1st South Aceh International Conference on Engineering and Technology (SAICOET 2018)

IOP Conference Series: Materials Science and Engineering
Volume 506

Kabupaten Aceh Selatan, Indonesia
8 – 9 December 2018

ISBN: 978-1-5108-8554-7
ISSN: 1757-8981

Printed from e-media with permission by:

Curran Associates, Inc.
57 Morehouse Lane
Red Hook, NY 12571

Some format issues inherent in the e-media version may also appear in this print version.

Printed by Curran Associates, Inc. (2019)

For permission requests, please contact the Institute of Physics at the address below.

Institute of Physics
Dirac House, Temple Back
Bristol BS1 6BE UK

Phone: 44 1 17 929 7481
Fax: 44 1 17 920 0979

techtracking@iop.org

Additional copies of this publication are available from:

Curran Associates, Inc.
57 Morehouse Lane
Red Hook, NY 12571 USA
Phone: 845-758-0400
Fax: 845-758-2633
Email: curran@proceedings.com
Web: www.proceedings.com

Table of contents

Volume 506

1st South Aceh International Conference on Engineering and Technology (SAICOET)

8–9 December 2018, Politeknik Aceh Selatan Campus, Indonesia

Accepted papers received: 7 March 2019
Published online: 15 April 2019

Preface

Papers

IOP Conf. Series: Materials Science and Engineering **506** (2019) 011001 doi:10.1088/1757-899X/506/1/011001

Preface The 1st SAICOET2018

South Aceh International Conference On Engineering and Technology (SAICOET) 2018 is the first international conference organized by Politeknik Aceh Selatan. The goal of the conference is to facilitate researchers around the world to publish and to share their current research. The 2018 edition of the conference will be held in Tapaktuan, Aceh Selatan, Indonesia on December, 8-9, 2018 at the Rumoh Agam, Aceh Selatan and Building B in Politeknik Aceh Selatan. These proceeding contains the selected scientific manuscripts submitted to the conference. It is with great pleasure to welcome you to the "1st South Aceh International Conference On Engineering and Technology (SAICOET) 2018 " that is held at Politeknik Aceh Selatan, Indonesia. The call for papers attracted submissions of over more than 50 abstracts from both local and international universities. Thanks to Universiti Malaysia Pahang (UMP) to become as Co-Organizer for this conference.

The scientific papers published in these proceeding have been revised and approved by the technical committee of the 1st SAICOET2018. All of the papers exhibit clear, concise, and precise expositions that appeal to a broad international readership interested in science, technology and engineering. The topics of the Conference were: (1) Mechanical Engineering; (2) Industrial Engineering; (3) Computer and Informatics Engineering; (4) Development and Environmental Engineering; (5) Development and Environmental Engineering; and (6) Agricultural Engineering.

The reports present original ideas or results of general significant supported by clear reasoning and compelling evidence, and employ methods, theories and practices relevant to the research. The authors state clearly the questions and the significance of their research to theory and practice, describe how the research contributes to new knowledge, and provide tables and figures that meaningfully add to the narrative. In this edition of SAICOET2018 representatives coming from academia, industry, governmental and private sectors. The plenary and invited speakers present, discuss, promote, and disseminate research in all fields of mechanical engineering in a constructive manner. Numerous social events that provide opportunities for the participants to renew old contacts and establish new ones were scheduled to accompany and follow the scientific activities.

Politeknik Aceh Selatan would like to thank the Organizing Committee, Co-Organizer and Sponsors supporting the conference and everyone who contributed along the conference and publication of these proceeding.

Darma Setiawan Putra, M.T
The Chairman of SAICOET2018

Assoc. Prof. Dr Mohd Shahrir Mohd Sani
The Chief Editor of SAICOET2018

IOP Conf. Series: Materials Science and Engineering **506** (2019) 011002 doi:10.1088/1757-899X/506/1/011002

Peer review statement

All papers published in this volume of *IOP Conference Series: Materials Science and Engineering* have been peer reviewed through processes administered by the proceedings Editors. Reviews were conducted by expert referees to the professional and scientific standards expected of a proceedings journal published by IOP Publishing.

IOP Conf. Series: Materials Science and Engineering **506** (2019) 012001 doi:10.1088/1757-899X/506/1/012001

Analysis of Bulletproof Vest Made from Fiber Carbon Composite and Hollow Glass Microsphere (HGM) in Absorbing Energy due to Projectile Impact

Muhammad Anhar Pulungan[1], Sutikno[2], M S M Sani[3]
[1]Mechanical Engineering Department, South Polytechnic Aceh (POLTAS), Aceh 23751, Indonesia
[2]Mechanical Engineering Department, Institut Teknologi Sepuluh November (ITS), Surabaya 60111, Indonesia
[3]Advanced Structural Integrity and Vibration Research (ASIVR), Faculty of Mechanichal Engineering, University Malaysia Pahang (UMP), 26600 Pekan, Pahang Malaysia

*Corresponding author: anhar.240205@gmail.com

Abstract. Bulletproof vest serves as a barrier and simultaneously absorbing the impact energy of a projectile shot from a firearm so it could not injure the users. Manufacture of lighter bulletproof vests with good absorbent of impact energy, is expected, because it supports the mobility and safety of its users. In this study, a composite composed of an epoxy matrix with a 16% Hollow Glass Microsphere (HGM) and carbon fiber reinforce to be implemented in a bulletproof vest. The objective of this research is to analyze the bullet-proof vests made of epoxy matrix composites with reinforcement in the form of HGM and carbon fiber through simulation with finite element method. Simulations with Ansys conducted in accordance with NIJ Standard 0101.06 from the U.S. Department of Justice, where the projectile initial velocity of 426 m/s for the category IIIA class weapon with a kinetic energy of projectile amounted to 528.37 Joules. The simulation with Ansys was performed by varying the thickness of bulletproof vests to obtain optimum thickness. The outcome of this research is a bulletproof vest that absorbs the impact energy of the projectile, so that the energy transmitted to the body is smaller than 170 Joules. Having obtaining the optimum thickness of bulletproof vest, an experimental verification will be performed to validate the simulation results. In the simulation results showed that a bullet proof vest with a thickness of 20 mm has been able to meet the standards of Major General Julian S. Hatcher, a U.S. Army ordnance expert with great energy generated at 138.77 Joules.

1. Introduction

Technology is evolving rapidly in increasing of time, this resulted in the need of amassed usage of an alternative material because of the unbalanced availability of material now days. In the industrial world, the science of metal is very important in all aspects. As aspects of the usage of appropriate materials costs, the safety factor manageable maintenance and efficiency [1]. In the automotive industry, aerospace and military response to the global issues of material crisis as a challenge, and therefore a lot of equipment are being made with high efficiency lightweight material [2]. One that needs to be developed is the body of a lightweight material. By reducing the weight of a product, the lower the energy consumption, so as to contribute to the issue of energy crisis being faced around the world.

One of the materials used in the manufacture of a bulletproof vest on the inside of the light is a composite material. The advantages of this material are the strength to weight ratio that is higher than the metal material. In addition of considering the physical properties such as low density, the mechanical properties of the composite material should also be considered because of the associated safety of users of the bullet-proof vests and able to provide protection against impact loads users from working [2-4]. Bulletproof vest is a protective clothing that is widely used in the military world. Bulletproof vest protects the body the chest, stomach and backbone. Backbone and chest are major parts of the body that protects the vital organs of man. Bulletproof safety vest is comfortable and has passed the requirements of Hatcher's Notebook (1962) by Major General Julian S. Hatcher, a U.S. Army ordnance expert [5-6].

Hollow Glass microspheres (HGM) is a glass ball microscale applied for; research, medical, and consumer products in various industries. The glass balls generally have diameters ranging from 1 micron to 1000 micron. Hollow Glass Microsphere have a range of diameters from 10 microns to 300 microns. HGM is usually used as a filler for composite materials, the advantages of this HGM among others:

- HGM has a density of small.
- Low thermal conductivity.
- The resistance to compression loads well.

HGM is a type of particle amplifier. HGM is a sphere made of glass with a certain thickness and has a hole in it containing an inert gas. HGM has many advantages, for example, has a density that excels because it has a hole so suitable for use to produce light material combinations (composite). HGM type IM30K is made of soda-lime-borosilicate glass with density of 0.6 g/cc, and an average diameter of each particle is 18 microns. HGM has high isostatic crush strength which is 28000 psi [7-9]. Table 1 shows the types of HGM IM30K that will be used in this study.

Table 1. Specifications HGM IM30K [10]

Property	**IM30K**	
Shape	*Hollow spheres with thin walls*	
Composition	*Soda-lime-borosilicate glass*	
Color, Unaided Eye	*White powder*	
Physical Properties	**IM30K**	**Test Method**
Isostatic Crush Strength	28000 psi	3M QCM 14.1.8
True Density	600 kg/m^3	3M QCM 14.24.1
Packing Factor (bulk density to true particle density)	63%	
Oil Absorption	33.5	ASTM D282-84 *of polymer additive*
Softening Point	600^{0}C	
Flotation Density	90%(in volume)	3M QCM 37.2
Volatile content (by weight)	0.5 % max	3M QCM 1.5.7
Alkalinity (miliequivalents)	0.5 mEq	3M QCM 55.19gr/max
pH(5% loading in water)	9.5 %	ASTM D3100-1982
Diameter	18 μ	3M QCM 193.0

Carbon fiber is one form of composite materials. Composite materials, which are taken from the English term composition materials or shortened to composite materials, is a material made of two or more materials making up mutually having different physical and chemical properties, that will produce

IOP Conf. Series: Materials Science and Engineering **506** (2019) 012001 doi:10.1088/1757-899X/506/1/012001

a material characteristic different from the materials constituent. Carbon fiber composites are one type of composite material using carbon fiber as one constituent [11-13].

This study aimed to make a bulletproof vest with a combination of carbon fiber and 16% HGM-epoxy that is able to absorb all the impact energy and reduce weight so as to produce a lightweight vests, which will increase movement while on duty, and of course to reduce the penetration of bullets in accordance with the testing standards NIJ Standard 0101.06 that is equal to 44 mm. This study will also analyze the effect of the addition of carbon fiber as reinforcement in reducing the impact waterwheel that will be accepted by the body, thereby reducing the level of injury that would be suffered by the user [14-15].

2. Research Methodology

Stages of research conducted is referring to the flow chart below which shows that the process occurs from several stages and test so that the best results are obtained as shown in the figure 1:

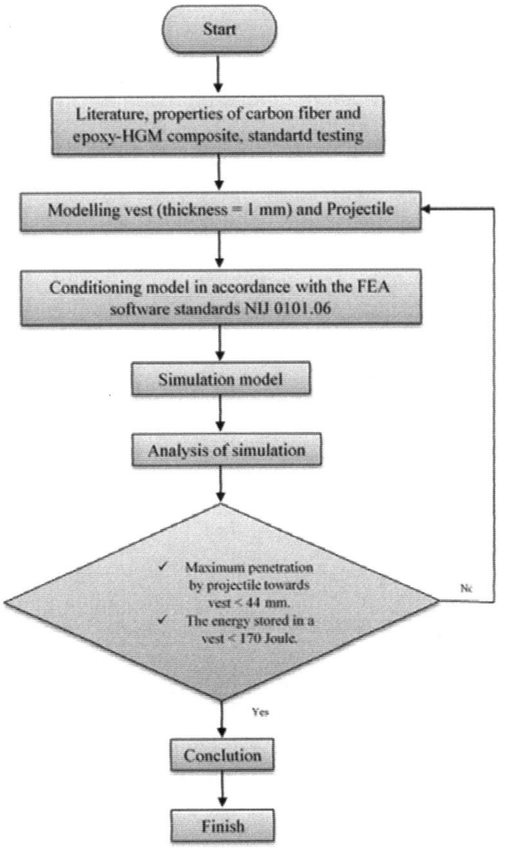

Figure 1. Flowchart research

In figure 2 shows the process that occurs in the ansys software starting from modeling to the results . In this procedure includes several stages [16-17], namely:

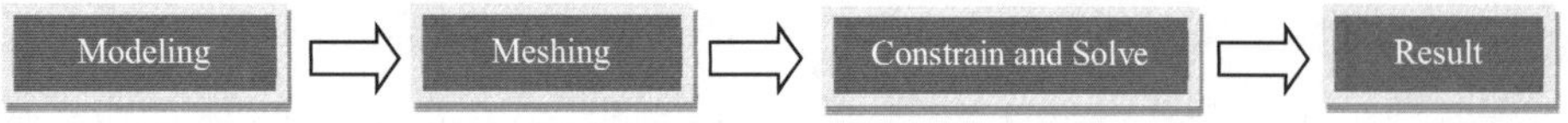

Figure 2. Simulation procedures

2.1 Conditioning of impact test and the absorption of kinetic energy
In figure 3, shows the selection of material types to be used in this study which aims to provide the right material properties of specimen of material used [18-19]

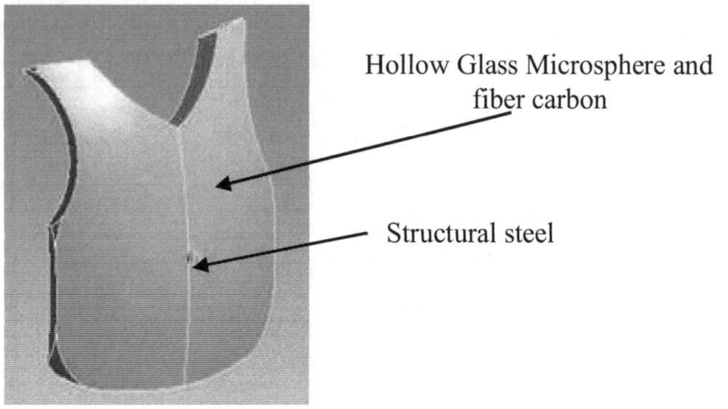

Figure 3. The choice of material in the vests and bullet models

2.2 Conditioning Meshing.
Figure 4 shows how the selection process is very important in using Ansys, in this study using the smallest meshing value is aim to get the better results.

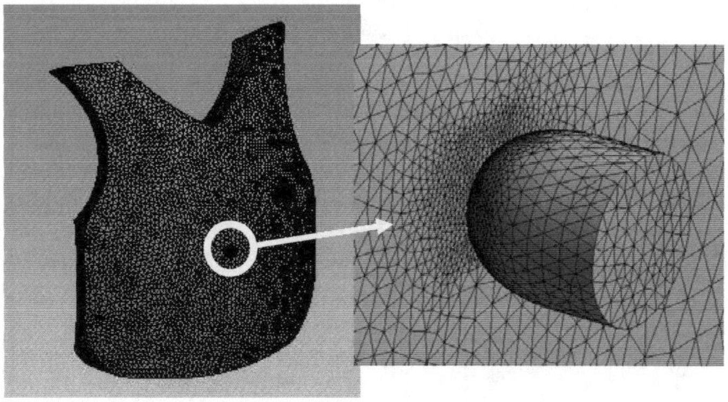

Figure 4. Meshing selection of the impact test

2.3 Conditioning of load and fixed support on the model
Figure 5 shows that the load in figure 5.a aims to give impact capability to the bullet by adding the speed of the bullet according to the standart test using the III-A type of weapon that is 426 m/s. Then in figure 5.b, it provides fixed support on both sides of the vest design which aims as a barrier so that during the collision the vest design remains in the desired position.

IOP Conf. Series: Materials Science and Engineering **506** (2019) 012001 doi:10.1088/1757-899X/506/1/012001

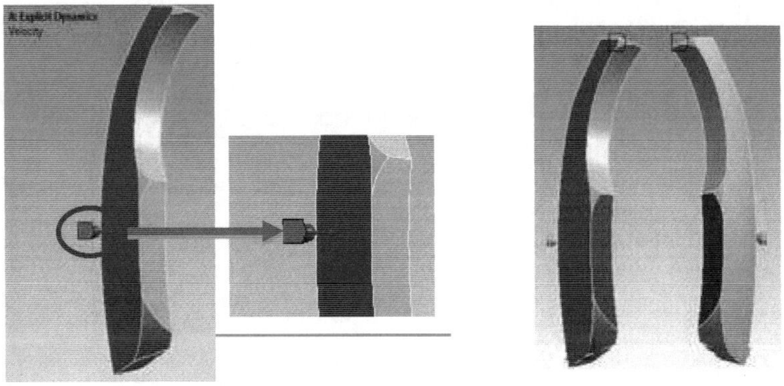

(a) (b)

Figure 5. Design of the bulletproof vest (a) Provision of the burden of a speeding bullet 426 m/s (b) fixed support

3. Results and Discussion

Bulletproof vest deforms when receiving impact and deformation varies between different thicknesses of vest. The smallest deformation happens to the vest with the highest thickness, which is 20 mm followed by a bulletproof vest thickness of 15 mm and the highest deformation is the vest with thickness of 1 mm, as shown in table 2 below.

Table 2. Specifications and Energy owned by the vest and the bullet.

No	Thickness (mm)	The depth penetration (mm)	Kinetic Energy of Bullets (J)	Kinetic energy Vests (J)	Internal energy Vests (J)
1	1	44.71	528.37	270.99	251.99
2	5	21.95	528.37	258.10	255.71
3	10	12.39	528.37	228.49	269.63
4	15	9.81	528.37	198.29	291.98
5	20	5.54	528.37	138.77	348.27

Figure 6 shows that with the increasing of thickness, the ability of a projectile to penetrate the vest is also decreases. NIJ Standard 0101.06 from the U.S. Department of Justice stated that maximum deformation allowed for a safe to use bulletproof vest is amounted to 44 mm. At a thickness of 1mm projectile capable of penetrating the vest as far as 44.71mm, while the thickness of 5 mm, 10 mm, 15 mm and 20 mm, the deformation's depth less than 44 mm is equal to 21.95 mm; 12.39 mm; 9.81 mm; and 5.54 mm respectively. It can be concluded that the vest with a thickness of 5 mm, 10 mm, 15 mm and 20 mm has met the standard of NIJ Standard 0101.06 from the U.S. Department of Justice.

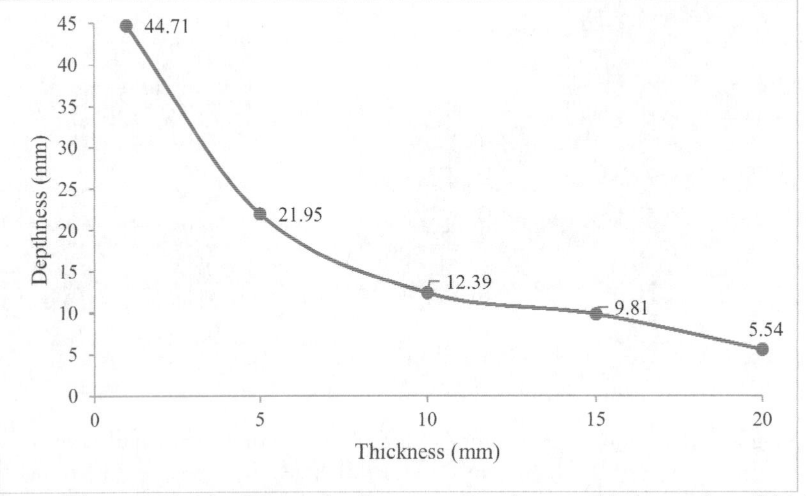

Figure 6. Graph of the projectile penetration decreased in increasing of thickness

In **Figure 7** below shows that with the increasing thickness of the vest, the bullet's kinetic energy that converted into the vest's kinetic energy is decreased. The kinetic energy received by the vest can still burden the user with load. In a bulletproof vest with a thickness of 1 mm, maximum kinetic energy accepted is 270.99 joules, a thickness of 5 mm maximum kinetic energy accepted is 258.10 joules, while for thickness 10 mm, 15 mm and 20 mm, the maximum kinetic energy accepted are 228.49 joules, 198.29 joules and 138.77 joules respectively. Hatcher's Notebook (1962) by Major General Julian S. Hatcher, a U.S. Army ordnance experts declared energy of 170 joules can cause paralysis in humans, so it can be concluded that the vest with a thickness of 20 mm has met these standards.

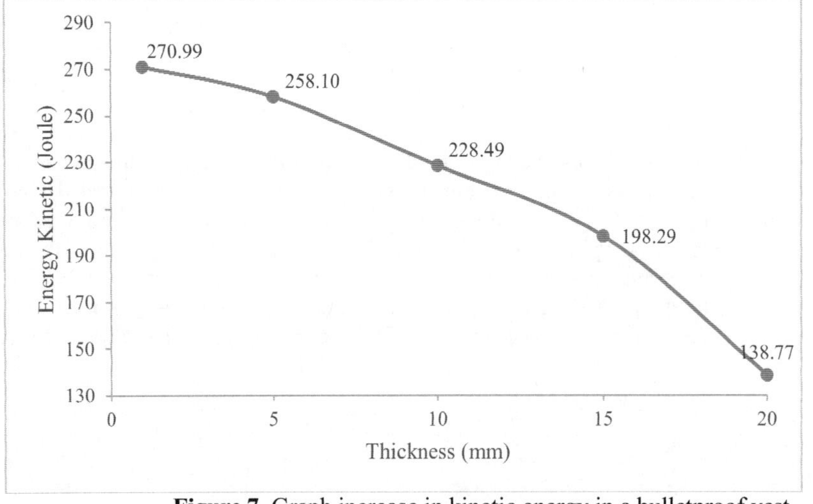

Figure 7. Graph increase in kinetic energy in a bulletproof vest

Figure 8 below shows that with increasing thickness, the kinetic energy of the bullet is converted into internal energy vests also increased. At vest with a thickness of 1mm the internal energy resulted id 251.99 joules, and the vest with a thickness of, 5 mm, 10 mm, and 20 mm, the internal energy resulted are, 255.71 joules, 269.63 joules, 291.98 joules, and 348.27 joules respectively.

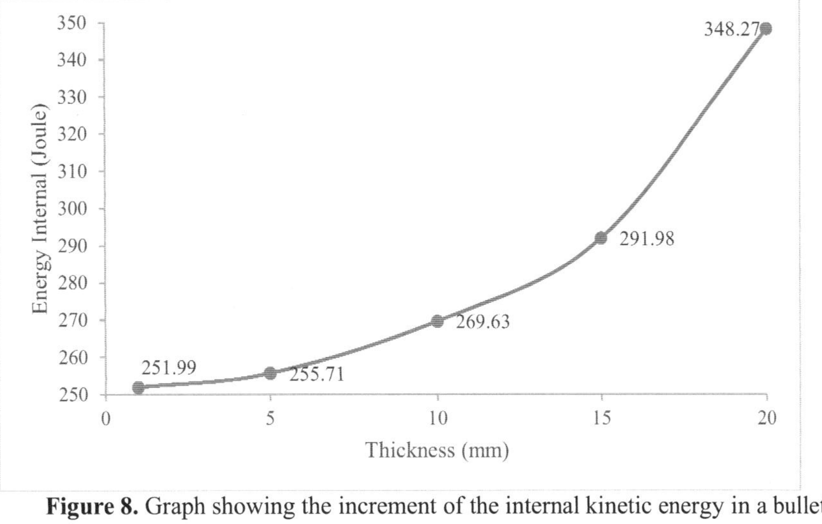

Figure 8. Graph showing the increment of the internal kinetic energy in a bulletproof vest.

4. Conclusions

Based on the simulation, it is concluded that:

i. The bulletproof vest with a thickness of 20 mm produces 138.77 joules of energy, which is safe to use. Based on the reports of Major General Julian S. Hatcher, a U.S. Army ordnance experts note that the total energy of 170.2 joules is capable of causing injury and able to paralyze the victim.
ii. The bulletproof vests with a thickness of 5, 10, 15, and 20 mm are safe to be used. Based on the NIJ standard 0101.06, penetration of the projectile against a bulletproof vest may not exceed 44 mm.
iii. A vest that is safe to be used is a vest with a thickness of 20 mm. Based on two of these references, (standard NIJ and Major General Julian s.)
iv. In this study, a bulletproof vest with a thickness of 20 mm is able to absorb the kinetic energy of the projectile in the amount of 138.77 joules of energy. The kinetic energy of the projectile will be transferred to the bullet-proof vests and is converted into kinetic energy and internal energy in a bulletproof vest.

References

[1] A Trofimov, L. Pleshkov, H. Back 2007 Hollow glass microsphere for high strength composite cores *Alchemie Technology* vol 50 (44-46) pp 48-50
[2] M S M Sani, N A Z Abdullah, S N Zahari, J P Siregar and M M Rahman 2016 Finite element model updating of natural fibre reinforced composite structure in structural dynamics *MATEC Web of Conferences* 83(03007)
[3] A Tasdemirci, G Tunusoglu and M Güden 2012 The effect of the interlayer on the ballistic

performance of ceramic/composite armors: Experimental and numerical study *International Journal of Impact Engineering* vol 44 pp1-9
[4] D Bürger, A Rocha de Faria, S F M de Almeida, F C L de Melo, Maurício V. Donadon 2012 Ballistic impact simulation of an armour-piercing projectile on hybrid ceramic/fiber reinforced composite armours *International Journal of Impact Engineering* vol 43 pp 63-77
[5] National Institute of Justice 2016 Development and evaluation agency of the United States Department of Justice, NIJ 0101.06
[6] US Patent Protective body armor garment shell *US5331683 A*
[7] F Y Arista 2013 Pengaruh Penambahan HGM terhadap sifat fisik dari komposit dengan matrix epoxy *Institut Teknologi Sepuluh Nopember*
[8] Hindun 2015 Analisa komposit matriks epoxy dengan penguat HGM untuk pembuatan bumper depan kendaraan *Institut Teknologi Sepuluh Nopember*
[9] Z Q Lutfianisa 2015 Analisa kemampuan rompi anti peluru yang terbuat dari komposit hgm 16% dalam menyerap energi akibat *impact* proyektil *Institut Teknologi Sepuluh Nopember*
[10] W Ritonga 2017 Pengaruh variasi fraksi volume, temperatur curing dan post-curing terhadap karakteristik tekan komposit epoxy - HGM IM30K *Jurnal Teknik Institut Teknologi Sepuluh Nopember* vol 6(1) pp 196-200
[11] H A Maples, S Wakefield, P Robinson and A Bismarcka 2014 High performance carbon fibre reinforced epoxy composites with controllable stiffness *Composites Science and Technology* vol 105 pp 134-143
[12] W A de Morais, J R M d'AlmeidaI, L B Godefroid 2003 Effect of the fiber reinforcement on the low energy impact behavior of fabric reinforced resin matrix composite materials *Journal of the Brazilian Society of Mechanical Sciences and Engineering* vol25(4) pp 325-328
[13] S K Mazumdar 2001 Composites Manufacturing: Materials, Product, and Process Engineering *CRC Press: United Kingdom* First Edition
[14] National Institute of Justice 2016 Development and evaluation agency of the United States Department of Justice, NIJ 0101.06.
[15] US Patent Cartridge for small arms 5094169
[16] M S M Fouzi, K M Jelani, N A Nazri and M S M Sani 2018 Finite Element Modelling and Updating of Welded Thin-Walled Beam *International Journal of Automotive and Engineering* vol 15(4) pp 5874-5889
[17] N A Abdullah, M S M Sani and I Zaman 2017 Correlation of Structural Modal Properties of Go-Kart Frame Structure Using Different Type of Joint in Finite Element Modeling *Advanced Science Letters* vol 23(1) pp 11315-11319
[18] M N M Merzuki, M R M Rejab, M S M Sani, B Zhang, M Quanjin and W Rafizi 2019 Investigation of modal analysis on glass fiber laminate aluminium reinforced polymer: An experimental study *IOP Conference Series: Materials Science and Engineering* vol 469 (013165)
[19] M S M Fouzi, M S M Sani and Y Muchlis 2019 Finite Element Modelling and updating of welded joint for dynamic study of exhaust structure *IOP Conference Series: Materials Science and Engineering* vol 469 (012099)

The Design of Flexible Rubber Tapping Tool with Settings the Depth and Thickness Control

H Susanto[1*], S Ali[1] and Hanif[2]
[1]Department of Mechanical Engineering, Faculty of Engineering, Teuku Umar University, West Aceh, Indonesia
[2]Department of Mechanical Engineering, Lhokseumawe State Polytechnic North Aceh, Indonesia

*Corresponding author: herdisusanto@utu.ac.id

Abstract. Road map of land for agricultural commodities in West Aceh is prioritized for rubber plantations. This commodity is the main source of livelihood and employment for the commodity. Furthermore, it is very difficult for a novice tap to be able to control the depth, thickness, and slope of tapping angle. Therefore, rubber tapping is created to increase the productivity of rubber crops in this region. This research conducted using software where the selection of component materials and manufacturing are tested on 9 rubber trees. The treatment is carried out according to the design plan; controlling the depth between 1-1.5 mm of cambium, tapping the thickness 1.5-2 mm, and tilting the angles of 35^0-60^0. Thus, the result of the functional testing shows that rubber tapping can function properly and the test data at interval 09:00-10:00 WIB for slope 60^0, 45^0, 35^0 obtained the average latex amount of 1.83 grams, 1.11 grams, and 0.74 grams. In addition, the rubber tapping capacity of 5-6 seconds per tree is better when compared to conventional rubber tapping between 6-8 seconds with 5.32 cm^3 rubber bark consumption.

Keywords: Latex extraction, equipment testing, tapping slope

1. Introduction

The rubber tree is one of the major economic crops in Southeast Asia and valued for its latex production and timber [1]. Indonesia has appropriate land for planting rubber trees. Some of which are in Sumatra and Kalimantan. In 2005, the area of rubber plantations was recorded at more than 3.2 million hectares and spread throughout Indonesia. Around 85% of rubber plantation land belongs to the society, about 7% is state-owned, and around 8% is privately owned. [2]

Road map for land use in the West Aceh farming community prioritized for rubber plants [3,4]. Thus, it makes rubber as one of the leading commodities with the plantation area cultivated by the society reaching 23,862.37 hectares with an average income of 52,091.24 tons per year [5]. This community is the main source of livelihood for society in West Aceh.

Rubber tapping is the process of extraction of latex from rubber trees. Rubber tree tapping is considered to be a skill-oriented job. During the tapping process, the taper has to make a downward half spiral incision on the tree bark to extract the white milky liquid called latex [6]. Rubber tapping technology has an important role in increasing rubber productivity. A rubber tapper must be able to control bark consumption and depth of tapping to protect tissue of the wood [7,8]. Tapping is performed on a very limited area of the stem, which raises the question of the spatial extent of the affected area with respect to its impact on radial growth. Rubber tree growers generally estimate total wood production or production deficit due to tapping on the basis of girth measurements at a particular height, which is assumed representative of the whole stem growth [9,10]. Skills in rubber tapping must be trained for years and the current conditions are increasingly difficult to get skilled workers in rubber tapping. In general, rubber planters often find difficulty to tap it in accordance with applicable regulations [6], This is due to the quality of tapping that is not qualified [11]. Based on the results of monitoring and interviews that have been carried out with several rubber planters, the researcher found that one of the main obstacles in tapping rubber is if the tapping technique in slicing rubber tree bark is not optimal, so the depth and thickness of slicing are irregular. In addition, if the cambium from the rubber tree is tapped, then, the bark of the rubber tree will be cured for a long time and even it can die. For this reason, the application of rubber tapping aids with depth, thickness, and tilt angle can be examined.

2. Research Methodology

2.1. Rubber Tapping Tool Design

The design of rubber tapping tools is designed and simulated using software. The components are arranged according to the specifications of good tapping techniques with the main consideration of control tapping depths of 1 - 1.5 mm from cambium and tapping thickness 1.5 - 2 mm and slope tapping grooves 30^0-40^0 [12,13]. The slice depth regulator can be adjusted according to the slices in the range of 1- 1.5 mm, depending on the age of the rubber plant; the older the rubber tree, the depth of the slices from the bark of the rubber tree will be deeper because of the more latex vessels [14]. In that case, blades are used using bolt joints so that it is easy to replace if blunt has occurred.

The track rail of the tapping device functions as a controller for the tilt angle of the tapping groove and it can be arranged in the range of 30^0-60^0. This rail serves as a tapping mall to save rubber tapping time. If all this time the garden farmers only use rollers and litters to form the tapping patterns, thus, the tapping rails are made to facilitate the tapping process and it is highly recommended for beginner tappers to homogenize the tilt angle. In addition, the tapping device can also be removed from the railroad tracks. For skilled tappers, this tool can be released from the railroad house and in the tapping process; it is permissible to use the tapping device only. In short, the rail house is only used for first tapping, and so on may not use it again.

2.2. The Selection of Rubber Tapping Tool Components

For material selection of rubber tapping components, frame construction using low carbon steel strip plate profiles, rail track components of the blade using ST37 profile steel, while tapping blades made of carbon steel (modified from manual blades) can be found in several markets in Meulaboh. Then, the handle of the pull and grip are made of wood, the depth controller of leather slices using pads, and the assembly using welding connections and bolts.

2.3. Manufacturing Rubber Tapping Tools

The process of manufacturing rubber tapping tools holds at the Teuku Umar University in Mechanical Engineering Workshop. The formation of tapping tool components is formed using a milling machine. The electric welding machines and acetylene is used to connect relatively small component parts and some parts are also needed as depth control; pads and blades for replacement if damaged or dried. Then, it is connected using a bolt for the formation and perforation of the tapping device component using a grinding and sitting drill.

IOP Conf. Series: Materials Science and Engineering **506** (2019) 012002 doi:10.1088/1757-899X/506/1/012002

2.4. Testing of Rubber Tapping Tools

The testing of rubber tapping tools with a depth control system, thickness, and slope of tapping grooves implemented at the rubber plantations of Sepakat Group, Paya Lumpat Village, Samatiga District, West Aceh Regency, Aceh Province. Additionally, the tapping testing process was applied to nine rubber trees as the objects in this research. Tapping angles of 35^0, 45^0, and 60^0 are arranged with combinations in three rubber trees for three taps testing with the same angle. Then, the same tree is tapped at a different angle. It means, for nine times of testing, each tree will experience three tests for each angle of 35^0, 45^0 and 60^0.

2.5. Data Analysis

Analysis of testing data using a rubber tapping tool that has been designed. The manufacture is determined from the results of field testing of rubber trees, and the ratio to the standard size of the slice depth 1-1.5 mm (from cambium), slice thickness 1.5-2 mm, and slope 35^0-60^0. If the measurement results are in that range, it can be stated that the tool is suitable for use and development. Afterward, rubber tapping test data is analyzed in graphical form to determine the optimal angle of latex production. Rubber tapping capacity, and bark consumption are solved using equation (1) and (2) respectively [15]:

$$KP = \frac{BPT}{t} \tag{1}$$

Where, KP denotes the tapping capacity (tree per hour); BPT denotes the number of trees tapped and t denotes the time (in hour).

$$KK = PL \times KS \times TS \tag{2}$$

Where KK is the bark consumption (in cm^3), PL is the length of track (in cm), KS is the depth of tapping (in cm) and TS is the tapping thickness (in cm).

3. Result and Discussion

3.1. Rubber Tapping Tool Design

The design of rubber tapping tools is designed using Autocad software, where the components are arranged according to the specifications of good tapping techniques with the main consideration of tapping depth 1-1.5 mm and tapping thickness 1.5-2 mm and tilt slope leads 30^0-60^0 [16,17]. The explanation of the components of the rubber tapping device to adjust the depth, thickness, and slope of the tapping angle are described in Figure 1.

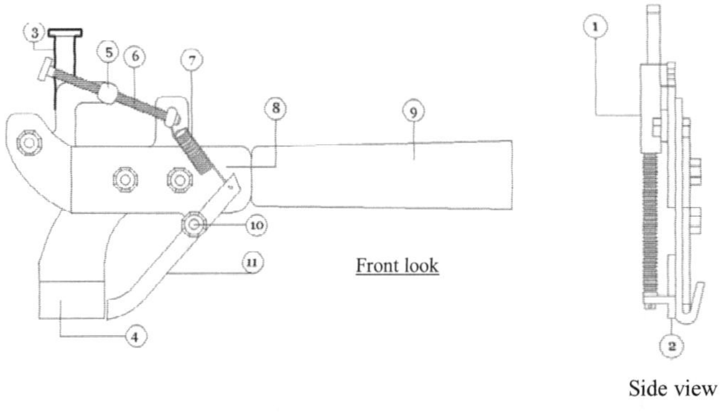

The Captions

1. Depth sensor
2. Depth control wheel
3. Swivel handle screw depth
4. Tapping blade
5. Steering sensor tapping position
6. Screw the tapping sensor position
7. Spring regulating the thickness of the tapping
8. Home tapping aids
9. Pull handle tapping tool
10. Tapping tool binder bolts

Figure 1. Design of rubber tapping tools

3.2. The Process of Making and Assembling Rubber Tapping Tools

The process of making rubber tapping tools using an iron plate and solid and formed by milling, turning and welding according to the design drawings that have been determined [17]. The manufactured rubber tapping equipment is shown in Figure 2.

Figure 2. Rubber tapping equipment that has been manufactured

3.3. Functional Test

The functional test aims to determine whether all components of the rubber tapping device can function properly in accordance with the design plan that has been determined. The test process is carried out in a mechanical engineering workshop by directly cutting the 2 m long rubber tree, after all rubber tapping aids are guaranteed to function properly, then continued with elementary tests.

3.4. Elementary Test

The initial field testing using a rubber tapping device that had been designed to be constructed was applied on nine rubber trees that were around 5 years old. Then, division for each corner; 35^0 angles are 3 trees, 45^0 angles are 3 trees and 60^0 angles are 3 trees, the thickness and depth of the tapping are set 1.5-2 mm with the same length of tapping grooves for all 38 cm trees. The procedures in field testing are as follows.

Before numbering is done on rubber trees that have not been tapped, the tree number and tilt angle are given first so that the tree is not swapped during data collection during measurement. The numbering of rubber trees from 1-9 trees shown in Figure 3.

Then, attach the tilt angle adjustment tool. This tool serves to homogenize all tapping angles. In this study, the angles chosen are angles of 35^0, 45^0 and 60^0 which function to determine the optimal angle in tapping rubber trees. The installation of tapping angle tilt aids is shown in Figure 4.

Figure 3. Rubber tree

Figure 4. Installation of regulatory aids tilt angle of the tapping

The next step is to make a tapping groove line. This is done to give a mark on the rubber tree before the tree is tapped. If it is finished, then the tilt angle tilt aids are released from the rubber tree. The process of scraping a rubber tree using a tapping angle tilt auxiliary tool is shown in Figure 5.

Rubber trees are ready to be tapped if there are scratches on the rubber tree. For the next tapping or the next day, the tapping angle tilt tool is no longer used. With the wool cloth, this tool is only used at the beginning of the tapping process and for the next process to follow the existing tapping path. The process of tapping rubber trees using a rubber tapping device is shown in Figure 6.

Figure 5. Marking scratches of tapping grooves

Figure 6. The process of tapping rubber trees

The things that must be considered in the tapping process of rubber trees using tapping tools are the thickness and depth of the tapping groove. They must be arranged in accordance with the thickness of the rubber tree bark and must not touch the wood from the rubber tree. The process that must be considered to adjust the thickness and depth of rubber tapping trees is shown in Figure 7.

The retrieval of the latex data from the tapping results is obtained when all rubber trees have been tapped according to the number and slope of the lead angle that has been determined at point 2. The next day, the latex in the holding bowl is taken and labeled so as not to be confused in the data collection. All labeled latex is put in a transparent plastic bag so that they are easily visible.

Figure 7. Setting depth and thickness

After the latex collection process is completed, the tapping process is carried out again on the same tree and the same angle for three days. Then, the tilt angle position of the rubber tree is replaced with the other position. Six days later, the tilt angle position was replaced again. Thus, all trees experience all tilt angles for three consecutive days.

Measurement of the results of rubber leads is measured by weighing every latex that has been tapped one by one. This weighing process is carried out using a digital scale with an accuracy of 0.01 grams.

3.5. Rubber Test Equipment Testing Results

Each tree is subjected to 3 tapping times with the same angle and the number of trees that are the object of testing the equipment is nine rubber trees. In other words, the tapping test is carried out on three rubber trees with the same angle, with the same thickness and tapping depth of 7 mm (from the outer bark), and the same tapping path length of 38 cm. The test results are shown in Figure 8.

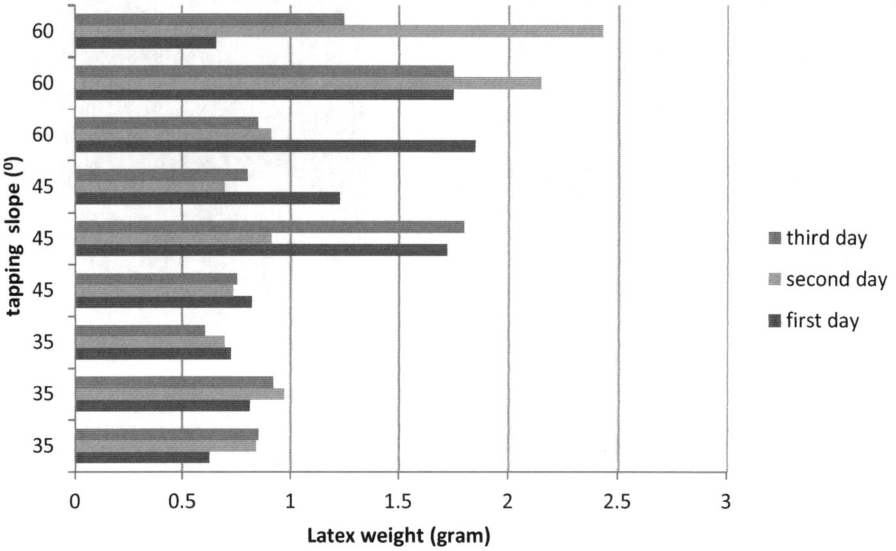

Figure 8. Test results of rubber tapping on the first, second and third day

Figure 8 shows that under the same tree conditions, with three tapping shows fluctuating results. In general, it indicates that the tapping process on the rubber trees with 60^0, 45^0 and 35^0 in the first, second and third day demonstrate that in 60^0 angles produce heavier latex (in average 1.51 grams) than tapping angle of 45^0 (in average 1.05 grams) and 35^0 (in average 0.78 grams)

Tests for the fourth, fifth and sixth days are done on the same tree, but the tilt angle is changed. If the first test uses variations 60^0, 45^0 and 35^0, then the second test uses variations of angles 45^0, 35^0 and 60^0. The second test results are shown in Figure 9.

Graph data from Figure 9 shows that in general the results of rubber leads are dominated by the tilt angle of 60^0 (in average 1.35 grams). Fluctuations in test results occur almost on average on all trees in each test, this can cause due to the quality of the rubber tree so that when tapping many latex vessels are closed, and the humidity around the tree also affects the amount of latex produced.

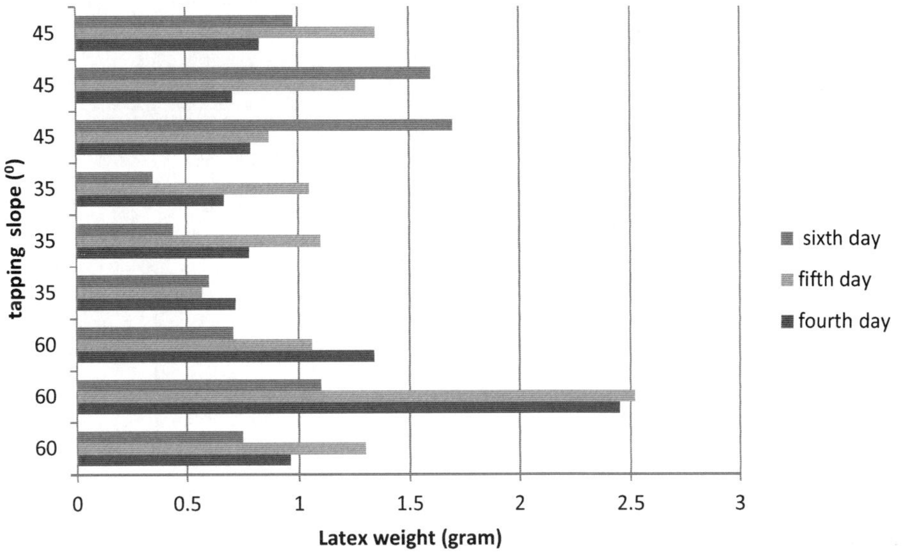

Figure 9. Test results of rubber tapping on the fourth, fifth and sixth day

The seventh day of testing with variations in angles of 35^0, 60^0 and 45^0. They are presented in Figure 10.

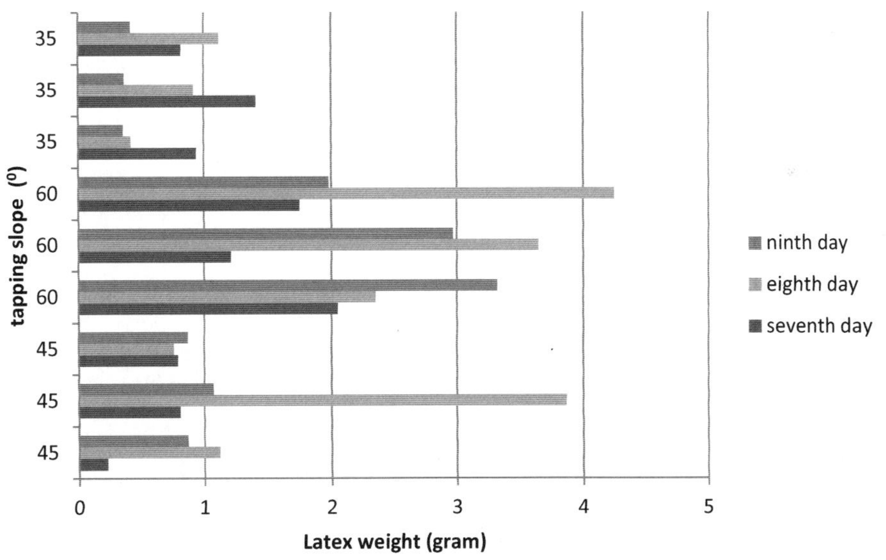

Figure 10. Test results of rubber tapping for the seventh, eighth and ninth day

The test results are shown in Figure 10 also show that the tilt angle of the lead is 60^0 (in average 2.62 grams) resulting in more latex when compared to the other angle, average 1.15 grams for angle slope 45^0 and angle slope 35^0 average latex of 0,75 grams.

If the data from the first day to the ninth day are classified according to the slope of the tapping angle, a graph will be generated as shown in Figure 11.

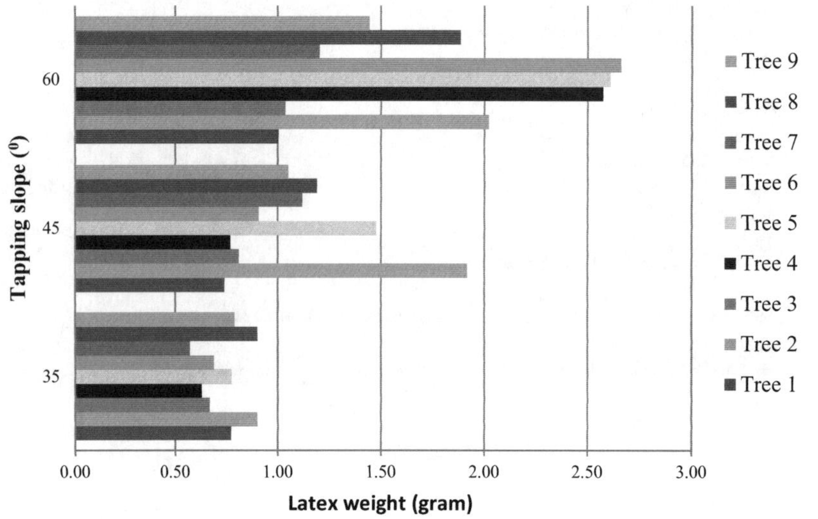

Figure 11. Grouping of latex production based on tilt angle

Figure 11 shows that the latex distribution for angle slope 60^0 produces an average latex of 1.83 grams, for the slope of the lead angle 45^0 produces an average latex of 1.11 grams, and with 35^0 tilt angles produces an average latex of 0.74 grams. From the data above, it shows that rubber tapping at intervals between 09.00-10.00 WIB should use a tapping angle of 60^0 because the angle is effective and fast to drain the latex on the tapping groove. The greater the slope of the tapping angle, the faster the flow of the latex in the tapping groove, but if the tilt angle of the tapping is greater, the latex will come out of the tapping groove and spill over the rubber stem. The slope of the 35^0 and 45^0 rubber lead angles is not suitable for use at the time interval above 9.00 WIB and the dry season. This is caused by a network of latex vessels that are quickly closed due to the slow flow of latex flowing in the tapping groove. In addition, this can also cause latex clotting.

The average rubber tapping time calculated on each rubber tree as many as nine trees per day is shown in Figure 12.

Figure 12 shows the total time needed for nine rubber trees to be tapped every day. Using equation (1), the average rubber tapping capacity for each tree is 5.59 seconds (tapping time between 5-6 seconds of application). Hence, tapping capacity using tapping aids is better when compared without using a tapping tool, which is between 6-8 seconds of application [7]

Bark consumption with tapping path length of 38 cm, tapping depth of 7 cm from outer bark and thickness of tapping 0.2 cm. Consequently, the consumption of rubber bark by using equation (2) is 5.32 cm^3

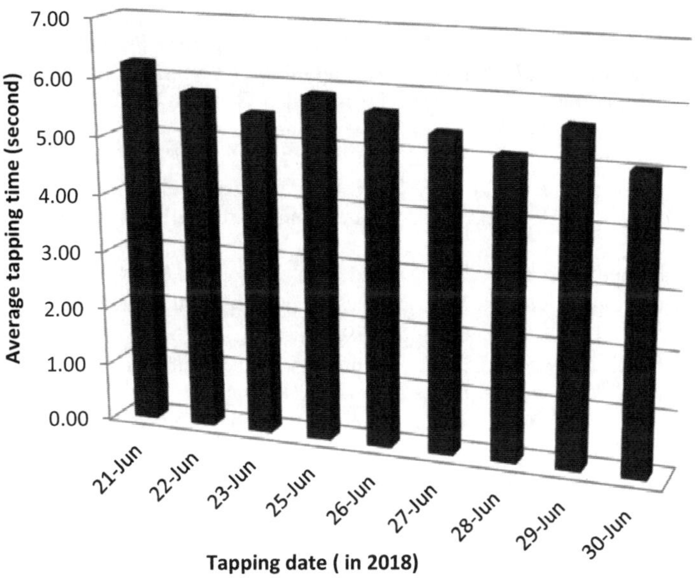

Figure 12. Average rubber tapping time

4. Conclusion

Based on the results discussed above, it can be concluded that:

i. Test results data at intervals between 09.00 - 10.00 WIB for the slope angle of 60^0, 45^0 and 35^0 obtained an average mass of latex of 1.83 grams, 1.11 grams, and 0.74 grams respectively.
ii. Rubber tapping capacity 5-6 seconds per tree is better when compared to rubber tapping devices without using a tapping tool, which is between 6-8 seconds with 5.32 cm^3 bark consumption.
iii. Rubber tapping at intervals between 09.00 - 10.00 WIB should use a tapping angle of 60^0 because the angle is effective to drain the latex on the tapping groove. The greater the slope of the tapping angle, the faster the flow of the latex in the tapping groove, but if the tilt of the tapping angle is greater, it can also cause the latex to come out of the tapping groove and spill over rubber.

References

[1] Kunjet S Thaler P Gay F Chuntuma P Sangkhasila K and Kasemsap P 2013 Effects of drought and tapping for latex production on water relations of Hevea brasiliensis trees *Kasetsart Journal Natural Science* vol 47 pp 506-515.

[2] Ginting R Siregar I and Ginting T. 2015 Perancangan Alat Penyadap Karet di Kabupaten Langkat Sumatera Utara dengan Metode Quality Function Deployment (QFD) dan Model Kano *Journal J@ti Undip* vol 10(1) pp 33-40

[3] Lacote R Gabla O Obouayeba S Gohet E Doumbia A Gnagne M and Eschbach J M 2006 Some Considerations Concerning The Panel Management in Rubber Tapping (Hevea Brasiliensis) *International Natural Rubber Conference* pp 3-15.

[4] Ritung S Wahyunto Agus F and Hidayat H 2007 Paduan Evaluasi Kesesuai Lahan (dengan contoh peta arahan penggunaan lahan Aceh Barat) *Balai Penelitian Tanah dan World Agroforestry Centre* chapter 3 pp 21

[5] Susanto H 2018 Desain dan Manufaktur Teknologi Tepat Guna Pedesaan *Bandar Publishing* chapter 2 pp 15

[6] Vinayaka N N Athul S N Madhuraj M N and Nandana S P 2017 Project Report on Automated Rubber Tapering Machine *Visvesvaraya Technological University* chapter 1 pp 2

[7] Wibowo SA 2011 Disain dan Kinerja Pisau Sadap Elektrik Untuk Tanaman Karet (hevea brasiliensis) *Institut Pertanian Bogor* chapter 2 page 4

[8] Sutardi 1991 Sistem sadap karet kearah atas *Pusat Penelitian Perkebunan Getas* pp 67

[9] Gohet E 1996 The production of latex by Heveabrasiliensis Relation with growth Influence of different factors: clonalorigin, hormonal stimulation, carbohydrate reserves *Université Montpellier II.* pp 343

[10] Unakorn S Philippe T Poonpipope K Andre L Arak C Boris A Eric G Sornprach T and Thierry A Effect of tapping activity on the dynamics of radial growth of Heveabrasiliensistrees *Tree Physiology Vol 26 Heron Publishing* pp 1579–1587

[11] Herlinawati E and Kuswanhadi 2012 Beberapa Aspek Penting pada Penyadapan Panel Atas Tanaman Karet *E-Journal Warta Perkaretan* vol 31pp 66-74

[12] Ritonga A I 2016 Tehnik Penyadapan Tradisional pada Tanaman Karet di Tapanuli Selatan *Jurnal Nasional Ecopedon* vol 3 pp 17-20

[13] Andry dan Suprianto J 2009 Teknologi Penyadapan Tanaman Karet *Balai Pengkajian Teknologi Pertanian Jambi* pp 16

[14] Ulfah D Thamrin G A R and Natanael T W 2015 Pengaruh Waktu Penyadapan dan Umur Tanaman Karet Terhadap Produksi Getah (Lateks) *Jurnal Hutan Tropis* vol 3 pp 247-252

[15] Wicaksono M S Ahmad H and Siswoyo S 2015 Modifikasi Alat Penyadap Karet (Lateks) Semi Mekanis *Berkala Ilmiah Pertanian* vol 4 pp 1-4

[16] Susanto H and Hanif 2017 Rancang Bangun Alat Sadap Karet dengan Pengaturan Kedalaman, Ketebalan dan Kemiringan Sudut Sadap *Seminar Nasional Sains dan Teknologi TM-009 Universitas Muhammadiyah Jakarta* pp 1-9

[17] Susanto H and Hanif 2017 Desiminasi Alat Sadap Karet dengan Pengaturan Kedalaman, Ketebalan dan Kemiringan Sudut Sadap *Seminar Nasional Teknologi Terapan Jilid II Universitas Gajah Mada* pp 54-63

Experimental Study of Bioethanol Production as Fuel from Salacca Zalacca Waste and Coconut Water Waste Combination

Muhammad Idris[1*] Uun Novalia[2]
[1] Mechanical Engineering of Universitas Harapan Medan
[2] Industrial Engineering of Universitas Harapan Medan

*Corresponding author: idrissurv@gmail.com

Abstract. The process of burning fuel from the engine produces exhaust emissions that have a negative impact on the environment. Bioenergy as biofuel needs to develop to reduce emissions. Biomass is one of the bioenergy sources and has high potential to be used as a liquid fuel; some of the available biomass is a waste of salacca zalacca fruit and coconut water waste. In the process of bioethanol production, several mechanical factors influence it, including the energy needed for the distillate process, bioethanol mass flow rate and the ethanol content it contains. This research aims to analyze the energy needed for the distillation process to reach 85 OC, calculate the mass flow rate and determine the ethanol content of this fuel. An acquisition parallax microcontroller device is installed on the distillation vessel and condenser. This tool is used to record changes in temperature during the distillation process. From the experiments that have been done, the total energy needed for the distillation process is 346.32 W, bioethanol mass flow rate is 0.0665 kg /s

Keywords: Bioethanol, Salacca Zalacca, coconut water, energy.

1. Introduction

Increased human activity in the field of industrial, fabrication, vehicles and other activities involving the combustion of petroleum fuels has led to a decrease in petroleum reserves in the world [1]. On the other hand, combustion of petroleum fuels either in external combustion or internal combustion engines causes the formation of exhaust emissions, CO, CO2, NO_x, HC and negatively impact the environment (global warming), health and global economy [2]. This has prompted humans to immediately switch the consumption of petroleum fuels to alternative fuels [3]. Biomass is an alternative fuel and can be as fuel in the gasoline engine after going through biochemical and thermochemical processes [4]. That the biomass potential undeveloped reach to 99%, especially in Indonesia [5].

Various studies have given a contribution produces some alternative fuel from various of biomass waste. It's clean and environmentally friendly and relatively cost effective [6]. There are several of biomass from agree cultural waste is abundant. The potato waste that becomes an environmental problem is used as raw material for bioethanol [7-8]. In addition to potatoes, sugar cane is also an option for bioethanol feedstock [9], some other biomass can be used as bioethanol feedstock such as: pineapple [10], alga [11], carbohydrates with syngas fermentation [12], hydrolysis [13], newspaper waste [14], and others.

There are two variations of liquid fuel that can be obtained from biomass derivatives such as biofuel for diesel engines such as fuel from palm oil [15], castor oil [16], and others [17]. There are also biofuel fuels for gasoline engines such as bioethanol, diethyl ether [18], hydrogen [19] and others. Based on several factors' considerations of these two fuels each has advantages and disadvantages. Among other factors are thecost production [20], value of caloric, octane and other factors [21].

Salacca zalacca and coconut water also a biomass that can grow easily in the tropics, as well as coconut that in fact still has a close relationship with salacca zalacca. Salacca zalacca and coconut harvest cycles also have similarities, both of which can continue to bear fruit throughout the year, but at certain times these two plants experience the harvest season and the crisis. At harvest time, salacca zalacca cannot be sold out all in the market, leaving a post-harvest problem. The harvest problem is not worth selling because the quality is low so increase the volume of waste. Also, the coconut water is obtained from a coconut seller who sells coconuts for food supplies for households. Every coconut is averaging 0.5 liters of water contained and disposed. Based on its phenomenon, the authors try to take and utilize to be processed into a mixture of fermentation of salacca zalacca waste and coconut water waste.

2. Research objectives:

During the distillation process, the phenomenon of convection heat transfer and the power needed to heat the solution to 85 OC on the distillation vessel is unknown. In addition, to analyze these phenomena, this aims of this study is to

i. Determine the characteristics of bioethanol from a mixture of the two ingredients in question. and also to determine:
ii. Analyze how much power is needed to heat the distillation chamber up to 85 ^{O}C and how much the convection heat transfer
iii. To study the characteristics of bioethanol, which is derived from various biomass waste materials

3. Methods

3.1. Experimental setup

Distillation is a phase separation process using the boiling point of the phase to be separated by using heat energy. The heat energy needed to increase the fluid temperature from 29.5 to 85 (^{O}C) theoretically is calculated by equation (1). The heat source is given through heating elements which are mounted on distillation with specifications as in table 1

$$q = hA\,(T_w - T_\infty) \qquad (1)$$

Table 1. Properties of heater element [22]

Heater element	Max. Power (W)	h (W/m^{2}C)	Diameter (m)	Length (m)	Area m^2
Aluminum	230	5.00E+03	1.50E-03	1.33E-01	6.24E-04

In this study, the authors build a distiller with a very simple design, but still consider its functions. This is done with the consideration that the distiller plan is intentionally built for in rural areas, resulting in minimal maintenance costs and operating costs [23].

3.2. Raw materials

Bioethanol is a mixture of 1:1 of zalacca and coconut water, which is mashed using a blender machine into a fermentor solution. The properties of the fermentor solution is the average amount of both materials as in table 2 below.

Table 2. Raw material of bioethanol

	Ingredient	ρ (kg/m³)	Alcohol content (%/vol)
1	Salacca Zalacca	1066.7	70
2	Coconut water	1030	10
3	mixture (average)	1048.35	40

3.3. Mass flow rate
The bioethanol mass flow rate is a condensed vapor mass per time change during the distillation process, mathematically written in equation (2).

$$\dot{m} = \frac{m_{cond}}{t}\left(\frac{kg}{s}\right) \quad (2)$$

3.4. Bioethanol content yield
In general, bioethanol content can be known by using a measuring instrument, in this study refractometer is used to determine the bioethanol content in units of %. The drop samples of bioethanol putted in the prism assembly and then look at the eyepiece. Refractometer seen in figure 1.

Figure 1. Refractometer

3.5. Measurement Tools and Experimental Apparatus
The measurement tools installed on the distillation chamber to record the distillation temperature process and connected by parallax acquisition microcontroller data PLX-DAQ using ATmel ATmega, it has specification showed in table 3.

Table 3. Specification of measurement tools

Device	Flash	EEPROM	RAM	General Purpose I/O pins	16 bits resolution PWM Channels	Serial USARTs	ADC Channels
ATmega2560	256KB	4KB	8KB	86	12	4	16

Five measurements of the k type thermocouple are allocated to records the temperature during the distillation process. The thermocouples range temperature of 0 - 1100 ^{o}C. Table 4 shows the identify of some of the variables recorded.

Table 4. the temperature measurement of the distillation process

Variable	Identify	Explanation
T1	V_{in}	Vapor into the condenser
T2	V_{out}	Vapor out the condenser
T3	W_{in}	Water into the condenser as cooling fluid
T4	W_{out}	Water out of the condenser after cooled the condenser
Tdist.	$T_{distillation}$	Temperature of distillation chamber

Experimental Apparatus in this study, especially to distillation proses is shown in the figure 2 belows.

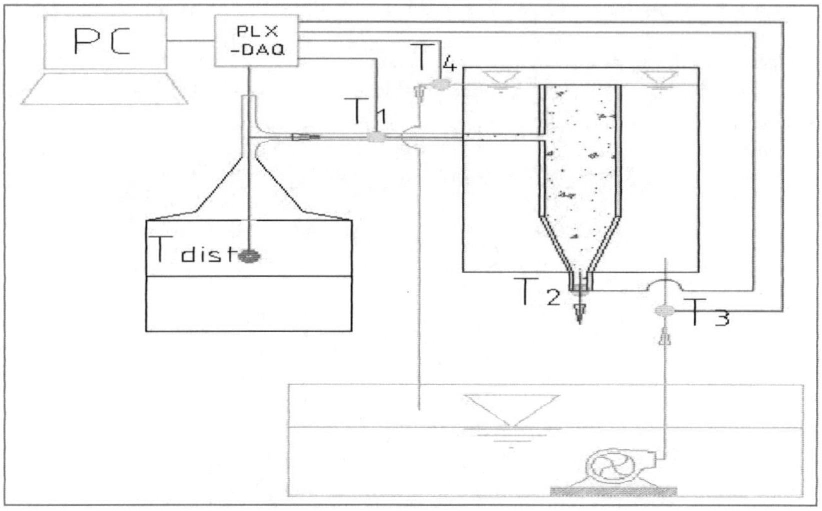

Figure 2. Experimental apparatus and measurement

Fermentation solution put into a distillation vessel of 500 mL, then heated for 1125 seconds 85 OC. The solution vapor flows through the pipe to the condenser and the vapor is condensed. Sensor temperature is installed at 5 measurement points which are at the distillation vessel, the vapor inlet into the condenser, vapor outlet from the condenser, cool water inlet into the condenser and cool water outlet from the condenser

4. Result and Discussion

After the experiment conducted, energy needed to heat up fermentation solution was calculated, temperature distillation during process was recorded by PLX-DAQ. The mass flow rates during the distillation process, quality and calorific of bioethanol were measured.

4.1. Energy to heat up

The energy needed for distillation to reach 85 ^{O}C is calculated using equation 1, then the result is 173.16 W. This is due to the two heater elements are installed in a distillation vessel so the total energy becomes 86.58 W each heater element. The detail can be seen in table 5.

Table 5. Calculation of distillation energy

h (W/m^{2}C)	Area (m^2)	Tw (^{O}C)	Ts (^{O}C)	Q (W)
5000	6.24E-04	85	29.5	173.16

4.2. Distillation process

The fermentation solution of 500 mL was put into a distillation vessel, then the distillation vessel was heated from 29.5 ^{O}C to 85 ^{O}C. Vapor flows through a 2.5E-2 m diameter pipe to a condenser at a temperature of 80 ^{O}C. temperature is recorded as long as the distillation process lasts 1125 seconds using the parallax microcontroller data acquisition (ATmel 2650), can be seen in figure 3.

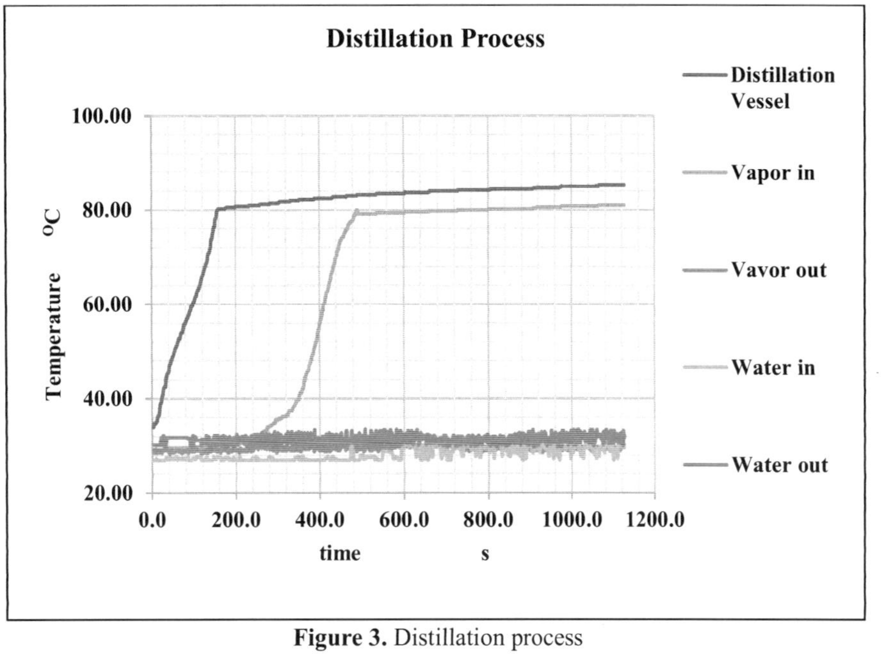

Figure 3. Distillation process

4.3. Mass flow rate

The bioethanol mass flow rate at the distillation process for 1125 seconds, calculated using equation (2), measured the mass of bioethanol directly by measuring the digital scales of time unity. Then the bioethanol mass flow rate is 0.0655 kg/ s.

4.4. Bioethanol content

After the distillation process is finished, the output of bioethanol fuel yielded and the bioethanol content determined by refractometer. The prism assembly shows of 40% bioethanol content. It can be seen in figure 4 that if we retest ethanol content in each raw material, the final results show a very good number compared to the ethanol content of a mixture of zalacca and coconut water.

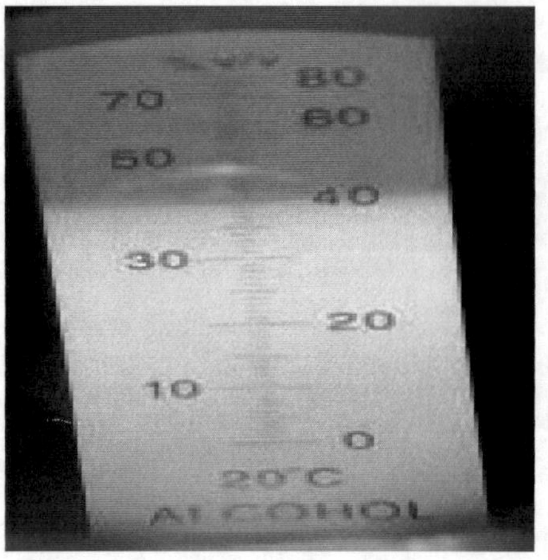

Figure 4. Ethanol content showed by refractometer

Therefore, for the next study is to reduce and measure the bioethanol heating value of these two ingredients. it showed in table 6.

Table 6. Comparison ethanol content

Ethanol content of salacca zallacca(%)	Ethanol content of coconut water (%)	Average (%)	Distillated (%)
70	**10**	**40**	**40**

5. Conclusion

Biomass has the big potential as a biofuel source, in addition to the abundant amount of raw material from agricultural waste it is also a source of environmentally friendly fuels, bioethanol is a liquid fuel that has physical properties such as fossil liquid fuels, so the user does not need to change engine geometry. In the bioethanol production process, there are several mechanical factors that influence it, the energy needed to heat up to 85 ^{O}C, the amount of mass flow rate and ethanol content that is owned by the fuel. The energy needed for distribution reaches 85 ^{O}C is 346.32 W, bioethanol mass flow rate is 0.0655 kg / s and ethanol content is obtained 40%

Acknowledgment

The authors would like to expresses his gratitude to the Director of Research and Society Service (DRSS), Ministry of Research and Higher Education for funding the Beginning Lecturer Research with Certificate Number 3 /E /KPT / 2018 with the period April to September 2018.

References

[1] S B Kumar U S Sushma L Chandrasagar V B Raju and V Devi 2017 Use of Waste Frying Oil as C.I. Engine Fuel-A Review *Open Access Library Journal* vol 4 pp 1-37

[2] A Mansouri R Rihani A N Loufi and M. Ozkan 2016 Production of Bioethanol From a Mixture of Agricultural Feedstocks: Biofuel Characterization *Fuel Science Direct* vol 185 pp 612-621

[3] M K Balki and C Sayin 2014 The effect of compression ratio on the performance, emissions and combustion of an SI (spark ignition) engine fueled with pure ethanol, methanol and unleaded gasoline *Energy* vol 71 pp 194-201

[4] P Basu 2010 Biomassa Gasification and Pyrolysis Practical Design and Theory, *Elsevier* (Burlington, USA)

[5] U.S. Department of Commerce, International Trade Administration 2010 Renewable Energy Market Report: Indonesia *U.S. Department of Commerce ; International Trade Administration* (Jakarta)
[6] S A Shahir H H Masjuki M A Kalam A Imran and A M Ashraful 2015 Performance and emission assessment of diesel–biodiesel–ethanol/ bioethanol blend as a fuel in diesel engines: A review *Renewable and Sustainable Energy Reviews* vol 48 pp 62-78
[7] S Rath A K Singh H Masih Y Kumar J K Peter P Singh and S K Mishra 2014 Bioethanol Production from Waste Potatoes as an Environmental Waste Management and Sustainable Energy by Using Cocultures Aspergillus Niger and Saccharomyces Cerevisiae *International Journal of Advanced Research* vol 2 pp 553-563
[8] J Joshi 2014 Enhanced Production Of Ethanol From Red Potatoes Grown In Hilly Regions Of Nepal Using Various Nitrogen Sources *International Journal of Applied Sciens and Biotechnology (IJASBT)* vol 2 (1) pp 41-44
[9] H Zabed G Faruq J N Sahu M S Azirun R Hashim and A N Boyce 2014 Bioethanol Production f rom Fermentable Sugar *The Scientific World Journal* vol 2014 pp 1-11
[10] A Tropea D Wilson L G L Torre R C B Lo P Saugman P T Davies G Dugo and K. W. Waldron 2014 Bioethanol Production From Pineapple Wastes *Journal of Food Research* vol 3 pp 60-70
[11] M Roy K Kundu and V Dahake 2015 Bioethanol Production From Indigenous Algae *International Journal of Environment* vol 4 pp 112-120
[12] N Shah and T Rehan 2014 Bioethanol Production from Biomass carbohydrate vs Syngas Fermentation *Journal of Chemistry and Biochemistry* vol 2 pp 161-167
[13] M El Harchi F F Kachkach and N. Elmtili 2015 Bioethanol Production From Acid Hydrolysate of Ulva Rigida C. Agardh by using Pachysolen Tannophilus and Zymomonas Mobilis *Moroccan Journal of Biology* vol 12 pp 27-34
[14] S A Byadgi and P B Kalburgi 2016 Production of Bioethanol from Waste Newspaper *Procedia Environmental Sciences* vol 35 pp 555-562
[15] M A Sukiran F Abnisa W M A W D Wan Daud and N A Bakar 2017 A review of torrefaction of oil palm solid wastes for biofuel production *Energy Conversion and Management* vol 149 pp 101-120
[16] A V Bueno M P B Pereira J V D Oliveira Pontes F M T de Luna and C L Cavalcante 2017 Performance and Emissions Characteristics of Castor Oil Biodiesel Fuel Blends *Applied Thermal Engineering* vol 125 pp 559-566
[17] M A Sukiran F Abnisa W M A D Wan Daud and N A Bakar 2017 A review of torrefaction of oil palm solid wastes for biofuel production *Energy Conversion and Management* vol 149 pp 101-120
[18] H K Noh and S Y No 2017 Effect of bioethanol on combustion and emissions in advanced CI engines: HCCI, PPC and GCI mode – A review *Applied Energy* vol 208 pp 782-802
[19] T Sinigaglia F Lewiski M E S Martins and J C M Siluk 2017 Production, storage, fuel stations of hydrogen and its utilization in automotive applications-a review *International Journal of Hydrogen Energy* vol 42 pp 24597-24611
[20] Megawati 2015 Bioetanol Generasi Kedua *Graha Ilmu* (Yogyakarta)
[21] T J K J E Berinyuy B and A N Okale 2011 Physico-chemical properties of bio-ethanol/gasoline blends and the qualitative effect of different blends on gasoline quality and engine performance *Journal of Petroleum Technology and Alternative Fuels* vol 2 pp 35-44
[22] J P Holman 2010 Heat Transfer Tenth Edition *McGraw-Hill* (NewYork)
[23] H Homma H Homma and M Idris 2014 Numerical Analysis on Wood Pyrolysis in Pre-Vacuum Chamber *Journal of Sstainable Bioenergy System* vol 4 pp 149-160

Investigation of knife quality by using forging and flame hardening methods

Balkhaya[1*], M P Anhar[1], Suwarno[2], M S M Sani[3]
[1]Mechanical Engineering Department, Politeknik Aceh Selatan, Indonesia
[2] Mechanical Engineering Department, Institut Teknologi Sepuluh Nopember Surabaya (ITS), Surabaya 60111, Indonesia,
[3]Automotive Engineering Centre (AEC), University Malaysia Pahang (UMP), Pekan Campus 26600 Pekan, Pahang Malaysia

[*]Corresponding author: balkhaya@poltas.ac.id

Abstract. The research was conducted to investigate the quality of knives the material and the forging process against wear resistance. The forged material is AISI 1050 steel, AISI 4340 steel, AISI L- 6 steel, and JIS SUP 9 steel. The manufacturing of the knives ware done by heating the material to a temperature ranging from 900-950°C then forged repeatedly manually until the temperature drop in between 650-675°C. Heating and forging are carried out several cycles to form a knife. Hardening was done by heating the knives to reach austenite temperature by flame hardening method, then quenching using water cooling media. research of wear resistance was done on the sharp side by using an actuator tribometer pin-on-plate. The results showed that wear and tear were influenced by the material and the treatment. The flame hardening process can be reduced the wear rate, the wear rate found on AISI 1050 steel knives is 5.439 x 10^{-4} mm^3/Nm after being forged, while the lowest wear rate was found on AISI L-6 steel knife which ware 2.44 x10^{-5} mm^3/Nm after flame hardening. The flame hardening process can reduce the wear rate, highest wear rate found on AISI 1050 steel knives ware 5.439 x 10^{-4} mm^3/Nm after being forged, while the lowest wear rate was found on AISI L-6 steel knife which is 2.44 x 10^{-5} mm^3/Nm after flame hardening. Therefore, it can be conclude that traditional knife quality especially the wear resistance can be improved by optimizing the heat treatment schedule.

1. Introduction

Cutting components are often found in agricultural machineryies and food processing. The ability of the cutting component was determined by the quality of the material and the manufacturing method used. The manufacturing process of the knife from steel bar, ingot, or plate is typically done by forging [1]. The knife making community knows two methods of making quality knives namely the Damascus method and pattern welding [2-3]. Both methods are the same as the methods used by traditional blacksmiths but it is differed in terms of the material used. traditional blacksmiths produce knives with limited metallurgy knowledge, so the knives were produced is low quality. In theory, the quality of traditional blacksmiths blades can be improved by selecting materials, manufacturing processes, heating treatment, and proper finishing

Damascus steel elements indicated that there are 1.60% C, 0.56% Mn, 0,17%P, 0.02% S, 0.048% Si, 0.012%Ni, 0.048% Cu, 0.01% V dan 0.002%Ti [4]. Damascus steel included in high carbon steel (ultra-high carbon (UHC)) with the chemical composition including hyper-eutectoid consisting of pearlite (lamellar cementite and ferrite). On the knife surface cementite sheets are not arranged in parallel but are corrugated, this matter was influenced by the forging process which is not uniformly but in the form of a damask pattern.

Steel AISI 1086, damascus steel, AISI 52100 steel and AEB-L steel also have different sharpness levels after forging, quenching and tempering treatment [5]. Steel AISI 52100 with a hardness of 61 HRC has a better sharpness than AISI 1086, However, the hardness value of 41 HRC with fine pearlite matrix or quenching results and tempering damascus steel has a slightly better side sharp edge than AISI 52100 steel, AISI 1086 steel, and AEB-L steel.

Theoretically knife's quality which produced by blacksmith can be improved with right material choice, good manufacturing process, and perfect finishing process. In this work, we study the traditional forging method done by a local blacksmith to make a knife from the various of the steel. Our goal of this research is to develop the process parameter to control the wear resistance of the knife.

2. Research method

2.1 Knife Making Method

Workpieces that have been cut then heated by using a blacksmith furnace with charcoal fuel. The time to reach the forging temperature depends on the material and dimensions used. The maximum temperature limit in the forging process is 50°C below the liquidus line the phase diagram Table 1 below shows the temperature and the heating temperature of each knife materials

Table 1 Temperature and heating time.

Materials	Preheating		Heating in the forging process	
	Temperature (°C)	Time (second)	Temperature (°C)	Time (second)
JIS SUP 9	34-850	195	390-850	56
AISI 1050	34-850	210	400-859	58
AISI L-6	34-957	265	625-957	79
AISI 4340	34-839	306	531-839	61

The material that has been heated using a blacksmith furnace then placed on the top of the anvil by using a clamp as a handle, then forged repeatedly using a hammer. In one heating, forging was carried out between 32-39 times per 10 cm with forging time ranging between 28-35 seconds. The forging cycle was carried out several times to forming the knife with forging temperature as shown in Table 2.

Table 2 Forging temperature and amount of forging.

Materials	Forging temperature (°C)	Forging time (second)	Max forging amount (Time)
JIS SUP 9	810-390	35	39
AISI 1050	819-400	35	38
AISI L-6	950-625	28	32
AISI 4340	830-531	34	38

2.2 Flame Hardening method

Manual flame hardening was done by heating the knife surface reach a temperature of 850°C. Heating was done by using a mixture of oxygen and liquid petroleum gas (LPG), Then the knife is quenched using cooling media form of water.

2.3 Wear resistance test

Before measuring the wear resistance, the knives was cut with dimensions of 10 mm, long and thickness adjusted to the dimensions of the knife. then the sharp side surface was smoothed using a griding tool machine as shown in Figure 1a. Next Weigh the mass of the specimen before and have tested the wear resistance in order to find out the lost mass of each knife specimen. then weigh the mass of the specimen before and after tested the wear resistance in order to find out the lost mass of each knife specimen.

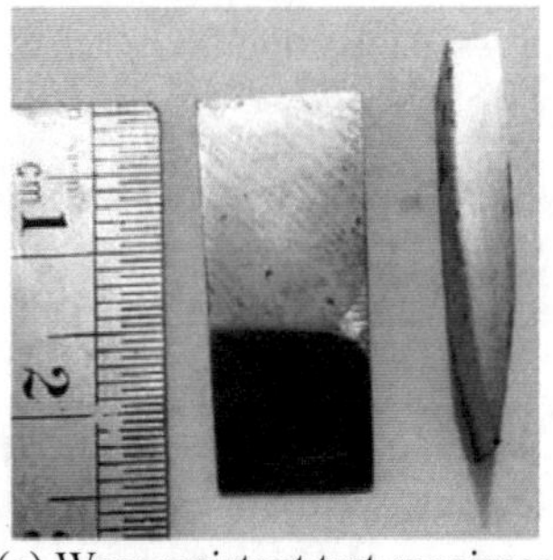

(a) Wear resistant test specimen

(b) Testing process is wear resistant

Figure 1 Testing is wear resistant

Wear resistance testing was carried out using a pin-on-plate tribometer actuator as shown in Figure 1b. In the test equipment was given a load of as 1290 g, then swiping the sharp side of the knife specimen on the plate that agitates back and forth for 25 minutes with a swipe plate speed of 0.4 m/s. Testing every condition was carried out two measurements then taken on average. The weight difference before and after the test is the wear data from the knife specimen. Specific wear rate is calculated using the following equation[6].

$$K = \frac{\Delta V}{F \cdot L} \quad (1)$$

Where K is the Spesific wear rate (mm^3/Nm), ΔV is the volume changes (mm^3), F is the force applied (N), L is the distance sliding (m).

3. Experiment Results

Knives after forged are shaped as shown in Figure 2. The results of visual observation showed that on each knife did not defect occur.

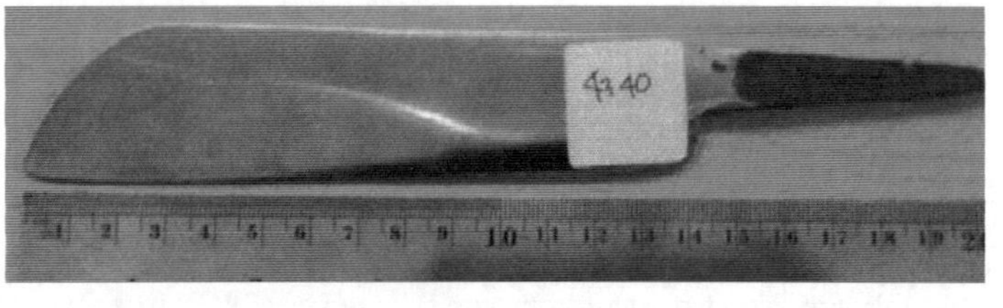

Figure 2 Steel knife 4340 manual forging results

Friction of the knife with other objects when cutting will cause a loss of sharp edges. In the end it causes the knife to wear out so that it cannot cut perfectly. Wear resistance defined as the weight of the knife specimen reduced after sharp side friction with hard objects. From the test results obtained the specific wear rate value on the knife material after being forged as show in Figure 3.

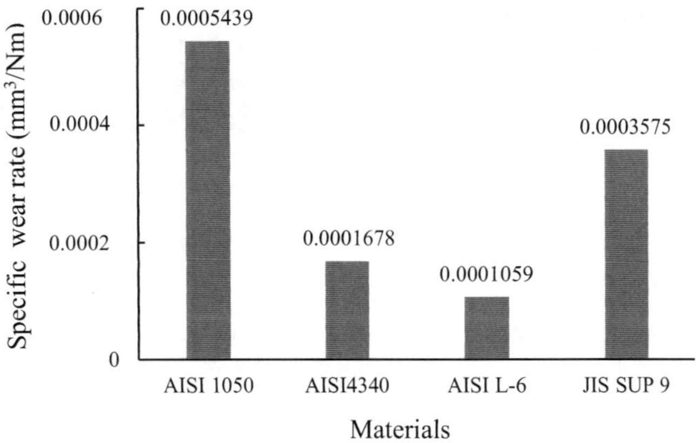

Figure 3 Graph the rate of wear resistance of the Knife after being forged

From Figure 3, It can be observed that the forging process, the highest wear resistance occurs in AISI 1050 steel knife in the amount of 5.439 x 10^{-4} mm^3/Nm. the lowest wear resistance of steel knife AISI L-6 is 1.059 x 10^{-4} mm^3/Nm. While the wear resistance of AISI 4340 steel blade is 1.678 x 10^{-4} mm^3/Nm lower than the wear resistance that occurs in steel knife JIS SUP 9 is 3.575 x 10^{-4} mm^3/Nm. After the hardening is carried out, there is an increase in wear resistance in each knife as shown in Figure 4.

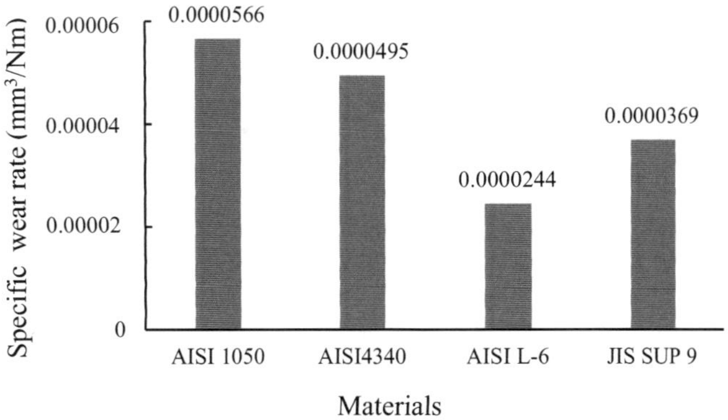

Figure 4 Graph the rate of wear resistance of the Knife after flame hardenig

On steel knife of AISI 1050 of wear resistance is 5.66 x 10^{-5} mm^3/Nm higher than AISI 4340 steel knife, AISI L-6 and JIS SUP 9. The lowest wear resistance occurs in steel knife AISI L-6 is 2.44 x 10^{-5} mm^3/Nm. From the type of material, it can be observed that the highest wear resistance occurs in AISI 1050 steel knife and the lowest is in AISI L-6 steel knife as shown in Figure 3-4, this is also due to differences in levels of violence [6-7]. The JIS SUP 9 knife material has the highest hardness value after being hardened. In addition, Cr levels of the material can be also kept wear resistance levels. Steel AISI L-6 has a Cr content of 1.13% while AISI 1050 has a Cr content of 0.0262%. Increased wear resistance is also caused by microstructure in the material [8-10].

IOP Conf. Series: Materials Science and Engineering **506** (2019) 012004 doi:10.1088/1757-899X/506/1/012004

4. Conclusion

The quality of the knife was affected by the type of material and the flame hardening process. The knife of the forging process has a higher wear rate compared to the knife of the flame hardening process. The highest wear rate obtained on AISI 1050 steel knife is 5.439 x 10^{-4} mm^3/Nm after the forging process. The lowest wear rate is 2.44 x 10^{-5} mm^3/Nm obtained on AISI L-6 steel material after the flame hardening process.

References

[1] J D Verhoeven 2001 The mystery of damascus blades *Scientific American* vol 284 (1) pp 74-79

[2] F Özkan, H Kirdan, I H Kara, F Hüsem, Y Akinay, H C Ug, Y Sun and H Ahlatic 2014 Mechanical propertis of pattern welding 1075-15N20 Steel *Mater test* vol 56 pp 897-900.

[3] Yoso, M., Takaiwa, T., Minagi, Y., Kanaizumi,T., Kubota, K., Hayashi, T., Morito, S., Ohba,T. 2013 Study of Javanes Sword from a Viewpoint of steel strength, *Alloys Compd.* vol 557 pp 690-694.

[4] F Fatollahi Fard. 2011 Damascus Steel, *MSE.* **121** pp 1-10.

[5] J D Verhoeven, A H Pendra and H F Clark 2008 Wear tests of steel knife blades *Wear* vol 265 pp 1093-1099.

[6] M M Khonsari and E R Booster 2001 Applied Tribology, Bearing Design and Lubrication *Wiley Publisher* New York.

[7] Balkhaya and Suwarno 2017 Effect of raw materials and hardening process on hardness of manually forged knife *AIP Conference Proceedings* vol 1855(030010)

[8] V F Silva, L F Canale, D Spinel, W W B Filho and O R Crnkovic 1999 Influence of retained austenite onshort fatique crack growth and wear resistance of case carburized steel *Journal of materials Engineering and Performance* vol 8(5) pp 543-548

[9] R N Penha, J Vatavuk, A A Couto, S A de L Pereira, S A de Sousa and L. de C F Canale 2015 Effect of chemical banding on the local hardenability in AISI 4340 steel bar *Engineering Failure Analysis* vol 55 pp 59-68

[10] V A González-mora, M Hoffmann, R Stroosnijder, E Espinar, J M Llamas, M Fernandez-Fairen, and F J Gil 2011 Influence of different cocrmo counterfaces on wear in ultra-high molecular weight polyethelene for rrtificial joints *J. Biomed. Sci. Eng.* vol 4 pp 375-382

IOP Conf. Series: Materials Science and Engineering **506** (2019) 012005 doi:10.1088/1757-899X/506/1/012005

Realizing Industry 4.0 Through STEM Education: But Why STEM Is Not Preferred?

Nor Samsinar Kamsi[1], R B Radin Firdaus[2*], Farrah Dina Abdul Razak[1] and M Ridha Siregar[2]

[1]Universiti Teknologi MARA Cawangan Perak, Kampus Tapah, 35400, Malaysia.
[2]School of Social Sciences, Universiti Sains Malaysia, 11800 USM, Malaysia.

*Corresponding author: radin@usm.my

Abstract. Since the 1760s, industrialization of the world has gone through four revolutions where the 4th revolution has expeditiously taken over from the traditional practices. The Industry 4.0 that embraces the computerization in smart factories requires improved capacities in science, technology, engineering, and mathematical (STEM) fields. Nevertheless, the present circumstances in Malaysia demonstrate that the numbers of youth who decide to follow STEM routes for tertiary education are far less encouraging. Thus, this study aimed to explore the factors affecting university students' decisions. Data used in this study was obtained from a survey of 102 Diploma in Accountancy students from one of the public universities in Malaysia, who were selected using purposive sampling method. This study employed the descriptive statistical method, principal component analysis (PCA) and multiple regressions analysis. The PCA on the questionnaire measurement items yielded factors accounting for 68.22 per cent of the total variance. In the multiple regression model, low morale attitude, learning experience and return on investment explained 53 per cent of the total variance of perception. This study concludes that the roles of national workforce that consists of education system, culture, and government must be strengthened to enhance STEM education system in Malaysia.

1. Introduction

Industry 4.0 consists of physical cyber, Internet of things (IOT), knowledge, innovation and talent. These are the factors that will enable a developing country like Malaysia to become a developed nation. According to the National Council for Scientific and Research Development, Malaysia requires at least 500,000 scientists and engineers by the year 2020 but the current statistics shows that there are only 70,000 registered engineers, which is only 17 per cent of the figure [1]. Thus, to meet the demands of human capital in the high-tech economy, the higher education system in a country should be tailored according to the needs of Industry 4.0.

Transformation in education system is inevitable to sustain a strong economic growth and spurring investment in a country [2]. The education market structures will shape the behavior of the participants in the market that give an effect towards achievement and productivity [3]. The issues around the world including Malaysia is while countries exerting to enhance their science, technology, engineering and mathematical (STEM) education, the students' interest in STEM is diminishing throughout their schooling period, resulting them to divert away from STEM field when pursuing higher education.

2. Literature Review

2.1. Perception and expectation of learning STEM

There are many factors that could shape the perception of students towards STEM courses in higher institution. Perceived internal barriers (confidence and preparation) are significantly related to a decrease in STEM career aspiration [4]. The internal barriers consist of structural barriers (the institution's capacity), psychological barriers (the level of STEM courses' difficulties) and perceived barriers of educational and career outcomes. Dustmann [5] found that the parental background such as parents' profession, working environment, and peer group of the father and mother seem to play important roles for education of the child.

2.2. Interest, morale and attitude

Wang and Degol [6] conducted an intensive review with regard to the links of intellectual aptitude and motivational beliefs to performance, educational, and career choices. The authors found that motivational beliefs, which are informed by aptitudes in math and science, competence beliefs, interest, occupational, and life values clearly play a role in the decision to pursue STEM versus non-STEM fields. For instance, in comparison between genders, females prioritize family when they have children thus more likely to choose careers outside of STEM fields due to more career options.

2.3. Learning experience

Students spend their time between six to nine hours at school, thus experience of school science has a large influence on students' choice to take science in the future [7]. Chong [8] criticized learning science in Malaysian classrooms of focusing on study of factual knowledge where students do not understand the nature of science. For Malaysian students, after studying mathematics for several years, the knowledge can only be applied after they enroll in science engineering stream at the age of 16. Therefore, there is a gap of time between knowledge and application which students fail to see how the knowledge should be integrated.

As for the teachers, Pimthong and Williams [9] explored the understanding of STEM among preservice teachers and found that the teachers are aware of the importance of STEM subjects but not able to explain how those subjects integrate. Hadi and Fred [10] suggested that mathematical and science subject should not be implicitly integrated. The curriculum should make mathematics more explicit, such as teaching mathematics into physics lessons through designed activities. Effective communication is vital in determining the efficacy of pedagogy and learning. Since 2003, Malaysia has implemented teaching science and mathematics in English. However, Thomas and Watter [11] found that the situation in Malaysia is contradicted with that of in India. Disadvantages among native Malay speaking teachers in delivering STEM knowledge in English resulted in failure of STEM education. Implementing constructive curriculum in Malaysia seems difficult as to change the traditional beliefs of the teachers [12].

In view of teacher effectiveness, Shirazi [7] highlighted that teachers' personality and teaching style are one of the top three factors. However, even though we do have highly qualified teachers, the interaction between education and production is fundamentally impossible if student trainees have no access to modern technology and equipment. The outcome must be varied across the nation in terms of urban versus rural areas. In Malaysia, class sizes are large (40–50 students) and poorly furnished [11]. The lack of access to science resources and facilities, high quality teachers, and advanced coursework would affect the quality of teaching STEM subjects [13].

2.4. Making Money Worth (Return on Investment Education)

In Malaysia, most students whose parents are not able to provide education fund will use educational loan from Perbadanan Tabung Pendidikan Tinggi Nasional (PTPTN) to pursue their study. Thus, students perceive their loan to be worthwhile investment for career opportunities as to outweigh their debt burdens [14]. By using 2007 Malaysian Household Income Survey (HIS), Husaina [15] found that due to current job market demands, completion of primary and lower secondary schooling does not result in higher returns to education in the market, resulting in the youths being further to the university level in order to capture higher returns to education.

3. Methodology

This study aims to examine factors that influence university students, who used to be a science stream student in secondary school, in opting for non-sciences field at university. This study employed quantitative approach using purposive sampling technique. The respondents were accounting students of Universiti Teknologi MARA (UiTM) Cawangan Perak who used to be a science stream student in secondary school. They were identified as "undergraduates who migrated from STEM routes to non-STEM routes". A total of 112 respondents were involved in this survey.

The descriptive questions for the first objective were designed based on Constantinou et al. [16], while the Likert-scale questions for the second objective were developed following Loren and Naltan [17]. In order to measure consistency of all items in the questionnaire, a reliability test using Cronbach's alpha was undertaken. We used Principal Component Analysis (PCA) to transform correlated variables into uncorrelated variables. A total of 20 questionnaire items were filtered using dimensionality reduction to extract relevant components. Based on these components, this study examines the relationship between independent variables (low morale attitude, learning experience, return on investment) and dependent variable (perception) by applying multiple regression analysis.

4. Results and discussion

From the analysis, 96.9 per cent respondents decided which courses they want to pursue. In investigating where they got the idea to pursue the course, figure 1 shows the channel of references. The respondents were asked to put ranks between those options and most of them chose their own opinion and ideas as the first rank. This indicates that their decision to pursue in non-STEM course is based on their own ideas and opinion (respondents' perception).

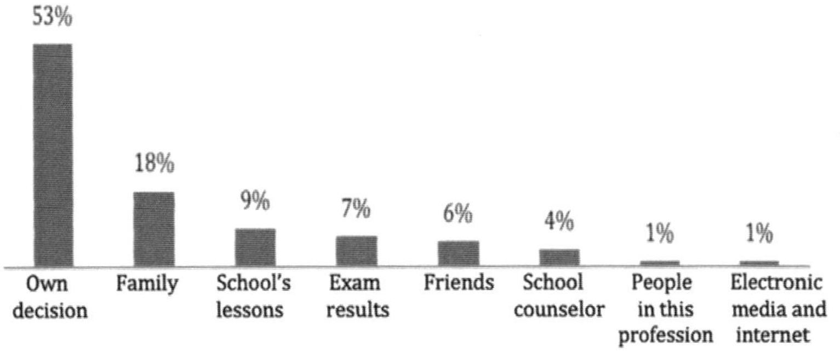

Figure 1. Source of references.

Figure 2 represents what are the criteria that influence them to pursue in accounting instead of STEM course. 84.6 percent believed that choosing accountancy would offer better future career prospects, followed by high salary. This is parallel with the range of salary for senior manager in accounting, and audit & tax services that is in the range of between RM10,444 to RM15,911 while the range salary for senior manager in consulting (IT, science, engineering & technical) is from around RM8,250 to RM11,000 [18].

In regard to their preferences in job scope, 55.4 per cent prefer to work in public sector, which gives better assurance. Meanwhile, there were 64.6 percent respondents who used to work during period of study but only 21.5 per cent of them claimed that work did influence their choice to pursue in accountancy.

In the observation of the roles of parents in the process of decision-making, figure 3 indicates that most of them do have frequent interaction with parents. However, figure 4 and figure 5 indicate that the roles of parents seem insignificant due to high percentage of parents who always accept the students' idea instead of assuming a guidance role to them. This situation reflects that there are weaknesses in culture of guiding the children and lack of awareness on how important to emphasize the science education as to fulfill the requirement of 4th industry revolution. Regardless of teaching approach, the general culture provided by parents as a role model would attribute to the school ethos [19].

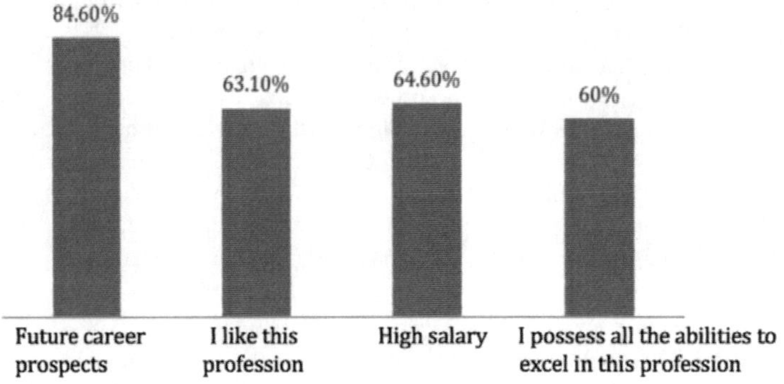

Figure 2. Criteria that influence respondents to choose accounting course.

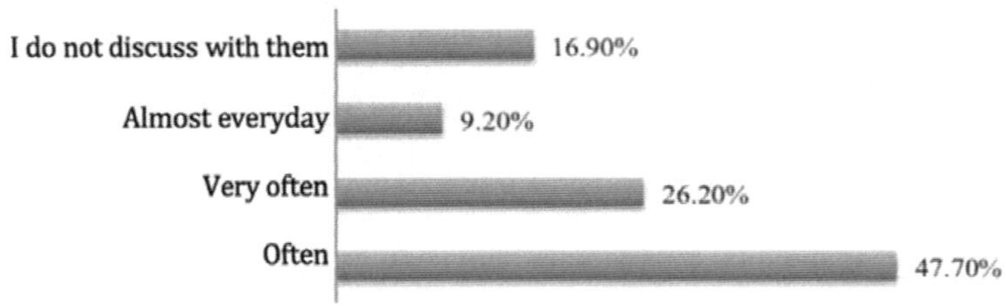

Figure 3. How often do you discuss with your parents about your studies and professional development?

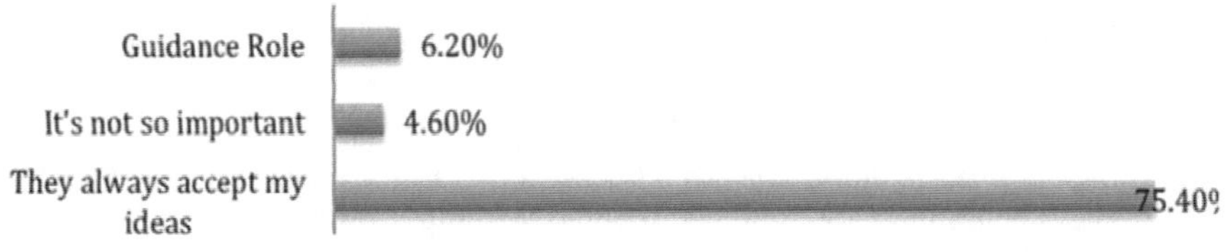

Figure 4. What is the role of your parents?

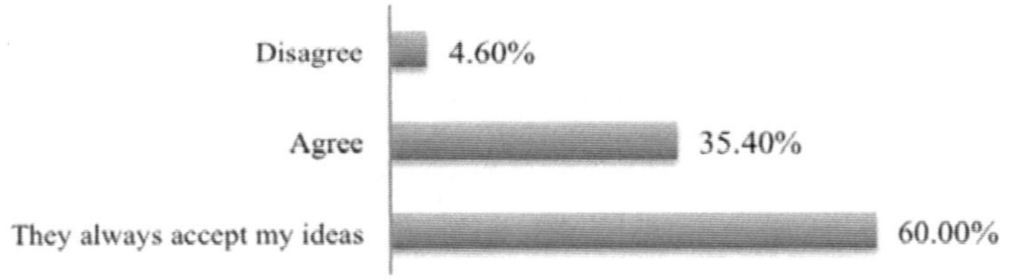

Figure 5. Did your parents disagree with the course you chose?

In Table 1, the values of the Cronbach's alpha, α for the first and second factor are strong, and the rest are acceptable (α > 0.5) to explain the consistency of statements in measuring every component.

In order to investigate which factors that have the most variance, principal components analysis (PCA) was used. Bartlett's test of sphericity was statistically significant ($p < .05$) and the Kaise Meyer Olkin (KMO) value was 0.544, indicating that the sample is adequate to explain the case. The PCA on the questionnaire measurement items yielded 6 factors with eigenvalues greater than 1, accounting for 68.22 per cent of the total variance and the first factor accounted for 24.04 per cent for the variance. Component 1 in Table 1 clearly represents 'Perception', Component 2 referring to 'Low Morale Attitude', Component 3 assigned as 'Learning Experience during Secondary School', and Component 4 as 'Returns on Investment'.

Table 1. Cronbach's Alpha test.

	Component			
	1	2	3	4
Cronbach's Alpha (α) =	0.757	0.706	0.604	0.565
Law, Accountancy and Business give more immediate work results	0.849			
The demand of labor from the job provider in STEM field is quite rigid since the expectation is very high.	0.808			
The level of syllabus will be tougher (it is highly elitist field).	0.604			
I have the inferior feeling of dropping out from university	0.432			
I might be interested in STEM courses if the teaching of subject is more engaging at young age		0.914		
I felt discourage		0.638		
I scored low marks in my entire test		0.792		
I had difficulties to understand the subjects		0.729		
I did not see the real picture of how things (the concept, theory, exercise in class) actually work.			0.827	
Teachers only talking at the front of a classroom while students taking notes and doing worksheet			0.619	

There were less practical science activities during lessons		0.445	
I want to make my money worth so I chose non-STEM courses			0.75
Taking non-STEM courses ensure me graduate on time			0.619

The highest is value of statement in perception, revealed that undergraduates choose this course as they want to get hired immediately right after completion of study. They also have an opinion that the demand for a worker in STEM field might be rigid due to the expectation being very high. Besides, they also think that the level of syllabus will be lot tougher as they think it is a highly elitist field and have the inferior feeling of dropping out from university. In contradiction, the respondents are actually still having interest in STEM field. This can be identified from the highest value in the second components, which belongs to the statement that "I might be interested in STEM courses if the teaching of subject is more engaging at young age". However, the weaknesses in teaching approach resulted in them feeling discouraged. Accompanied by low morale, these have led the students to score low marks due to the difficulties to understand the subject. Recalling the experience of learning STEM subjects during secondary school, the highest score stated that they actually did not see the real picture of how things actually work and teachers were still using traditional method and less practical science activities during lesson. Student are also concerned on the return on investment for assurance of their money's worth by choosing the right track and able to grad on time.

From the regression analysis in Table 2, the concern on return on investment is the most driven factor that influences undergraduates to migrate from STEM courses. Most students are using educational loan to pursue their study. Given the rising cost and economics condition, the uncertainties of getting a job from STEM field could be a high risk for a borrower in repayment default [14].

Amongst of the barriers for a nation to accumulate the talents in STEM field are the internal barriers that exist due to weak learning experience, and fragile attitude. This is supported by Boon [20] who stated that our science lessons are no more focusing on hands-on scientific investigations as there is reduction in science practical works ever since the science practical examination was scrapped in the early 2000s.

Table 2. Regression analysis.

Variables	B	Std. Error	*t*	Sig.
Low Morale Attitude	0.165	0.098	1.679	0.098
Learning Experience	0.239	0.094	2.538	0.014
Return on Investment	0.289	0.081	3.552	0.001

R^2= 0.53%

5. Conclusions

The study result implies that the roles of national workforce that consists of education system, culture, and government must be strengthened to enhance STEM education system in Malaysia. This is important in order to cultivate and spur the interest in STEM in every young generation. Joseph [13], in his model of human capital for Smart Manufacturing and Industry 4.0 revolution, stated that the roles of national workforce which consist of education system, culture, and government are vital in forming the human capital equipped with education, knowledge, experience, skills, creativity, and motivation. The role of government is to provide the support system especially in educational fund for the STEM fields' students. By providing financial aid, it could lower the educational cost and will increase the return on investment since the market salary offered by the industry still do not outweigh the burden of cost studying since the salary range is somehow similar with the non-STEM groups' income. At institutional level, finding the right formula of teaching STEM subjects could be the key factor. Even if numerous policies were formulated; the implementation at school level still requires interpretation and the ingenuity of the teachers. In terms of flourishing the STEM field in societies' culture, parents must implicitly and explicitly value the science education. The priority is to create the worth of STEM field environment and children can choose their interest based on what they have experienced.

Acknowledgements
This research has been funded by Universiti Sains Malaysia through Short Term Grant (304/PSOSIAL/6313317).

References

[1] Audrey V 2018 Malaysia Needs 500,000 Scientist and Engineers by 2020 *New Straits Times*
[2] Boev S 2017 *Problems of Economic Transition* **59**(7–9) 576–599.
[3] Caroline M H 2003 The Economics of School Choice *National Bureau of Economic Research* Available from: https://www.nber.org/chapters/c10083.pdf [Accessed 3rd August 2018]
[4] Allison S and Alexis M 2014 Perceived Barriers to Higher Education in STEM among High-Achieving Underrepresented High School Students of Color
[5] Dustmann C 2004 *Oxford Economic Papers* **56**(2) 209-203.
[6] Wang M T and Degol J 2013 *Developmental Review* **33**(4) 304-340.
[7] Shirazi S 2017 *International Journal of Science Education* **39**(14) 1891-1912.
[8] Chong B P 2005 Understanding of the Nature of Science and Its Relationship with Academic Achievement and Gender Among Form 6 Students, Unpublished M. Ed Project Paper *University Malaya, Kuala Lumpur, Malaysia.*
[9] Pimthong P and Williams J 2018. *Kasetsart Journal of Social Sciences*
[10] Hadi M A and Fred F 2016 STEM Education in Malaysia: Reviewing the current physics curriculum. *International Conference of Association for Science Teacher Education,* Reno, Nevada, United States.
[11] Thomas B and Watters J J 2015I *International Journal of Educational Development* **45** 42-53.
[12] Yunus H M, Ismail Z and Raper G 2004 *Journal of Science and Mathematics Education in S.E. Asia* **27**(1) 166-203.
[13] Joseph E A 2018 Human capital in the smart manufacturing and industry 4.0 revolution. *Digital Transformation in Smart Manufacturing*
[14] Baum S and O'Malley M 2003 *Journal of Student Financial Aid* **33**(3) 7-19
[15] Husaina B K 2013. *International Journal of Educational Development* **33**(4) 380-393.
[16] Constantinou A, Michailidis, M P and Kyriakidou N 2011 Career guidance: A study of factors that influence students' choices of undergraduate studies *European Career Guidance Counselling Conference 2011*
[17] Loren A and Naltan L 2014 *Journal of Education and Human Development*, **3**(2) 391 – 404.
[18] Jobstreet.com 2017 *Salary Report 2017*
[19] Jayantee N 2011 Factors Affecting the Choice of Science Subject Among Girls at Secondary Level in Mauritius (Ph.D. Thesis) *Brunel University*
[20] Boon N S 2014 Malaysian School Science Education: Challenges and the Way Forward. *Curriculum Development Division, Ministry of Education Malaysia*

The Substitution of Fine Aggregate with Interlocking Material of Stone Quarry Dust in High Strength Concrete

K C S Zein [1], Meillyta [1], A Wahyuni[1]
[1]Civil Engineering Department, Faculty of Engineering, Universitas Muhammadiyah Aceh, 23245 Banda Aceh, Indonesia

*Corresponding author: meillyta@gmail.com

Abstract. Some of the parameters for improving the quality of concrete are the selection of its constituent materials and compaction process to increase the interlocking between materials. The selection of cubical agregate as fine agregate in the form of stone dust origin from quarry stone residue was chosen as a research object as it is suspected can increase the interlocking material due to its sharp edge. This study aimed to improve the interlocking between material in concrete so as to improve the compressive strength of high strength concrete. The research method is using the standard in SNI 03-6468-2000 (Pd T-18-1999-03) with water cement ratio of 0.3 and 25-50 mm planned slump. Sica viscrocrete – 10 and silica fume was added as additive. The percentage of sica viscrocrete – 10 and silica fume respectively is 1.5% and 10% from cement weight. Dust is used as the substitution of fine agregate. Compressive strength test in this study was performed using 30 cm height and 15 cm width concrete cylinder specimen at 7 and 28 days old. From the test, it was obtained that the compressive strength at 7 days is 53.14 MPa for concrete using natural sand and 52.04 Mpa for concrete using stone dust and at 28 days old is 68.19 MPa for concrete using natural sand and 82.52 MPa for concrete using stone dust. It can be concluded that stone dust as fine agregate substitution can boost more compressive strength of high strength concrete as it has a cubical and sharp plane edge which can help to increase the interlocking between material during the compressive strength test.

Keywords: Quality improvement, high strength concrete, compressive strength test

1. Introduction

High strength concrete is one of concrete types which has high compressive strength. Generally, its required compressive strength is greater than 41.4 MPa [1]. Most of the concrete constituent material is aggregate which tied with cement paste. The commonly used agregate is come from nature, for example river stone and sand. Natural agregate caused the presence of porosity which is becoming one of the most influential problems of compressive strength in high quality concrete. The use of natural river sand and uniform granule size which has round form can provide airspace in concrete which result in smaller compressive strength.

Meanwhile, angular aggregate particles with the sharp corner can resist the possibility of aggregate movements during compressive strength test. This is because it has more than one split plane that will give more extra bonds between concrete constituent, reduce the percentage of porosity in concrete and increase its workability.

Crushed stone dust is chosen as natural fine agregate replacement because its cubical form is suspected can provide better interlocking between material. Concrete with better material interlock will increase its strength. The aim of the study is to determine the compressive strength of high strength concrete at 7 days and 28 days old using crushed stone dust as natural fine aggregate replacement. This research was performed at Construction and Building Materials Laboratory Civil engineering Departement Universitas Muhammadiyah Aceh. Mix design method was using standard in SNI 03-6468-2000 (Pd T-18-1999-03) with water cement ratio of 0.30 and slump of 25-50 mm. Concrete cylinder of 15 cm diameter and 30 cm height was used as specimen.

2. High Strength Concrete

High strength concrete can be interpreted as a high performance concrete that takes into account the durability as well as workability of concrete. In high strength concrete, the decreasing of particle and pore size will increase the material strength [2]. Generally high strength concrete has a required strength of 40–80 MPa, and nowaday increasing accordance to its application.

High strength concrete has been used widely in high rise structure as coloumn, beam and foundations. It is also use for bridges structures and highway pavement. Its application in structures reduce the dimension of structural element so the dead load also reduces. The characteristic of high strength concrete is its high early strength, high modulus elasticity, durability and compaction without segregration [3].

3. Aggregate

Aggregate is one of the important ingredients in concrete, which occupies 70% -75% of the concrete volume, so that the aggregate is very influential on the properties of concrete [4]. Once the aggregate is used according to the requirements then the concrete may have good workability, strong, durable and economical value. In the use as concrete material, aggregate is divided into 2 i.e fine aggregate and coarse aggregate, which is usually, comes from the river.

The shape of aggregate may affect the workability of the concrete mix during the mixing period in terms of the compacting energy required for the mixture to be set and the strength of structural concrete during its service life [5]. Moreover, the strength of the aggregate will also affect the compressive strength of the concrete. Main factors affecting concrete strength are aggregate interlocking and porosity [5]. Aggregate with good gradation and sharp edge is more likely to have better interlocking.

Stone dust is a side product of a quarry which has a sharp corner. It is resulted from the process of breaking stone into pieces which has a granular size of 0 - 5 mm and shaped like a pebble but has a smaller diameter which is less than 5mm [6]. It is produced from the location of the stone crusher contain approximately 17% to 25% dust fraction [6]. Therefore, stone dust has a potential volume to be used in other purpose for example to replace the use of fine aggrefate from the river.

4. Compressive Strength

Compressive strength of concrete is obtained using a test machine, by providing a multilevel press load with a certain incremental load speed on the cylindrical test object until it crushed [7]. Generally, the testing procedures is using ASTM C39-86 standard. The compressive strength is measured using standard cylinder specimen of 150 mm diameter and 300 mm height at 28 days old.

The compressive strength of each specimen is determined by the compressive stress (f'c) achieved at the 28-day test due to the experimental load. Based on the collapsed loads that can be accepted by the specimen, the value of the compressive strength of the structural concrete can be calculated using equation (1) below.

$$fc' = \frac{P}{A} \tag{1}$$

Where fc' is the compressive stress (MPa), P is the load (N), and A is the cross sectional area (mm^2).

5. Research Methodology

Method used in this research is according to [8]. This procedure applies only for high strength concrete produced using conventional materials and production methods. The research process was performed according to the figure 1 below.

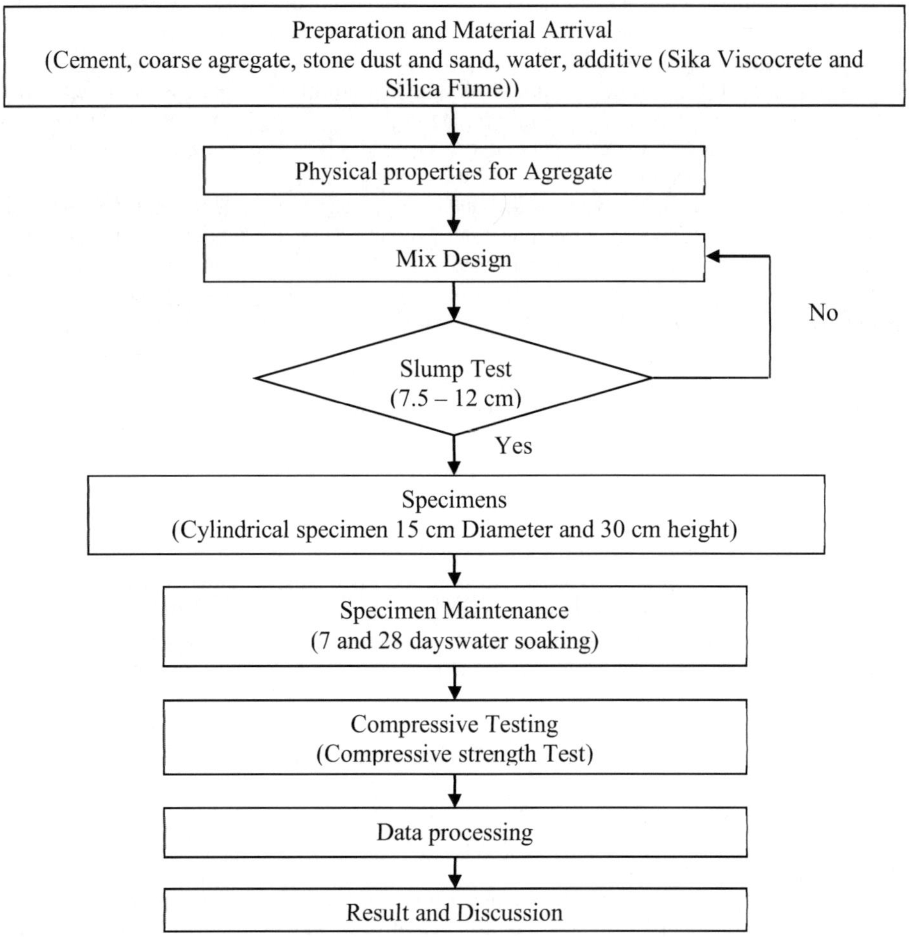

Figure 1. Research Procedure

Variaton and the number of specimen can be seen in table 1. The use of water in high strength concrete is very little which results on mortar that is difficult to work with. A high level of workability for mortar will be obtained by adding superplastizer to achieve the appropriate slump value without bleeding and segregation. In this study, Sika Viscocrete 10 was chosen as an additive. It was added in the same amount for each variation which was 1.5%.

IOP Conf. Series: Materials Science and Engineering **506** (2019) 012006 doi:10.1088/1757-899X/506/1/012006

Table 1. Variation and Number of Specimen

Variation of Fine Agregate	Days		Number	Note
	7	28		
Dust 100% + 1,5% Sika Viscocrete + 10% Silica Fume	10	10	20	
Pasir 100% + 1,5% Sika Viscocrete + 10% Silica Fume	10	10	20	comparison
	Total Number		40	

6. Results and Discussion

Based on the result of mix design in the form of water, cement, fine sand, coarse sand, dust and additive material (silica fume den sika viscocrete-10) using SNI 03-6468-2000 method (Pd T-18-1999-03) are summarized in table 2 in the form of a proportion of concrete mixtures for each variation per 1 m3

Table 2 Proportion of Concrete Mixture for each variation per 1 m^3

Material type	Material Weigth(kg)	
	Stone Dust 100%	Sand 100%
Water	159	167
Cement	521	548
Split	1144	1144
Stone Dust	574	-
Sand	-	553
Silica Fume	52,1	57
Sika Viscocrete	7,8	8,5

From the calculation of the proportion of concrete mixture, it can be seen the use of cement in concrete mixture using stone dust is less compared to the one that used natural river sand. This is because in the aggregate physical properties of the two types of aggregates are obtained different bulk density which affect the required water content.

From the slump test, it is obtained average slump of 19 – 21 cm, which meets the required slump of 25 -50 cm. Slump test is shown in table 3 below.

Table 3. Slump Test

No	Variasi Pengujian	Slump (cm)		
		10 specimens	10 Specimens	Rata-rata
1	Stone Dust 100%	19	21	20
2	Sand 100%	21	21	21

The high slump value that can be seen in the table above due to the addition of superplasticizer to obtain high workability in concrete specimen.

From the test results and data processing performed statistically, concrete compressive strength values obtained at 7 days and 28 days old concrete specimen are shown in table 4.

Table 4. Test Result and Data Processing

Specimens Variation		Average f'c (MPa)	Sd	f'c	f'c	f'c (corrected)(MPa)
Life Factor (day)	Fine Agregate			min	max	
7	Pasir (100 %)	52.13	3.68	47.2	57.05	53.14
	Dust (100 %)	52.07	2.5	48.72	55.42	52.04
28	Pasir (100 %)	66.69	5	59.99	73.39	68.19
	Dust (100 %)	83.94	6.27	75.55	92.34	82.52

The ratio of compressive strength of concrete of stone dust and natural sand mixture at 7 days and 28 days old test is shown in bar chart at figure 2. The result shows that concrete mixture using stone dust as fine aggregate, the compressive strength corrected at 28 days old reach 82.53 Mpa. It is found that with the substitution of stone dust as fine agregate can boost the compressive strength of high strength concrete. Moreover the obtained compressive strength suit the minimal compressive strength of high strength concrete requirement in SNI 03-6468-2000 (Pd T-18-1999-03) which is 41.4 MPa. The compressive strength of concrete using stone dust is increasing 50.1% from the design.

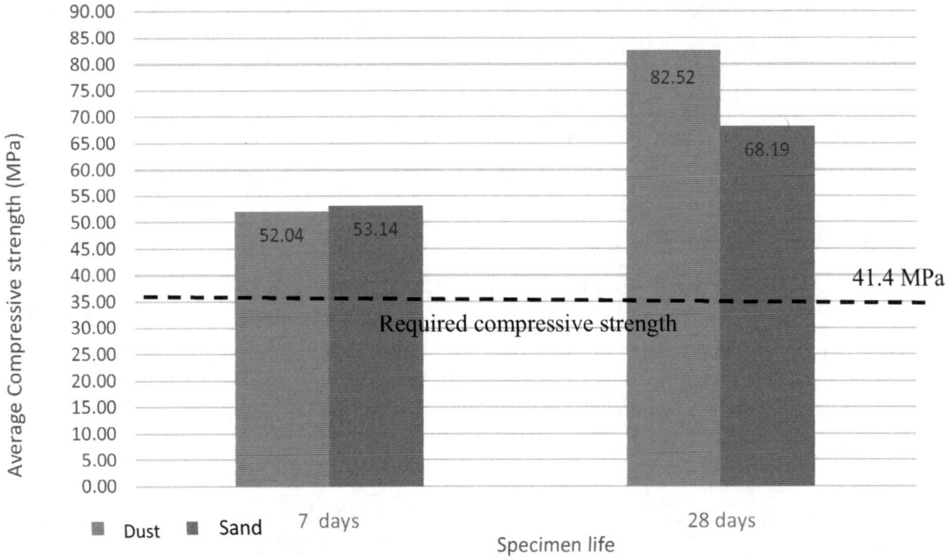

Figure 2. Comparison Graph of Compressive Strength for each Variation

7. Conclusion

Based on the result, it can be concluded that the use of quarry stone dust as fine agregate substitution can increase the compressive strength og high strength concrete. It is proven that stone dust granule which has angular aggregate particle with sharp corner gives better agregate interlocking which can prevent the possibility of agregate movement in the concrete with its higher result of compressive strength. Moreover, higher compressive strength at the earlier age of concrete reaches 52.04 Mpa which is more than the required strength, confirm the theory that high strength concrete has an high early strength.

Acknowledgments
This research was supported by Universitas Muhammadiyah Aceh. We appreciate and thanks to all colleagues and all those who have contributed from research conducted until this paper was published.

References

[1] Standar National Indonesia Pd-T-04-2004-C 2004 Tata Cara Pembuatan dan Pelaksanaan Beton Bermutu Tinggi, Jakarta *Badan Standardisasi Nasional*
[2] Nugraha and Antoni 2007 Teknologi Beton dari Material Pembuatan Beton Kinerja Tinggi (Yogyakarta)
[3] P C Aitcin 2004 High Performance Concrete *E & FN Spon*
[4] O E Gjørv 2008 Development in the Formulation and Reinforcement in Concrete *Mindness S. Woodhead Publishing Limited* p79
[5] Alizar 2010 Teknologi Bahan Konstruksi *Pusat Pengembangan Bahan Ajar – UMB* (Jakarta)
[6] Celik T and Marar K 1996 Effects of Crushed Stone Dust on Some Properties of Concrete *Journal of Cement and Concrete Research* (Pergamon) vol 26(7)
[7] Dipohusodo I Struktur Beton Bertulang *PT Gramedia Pustaka Utama Publisher* (Jakarta)
[8] Standar National Indonesia 2000 Tata Cara Perhitungan Campuran Beton Berkekuatan Tinggi *Yayasan Penyelidik Masalah Bangunan* (Bandung) SK SNI 03-6468-2000 (Pd T-18-1999-03)

Spatial Distribution of Anemia Among Adolescents In Banda Aceh, Indonesia

K M Zurnila[1*], Hizir[1], R Latifah[1] and U Farazilla[1]
[1]Statistics Department, Faculty of Mathematics and Natural Science, Syiah Kuala University, Banda Aceh, Indonesia

[*]Corresponding author:kesumaku@yahoo.com

Abstract: Adolescents are an important assets to determine whether a country will be advance in the future. A good nutrient intake is needed to ensure their growth and development. Lack of nutrients will cause various health problems, one of them is anemia. Anemia is influenced by many factors, including spatial factors. The authors examine the relationship between anemia and variables that affect it. The relationship may be influenced by regional or spatial factors. Total data 602 of adolescents in nine sub-districts in Banda Aceh were analyzed using GeoDa software. The results of this study revealed that the number of cases of anemia in adolescents in Banda Aceh is no correlated with nutritional status, nutrient intake, breakfast habits, anemia knowledge, parental income, and the number of family covered, except in Kuta Alam and Lueng Bata Sub-districts. The numbers of cases of anemia in adolescents in Kuta Alam Sub-district are more correlated with nutritional status, nutrient intake, breakfast habits, anemia knowledges, parent's income, and numbers of family covered.The number of cases of anemia in adolescents in Lueng Bata Sub-district is more correlated with nutritional status, and the number of family covered. Thus, for nutrient intake, anemia knowledge and parental income, they are no correlated with the number of anemia cases.

Keywords: adolescent, anemia status, spatial, region

1. Introduction

Anemia is a condition where amount of oxygen-carrying red blood cells is insufficient to meet the physiological needs, which vary according to age, sex, smoking habits, and pregnancy status. Physiologically, anemia occurs when the levels of hemoglobin which carries oxygen to the body's tissues is reduced [1]. The normal hemoglobin level for women is higher than12 gr/dl and more than 13 gr/dl for men. According to World Health Organization (WHO), global prevalence of anemia was 24.8%, which means that about 1.62 billion people in the world suffer from anemia. In India, nutritional status measured by anthropometry has a significant relationship to anemia[2]. Another survey in Indonesia showed that the prevalence of anemia in Indonesia is still high at 21.7% with 5-14 years old anemia sufferers was 26.4%, and 15-24 years old anemia sufferers was 18.4% [3]. Adolescents who do not like breakfast and eat less foods containing iron are among the factors that can increase the risk of anemia [4]. Although the number og anemia is high, but there has been no special attention to this case, information about anemia in adolescents has not been listed on Indonesian's Health Profile.

Banda Aceh is the capital of Aceh province consists of nine subdistricts. In 2015, Banda Aceh'spopulation was 250.303 with adolescents's proportion is 15.96% or 39.948 people [5]. Based on adolescent nutrition survey in Banda Aceh 2017, 46.35% adolescents in Banda Aceh got anemia [6]. It is indicate that anemia is a problem needs to get attention from government.

In this paper, the authors examine the relationship between anemia and variables that affect it. The relationship may be influenced by regional or spatial factors.

2. Research Methodology

In spatial data analysis, it is possible that one region depends on another region. Meaning that observations in a region can be influenced by observations in other areas.

The sampling technique of this research is cluster random sampling, where primary school, junior high school and senior high school are used as clusters. List of all schools and the number of population for each cluster in each subdistrict obtained from Badan Pusat Statistik (BPS), then the cumulative frequency is calculated. For each cluster, 30 schools were selected using Probability Propotional to size (PPS) [6] The number of data in nine sub-district in Banda Aceh are 603 which 239 adolescents classified as anemia. Data were computerized using Epidata and analyzed with GeoDa software [7]. Geoda software was created and developed by Dr.Luc Anselin which was originally only used at Spatial Analysis Laboratory (SAL), University of Illinois Urbana-Campaign in 2003 [8].

Figure 1 shows the distribution of adolescent anemia data in Banda Aceh. The number of cases of anemia in adolescents is most prevalent in Kuta Alam Sub-district , which is as many as 48 people. Sub-district with the second case of anemia is Baiturrahman Sub-district. While the fewest cases found in Lueng Bata Sub-district (12 people).

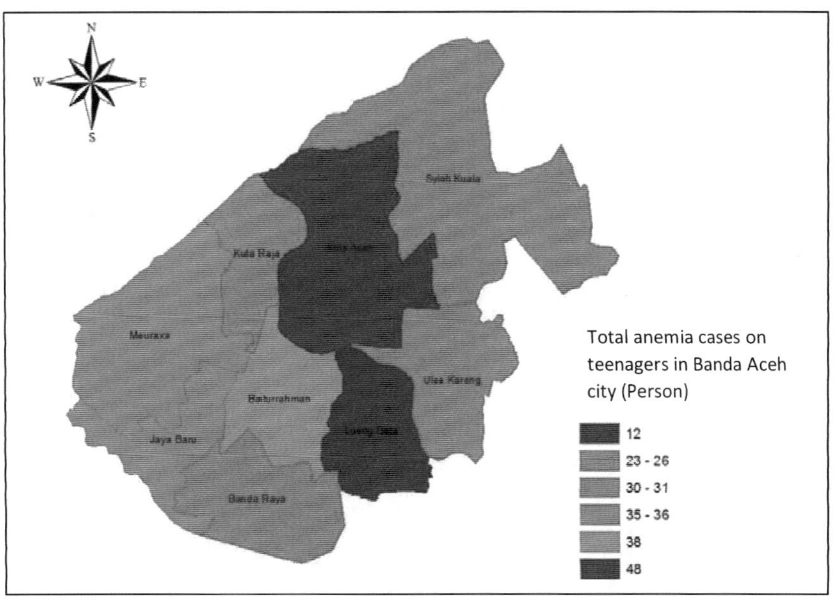

Figure 1. Spatial data distribution for adolescent anemia in Banda Aceh

3. Result and Discussion

Anemia cases in adolescents is one of the nutritional problems in Banda Aceh which is not currently receive special attention. The pattern of anemia incidence is need to know so that the government can give priority in handling the case. By using Analysis of variance (ANOVA), we compared anemia status among the nine Sub-district in Banda Aceh. Here, the factor sub-district and the response variabel is anemia status.

Table 1 showsthat P-value is less than 0.05, the test results statistically significant at 5% level. We conclude that the average of anemia status among nine sub-district in Banda Aceh are different

Table 1. ANOVA forAnemia status among nine sub-districts in Banda Aceh

Haemoglobin	Sum of Squares	df	Mean Square	F	Sig.
Between Groups	31.965	8	3.996	2.746	.006
Within Groups	391.432	269	1.455		
Total	423.398	277			

4. Thematic Map

The thematic map depicted using GeoDa Software, in this study, an illustration of the relationship of anemia with combination of two factors. Thematic maps are maps that show qualitative and quantitative information on an element that has to do with important topographic details [9]. Sub-district areas in each thematic map refer to Figure 2.The horizontal and vertical axis on the map shows the quantity of independent variables.While the number of occurrences of anemia is indicated by the color on the thematic map.If the colors on the map close to red, then the number of anemia events in certain districts will be more and more. If the color on the map is close to blue, then the number of occurrences of anemia is less.

Figure. 2 shows the relationship between 1 nutritional status and nutrient intake on the number of cases of anemia in a region. The horizontal axis shows the quantity of adolescents with abnormal nutritional status.While the vertical axis shows the quantity of adolescents who consume foods that do not meet the minimum nutrient intake. Figure. 2 (a) shows that the more teenagers who have abnormal nutritional status and the more teenagers who do not consume nutritious food, the fewer teenagers are anemia. Different pattern appear in the Kuta Alam and Lueng Bata Districts.In Kuta Alam, The conditions are number of adolescents with anemia, adolescents who have abnormal nutritional status, and the ones who do not eat nutritious foods are high. In Lueng Bata District, the number of adolescents with anemia is low, the number of adolescents with abnormal nutritional status are low and the number of adolescents who do not eat high nutritious food are high.

Figure 2 (b) shows the relationship between nutritional status and breakfast habits with anemia in adolescents.The horizontal axis shows the quantity of adolescents who have abnormal nutritional status, while the vertical axis shows the quantity of adolescents who breakfast less than 5 times a week. It appears that the more teenagers with abnormal nutritional status and the more adolescents who eat breakfast less than 5 times a week, then the fewer adolescents with anemia.But different things are seen in Kuta Alam and Lueng Bata Subdistricts. In Kuta Alam, number of adolescents with anemia are high, in line with adolescents who have abnormal nutritional status, and adolescents who take breakfast less than 5 times a week. In Lueng Bata, the number of adolescents with anemia is low, the number of adolescents with abnormal nutritional status is also low and those who eat breakfast less than 5 times a week are in the moderate category.

Figure 2 (c) shows the relationship of nutritional status variables and anemia knowledge variables with the number of cases of anemia. nutritional status, which has the definition of the number of adolescents who have abnormal nutritional status, is indicated by the horizontal axis. While knowledge of anemia, which has the definition of the number of adolescents who have low knowledge about anemia, is shown by the vertical axis.Based on Figure 2 (c) it can be seen that the more teenagers who have abnormal nutritional status and the more teenagers who have low knowledge about anemia, the fewer cases of anemia. However, in Kuta Alam, number of adolescents with anemia are high, number of adolescents who have abnormal nutritional status and the number of adolescents with low knowledge of anemia is also high. In Lueng Bata, all categories are low. The number of adolescents who have anemia, the number of adolescents who have abnormal nutritional status and those who have knowledge about anemia are low.

Figure 2(d)shows the relationship between nutritional status and parent's income on the number of cases of anemia. The horizontal axis shows the quantity of adolescents with abnormal nutritional status, whereas the vertical axis shows the quantity of adolescents with low-income parents.The more teenagers who have abnormal nutritional statusthey have low-income parents, the fewer cases of anemia. In contrast, Kuta Alam, The number of adolescents suffering from anemia is high with the number of adolescents who have abnormal nutritional status and the number of adolescents who have low income parents is also high. While in Lueng Bata, the number of adolescents who have anemia are low with the number of adolescents who have abnormal nutritional status are also low and the number of adolescents who have low-income parents are high.

Figure 2 (e) shows the relationship between nutritional status and the numbers of family covered with the number of cases of anemia in adolescents. Horizontal axis shows the quantity of adolescents who have abnormal nutritional status. The vertical axis shows the quantity of adolescents who live with parents / carers, and covered more than 4 people. Based on Figure 2 (e) it can be seen that the relationship pattern of nutritional status and the number of dependents on cases of anemia can not be determined. For example if the number of adolescents who have abnormal nutritional status is few and the number of adolescents who live with the parents that covered more than 4 people also few, the number of occurrences of anemia in some districts is high or low in others.

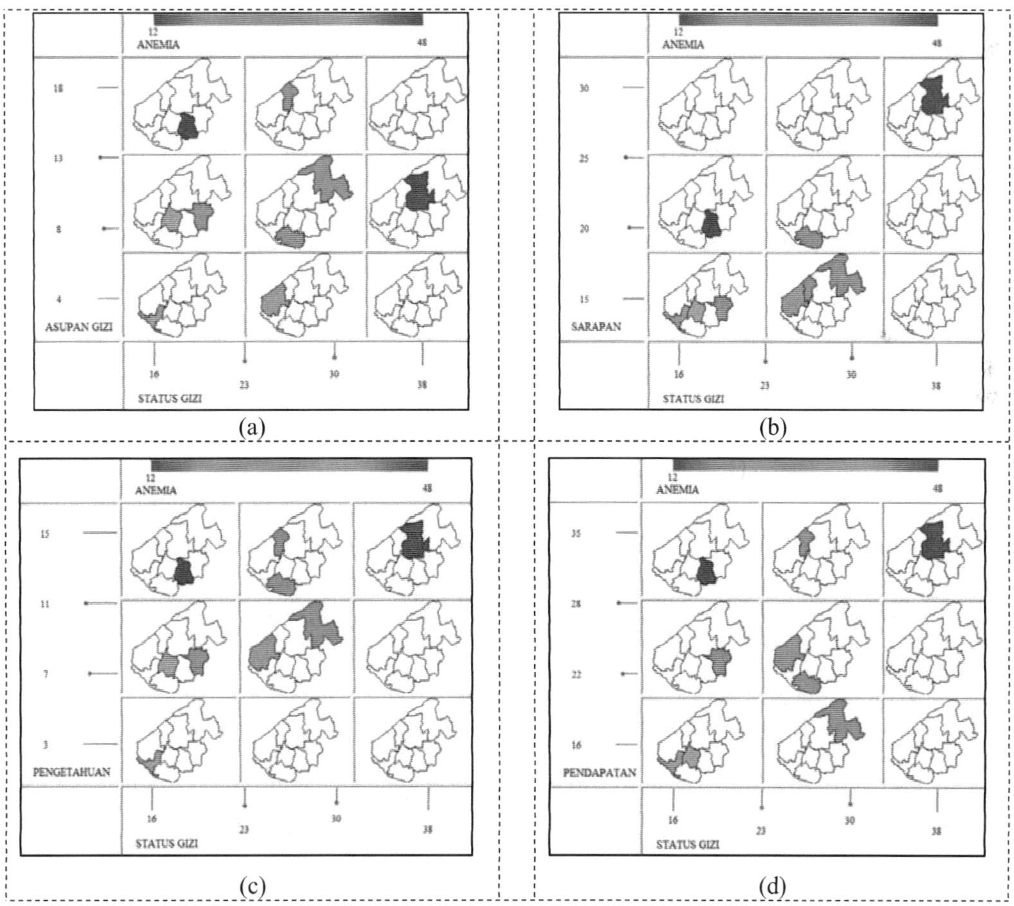

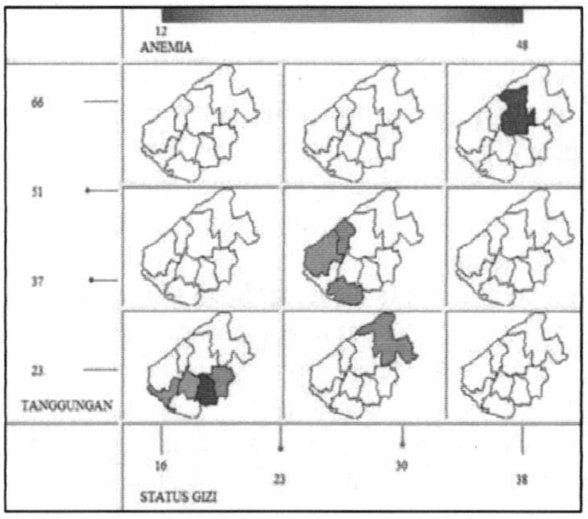

(e)

Figure 2. The association of anemia with combination of nutritional status and other factors (nutritional intake (a), breakfast (b), knowledge (c), income (d) and the number of families covered (e).

5. Conclusion

Based on the results, the conclusions of this study are as follows:

i. The number of cases of anemia in adolescents in Banda Aceh is no correlated with nutritional status, nutrient intake, breakfast habits, anemia knowledge, parental income, and the number of family covered, except in Kuta Alam and Lueng Bata. The number of cases of anemia in adolescents in Kecamatan Kuta Alam are more correlated with nutritional status, nutrient intake, breakfast habits, anemia knowledges, parent's income, and numbers of family covered

ii. The number of cases of anemia in adolescents in Lueng Bata is more correlated with nutritional status, and the number of family covered. For nutrient intake, anemia knowledge and parental income, they are no correlated with the number anemia cases

Acknowledgments

The authors would like to thank our colleague Fitriana, AR for helped us out in proofreading our English before submitting a paper.

References

[1] World Health Organization 2004 Focusing on Anaemia: Towards an Integrated Approach for Effective Anaemia Control World Health Organization Geneva

[2] T Premalatha S Valarmathi Parameshwari S S S Jasmine and S Kalpana 2012 Prevalence of anemia and its associated factors among adolescent school girls in Chennai, Tamil Nadu, INDIA *Epidemol* vol. 2 pp 118

[3] Health Department of Republic Indonesia 2013 Basic Health Research (Riskesdas) Health Department of Republic Indonesia

[4] M Arisman 2010 Nutrition in the life cycle second edition (Jakarta) *EGC*

[5] Banda Aceh Bureau of Statistics 2016 Banda Aceh in Numbers Banda Aceh Bureau of Statistics

[6] K M Zurnila and R Latifah 2017 Identification of adolescent Nutritional Status in Banda Aceh, Indonesia *Statistika* vol. 17 pp. 63–69

[7] K M Zurnila and C Virasakdi 2015 Utilization of the Local Government Health Insurance Scheme (JKA) for Maternal Health Services Among Women Living in Underdeveloped Areas of Aceh Province, Indonesia *Asia-Pasific J. Public Health* vol. 27 pp. 348–359
[8] Luc Anselin 2003 *GeoDa(TM) 0.9 User's Guide* University of Illions: Urbana-Campaign
[9] I. Indradari dan S. Tullus 2014 *Kartografi* Yogyakarta: National Land College

Rapid and Simultaneous Detection of Hazardous Heavy Metals Contamination in Agricultural Soil Using Infrared Reflectance Spectroscopy

Devianti[1], Sufardi[2], Zulfahrizal[1], Agus Arip Munawar[1]*
[1]Department of Agricultural Engineering, Syiah Kuala University, Banda Aceh - Indonesia.
[2]Department of Soil Science, Syiah Kuala University, Banda Aceh - Indonesia.

*Corresponding author: aamunawar@unsyiah.ac.id

Abstract. Soil function can be disrupted by highly accumulated contaminations in soils, due to several human activities such as building construction, mining, sewage disposal, transportation, and fertilizing. These activities may cause contaminations especially by heavy metals accumulated in soil and surely will affects growing plants on it. A real time, quick, robust and simultaneous method is therefore required to rapidly detect heavy metal contamination in agricultural soil. Infrared technology has been widely employed and applied in many field areas including in agriculture and soil science. The main aim of this research is to study the feasibility of infrared spectroscopy as a fast and robust method in detecting hazardous contaminations caused by accumulated heavy metals namely Zn and Pb in agricultural soil. Diffuse reflectance spectral data were acquired for a total of 8 soil samples from four different site locations in Aceh province. Spectra data, in wavelength range of 1000-2500 nm, were enhanced using multiplicative scatter correction (MSC). Hazardous detection was performed using principal component analysis (PCA) and accumulated heavy metals were predicted using partial least square regression (PLSR) method. The results showed that IR spectroscopy can clearly distinguish contaminated soil and un-contaminated ones, with total explained variance derived from PCA was 97%. Both accumulated heavy metals (Zn and Pb) can be predicted simultaneously with coefficient correlation (r) were 0.98 for Zn and 0.98 for Pb, whereas residual predictive deviation (RPD) indexed were 3.85 for Zn and 3.91 for Pb. It may conclude that infrared spectroscopy can be used as a fast, robust and accurate method in detecting and predicting heavy metal contaminations in agricultural soil.

Keyword : *infrared reflectance spectroscopy, principle component analysis, partial least square regression*

1. Introduction

Soil is the main media for growing plants, whether it is food crops or plantations. In fact, plants can grow ideally on healthy soil, if the soil has physical and chemical properties that are suitable with plant growth. Healthy soil conditions can be seen visually by looking its texture, structure and humidity. Soil chemical properties are related to the amount of nutrients needed by plants, with the amount needed will be different for each growing phase [1]. Soil fertility is largely determined by the presence of nutrients in the soil, both macro nutrients, secondary, and micro nutrients. Nutrients that are lacking in the soil will obviously affect plant growth. Growth rate will be very disturbed and susceptible to disease because soil nutrients do not meet the minimum requirements needed. Conversely, the amount of excessive nutrients in the soil will also affect plant growth and the surrounding environment [2].

Human activities, such as mining, transportation, sewage disposal and fertilizing, have been posing an ongoing threat to the soil health [3]. Moreover, the consumption of metal-polluted crops (e.g., rice, corn and soybean) grown in agricultural soils greatly raises the potential risks of food security and human health [2, 4]. In precision agriculture practice, fertilization and excessive use of fertilizer will cause pollution to the environment because it can cause artificial nutrient deposits that are not utilized by plants. This of course will accumulate and make new problems especially pollution due to the use of pesticides, fertilizers and fertilizers that are above the normal amount [3].

The chemical, biological, and physical unbalance caused by soil contamination by hazardous materials may be detrimental to plant, animal, and human health. For example, the symptoms of reduced root growth, reduced seed sprouting, and seedling stunting, necrosis, and chlorosis appear susceptible plants growing in soils contaminated with heavy metals [5]. Agricultural crops (fruits, grains, and vegetables) for livestock human consumption, growing on contaminated soil, can potentially uptake and accumulate hazardous materials in their edible plant parts, and may be harmful to animal and human health through the food chains [6].

From this point of view, the determination of hazardous contamination on agricultural soils is necessary to monitor the health of agricultural soils and further to take preventative measures to avoid soil contamination. A reliable and environmentally friendly method is therefore needed to rapidly detect and survey the spatial distribution soil contaminations, to diagnose suspected contaminated areas as well as control the rehabilitation processes [2, 3, 7].

During the last few decades, infrared technology has been widely used and become most promising methods of analysis in many field areas including in soil science and agriculture due to its advantage; simple sample preparation, rapid, and environmental-friendly since no chemicals are used. More importantly, it has the potential ability to determine multiple parameters simultaneously [3, 6, 8, 9].

The overall research findings of numerous studies [9, 10, 11] on the application of near infrared spectroscopy (NIRS), shows that NIRS was feasible to be applied as a rapid and non-destructive tool for quality attributes prediction in agricultural sectors. The prediction model performance was sufficiently robust and accurate with correlation coefficient (r) range of 0.93 – 0.99 and residual predictive deviation (RPD) index was 1.53 – 4.68 which is categorized as coarse, sufficient and excellent prediction models respectively.

Therefore, based on advantages and excellence of NIRS, we attempted to apply the NIRS method in predicting soil contaminations by heavy metals (Pb, Zn and Cu). We hope that we will achieve satisfactory results from which NIRS can detect and classify soil condition. This study will describe the feasibility of NIRS technology in detecting hazardous contaminations on agricultural soil rapidly and develop the prediction model to quantify the amount of accumulated heavy metals on agricultural soil. The prediction model was established based on IR spectroscopic data after scatter correction using principal component analysis (PCA) and partial least square regression (PLSR).

2. Materials and Methods

2.1. Samples

A bulk of soil samples from four different site locations in Banda Aceh and Aceh Besar area were taken and stored for two days to equilibrate before spectra acquisition and further chemical analysis. Soil contaminations were made manually by injecting 250, 500, 750 and 1000 ppm heavy metals liquid namely Zn and Pb.

2.2. Spectra data acquisition

Infrared spectral data in form of diffuse reflectance spectrum were taken of all soil samples using infrared instrument (FTIR, Thermo Nicolet Antaris II MDS). The basic measurement with probe detector was chosen as a basic measurement in high resolution format. The infrared spectrum was collected and recorded in wavenumbers range from 4000 to 10 000 cm^{-1} with co-added 32 scans and averaged. Spectra data were stored in local computer with three different file formats (*.SPA, *.JDX and *.CSV) [5].

2.3. Spectra data correction
In order to obtain accurate and robust prediction results, infrared spectra data were enhanced and corrected using multiplicative scatter correction (MSC) [6].

2.4. Soil contaminant detection
Once spectra correction was completed, soil contamination was detected rapidly using corrected spectra data. These data were projected onto principal component analysis (PCA) in order to classify and distinguish soil samples between contaminated and un-contaminated [5]. On the other hand, soil contamination prediction models were developed using partial least square regression (PLSR) and validated using leverage validation method [5].

2.5. Prediction model performance
Soil organic prediction performances were judged for their accuracies and robustness using several statistical indicators: coefficient of determination (R^2), correlation coefficient (r), root mean square error (RMSE) and the residual predictive deviation (RPD) defined as the ratio between the standard deviation and the root mean square error. The higher the RPD, the greater and robust the model to predict heavy metal contamination on soil samples [5, 6]. It is obvious that the good model should have high R^2 and r coefficient, low value of RMSE and few number latent variables of PLSR [5].

3. Result and discussion
3.1. Typical spectra of soil sample
Typical diffuse reflectance spectra for soil samples was presented in Figure 1. The result shows several peaks represent the vibration of molecular bonds of C-C, O-H, N-H, C-H-O and C-H. Original spectra data before correction were still interference due to noise resulted from light scattering. These noises were corrected using several pre-treatment methods namely multiplicative scatter correction (MSC). As shown in Figure 2, these spectra correction methods clearly enhance spectra appearance and remove some noises due to light scattering.

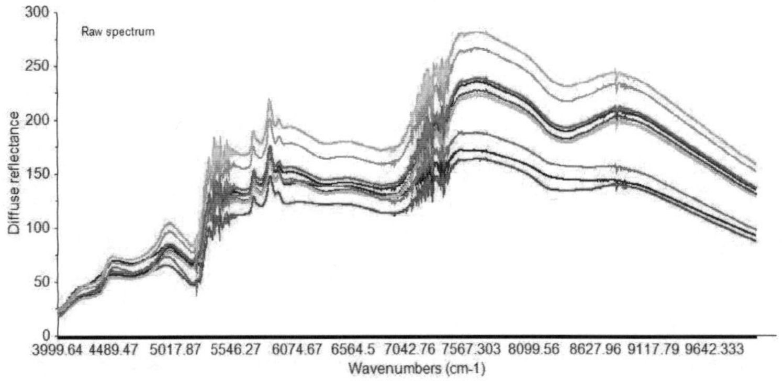

Figure 1. Typical diffuse reflectance spectra data of soil samples before correction.

Spectral data acquired from the near infrared instrument generally contain background information and noises which are interfered and affected desired relevant soil quality information such as C-organic or even soil contaminations. Interfering spectral parameters, such as light scattering, path length variations and random noise resulted from variable physical sample properties or instrumental effects need to be removed or minimized in order to obtain accurate, robust and stable calibration models. Thus, it is necessary to enhance and correct spectral data before further analysis. Spectra enhancement was conducted using multiplicative scatter correction (MSC) as presented in Figure 2.

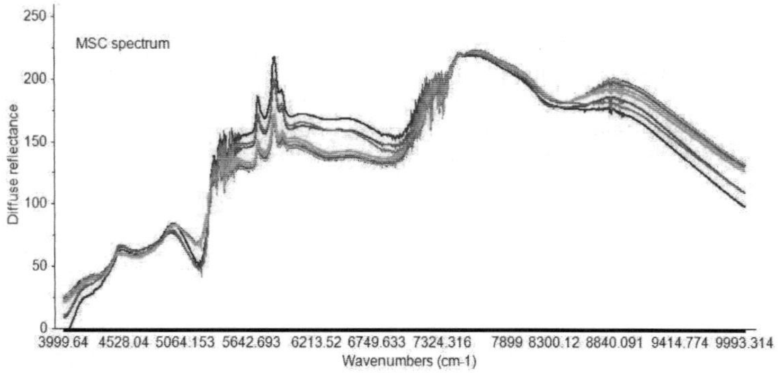

Figure 2. Diffuse reflectance spectra after correction using MSC.

As shown in Figure 2, spectra data were obviously enhanced and corrected compare to un-corrected one. In this study, NIR spectra data were enhanced using MSC, de-trending (DT) and combination of DT+MSC. The de-trending pre-treatment method tends to remove nonlinear trends in spectroscopic data, while MSC method used to reduce amplification due to light scattering and offset due to additive chemical effects in near infrared spectra

3.2. Soil classification based on spectra data
The PCA result as for soil contamination detection was shown in Figure 3. It shows a significantly cluster based on soil characteristics. The first principal component accounts for as much of the variability in the data as possible, and each succeeding component accounts for as much of the remaining variability as possible. PCA is used as a tool for screening, extracting, compressing and discriminating samples based on their similarities or dissimilarities of multivariate data. The first PC (PC1) accounted 91% of all spectra data while remaining 7% was explained by the second component (PC2).

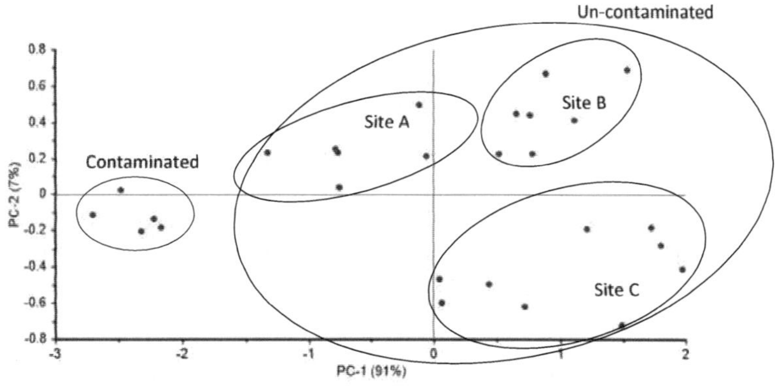

Figure 3. PCA based on infrared spectra for soil contamination detection

The differentiation of showed PCA classification is probably due to nutrient content and heavy metals contamination of related soil. Based on principal component analysis (PCA) results, it was obvious that NIRS method was able to distinguish and classify soil based on heavy metal contaminations. Soil spectral features in the infrared wavebands are highly correlated to the vibration modes of functional groups like the chemical bond of H and C, N, and O.

Moreover, the NIR technology applied in this study, was able to predict soil contaminations caused by heavy metals namely Zn and Pb. The best and most accurate prediction results for both heavy metals contaminants can be predicted by using MSC spectra data.

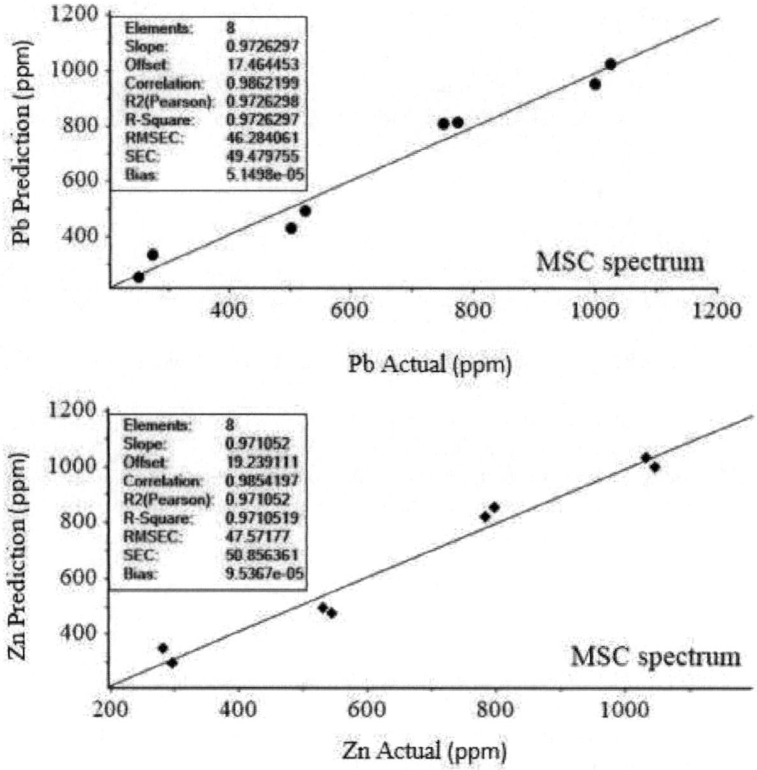

Figure 4. Pb and Zn prediction of soil samples

As shown in Figure 4, for Zn prediction, the NIR technique achieved excellent prediction result with correlation coefficient (r) was 0.98 and residual predictive deviation (RPD) index was 4.57. Furthermore, for Pb prediction, the SNV spectra correction method gave also excellent prediction result with r coefficient was 0.98 and RPD index was 4.25.

4. Conclusion

Based on obtained results, we may conclude that near infrared spectroscopy can be used and employed as a rapid and environmental-friendly method used to predict soil contaminations. Achieved present study shows that infrared technology was feasible to use as a rapid, innovative and non-destructive method in soils quality properties classification and evaluation based on their contamination. Moreover, this technology can also predict heavy metals contamination with excellent satisfactory result: r = 0.98 and RPD = 4.57 for Zn prediction, and r = 0.98 with RPD = 4.25 for Pb prediction.

Acknowledgements

We sincerely acknowledge to DRPM Kemristek-Dikti for providing research grant and funding through Hibah Penelitian Fundamental scheme 2018.

References

[1] Knadel M Gilsum R Hermansen C Peng Y Moldrup P de Jonge L W & Greve M. H. 2017 Comparing predictive ability of laser-induced breakdown spectroscopy to visible near-infrared spectroscopy for soil property determination. Biosystem Engineering vol 156 pp 157-172

[2] Mouazen A Kuang B De Baerdemaeker J Ramon H 2010 Comparison among principal component partial least squares and back propagation neural network analyses for accuracy of

measurement of selected soil properties with visible and near infrared spectroscopy. Geoderma vol 158 pp 23–31

[3] Moros J Vallejuelo S F O Gredilla A Diego A Madariaga J.M Garrigues S & Guardia M 2009 Use of reflectance infrared spectroscopy for monitoring the metal content of the estuarine sediments of the Nerbioi–Ibaizabal River (Metropolitan Bilbao, Bay of Biscay, Basque Country). *Environ Sci Technology* 43 (24), pp 9314–9320

[4] Ren H.Y Zhuang D.F Singh A.N Pan J J Qiu D.S Shi R.H 2009 Estimation of As and Cu contamination in agricultural soils around a mining area by reflectance spectroscopy: a case study Pedosphere vol 19 pp 719–726

[5] Munawar A A Hörsten D v Mörlein D Pawelzik E & Wegener J K 2016 Rapid and non-destructive prediction of mango quality attributes using Fourier transform near infrared spectroscopy and chemometrics *Engineering in Agriculture, Environment and Food* 9(1)

[6] Munawar A.A D.v Hoersten J K Wegener E Pawelzik & D Moerlein 2013 Rapid and non-destructive prediction of mango sweetness and acidity using near infrared spectroscopy Lectures note in Informatics GIL jahrestagung Potsdam

[7] Shi T Yiyun C Yaolin L & Guofeng L 2014 Visible and near-infrared reflectance spectroscopy: An alternative for monitoring soil contamination by heavy metals. *Journal of Hazardous Materials* vol 265 pp 166–176

[8] Salazar M J.H Rodriguez G L Nieto M.L Pignata 2012 Effects of heavy metal concentrations (Cd, Zn and Pb) in agricultural soils near different emission sources on quality, accumulation and food safety in soybean, J. Hazard. Mater pp 244–253

[9] Wang J Lijuan C Wenxiu C Tiezhu S Yiyun C & Yin G 2014 Prediction of low heavy metal concentrations in agricultural soils using visible and near-infrared reflectance spectroscopy. Geoderma, vol 216 pp 1-9

[10] Senesi G & Senesi N 2016 Laser-induced breakdown spectroscopy to measure quantitatively soil carbon with emphasis on soil organic carbon A review. Analytica Chimica Acta vol 938 pp 7-17

[11] Li Q S Y Y Chen H B Fu Z H Cui L Shi L L Wang & Z.F Liu 2012 Health risk of heavy metals in food crops grown on reclaimed tidal flat soil in the Pearl River Estuary China J Hazard. Mater vol 22 pp 148–154

IOP Conf. Series: Materials Science and Engineering **506** (2019) 012009 doi:10.1088/1757-899X/506/1/012009

Investigation into Joint Stiffness for Finite Element Modelling of the Dynamic Behaviour of a Structure with Bolted Joints

R Omar[1,2], M N Abdul Rani[1*], M A Yunus[1], W I I Wan Iskandar Mirza[1] and M S Mohd Zin[1]

[1] Structural Dynamics Analysis and Validation, Faculty of Mechanical Engineering, Universiti Teknologi MARA (UiTM), 40000 Shah Alam, Selangor, Malaysia
[2] Machinery Technology Centre, SIRIM Industrial Research, 44200 Hulu Selangor, Selangor, Malaysia
E-mail: mnarani@salam.uitm.edu.my

Abstract. Accurate representation of the joint stiffness of bolted joints in the finite element (FE) model of a jointed structure is vital to ensure the accuracy and reliability of the predicted dynamic behaviour of the structure. The aim of this work was to put forward an efficient FE modelling method embodying the representation of the joint stiffness for the investigation of an assembled structure with bolted joints. The predicted FE results were compared with those obtained from the experimental modal analysis (EMA). EMA was performed to measure the natural frequencies and mode shapes of the structure with bolted joints. The calculated total error from the comparison of the natural frequencies between the FE and EMA were recorded. The results of the investigation showed that there is a noticeable contribution of the joint stiffness in the representation of the structure with bolted joints in developing the efficient FE modelling method.

1. Introduction

Bolted joint is one of the common elements used in an assembled structure. Typically, a structure with bolted joints consists of several structural components or members jointed with a series of bolts and nuts. However, to the best of authors knowledge, there has been very little discussion about the appropriate representation of the joint stiffness for the investigation of the dynamic behaviour of the structure with bolted joints.

There are two typical components in structures with bolted joints which are the members and bolts. The member stiffness is determined by considering the effective spring effect of each structural member. Structure with multiple members shall has the overall member stiffness represented by several springs connected in series [1], [2]. In this work, the bolted joints consisting of 2 members, hence the overall member stiffness, k_m, is calculated by using equation (1) where k_1 and k_2 represent stiffness for the 1st member and the 2nd member respectively.

$$\frac{1}{k_m} = \frac{1}{k_1} + \frac{1}{k_2} \tag{1}$$

Motosh [2], [3] proposed the conical pressure distribution envelopes for the highly stressed portion of the joint as a technique to estimate the member stiffness of the bolted joint model. This stiffness estimating method was also used by other researchers to determine the stiffness of jointed members of a structure [1], [2], [4]. As shown in Figure 1, the jointed region of the assembled structure of this work is divided into 2 frustums which fall off transversely from the bolt centreline, replicating the proposed method used in previous researches.

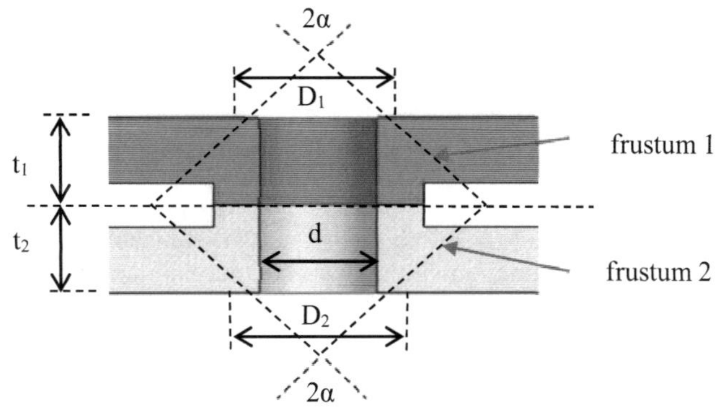

Figure 1. Conical pressure distribution envelope and frustums.

An analytical equation was originally proposed by Lehnhoff et al. [1], [2], [4] to calculate the stiffness of the member with the assumption of uniform pressure with conical envelope under the bolt's head. Subsequently, Shigley and Mischke modified this equation with a fixed standard cone angle (α) at 30° to obtain a simpler expression for the calculation of the member stiffness as in equation (2) [1], [2], [5]. In this work, equation (2) was used to calculate the stiffness of each member of the structure with bolted joints which denoted by k_1 for frustum 1 and k_2 for frustum 2.

$$k = \frac{0.577\pi E d}{ln\frac{(1.15t + D - d)(D + d)}{(1.15t + D + d)(D - d)}} \tag{2}$$

The stiffness values of the overall members and bolts are used in the FE model of the assembled structure with bolted joints for the investigation of its dynamic behaviour. The predicted results are compared with the counterparts obtained from experimental modal analysis. This paper proposes an efficient FE modelling method embodying the representation of the joint stiffness for the investigation of the dynamic behaviour of an assembled structure with bolted joints.

2. Modelling of Joints Stiffnesses

2.1. Modelling of Bolted Joints

A simple lap joint is used to study the dynamic behaviour of the structure with bolted joints. The lap joint comprises 2 identical steel plates with stainless-steel bolts and nuts. Each plate comes with the length of 380 mm, width of 45 mm and thickness of 6 mm. The 3D CAD models of the components were developed and designated as Plate 1 and Plate 2. Both plates are assembled using bolts and nuts in a configuration shown in Figure 2. For the purpose of reducing the assembly contact zones of the assembled plates, two cylindrical bosses were introduced in each plate with the diameter of 18 mm and thickness of 2 mm as shown in Figure 3 [6].

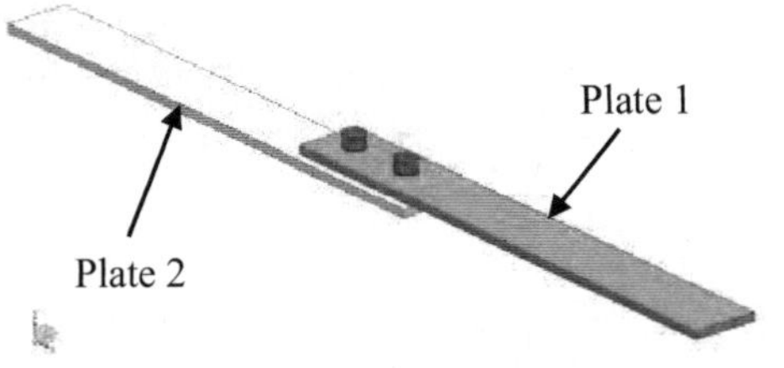

Figure 2. The structure with bolted joints.

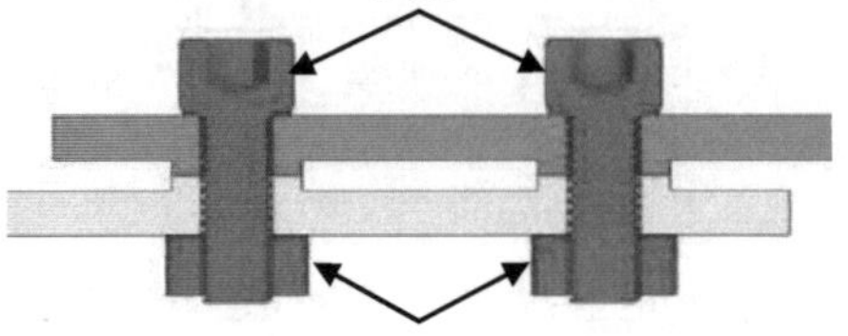

Figure 3. Cross-section of the bolted joints.

The material properties of Plate 1, Plate 2, bolts and nuts [7] are tabulated in Table 1. The values of the material properties of each plate are slightly different from the standard values. The difference is because those values have been updated using the FE model updating procedures [8], [9]. The procedures provide an accurate modelling of the Plate 1 and Plate 2 before assembling the plates to form an assembled structure.

Table 1. Material properties of Plate 1, Plate 2, bolts and nuts.

Component	Property	Value	Unit
Plate 1	Young's Modulus	2.064×10^{11}	N/m^2
	Poisson's Ratio	0.302	Unitless
	Mass Density	7.458×10^{3}	kg/m^3
Plate 2	Young's Modulus	2.094×10^{11}	N/m^2
	Poisson's Ratio	0.316	Unitless
	Mass Density	7.45×10^{3}	kg/m^3
Bolts and nuts	Young's Modulus	1.93×10^{11}	N/m^2
	Shear Modulus	0.75×10^{11}	N/m^2
	Poisson's Ratio	0.27	Unitless
	Mass Density	7.86×10^{3}	kg/m^3

2.2. Stiffness of the Members

Stiffness components of the 1st member and the 2nd member, which are Plate 1 and Plate 2 respectively, were calculated using equation (2). These stiffness values, named as k_1 and k_2 correspondingly, were used as the input to calculate the overall member stiffness, k_m, by using equation (1). Table 2 shows the calculated stiffness values of k_1, k_2 and k_m. The k_m value was then used as the translational stiffness component of the CELAS element representing the jointed members in the FE modelling of the bolted joints.

Table 2. Calculated members' stiffnesses.

Stiffness components	Value (N/m)
1st member stiffness, k_1	6.642×10^{9}
2nd member stiffness, k_2	6.954×10^{9}
Overall member stiffness, k_m	3.397×10^{9}

2.3. Stiffness of the Bolts

In this work, CBUSH element was used to represent the bolts in the FE model of the assembled structure. The CBUSH stiffness components which consist of axial stiffness (K_1), shear stiffness (K_2, K_3) and rotational stiffness (K_4, K_5, K_6) were defined by using the stiffness with Swift's flexibility formulae as used in the previous studies [10], [11]. The calculated stiffness components for CBUSH element are tabulated in Table 3.

Table 3. Value of stiffness components of CBUSH element.

Stiffness components	Value (N/m)
Axial stiffness, K_1	9.474×10^8
Shear Stiffness, $K_{2,3}$	5.403×10^8
Rotational Stiffness, K_4	1.000×10^5
Rotational Stiffness, $K_{5,6}$	3.458×10^9

3. FE Modelling and Analysis of the Bolted Joints

The FE model of the structure with bolted joints was developed as shown in Figure 4 and Figure 5. The model was discretised into a combination of 1977 CQUAD8 and 30 CTRI6 elements with the size of 5 mm. The size was chosen based on the results of convergent tests carried out [12]. Total number of nodes is 6745 with a total number of 40470 degrees of freedom. In each bolted joint model, CELAS elements were used to represent the jointed members' stiffness, CBUSH elements were used to represent the bolts, while RBE3 elements were used to represent the bolts' heads and nuts.

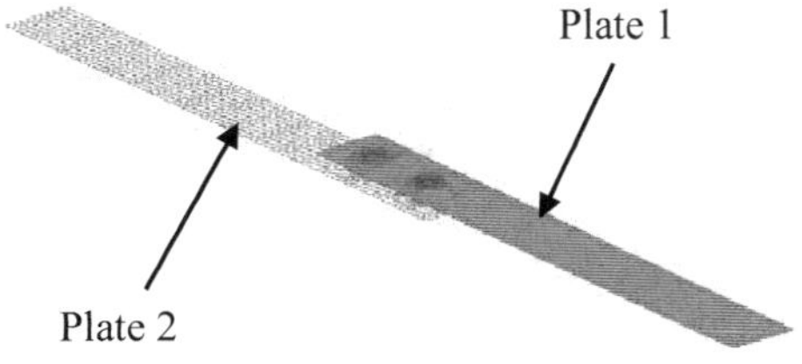

Figure 4. FE model of the assembled structure with bolted joints.

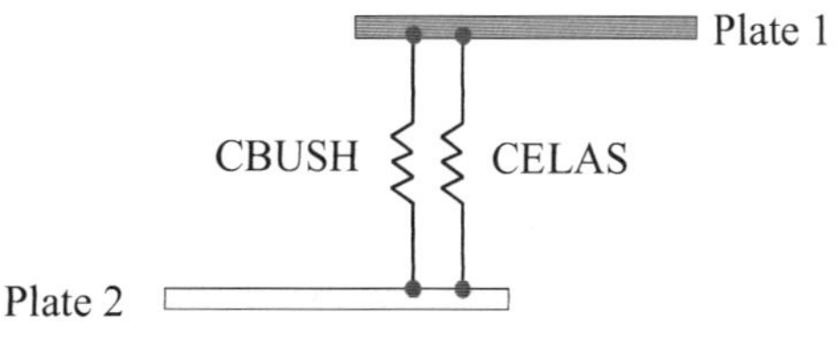

Figure 5. Representation of the members and bolts stiffnesses by CELAS and CBUSH respectively.

In this work, normal modes analysis for the developed FE model of the assembled structure with bolted joints was performed using the NX11 software packages. The analysis was performed to predict the first ten natural frequencies and mode shapes of the assembled structure. The predicted results were then compared with the measured results of the assembled structure obtained from experimental modal analysis.

4. Experimental Modal Analysis (EMA)

The modal parameters, which are the natural frequencies and mode shapes, of the structure with bolted joints were characterised by performing EMA. The equipment packages used to characterise the modal parameters were the LMS system, LMS Test.Lab software, impact hammer & accelerometers as shown in Figure 6. The assembled structure consisting of 2 identical plates namely Plate 1 and Plate 2, was made from steel beams. The experimental setup for the assembled structure is shown in Figure 7. Plate 1 and Plate 2 were joined by using the stainless steel M10 bolts and nuts. The assembled structure under the test was suspended from the test rig by using rubber bands to simulate free-free boundary conditions.

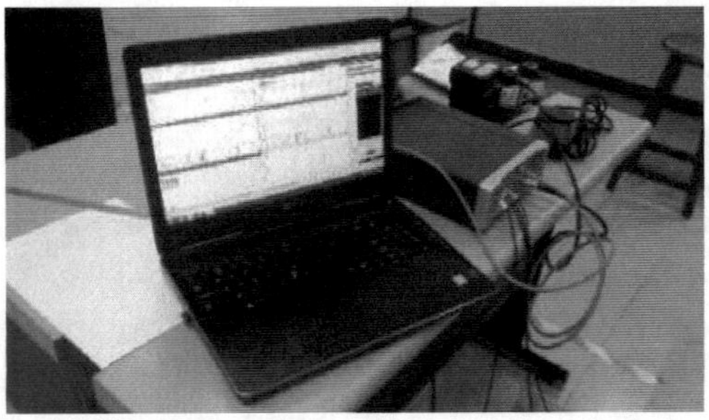

Figure 6. Testing equipment for EMA.

Figure 7. Experimental set up for the structure with bolted joints.

5. Results and Discussion

The predicted natural frequencies were compared with the measured counterparts to compute discrepancies in terms of total error. The comparison of the first 10 natural frequencies between the predicted and measured results of the assembled structure is tabulated in Table 4. The comparison of the results show that the total error recorded is 14.573%, while the modal assurance criterion (MAC) values show a very good correlation between the predicted and measured mode shapes. The results also revealed that the proposed FE modelling method embodying the representation of the members' and bolts' stiffnesses has contributed significantly towards an accurate modelling of the assembled structure with bolted joints, especially for the first 3 modes.

Table 4. Comparison between the measured and predicted results for the assembled structure.

Mode	I. EMA (Hz)	II. FE (Hz)	III. Error between I & II (%)	IV. MAC	V. Mode of vibration
1	75.08	75.28	0.266	0.982	Bending z-direction
2	200.77	202.44	0.832	0.940	Bending z-direction
3	404.92	401.63	0.813	0.992	Bending z-direction
4	469.30	446.93	4.767	0.945	Bending y-direction & torsion
5	630.77	632.62	0.293	0.973	Bending. z-direction
6	674.39	661.27	1.945	0.960	Torsion
7	1034.64	1019.16	1.496	0.944	Bending z-direction
8	1053.23	1067.47	1.352	0.910	Torsion
9	1276.95	1278.41	0.114	0.957	Bending z-direction
10	1435.09	1473.76	2.695	0.937	Bending y-direction & torsion
	Total Error		14.573		

6. Conclusions

The proposed FE modelling method incorporating the contribution of the joints stiffnesses into the investigation of the dynamic behaviour of the assembled structure with bolted joints is presented. The comparison between the predicted and measured modal parameters revealed that the members' and bolts' stiffness values have played a vital role in improving the accuracy of the prediction of the dynamic behaviour of the assembled structure with bolted joints. The accuracy and reliability of the predicted results of the FE model could be further improved by using the FE model updating method.

IOP Conf. Series: Materials Science and Engineering **506** (2019) 012009 doi:10.1088/1757-899X/506/1/012009

Acknowledgments
The authors gratefully acknowledge the MyBrain15 and FRGS (600-RM1/FRGS 5/3 (96/2016) for providing financial support.

References
[1] Haidar N, Obeed S and Jawad M 2011 *J. Mech. Sci. Technol.* **25** 2827
[2] Nassar S A and Abboud A 2009 *J. Mech. Des.* **131** 121001
[3] Motosh N 1976 *J. Eng. Ind.* **98** 858–61
[4] Lehnhoff T F, Il Ko K and McKay M L 1994 *J. Mech. Des.* **116** 550
[5] Shigley J E and Mischke C R 2006 *Mechanical Engineering Design* (New York, USA: McGraw-Hill)
[6] Omar R, Abdul Rani M N, Yunus M A, Wan Iskandar Mirza W I I and Mohd Zin M S 2018 *AIP Conf. Proc.* **1952** 020082
[7] Hibbeler R C 2011 *Mechanics of Materials* (Prentice Hall)
[8] Abdul Rani M N, Kasolang S, Othman M H, Yunus M A, Wan Iskandar Mirza W I I and Ouyang H 2016 *23rd International Congress on Sound and Vibration* pp 315
[9] Omar R, Abdul Rani M N, Wan Iskandar Mirza W I I, Yunus M A and Othman M H 2017 *J. Mech. Eng.* **SI 4** 202
[10] Chandregowda S and Ganga Reddy C R 2018 *AIP Conf. Proc.* **1943** 020001
[11] Whitman Z L 2012 (University of Utah)
[12] Omar R, Abdul Rani M N, Yunus M A, Mat Isa A A, Wan Iskandar Mirza W I I and Mohd Zin M S and Roslan L 2018 *Int. J. Automot. Mech. Eng.* **15** 5695

Finite Element Model Updating of Laser Stitch Welded Structure

M.A.S. Aziz Shah[1], M.A. Yunus[1], A.M. Saman[1], M.N. Abdul Rani[1], A.R. Bahari[1] and C. Peter[1]

[1]Structural Dynamics Analysis and Validation (SDAV),Faculty of Mechanical Engineering, Universiti Teknologi MARA (UiTM), 40450 Shah Alam, Selangor Malaysia.

mayunus@salam.uitm.edu.my

Abstract. The applications of laser stitch welding are increasing used in assemblies engineering products such as automotive structures due to its efficiency and re-liability in connecting a light weight material. In order to understand the vibration effect of the laser stitch weld towards the assembled structure, it is important to analyse its dynamic behaviour. However, they are relatively difficult to be analyse numerically as the laser stitch welded joints are found to be complex and have many uncertainties. This paper presents the investigation of finite element model for laser stitch welded structure. The reconciliation method of finite element model updating is used to improve the dynamic behaviour of the initial finite element model as close to experimental data. In this work, the finite element model of the laser stitch welded is developed by employing area contact model (ACM2) connector. Meanwhile, the experiment dynamic behaviour is obtained using an impact hammer and roving accelerometer with free-free boundary conditions. Later, the finite element model updating method is carried out in order to reduce the discrepancies of initial predicted models based on the experiment data. Based on the result, it was found that the ACM2 connector can be used to represent as la-ser stitch welds accurately by updating the most sensitive selected parameters.

Keywords: Laser Stitch Weld, Dynamic Behaviour, Experiment Modal Analysis, Finite Element Analysis, Model Updating.

1. Introduction

In automotive industries, a typical car body-in-white (BIW) is usually consists of thousands of small parts which are made from thin steel sheets and they are connected together by numerous types of joints. However, in order to fulfil the need of producing a light weight car structure and to improve the quality, the application of laser stitch welding has become more preferable.

The used of laser stitch weld joints that used in car BIW are not only form connections between thin steel sheets but also provide significant contribution to the stability and the rigidity of car [1]. Therefore, an accurate finite element (FE) model must be developed in order to predict dynamic behaviour of structure meanwhile experimental work is required as to determine the level of accuracy of the FE model.

Computerized analysis package such as FE analysis is found to be more beneficial to predict the dynamic behaviour of the structure. However, it is difficult and cumbersome to model the laser stitch weld joint itself. It is particularly be-cause of the local effects that occur due to welding process are realized to be complex to be included in the FE model and also influenced by many uncertain parameters such as geometry irregularities, heat affected zone and residual stress [2], [3].

Previous studies reported that there are few types of element connectors that are widely used to represent laser welded joints such as Rigid Body Element (RBE2), Area Contact Model (ACM2) and spot weld element such as CWELD format [4]. There are several works that related to the application of CQUAD4, CHEXA, RBE2, CWELD and ACM2 format for welded joints [5], [6]. Kuppuswamy et al. studied modelling techniques for laser welded joints with the accuracy of model were depended on the like sheet thickness, static and dynamic strength of the joint [7]. Ha et al. proposed a guideline for accurate finite element modelling of laser welded region for the crash analysis of vehicles [8].

In prediction stage, it is found that the initial FE models of structures are not always in good agreement with the measured counterparts. This is due to the in-valid assumptions made through the models. Usually, the invalid assumptions in structural dynamics cases are related to the material properties, joints and boundary conditions [9]. For instance, study showed that the prediction models of welded structure made from thin steel sheets that influential by initial stress can be successfully improve using model updating method [10]. Furthermore, other research study showed that, selecting the right updating parameters were really crucial in order to guarantee the reliability of FE models [11], [12]. Therefore, it is important to improve the initial FE models systematically using model updating approach as close as experiment data.

This paper is concerned with the development of the FE modelling with ACM2 connector is used as laser stitch welded joints. Further, this paper also discussed about the procedure of model updating of the dynamic behaviour of a laser stitch welded structure.

2. Finite Element Modelling of Laser Stitch Welded Structure

Finite element (FE) analysis is a numerical method that involve with discretizing of continuous system into small elements (discrete). There are lot of software packages for modelling structural dynamic models and the choices are highly depends on researcher preference. In this work, the structure is designed as hat-plate structure connected by laser stitch welds (as shown in Figure 1) in order to replicate the substructure of typical car BIW.

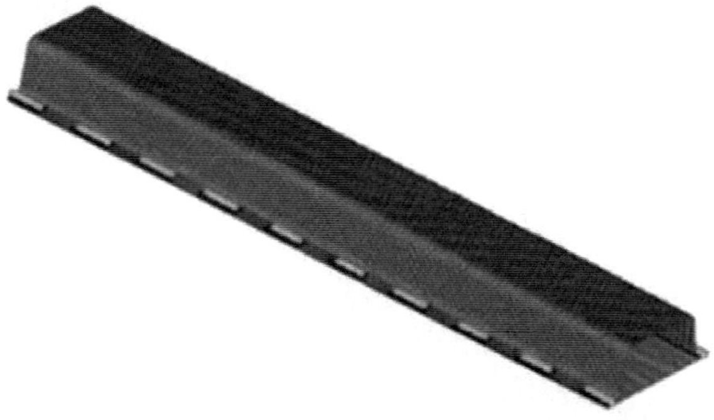

Figure 1. FE model of laser stitch welded structure.

The hat-plate structures were discretizing into a CQUAD4 shell elements with 13090 elements and 13348 nodes. Meanwhile, the modelling of the laser stitch welds in the hat-plate structure was prepared using ACM2 connector since it has capability to represent as laser stitch welds.

The natural frequencies and mode shapes of finite element model of hat-plate structure were predicted using normal mode analysis by solving the Eq. (1).

$$(\boldsymbol{K} - \omega^2 \boldsymbol{M})\phi \qquad (1)$$

IOP Conf. Series: Materials Science and Engineering **506** (2019) 012010 doi:10.1088/1757-899X/506/1/012010

where, **K** and **M** are the stiffness and mass matrices respectively, while ω and ϕ represent natural frequency and eigenvector respectively. In the normal mode analysis, the frequency of interest was set starting from 0 to 700 Hz meanwhile the nominal material values of mild steel were used as follow; Young's modulus: 210 GPa, Poisson's ratio: 0.30, density: 7900 kg/mm^3.

3. Experimental Modal Analysis of Laser Stitch Welded Structure

Experimental modal analysis is a technique to define the dynamic behaviour of the structure from a vibration testing, in terms of natural frequencies, mode shapes and damping ratios [13]. The result from the experimental data will be compared with finite element (FE) model to identify the correlation between its results. If the results from the test structures are shown a good agreement with predicted model, and then FE model can be used in further analysis with higher degrees of confidence.

In this paper, dynamic behaviour of the test structure was obtained using impact hammer and roving accelerometers under free-free boundary condition as presented in Figure 2. An impact hammer was used to excite the structure with one fixed reference accelerometer as shown in Figure 3 and Figure 4. Furthermore, to replicate the free-free boundary condition, four sets of nylon strings and springs were used to hang the structure from the clamps. Meanwhile The frequency bandwidth of interest of the test structure was 0 Hz to 700 Hz. Finally, LMS SCADAS system was used to interpret the load and signal produced by test structure.

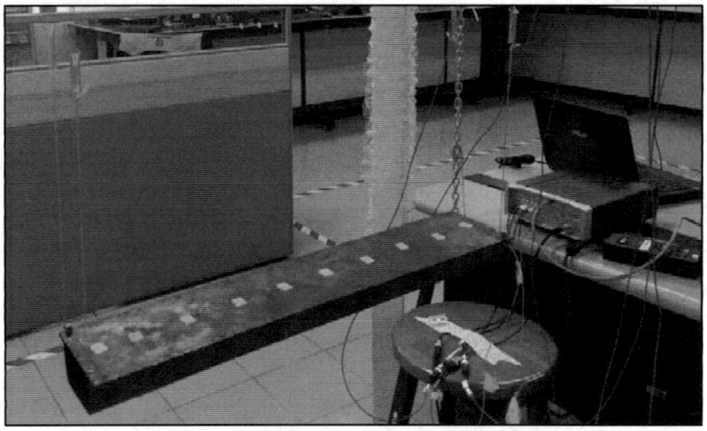

Figure 2. Experimental setup of the laser stitch welded structure.

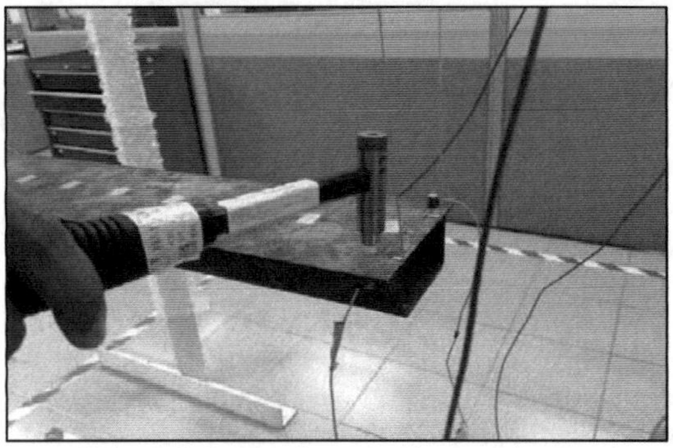

Figure 3. Impact hammer in experiment.

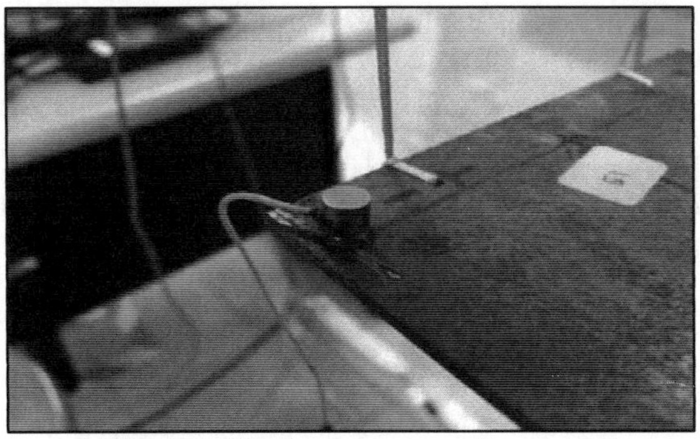

Figure 4. Fixed reference accelerometer.

4. Model Updating of the Laser Stitch Welded Structure

Model updating is an approach to improve the correlation of finite element (FE) models and the test structure by correcting the invalid assumptions of the model to an acceptable level of accuracy [14]. This is carried out by tuning some uncertain model parameters systematically. However, it is difficult to select the right parameters that sensitive to the change of modal properties. Therefore, selection of the most influential parameters was performed using sensitivity analysis.

In order to perform model updating, the NASTRAN optimization code (SOL 200) is used in which it allows the objective function (J) to be minimized by adjusting the eigenvalues of the initial FE model until the objective function is converged. The objective function can be expressed as

$$J = \sum_{i=1}^{n} W_i \left(\frac{\lambda_i^{\mathrm{fe}}}{\lambda_i^{\mathrm{exp}}} - 1 \right)^2 \tag{2}$$

where, λ_i^{exp} is the i-th experimental eigenvalue and λ_i^{fe} is the i-th predicted eigenvalue from the FE model and n is the number of eigenvalues involved in the updating procedure. In this process, the selected parameters are systematically updated until the minimum errors between FE and experiment results are achieved. However, it is important to pair the mode shapes of FE and experiment correctly as the equation (2) are only calculated the eigenvalue of FE and experiment from the same modes.

5. Results and Discussion

In this study, the dynamic behaviour of laser stitch welded structures which are natural frequencies and mode shape were obtained numerically and experimentally. Results obtained were showed in Table 1.

The experimental natural frequencies and mode shapes were used for the comparison with the initial FE counterparts. It can be seen that in Table 1, the total error of the first six modes which is 28.59 percent (Column III) and the average MAC value is above 0.68 (Column VI). Furthermore, the comparison also reveals that the largest error is in the 4th and 5th mode with 6.68 and 6.63 percent (Column III). These results suggest that it is obligatory to applied FE model updating method to the initial FE model in order to reduce it discrepancies against experimental result.

In order to identify the sensitivity parameters of the initial FE model, several parameters such as Young's modulus, Poisson's ratio, and density of the structure were investigated in sensitivity analysis. It was found that by analysing the global material properties of the structure alone were not enough to improve the initial result because welded structure like laser stitch welded structure were always tending to experience localized error consequences from the welding process. Therefore, it is necessary to include parameters such as material properties of laser stitch welds and heat affected zone (HAZ), and residual stress regions.

Table 1 shows (Column VI) the discrepancies between the experimental and the updated natural frequencies of the laser stitch welded structure. It can be seen that the discrepancies of the first six modes were reduced from 28.59 to 6.35 percent. Meanwhile, Table 2 shows the sensitivity analysis results of the structure. In this analysis, the parameters such as Young's modulus hat (Ehat), Young's modulus plate (Eplate) and residual stress hat (RShat) were allowed to vary in a range of material nominal values. This is because to guarantee the physical meaning of updated parameters. However, due to the high uncertainties level, the parameters such as Young's modulus laser stitch welds (Eweld), Young's modulus HAZ on hat (EHAZh) and Young's modulus HAZ on plate (EHAZp) were allowed to have large variation but in reasonable amount in order to justify the rigidity of the weld [15].

In this work, it was found that, it is important to include the local parameters such as laser stitch welds, HAZ and residual stress in FE model updating of laser stitch welded structure. It is essential to accurately identify the source of error that may affects the dynamic behaviour of the finite element model.

Table 1. Comparison between experiment, FE and updated natural frequencies of laser stitch welded structure

Mode	I Experiment (Hz)	II Initial FE (Hz)	III Error (%) [I-II/I]	IV MAC	V Updated FE (Hz)	VI Error (%) [I-V/I]	VII MAC
1	521.52	503.51	3.45	0.95	521.77	0.05	0.98
2	591.17	569.27	3.70	0.94	590.22	0.16	0.98
3	595.44	571.93	3.95	0.90	598.84	0.57	0.97
4	674.67	629.59	6.68	0.81	657.69	2.52	0.85
5	694.64	636.39	6.63	0.76	664.49	2.51	0.77
6	726.47	665.67	4.17	0.68	690.84	0.55	0.69
Total Error			**28.59**			**6.35**	

Table 2. Updated parameters of the laser stitch welded structure

No	Parameters	Initial value	Updated value	Percentage Different
1	Young's modulus Hat (Ehat)	210 GPa	200 GPa	4.76 %
2	Residual Stress Hat (RShat)	1.0	1.16	16 %
3	Young's modulus Plate (Eplate)	210 GPa	220 GPa	4.76 %
4	Young's modulus LSW (Eweld)	320 GPa	335 GPa	4.68 %
5	Young's modulus HAZ on Hat (EHAZh)	270 GPa	283 GPa	4.81 %
6	Young's modulus HAZ on Plate (EHAZp)	270 GPa	268 GPa	0.74 %

6. Conclusions

This work was conducted to investigate the dynamic behaviour of the laser stitch welded structure numerically by using ACM2 element connector as laser stitch weld connectors. The comparison of results reveals that ACM2 element connect-or can be used to represent as laser stitch welds. However, in order to improve the initial model, it is essential to apply the FE model updating method in light of the experimental results. The approach indicates that the right selection of updating parameters is important in reducing the errors in the finite element models. In addition, the inclusion of local parameters such as laser stitch welds, HAZ and residual stress as updating parameters has led to improvement towards the FE model.

Acknowledgments

The research has been funded by the Malaysian Ministry of Higher Education (MOHE) and Research Management Institute (RMI) of Universiti Teknologi MARA (UiTM) for Research Grant with the file number 600-IRMI/PERDANA 5/3/BESTARI (073/2018). Besides, the authors would like to express thankful-ness to the members of the Structural Dynamics Analysis and Validation (SDAV) UiTM for their technical assistance and laboratory facilities.

References

[1] R. A. Ibrahim and C. L. Pettit, "Uncertainties and dynamic problems of bolted joints and other fasteners," J. Sound Vib., vol. 279, no. 3–5, pp. 857–936, Jan. 2005.
[2] K. Sripichai, K. Asim, and J. Pan, "Stress intensity factor solutions for estimation of fatigue lives of laser welds in lap-shear specimens," Eng. Fract. Mech., vol. 78, no. 7, pp. 1424–1440, May 2011.
[3] K. Asim, K. Sripichai, and J. Pan, "Fatigue Failure of Laser Welds in Lap-Shear Specimens of High Strength Low Alloy (HSLA) Steels under Cyclic Loading Conditions," SAE Int. J. Mater. Manuf., vol. 4, no. 1, pp. 2011-01–0473, Apr. 2011.
[4] M. Norhisham, A. Rani, S. Kasolang, and M. H. Othman, "Finite element modelling and modal based updating for the dynamic behaviour of a laser stitch welded structure," 23th Int. Congr. Sound Vib. 2016, ICSV 2016, pp. 1–8, 2016.
[5] F. Kuratani, M. Okuyama, T. Yamauchi, S. Washior, and S. Washio, "Finite Element Modeling of Spot Welds for Vibration Analysis," 5th Asian Conf. Multibody Dyn., 2010.
[6] M. A. Yunus, M. A. S. Aziz Shah, M. N. A. Rani, W. I. I. Wan Iskandar Mirza, and S. Khushairi, "Experimental and Finite Element Analysis of Laser Stitch Welded Structure," J. Mech. Eng., 2017.
[7] N. Kuppuswamy, R. Schmidt, F. Seeger, and S. Zhang, "Finite element modeling of impact strength of laser welds for automotive applications," in Safety and Security Engineering II, 2007, vol. I, pp. 375–384.
[8] J. Ha, H. Huh, Y. An, and C. Park, "Compatible finite element modelling of laser welded region for crash analysis of autobody assemblies," Mater. Res. Innov., vol. 15, no. sup1, pp. s412–s416, Feb. 2011.
[9] J. E. Mottershead, M. I. Friswell, G. H. T. Ng, and J. a Brandon, "Geometric Parameters for Finite Element Model Updating of Joints and Constraints," Mech. Syst. Signal Process., vol. 10, no. 2, pp. 171–182, 1996.
[10] M. N. Abdul Rani, D. Stancioiu, M. A. Yunus, H. J. Ouyang, H. Deng, and S. James, "Model Updating for a Welded Structure Made from Thin Steel Sheets," Appl. Mech. Mater., vol. 70, pp. 117–122, Aug. 2011.
[11] M. A. Yunus, H. Ouyang, M. N. A. Rani, and A. A. M. Isa, "Finite element modelling and updating of bolted joints in a thin sheet metal structure," in 20th International Congress on Sound and Vibration 2013, ICSV 2013, 2013, vol. 2, pp. 1146–1152.
[12] M. A. Yunus, H. Ouyang, M. N. A. Rani, and A. A. Misa, "Modal test and model updating for a welded structure made from thin steel sheets," in 20th International Congress on Sound and Vibration 2013, ICSV 2013, 2013, vol. 2, pp. 1161–1167.
[13] M. A. S. Aziz Shah, M. A. Yunus, M. N. Abdul Rani, N. A. Tormidi, A. A. Mat Isa, and W. I. I. Wan Iskandar Mirza, "Closed Mode Separation of The Symmetrical Structure," J. Mech. Eng., 2018.
[14] J. E. Mottershead and M. I. Friswell, "Model Updating in Structural Dynamics: A Survey," Journal of Sound and Vibration, vol. 167, no. 2. pp. 347–375, 1993.
[15] N. A. Husain, H. H. Khodaparast, A. Snaylam, S. James, G. Dearden, and H. Ouyang, "Finite-element modelling and updating of laser spot weld joints in a top-hat structure for dynamic analysis," Proc. Inst. Mech. Eng. Part C J. Mech. Eng. Sci., vol. 224, no. 4, pp. 851–861, Apr. 2010.

The Prediction of the Dynamic Behaviour of a Structure Using Model Updating and Frequency Based Substructuring

W I I Wan Iskandar Mirza[1], M N Abdul Rani[1*], M A Yunus[1], M S M Sani[2] and M S Mohd Zin[1]

[1] Structural Dynamics Analysis & Validation (SDAV), Faculty of Mechanical Engineering, Universiti Teknologi MARA (UiTM), 40450 Shah Alam, Selangor, Malaysia

[2] Advanced Structural Integrity and Vibration Research (ASIVR), Faculty of Mechanical Engineering, Universiti Malaysia Pahang, 26600 Pekan, Pahang, Malaysia

[*]Corresponding author: mnarani@salam.uitm.edu.my

Abstract. Combining analytical with experimental data to predict the dynamic behaviour of an assembled structure via the frequency based substructuring (FBS) method, is a common practice in the field of structural dynamics. However, the accuracy of the dynamic behaviour predicted from the FBS method relies heavily on the quality of experimental frequency response function (FRF) of the interfaces on which, in practice, it is very difficult to obtain. In addition, the accuracy of the FBS method is highly dependent on experimental rotational degrees of freedom (DOF) which are always found to be very difficult to measure accurately. Therefore, this paper proposes a new frequency response function (FRF) coupling scheme that may uniquely address the difficulties and improve the quality of predicted results of the FBS method. The scheme is formulated based on the finite element method, model updating technique and experimental modal analysis. A simplified finite element model of a physical test substructure is developed to generate rotational FRF data by reconciling the initial FE model using the model updating technique. It was found that the scheme adopted allows generating full translational and rotational degrees of freedom data and leads to a significant improvement in the FRF coupling process between the analytical and experimental model in the FBS method.

Keywords: Structural Dynamics, Frequency Based Substructuring, Model Updating, Experimental Modal Analysis

1. Introduction

Dynamicists tend to combine analytical model with experimental model than combining models derived from analytical work for the investigation of the dynamic behaviour of engineering structures. Usually, analytical models are developed from the finite element method. However, the developed FE models are often found to be not in good agreement with experimental models due to the invalid assumptions of the model properties used in the FE models [1]. The model updating method is perceived to be the most efficient method to improve the accuracy of the FE models [2-3]. Adopting the updating method, the updated FE models has better representation of the dynamic behaviour of the physical test structures.

The dynamic behaviour of a complex built-up structure which consists of several substructures usually tends to be computationally expensive and highly problematic to be determined analytically [4-5]. Therefore, dynamic substructuring methods, which allow the dynamic behaviour of the complex structure being investigated economically and efficiently, are seen to have the edge over other methods.

IOP Conf. Series: Materials Science and Engineering **506** (2019) 012011 doi:10.1088/1757-899X/506/1/012011

One of the popular dynamic substructuring methods is the frequency based substructring (FBS) method. This method has outstanding versatility in assembling the frequency response function (FRF) of substructures which can be analytically derived or experimentally measured [6].

The review, history and general framework of FBS are available in De Klerk [7]. However, the method has been reported to be suffer from several issues, which mainly due to the rotational FRF of the experimental model [8-10]. This is because the rotational FRFs are very difficult to be measured and neglecting them during the process of FRF coupling might greatly affect the accuracy of the coupled FRF [11-12]. Mathematically, considered a single node system is described in matrix as follows:

$$\begin{bmatrix} x \\ y \\ z \\ \phi_x \\ \phi_y \\ \phi_z \end{bmatrix} = \begin{bmatrix} Y_{xx} & Y_{xy} & Y_{xz} & Y_{x\phi_x} & Y_{x\phi_y} & Y_{x\phi_z} \\ Y_{yx} & Y_{yy} & Y_{yz} & Y_{x\phi_x} & Y_{x\phi_y} & Y_{x\phi_z} \\ Y_{zx} & Y_{zy} & Y_{zz} & Y_{x\phi_x} & Y_{x\phi_y} & Y_{x\phi_z} \\ Y_{\phi_x x} & Y_{\phi_x y} & Y_{\phi_x z} & Y_{\phi_x x} & Y_{\phi_x \phi_y} & Y_{\phi_x \phi_z} \\ Y_{\phi_y x} & Y_{\phi_y y} & Y_{\phi_y z} & Y_{\phi_y x} & Y_{\phi_y \phi_y} & Y_{x\phi_z} \\ Y_{\phi_z x} & Y_{\phi_z y} & Y_{\phi_z z} & Y_{\phi_z x} & Y_{\phi_z \phi_y} & Y_{\phi_z \phi_z} \end{bmatrix} \begin{bmatrix} F_x \\ F_y \\ F_z \\ M_x \\ M_y \\ M_z \end{bmatrix} \tag{1}$$

where $[x,y,z]$ and $[\phi_x \phi_y \phi_z]$ represent the translational and rotational DOFs respectively. The abbreviation of Y, F, and M describe the FRF matrix, force-excite and moment-excite respectively. The matrix as presented in the above equation clearly shows that 75% of the single node matrix is built based on rotational information. Therefore, neglecting the rotational DOFs at the interface during the process of coupling will definitely lead to a reduction in the interface stiffness up to 75%. In other words, the neglect of the DOFs may result in a dramatic reduction in the accuracy of the prediction of FRFs.

Previous studies reveal that including the rotational DOFs in the process of coupling has led to successful and accurate coupled FRF between FE-experimental dynamic substructuring [9,13-14]. Recently, dynamicists tend to implement several modal expansion methods to derive the rotational FRF based on the measured translational FRF. However, the expanded rotational FRF relies heavily on the quality of the measured translational FRF that might contaminated with noise. Therefore this paper aims to propose a new coupling scheme that may uniquely address the difficulties and improve the quality of predicted results of the FBS method. The scheme is formulated based on the finite element method, model updating technique and experimental modal analysis.

In this work, the high dependency on the experimental data on which the FBS method relies largely, can be significantly reduced by using a simplified finite element (SFE) model which is correlated with the translational test data to obtain rotational information and a high quality derived FRF.

2. Description of the Jointed Structure

In this study, the applicability, capability, and the accuracy of the proposed scheme for coupling the finite element and test FRFs via the FBS method were demonstrated using a jointed structure consisting of a simple beam and an irregular plate structure as shown in Figure 1. The jointed structure was 200 mm in wide and 500mm in length. Both substructures were assembled by using two bolted joints. The dynamic behaviour investigation is focused on the frequency bandwidth between 0 to 2000 Hz which contains ten modes of interest. It was found that there were several bending and torsional modes at the connection interfaces.

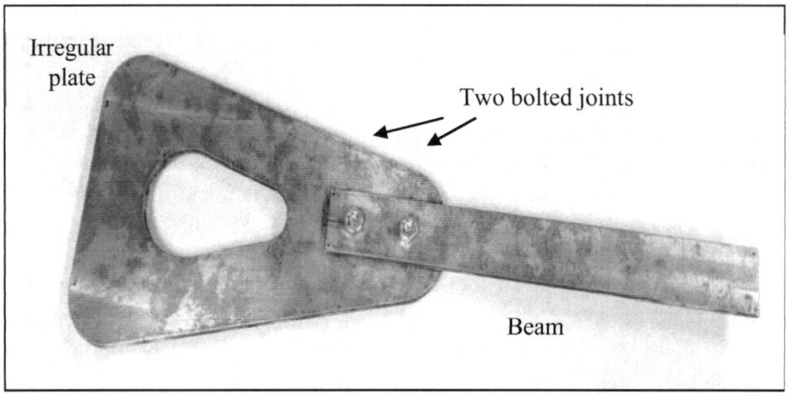

Figure 1. The jointed structure

Prior to performing the FBS analysis, the FRF of the physical jointed structure was measured for a benchmarking purpose using the experimental modal analysis by referring to the previous work presented in [6,13,15] . The structure was tested under free-free boundary conditions by using impact testing before disassembling it into two substructures, which are the beam (finite element model) and the irregular plate (experimental model). Three uniaxial accelerometers with 10mV/g sensitivity was used to acquire the response and the measured data was analysed by (Leuven Measurement System) LMS Test Lab.

The structure was suspended with soft springs to simulate free-free boundary conditions and was excited ten times to obtain an acceptable level of accuracy for the measured data. The finite element model of the beam was developed by using 3D elements. The dynamic behaviour of the FE model was calculated by using normal modes analysis.

3. The Development of the Simplified Finite Element (SFE) Model

Previous works [9-10] revealed that there was a possibility of deriving directly moment- excited FRFs from experimental models. However, in most cases, dynamicists tend to perform the modal expansion method to derive the moment-excited FRFs to provide a full FRF matrix which usually leads to successful FRF coupling [13-15].

In this work, a new coupling scheme was developed using the model updating method in the light of experimental FRF in which the model updating method was used to improve the finite element model of the physical irregular model. The updated FE model was then used to extract and derive the incomplete rotational FRF data required. It is worth noting that constructing an accurate finite element model of large structures such as a car body in white and helicopter, which usually consist of several complex components, is very difficult, challenging and time consuming. Therefore, instead of using the actual experimental model (the irregular plate structure), a simplified finite element (SFE) model is introduced in the work.

The SFE model of the test structure was developed by constructing a 2D finite element model with a much simpler structure without any curves or slopes. Figure 2 shows the developed SFE model for the irregular shape structure. The developed SFE model has multiple parameters that can be reconciled by referring to the experimentally obtained natural frequencies. By adopting the SFE, the derived FRF from the updated SFE model can be matched as close as possible to the actual experimental FRF. The updated SFE model was then used to derive a new FRF matrix which contained both translational and rotational FRF

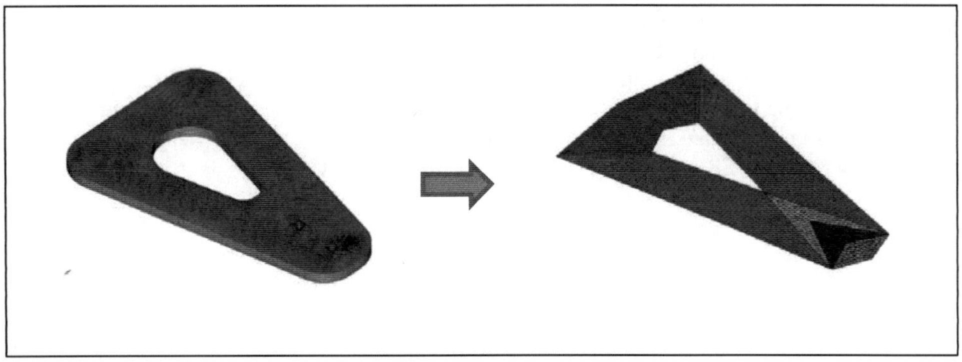

Figure 2. The full finite element and SFE model

4. The Finite Element Model Updating for SFE Model

In structural dynamics, model updating methods are used to improve the accuracy of finite element models by improving systematically and iteratively the design parameters of the models to achieve a better correlation with the experimentally measured dynamic behavior [16-17]. In this work, the finite element model updating method was adopted to improve the initial SFE model by updating the material properties and thickness of the model to match that of the physical irregular plate. Updating the parameters was performed by using MSC NASTRAN Solution 200 with the same objective function used in [17-18].

Table 1 presents the natural frequencies obtained from three different sources which are the experimental modal analysis (EMA), initial and updated SFE models. The total errors recorded were calculated from the direct comparisons between the natural frequencies of EMA, the initial FE and the updated FE. From the table, it can be observed that the total error recorded in the initial FE was successfully reduced from 77.4% to 2.95%. Therefore, it is imperative to note that the SFE model is reliable to be used in predicting the dynamic behaviour of the physical irregular shape.

Table 1. Comparison of the natural frequencies of the measured (EMA), initial and updated SFE models.

		Initial		Updated	
Mode	EMA	FE Hz	Diff. (%)	FE Hz	Diff. (%)
1	352.19	368.09	4.51	351.11	0.31
2	474.80	553.38	16.55	474.53	0.06
3	876.62	973.25	11.02	884.01	0.84
4	1038.00	1131.70	9.03	1035.40	0.24
5	1128.00	1229.60	9.01	1128.40	0.04
6	1886.70	2006.20	6.33	1886.60	0.00
7	1935.20	2026.10	4.70	1931.40	0.20
8	2160.50	2272.80	5.20	2173.60	0.61
9	2258.40	2508.00	11.05	2243.70	0.65
			77.4		2.95

IOP Conf. Series: Materials Science and Engineering **506** (2019) 012011 doi:10.1088/1757-899X/506/1/012011

5. Frequency Based Substructuring using Updated SFE Model

To evaluate the capability and accuracy of the proposed coupling scheme, the FRFs derived from the finite element model of the beam and the updated SFE model of the irregular plate were coupled as shown in Figure 3. The FRFs from both finite element models were derived by using the FRF synthesis method based on the calculated natural frequencies and mode shapes. Later, the FBS method was employed to calculate the FRF of the complete jointed structure by using rigid type of connections.

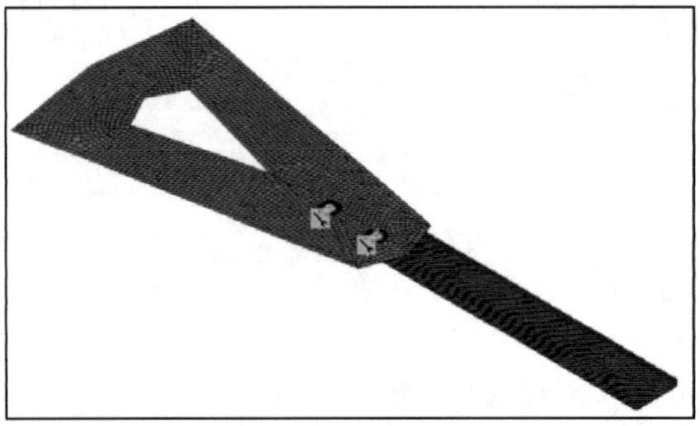

Fig. 3. The assembly of the FE model of the beam and SFE model of the irregular plate

The coupled FRF calculated from the FBS method was validated and compared with the measured FRF. Figure 4 shows the comparison of both FRFs. From the figure, there are eight visible resonance peaks and seven anti-resonances of the experimental and coupled FRF within the frequency bandwidth between 0-2000 Hz. The results obtained from the comparison indicates that the proposed FRF coupling scheme has been successfully used to calculate the FRF of the jointed structure.

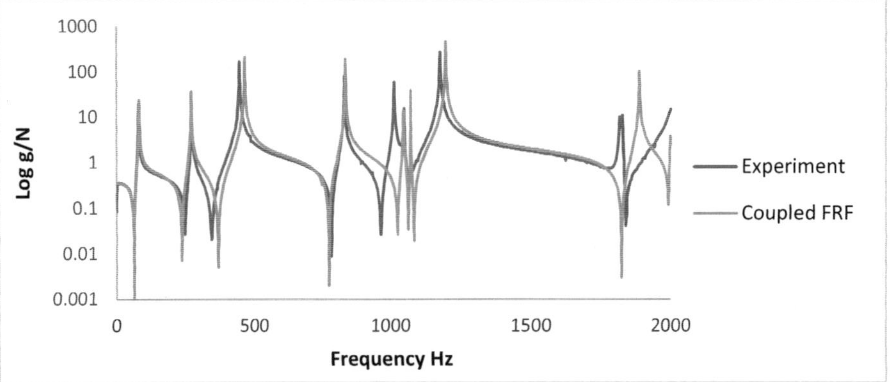

Figure 4. The comparison of coupled and experimental FRF of the assembled structure

It was also found that for the first four modes, the coupled FRF matches well with the measured one with having accurate patents in the resonance frequencies as well as anti-resonance. This achievement suggests that using the proposed coupling scheme can lead to a successful prediction of the coupled FRF for a low frequency bandwidth. A good correlation obtained between the predicted and measured FRF indicates that introducing the derived rotational FRF from the SFE model during the process of coupling has enhanced the flexibility of the interfaces of the model. From the engineering point of view, including the 75% of the rotational information has improved the rigidity of the interface, which has led to a successful FRF coupling process. This achievement is also in line with that of presented in the previous work [9,13,19]. Therefore, it is worth noting that the proposed FRF coupling scheme offers a great capability to accurately predict the dynamic behaviour of the jointed structure via the FBS method.

Another striking point of the comparison is that the coupled FRF has started to differ from the experimental FRF for higher modes. The discrepancies occurred were suspected from the invalid selection of coupling types which may be unable to accurately represent the physical properties of the two bolts. This is a normal phenomenon encountered in the field of structural dynamics [4-5] because the joints are very difficult to be modelled in detail [2].

6. Conclusions

The dynamic behavior of the jointed structure using the FBS method has been investigated. The new coupling scheme is proposed to effectively and efficiently couple the two substructures. The scheme is performed by implementing the model updating method for the SFE model in the light of the experimental data. The evaluation of the capability and accuracy of the proposed scheme has been presented and discussed. The results in the previous section demonstrated that the proposed scheme has been successfully used to calculate the FRF of the assembled structure. Therefore, it is worth noting that the updated SFE model has a great capability to be used for the FBS analysis. This remarkable achievement suggests that the proposed scheme is more efficient and effective over the conventional modal expansion method in order to obtain the full FRF matrix of the connection interfaces.

Acknowledgements

The authors are gratefully indebted to the Malaysian Ministry of Higher Education for providing financial support for this study through the fundamental research grant scheme (FRGS) (600-RM1/FRGS 5/3 (96/2016). They would also like to express their gratitude for the help and support given by Prof. Ir. Dr. Hj. Abdul Rahman Omar, Prof. Dr. Hadariah Bahron, and SDAV group members.

References

[1] K Yuen 2012 Updating large models for mechanical systems using incomplete modal measurement *Mech. Syst. Signal Process.* vol 28 pp 297–308

[2] W I Wan Iskandar Mirza, M N Abdul Rani, M H Othman, S Kasolang and M A Yunus 2016 Reduced order model for model updating of a jointed structure *J. Eng. Appl. Sci.* vol 11 pp 2383–2386

[3] J E Mottershead, M Link and M I Friswell 2011 The sensitivity method in finite element model updating: A tutorial *Mech. Syst. Signal Process.* vol 25 pp 2275–2296

[4] M N Abdul Rani, D Stancioiu, M A Yunus, H Ouyang, H Deng and S James 2011 Model Updating for a Welded Structure Made from Thin Steel Sheets *Appl. Mech. Mater.* vol 70 pp 117–122

[5] M N Abdul Rani, S Kasolang, M H Othman, M A Yunus and W I I Wan Iskandar Mirza 2016 Finite Element Modelling and Modal Based Up- Dating for The Dynamic Behaviour of a Laser Spot Welded Structure *ICSV 2016 - 23rd International Congress on Sound and Vibration: From Ancient to Modern Acoustics* pp 1–8

[6] W I I Wan Iskandar Mirza, M N Abdul Rani, M A Yunus, L Roslan and A B Ahmad Basri 2016 The investigation of the dynamic behaviour of a complex assembled structure using the frequency response *MATEC Web Conf.* vol 90 pp 1–10

[7] D De Klerk, D J Rixen and S N Voormeeren 2008 General Framework for Dynamic Substructuring: History, Review and Classification of Techniques *AIAA J.* vol 46 pp 1169–1181

[8] D Nicgorski and P Avitabile 2010 Experimental issues related to frequency response function measurements for frequency-based substructuring *Mech. Syst. Signal Process* vol 24 pp 1324–1337

[9] J E Mottershead, M Ghandchi Tehrani, D Stancioiu, S James and H Shahverdi 2006 Structural modification of a helicopter tailcone *J. Sound Vib.* vol 298 pp 366–384

[10] D de Klerk, D J Rixen, S N Voormeeren and F Pasteuning 2008 Solving the RDoF Problem in Experimental Dynamic Substructuring *Proc. 26th Int. Modal Anal. Conf. - IMAC*

IOP Conf. Series: Materials Science and Engineering **506** (2019) 012011 doi:10.1088/1757-899X/506/1/012011

[11] M S Allen, R L Mayes and E J Bergman 2010 Experimental modal substructuring to couple and uncouple substructures with flexible fixtures and multi-point connections *J. Sound Vib.* vol 329 pp 4891–4906

[12] R L Mayes, P S Hunter, T W Simmermacher and M S Allen 2008 Combining Experimental and Analytical Substructures with Multiple Connections *26th IMAC: Conference and Exposition on Structural Dynamics: IMAC XXVI* (Orlando, Florida)

[13] A Drozg, G Čepon and M Boltežar 2018 Full-degrees-of-freedom frequency based substructuring *Mech. Syst. Signal Process.* vol 98 pp 570–579

[14] A Williams, C Chipman and P Avitabile 2008 Modal and frequency based substructuring using rotational DOF conidetations *26th Int. Modal Anal. Conf. (IMAC XXVI)*

[15] P Peeters, S Manzato, T Tamarozzi and W Desmet 2018 Reducing the impact of measurement errors in FRF-based substructure decoupling using a modal model *Mech. Syst. Signal Process.* vol 99 pp 384–402

[16] N A Z Abdullah, M S M Sani, N A Husain, M M Rahman and I Zaman 2017 Dynamics properties of a Go-kart chassis structure and its prediction improvement using model updating approach *Int. J. Automotive and Mech. Eng.* vol 14 pp 3887–3397

[17] M S M Fouzi, K M Jelani, N A Nazri and M S M Sani 2018 Finite Element Modelling and Updating of Welded Thin-Walled Beam *Int. J. Automotive and Mech. Eng* vol 15(4) pp 5874-5889

[18] M N Abdul Rani, H Ouyang, M A Yunus and B A Aminudin 2013 Model Updating for a Thin Steel Sheet Welded Structure *20th Int. Congr. Sound Vib. 2013 (ICSV 2013)* **2** 1138–45

[19] S Manzato, E Risaliti, C Napoli, T Tamarozzi and B Peeters 2015 A Review of Frequency-Based Substructuring Methods and Their Applicability to Engineering Structures *Proceeding of International Conference of Structural Engineering Dynamics (ICEDyn)* (Lagos)

Frequency Response Function Based Updating of a Laser Spot Welded Structure using Synthesised Frequency Response Function

M S Mohd Zin[1], M N Abdul Rani[1*], M A Yunus[1], W I I Wan Iskandar Mirza[1], M A Ayub[1] and R Omar[2]

[1]Structural Dynamics Analysis & Validation (SDAV), Faculty of Mechanical Engineering, Universiti Teknologi MARA (UiTM), 40450 Shah Alam, Selangor, Malaysia
[2]Machinery Technology Centre, SIRIM Industrial Research, 44200 Hulu Selangor, Selangor, Malaysia

[*]Corresponding author mnarani@salam.uitm.edu.my

Abstract. Accurate analytical models of engineering structures are of paramount importance for dynamicists to be used in predicting the dynamic behaviour (frequency response function) of the structures. The finite element method (FEM) and the experimental modal analysis (EMA) have been known as powerful and useful methods that can be used to determine the frequency response function (FRF) of the structures. However, the finite element FRF are often not in good agreement with experimental FRF due to assumption properties in finite element (FE) model. Therefore, in order to have a reliable FE model of a structure, measured FRF obtained from the experimental modal analysis can be integrated with finite element FRF to reconcile the FE model and the procedures involved in the reconciliation is a model updating process. Prior performing the updating process, the experimental FRF must be in good quality to obtain accurate FE model results. One way to reduce noise data and eliminate suspicious modes in FRF data is to use the FRF synthesised method. The main goal of this study is to use frequency response function based updating to minimise the error of a laser spot welded structure by using FRF synthesised data. In this work, MSC Software and LMS Test Lab were used to predict and measured the FRF of laser spot welded structure. The results revealed that FRF based updating method is successfully reduced the correlation gap between synthesised FRF and predicted FRF.

Keywords: frequency response function, model updating, synthesised FRF, laser spot weld.

1. Introduction

A precise description of the dynamic behaviour of engineering structures is of paramount importance to the dynamicist community. Usually, there is a great deal of uncertainty about the selection of suitable methods to be used for the analysis. This is because the time to market products is shrinking. Experimental analyses usually are very costly, and long-time consume to be performed as compared to numerical simulations using the finite element method which is one of the most versatile numerical methods [1-2]. However, it is found that the finite element results are frequently not in good agreement with experimental counterparts due to the invalid assumptions made in the finite element modelling.

One way to refine, correct or update the finite element model through which the dynamic behaviour of a structure is predicted using model updating methods [3-9].

The subject of model updating methods has received much attention from many researchers. In the modal based updating method, [10-12] have used natural frequency values as a reference for updating the finite element model. Furthermore, [13-14] successfully corrected the initial FE model of engineering structure by using frequency response function data. Although the FRF based updating method required more computational time and space to solve problems, this method is more reliable than modal based updating method as claimed by [15-16].

One of the major drawbacks of the FRF based updating method is the noisy data in experimental FRF that will usually affect the accuracy of the finite element model updating results. The FRF that are contaminated with noise may cause the numerical solutions are not able to differentiate between the actual resonance peaks and the spurious modes or the noisy amplitudes. Hence, this scenario has driven the authors to implement the FRF synthesized method to eliminate the noisy data, that can efficiently be used in FRF based model updating. This paper presents a method that can be used to minimise the error between finite element FRF and synthesised FRF of a laser spot welded structure using the FRF based updating method. The updating will be applied to the initial finite element model of the welded structure in light of the measured results. The updated frequency response function is validated with those obtained experimentally to evaluate the accuracy of the FE model.

2. Finite element modelling of the laser spot welded structure

The finite element model of the laser spot welded structure was constructed using commercial finite element software as shown in Figure 1. In this work, the structure was modelled using QUAD4 shell elements with thousands of elements and nodes. Convergence tests were performed to determine the appropriate size of the elements and the element type used in the development of the FE model. The hat shape plate and flat shape plate were connected using ACM2 elements connectors that are available in the software to represent laser spot welded joints. The structure is thin steel sheets which have the following properties: the Young's modulus, E = 210 GPa, Poisson's ratio, n = 0.30 and density, r = 7500 kg/m^3. The frequency response function of the finite element model of the structure was calculated using MSC. NASTRAN solver. In this research, the frequency of interest was starting from 0 to 900 Hz.

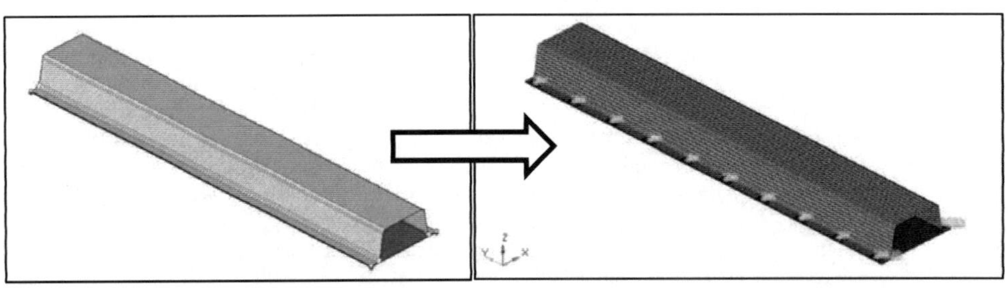

Figure 1. Geometric and finite element model of laser spot welded structure

3. Experimental work of the laser spot welded structure

In this study, the experimental modal analysis was performed on the laser spot welded structure with the nominal thickness of 1.5 mm, length of 564 mm and width of 110 mm. The structure was suspended by using soft rubber bands and strings to simulate free-free boundary conditions. This is because there were no constrains involved in the finite element model . The high sensitivity of 10.73 mV/N impact hammer and three accelerometers with 10mV/g sensitivity were used to measure the frequency response function (FRF), natural frequencies and mode shapes of the laser spot welded structure. The responses from the impact hammer and the accelerometers were acquired and analyzed by using Leuven Measurement System (LMS) front-end. The selection of the excitation points and the locations of measurement points were decided based on the modal participation factor (MPF) calculated in the initial prediction of the dynamic behaviour of the structure. The frequency bandwidth of interest was 0 to 900 Hz. The details of the configuration of the components and the test set up are available in [10]. In order to obtain an acceptable level of accuracy and minimize the noise in the data, the RMS averaging method was performed by exciting and testing the structure ten times.

3.1. *The synthesised frequency response function*

One way to determine the frequency response function (FRF) of a component is by calculating an FRF synthesis based on the number of mode shapes and natural frequencies of the components. The equation of the synthesised FRF [13] is expressed by

$$\mathrm{H_{syn}}(\omega_k) = \sum_{i=1}^{N}\left(\frac{\{\emptyset\}_i\{\emptyset\}_i^T}{(\omega_{n_i}^2 - \omega_k^2) + j^2\xi_i\omega_{n_i}\omega_k}\right) \tag{1}$$

where N is the number of calculated modes, $\{\emptyset\}_i$ is the ith mass normalized mode shape, ω_{n_i} is the ith natural frequency and ξ_i is the ith modal damping ratio. In this research, the synthesised FRF was used as a reference data in model updating method.

4. FRF based model updating

The updating of the initial finite element model was attempted using NASTRAN Solution 200 software. The objective of FRF based updating is to minimise the discrepancies between the finite element FRF and experimental FRFs by improving the parameters of the FE model in light of measured data to an acceptable level of accuracy. The finite element frequency response function must be corrected using the min-imisation of output residue at any frequencies using the equation below;

$$\min\sum_{j=1}^{m}\left(\frac{\mathbf{T}_i^{\mathrm{FE}}}{\mathbf{T}_i^{\mathrm{exp}}} - 1\right)^2 \tag{2}$$

where T_i^{FE} is the ith predicted frequency response function and T_i^{exp} represents the ith measured frequency response function.

5. Results and discussion

The objective of this paper is to minimise the error between finite element FRF and synthesised FRF using FRF based updating method. The synthesised FRF was calculated using Equation 2 starting from 0 to 900 Hz which contains ten modes. The comparison of synthesised FRF and original experimental FRF are presented in Figure 2 below. The synthesised FRF has successfully removed spurious mode (marked with red circle) between mode seven and mode eight. In addition, the noisy data in the frequency less than 400 Hz has also been successfully reduced. Hence, the synthesised FRF will be used as a reference data to correct the initial finite element FRF.

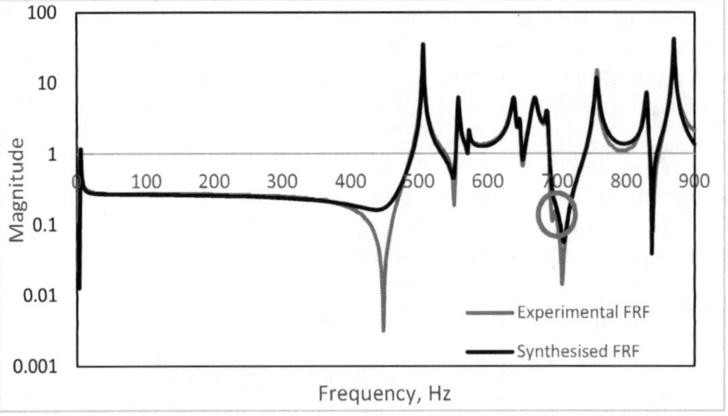

Figure 2. Graph of comparison between original experimental FRF and synthesised FRF

The computed initial finite element FRF as shown in Figure 3 shows huge discrepancies as against the synthesised FRF. The discrepancies may arise from erroneous assumptions about the model properties used in the initial model. In addition, the discrepancies may come from the simplification in modelling of the joints to represent the actual joints without considering the effects of the process of assembling the substructures. Therefore, the initial finite element model needs to be corrected to improve the accuracy of the model. However, it seems to be impossible to manually adjust the predefined model properties of the initial finite element model without using a systematic updating method. This study aimed to systematically correct the predefined model properties of the initial finite element model in the light of experimental data using the frequency response function (FRF) based model updating method.

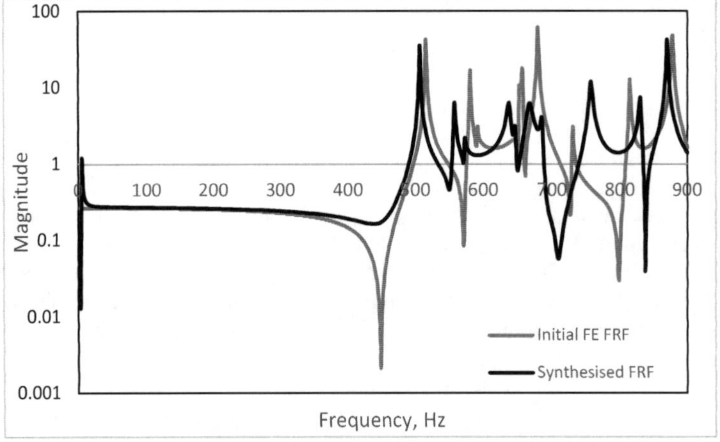

Figure 3. Graph of comparison between initial FE FRF and synthesised FRF

The potential updating parameters were selected from the finite element model to systematically modify the parameters until the predicted FRF is closed to measure FRF. Prior performing model updating process, the sensitivity method is carried out using Nastran codes. It was found that the responses which are the FRF are more sensitive to the Young's modulus and Poisson's ratio of the hat and plate as well as all model properties of the heat affected zones. The minimisation process of the discrepancies in the initial finite element model is calculated using the objective function as mentioned in Equation 2.

The comparison between the updated FE FRF and synthesised FRF is shown in Figure 4. The comparison of the FRF revealed that the updated FE FRF shows great improvement compared to initial FE FRF especially in the frequencies below 650 Hz. However, the FE FRF between 650 Hz to 800 Hz are not in good correlation due to uncertainties of stiffness in the FE model. In addition, the surrounding welded areas of the structure may be experienced initial stresses, that are very difficult to be measured [11]. However, in practice, the stiffness is usually assigned to be homogenous all over the finite element model of the laser spot welded structure. Based on the engineering judgement and observation, the updated results have shown an obvious improvement and are in good agreement with the measured one.

The parameter changes of the finite element model are tabulated in Table 1. From the results, Young's Modulus of heat affected zone shows huge difference from the actual value which is from 210 GPa to 241.4 GPa. This means that the HAZ regions are harder than any other regions. All updated parameter values are within the acceptable level of steel properties. In addition, the updated values are in line with the results presented in the previous studies [9,10,13].

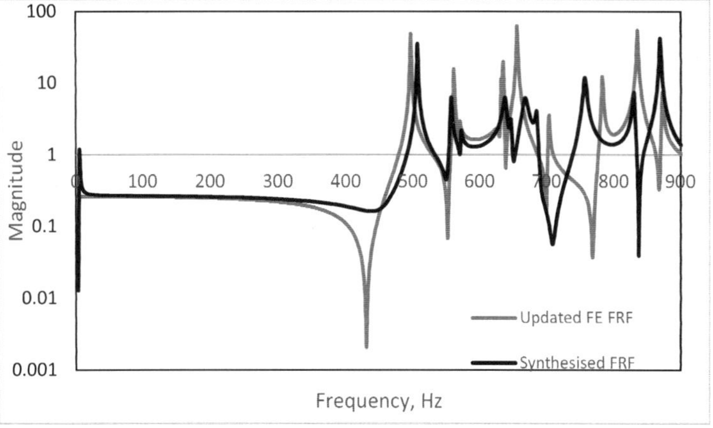

Figure 4. Graph of comparison between updated FE FRF and synthesised FRF

Table 1. Parameter changes of laser spot welded structure

Type	Elem/Set	Old	Actual
E	Hat	210000	178500
E	Plate	210000	187251
E	ACM2	210000	204591
E	HAZ	210000	241500
NU	Hat	0.30	0.33
NU	Plate	0.30	0.33
NU	ACM2	0.30	0.30
NU	HAZ	0.30	0.30

6. Conclusions

The attempt to predict and correct the frequency response function (FRF) of the laser spot welded structure using the FRF based updating method have been discussed. The results of updating the initial FE model of the structure are presented. The finite element method and experimental modal analysis have successfully used to determine the FRF of the laser spot welded structure. In addition, the synthesised FRF was successfully used to replace experimental FRF in the FRF based model updating method. Using the synthesized FRF in the FRF based model updating has proven to be successful in significantly improving the quality of the FE FRF against the measured FRF. Therefore, this method has a high potential to be used for various applications of FRF based model updating with noisy data and spurious modes are of great concern to the dynamicist community.

Acknowledgments

The authors wish to acknowledge the Malaysia Ministry of Higher Education (MOHE), Research Management Institute (RMI) and Institute of Graduate Studies of Universiti Teknologi MARA (UiTM) for providing financial support for this study through the fundamental research grant scheme (FRGS) (600-RMI/FRGS 5/3 (0096/2016)). They would also like to extend their sincere gratitude for the support and help given by Mr David Starbuck and Professor Hadariah Bahron and all the SDAV members.

References

[1] S S Bhavikatti 2005 Finite Element Analysis *New Age International*

[2] N S Gokhale, S S Deshpande and S V Bedekar 2008 Practical Finite Element Analysis Finite *To Infinite Publisher*

[3] M I Friswell and J E Mottershead 1995 Finite Element Model Updating in Structural Dynamics *Illustrate. Netherlands: Springer Science & Business Media*

[4] S V Modak, T K Kundra and B C Nakra 2002 Comparative study of model updating methods using simulated experimental data *Computers and Structure* vol 80 pp 437–447

[5] M Kurt, M Eriten, D M McFarland, L A Bergman, and A F Vakakis 2015 Methodology for Model Updating of Mechanical Components with Local Nonlinearities *Journal Sound and Vibration* vol 357 pp 331–348

[6] A Batou 2015 Model updating in structural dynamics - Uncertainties on the position and orientation of sensors and actuators *Journal Sound and Vibration* vol. 354 pp 47–64

[7] H Ahmadian, G M L Gladwell and F Ismail 1997 Parameter Selection Strategies in Finite Element Model Updating *Journal Vibration Acoustic* vol. 119(1) pp 37–45

[8] M I Friswell, J E Mottershead and H Ahmadian 2001 Finite-element model updating using experimental test data: parametrization and regularization *Philos. Trans. Math. Phys. Eng. Sci.*, vol 359 (1778) pp 169–186

[9] M S Mohd Zin, M N Abdul Rani, M A Yunus, W I I Wan Iskandar Mirza, A A Mat Isa and Z Mohamed 2017 Modal and FRF Based Updating Methods for the Investigation of the Dynamic Behaviour of a Plate *J. Mech. Eng.* vol 4(3) pp 175–189

[10] M N Abdul Rani, S Kasolang and M H Othman 2016 Finite Element Modelling and Modal Based Updating for the Dynamic Behaviour of a Laser Spot Welded Structure in *23rd International Congress on Sound & Vibration (ICSV23)* pp 1–8

[11] M N Abdul Rani, H Ouyang, M A Yunus and B A Aminudin 2013 Model Updating for a Thin Steel Sheet Welded Structure in *20th International Congress on Sound and Vibration (ICSV 20)* pp 1–8

[12] N A Z Abdullah, M S M Sani, N A Husain, M M Rahman and I Zaman 2017 Dynamics properties of a Go-kart chassis structure and its prediction improvement using model updating approach *Int. J. Automotive and Mech. Eng.* vol 14 pp 3887–3397

[13] M S Mohd Zin, M N Abdul Rani, M A Yunus, M S M Sani, W I I Wan Iskandar Mirza, and A A Mat Isa 2018 Frequency response function (FRF) based updating of a laser spot welded structure in *AIP Conference Proceedings* vol 020055 p. 020055

[14] J T Wang, C J Wang, and J P Zhao 2017 Frequency response function-based model updating using Kriging model *Mech. Syst. Signal Process.* vol 87 pp 218–228

[15] V Arora, S P Singh, and T K Kundra 2009 Comparative study of damped FE model updating methods *Mech. Syst. Signal Process.* vol 23 (7) pp 2113–2129

[16] V Arora 2015 Direct structural damping identification method using complex FRFs *J. Sound Vib.* vol 339 pp 304–323

IOP Conf. Series: Materials Science and Engineering **506** (2019) 012013 doi:10.1088/1757-899X/506/1/012013

Macroergonomics Conceptual Assessment of a Local SME in Banda Aceh using Causal Loop Diagram (CLD) and System Archetype: a Case Study in Banda Aceh

P D Sentia[1], T Mulyati[1*], Suhendrianto[1] and N Zuelda[1]

[1]Industrial Engineering Department, University of Syiah Kuala, Darusalam, Banda Aceh 23114, Indonesia

[*]Corresponding author: trinamulyati@gmail.com

Abstract. Macroergonomics is a sociotechnical approach that sees the relationship between one element and another in optimizing a work system. MP is one of Small Medium Enterprse (SME) that produces perfume in Banda Aceh, it often experiences work system problems such as uncomfortable physical environment, a small workstation, tentative work schedule and a work demand, broken machine and lack of skilled human. In order to reduce the problems found in this SME, macroergonomics approach is implemented to improve the work system by selecting important elements based on macroergonomics factor and elements. The result consists of 15 elements are the most influential element and analyzed to find the relationship of one variable to another by using system thinking or the Causal Loop Diagram (CLD). The result are success to the successful and fix that fail system archetype were obtained from the final CLD and interview. The result shows that, SME must spent macroergonomics investment by giving a training to the employees therefore they can understand and capable in manufacturing technology and tools, and decrease the number of evaluation, from twice a day becomes once a day and once a week for meeting.

Keywords: SME, CLD, Macroergonomics, Systems thinking, System archetype

1. Introduction

The limitation of technology and skilled human resources is the main problem facing SMEs when it comes to the production process. MP is one of SME produced Aceh's fragrance product which is expected to become the locomotive of national perfume. MP has been distributed throughout Indonesia, now it can be enjoyed by residents from Sabang to Marauke. This perfume is also often being a souvenir for foreign tourists who visit Aceh. MP is located in Lam Ara, Banda Raya, Banda Aceh. It hires many local manpower especially in Banda Aceh.

MP often faced production problems such as the work system. This SME has broken machine, and it makes the production system must be postponed, the physical environment of this SME is not comfortable for example a small workstation, tentative workschedule and such as the poor production processes such the number of production target never be reached by this SME, and the physical environment of this SME is not comfortable for example a small workstation, tentative workschedule, work demand, lack of skilled human. Macroergonomics is the appropriate approach to solve the MPs'

IOP Conf. Series: Materials Science and Engineering **506** (2019) 012013 doi:10.1088/1757-899X/506/1/012013

problem by considering its five factors; person, organisation, technology and tools, task and environment.

The problems also caused by the design of the system that is less effective, therefore there is an absence of a harmonious system. According to Kleiner [1] macroergonomics is a work system design that focuses on the interaction of organizations and systems, and it is a sociotechnical approach from the top down to create harmonious work system design through five factors and its 25 elements [2]. By using macroergonomics approach, there has been a emphasis on ergonomics in terms of achieving large scale organizational improvement. Macroergonomics can change an organization's culture and can achieve 60% to 90% performance improvements [1,3].

Macroergonomics is often used in a health sector to identify about work system design in a hospital [4-7] it also decreases the risk by involving managerial staff, workers or organization to attain a new system design which will be more effective [3,8-9]. The goal of macroergonomics is achieved through a general approach to design which consider interrelated subsystem [1,10].

The problem that is interconnected with other variables that can be said as a systemic problem. Therefore this problem is necessary to be solved by using system thinking. The system thinking is a tool to enhance a complex problem from various perspectives, to understand connectedness and interactions [11].

System thinking is used to see the behaviour of the variable in a complex system, shows the link one to another variables, understand the system and its structure [12]. This study integrates macroergonomics and sytem thinking to find and predict a future trend of variables. Based on the explanation of the problems that occurred in MP, using macroergonomics approach and Causal Loop Diagram (CLD) and System Archetype is the appropriate to solve the problem.

2. Research Methodology

The process of selecting primary data is done using in-depth interview (IDI) to collect the main variables and Focus Group Discussion (FGD) to verifivied the conceptual model, the interview result will eliminate unimportant element. The table 1 will show macroergonomics factor and its elements. All of these elements will be asked to the respondent in the interview process. The secondary data will be number of production from April 2015 to April 2018, then the CLD is developed through three steps, namely preliminary CLD, Working CLD and Final CLD.

The final CLD will be analyzed and based on the field observation and also based on the information obtained from the interview. in FGD the researcher also gives the respondents several graph about behaviour over time for each system archetype, the respondents are asked to choose one the element performance through the graph for the elements which do not have any data, but for the elements which have the data, the researcher find the behaviour over time through its data that will show particular system archetype pattern. Then, the prescriptive action is made by stakeholders, based on the future trends that has been explained.

3. Macroergonomics Factor and Element

The interview with seven people in MP was completed. The System Engineering Initiative for Patient Safety (SEIPS) model helps to capture themes indicated the important factors in the work system. The interview is able to show how the problem can be solved through barriers, physical work design and personnel issues. These next five factors describes its elements in detail then the relationship were identified in each factor with verbatim employee comments example and research interpretation.

3.1 Person

There are four elements in person factor and employees metioned. Three among them are important, and they believe motivation and need is not important element that affect work system peformance.

3.2 Organisations

There are eight elements in organisation factor, employees consider three among them are not important such as coordination, colaboration and communication, social relationship and supervisor and management style.

3.3 Technology and Tool

Among three elements in technology and tools factor based on the interview, the employees claims information and technology is not important.

3.4 Task

Based on the interview with employees among six elements in task factor, they consider the only two elements are important.There are task variety and work demand.

3.5 Physical Environment

The employees believe among five elements in physical, distribution only an element that considered as not important element.

The finding in table 1 shows that the organisational culture was the most statements during the interview. The employees believe that the organisational culture is the most important element in manufacturing process to complete their work.

Table 1: Macroergonomics Element Selection.

No	Factor	Element	Total Statement
1	Person	Education, Knowledge and Skill	7
		Physical Characteristics	4
		Psychological Characteristics	7
		Motivation and Need	0
2	Organisation	Teamwork	16
		Coordination, Collaboration, Communication	0
		Organisational culture	17
		Culture	0
		Work Schedule	6
		Social Relationship	0
		Supervisor and Management Style	0
		Evaluation and Rules	11
3	Technologies and Tools	Information and Technology	0
		Manufacturing technology and Tools	14
		Human Resource Characteristics in Technology and Tools	5
4	Task	Task Variety	3

		Job Content	0
		Challenge and use of skills	0
		Autonomy, Job Control and Participation	0
		Work Demand	4
		Distribution	0
5	Physical Environment	Noise	10
		Lighting	7
		Temperature, Humidity and Air Quality	5
		Workstation and Layout	10

This results are consistent with the studies that found the organisational culture is the key of the system running well, and this is in line with the studies that show the significant relationship between organizational culture and employees' innovation. In this SMEs, the employees stated that they felt comfort because their organizational culture, namely Trust Care Love (TCL) and they felt home when they are working in the office. This SME is quite different; they implement the culture of the Islamic too. All of employees are obliged to go to mosque performing Jemaah. The teamwork element is the second highest rank in the interview. The employees acknowledge the teamwork makes them feel respected thereby increasing the job satisfaction. The manufacturing technology and tool element is the third highest rank in the macrorgonomics elements. The employees realize that the manufacturing technology is really matter for the manufacturing system now due to make their activity easier, but the advance technology will spend so much cost and need a serious investment, and if the demand remain stable the employee will need the advanced manufacturing technology to help them in manufacturing process because there are several machines that often broken and it makes the employee works semi manually. The lowest rank in the 25 elements is the task variety elements because they believe that they still can cover all of the manufacturing system even one person hold two responsibilities. It does not really matter for them as long as they can manage their time as well as possible. To sum up this discussion, every factor do not have difference significantly employees agree that all of these five factors are affected the manufacturing system, therefore these 15 elemets chosen become the main key variables to build an preliminary CLD.

Table 2 shows the main key variables of the Causal Loop Diagram, this variables are obtained from through the interview based on table 1. The variables which is considered as unimportant element has been eliminated. The main key variables consists of 15 macroergonomics elements and two additional elements.

IOP Conf. Series: Materials Science and Engineering **506** (2019) 012013 doi:10.1088/1757-899X/506/1/012013

Table 2. The Main Key Variables.

No	Key Variables	CLD variable
1	Education, Knowledge and Skill	EKK
2	Physical Characteristics	Physical Characteristics
3	Psychological Characteristics	Psychological Characteristics
4	Teamwork	Teamwork
5	Organisational Culture	Organisational Culture
6	Work Schedule	Work Schedule
7	Evaluation and Rules	Evaluation and Rules
8	Manufacturing Technology and Tools	Manufacturing technology and tool
9	Human Resource Characteristics in Technology and Tools	HRCT
10	Task Variety	Task Variety
11	Work Demand	Work Demand
12	Noise	Noise
13	Lighting	Lighting
14	Temperature, Humidity and Air Quality	THAQ
15	Workstation and Layout	WL
16	Sales	Sales
17	Production	Production

4. Result and Discussion

4.1 Causal Loop Diagram (CLD)

The preliminary CLD based on the SEIPS model without the intervention of researcher's logic and FGD. The preliminary CLD shown in figure 1 has been used for examining different component of work system. It has five key components or five variables, namely macroergonomic factor. This preliminary CLD was adopted from pharmacy work system. All of variables has been connected each other. It contains nine reinforcing loop (R) and there is no balancing loop in this preliminary CLD or no loop that encounter the system. The model is also used to see the how the interrelationship between structural variables to influence the work system process as the outcomes of the work system. The preliminary CLD will be a basic information to build an working CLD. This working CLD is built based on MP case, researchers' logic and field observation without any intervention by stakeholders.

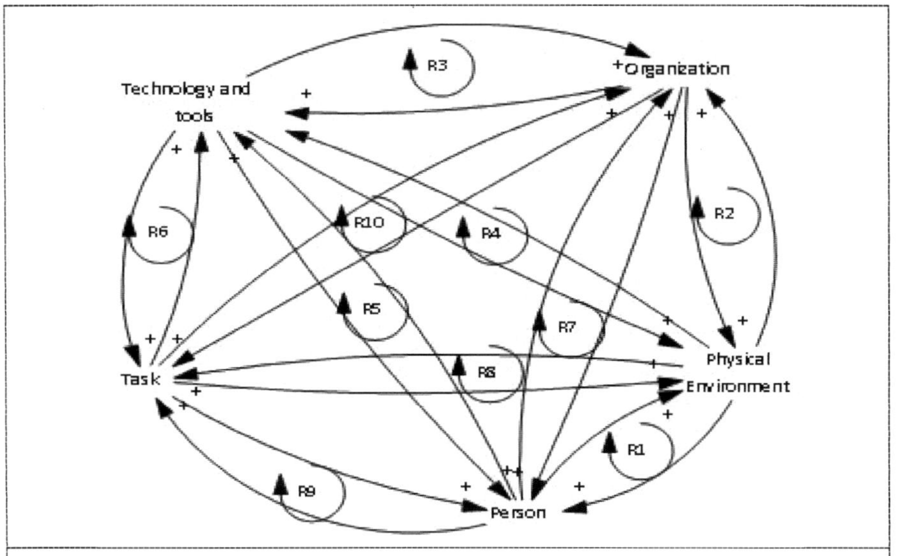

Figure 1. Preliminary CLD

Figure 2 is the working CLD which is built based on MP case, researchers' logic and field observation without any intervention by stakeholders. Many variables are added in the CLD such as task, teamwork, HRCT, production, slaes and evaluation and rules. This woking CLD contains two feedback loops, namely two reinforcing loops, both of these loops is affecting the work system.

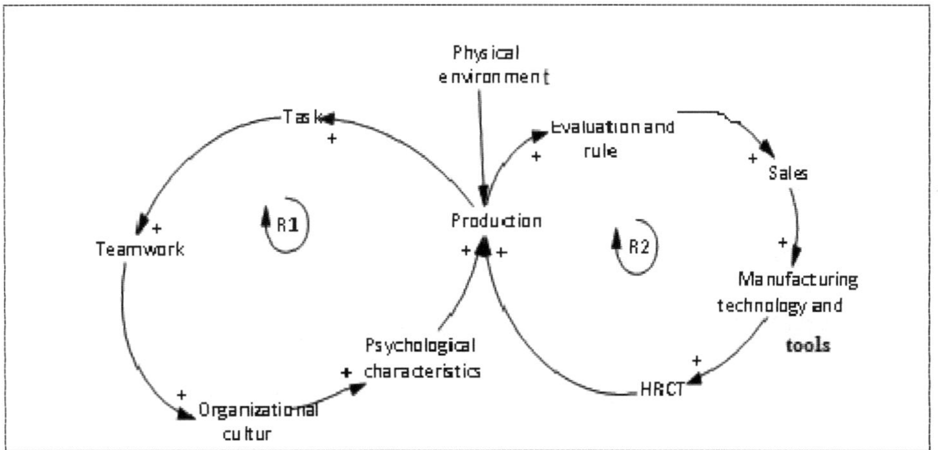

Figure 2. Working CLD

Figure 3 shows the final CLD which is built from Focus Group Discussion (FGD) with the manager of SME as an expert is shown in figure. There are three feedback loops, they are R1, R2, and R3.

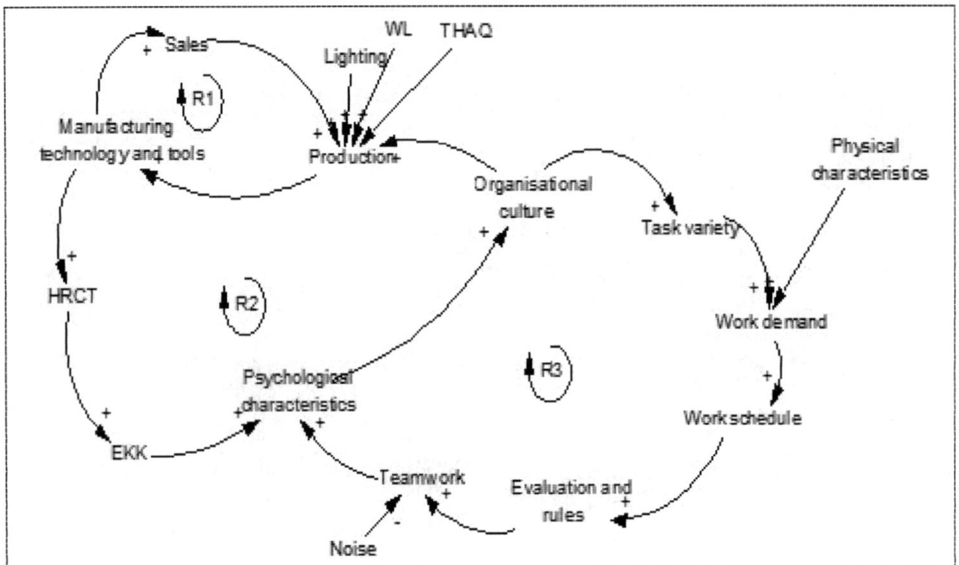

Figure 3. Final CLD

4.2 System Archetype

The most obvious system archetype seen in the conceptual model is success to the success to the successful and fix that fail. These system archetypes are obtained from final CLD by seeing the dominant loop generally and based on the historical data that construct the graph. Figure 4 is an archetype which presents the improvements in R1 group is given more resources than another group namely manufacturing technology and tools loop or R2, that can make teamwork performance continues to improve and manufacturing technology and tools loop will erode. In the case of SME MP, to increase the number of production there are two ways, namely improving the teamwork or improving manufacturing technology and tools or do this both with the same effort. Below is the explanation about both of loops:

*R1.*When the production is getting higher, the variety of task will increase too, and the work demand of employees also increase, it needs a good teamwork which require a good psychological characteristics of the employees and can increase the production again.

R2. When the production is getting increase, this SME keep buying the manufacturing technology, but do not have any kind of human resource in technology and tool (HRCT), therefore the production is going to erode.

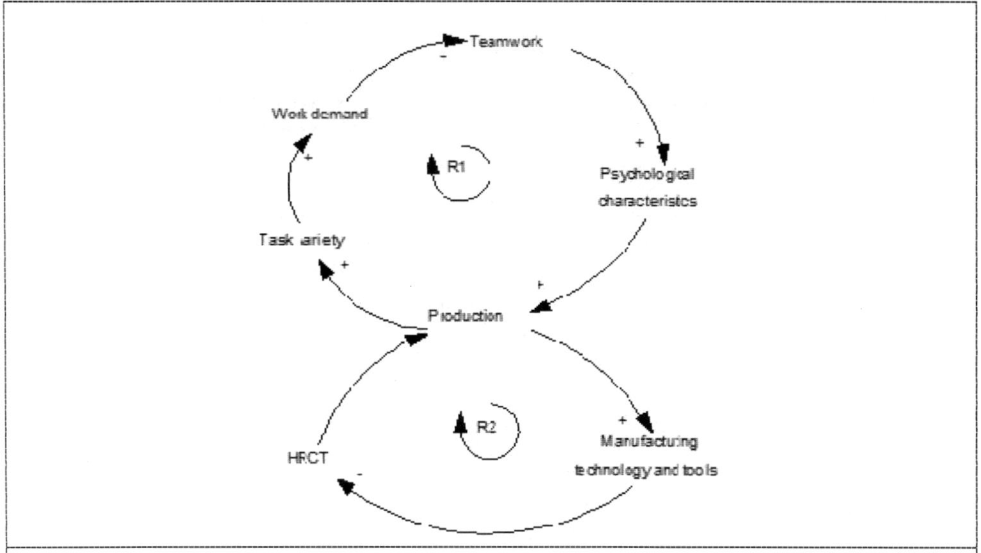

Figure 4. Success to the successful archetype

In this case, the teamwork loop is given more effort to spent the personal bonding between employees are getting higher day by day. Meanwhile, the manufacturing technology and tools need more effort to support production. In the field study, there are five machines in the SME but, the operator is not there and the machine does not work anymore. Therefore, it can decrease the number of production too. Hence, the success to the successsful archetype can be represented by behaviour over time.

Figure 5 is the Behaviour Over Time (BOT) of Success to The Successful archetype. This archetype can be identified through its trended data, by looking for diverging variables are examined. As a group that has been given more effort, the success improves even more. In the other hand, other partys' performance will continued to erode. The second archetype appears in the CLD is fix that fail.

Figure 5. BOT of Success to the Successful

Figure 6 shows the second archetype, namely fix that fail. This archetype consists of two loops which is balancing loop and reinforcing loop. Below is an explanation about each loop:

R1. This archetype shows the condition where the SME is very busy to improve the organisational culture performance by improving the teamwork of the employees as the short term solution that SME expected will increase the organisational culture performance.

R2. Teamwork is the short time solution and also can be a cause toward the unintended consequences. However, the teamwork will increase the number of rules and evaluation. The meeting will be often than usual and it spends more employees' time to attend the meeting and absolutely it increase the work schedule which can make employees feel bored. Meanwhile the meeting keep facing the same problems every day and it will reduce the employees' satisfaction to work which can cause the uncomfortable feeling toward the employees and at the end can decrease the organisational culture performance.

The organisational peformance data obtained through the FGD by giving behaviour over time of 10 system archetype and the employees choose the fix that fail archetype which similar to their organisational culture performance can be seen in figure 6. The unintended consequence is a work schedule that makes employees do not statisfied with their job and can impact the production due to many evaluations that has been made by the SME. The employees will feel tired and their performance in work will be decreased. The trend of organization as the quick fix solution, at the first place it can increase the peformance, but over time it continue to decrease and almost reach the unintended qonsequences.

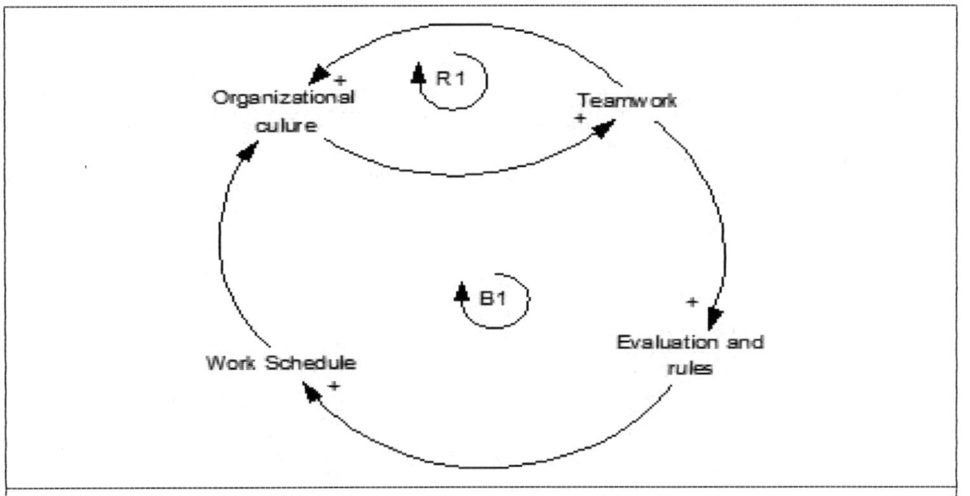

Figure 6. Fix that Fail Archetype

4.3 Prescriptive Action

4.3.1 Success to the Successful. First, the measurement or the result of this study shows that the improvement in organisational culture continually do not set up to favor the SME's performance or work system. The SME need to stop taking a look about the organizational culture, because it has been being a custom of the SME and the members will apply it unconciously.

The SMEs' goal is to improve their work system through the production. This can be achieved if this SME spent macroergonomics investment to the manufacturing technology and tool. This SME should give a training toward their employee and at the end can be able to fix and understand the manufacturing machine, to prevent the same problems recure over time such as after the machine is brought to the SME, the SME cannot use it to support their production.

Then, investigate the initial condition of the manufacturing technology and tool. Later, evaluate the current measurement of manufacturing technology whether or not the technology or organizational cultures really support the work system. After that, ask the outsiders or the expert for the alternative strategy. The next step is assess the effects of the new alternatives after hiring the human resouce characteristics in technology and tool. Last but not least, keep scaning the gaps and areas for the continous improvement.

4.3.2 Fix that Fail. The second archetype is about the management of this system archetype is must focusing on the fundamental cause and identifying a problem symptom. Here is the explanation about the prescriptive action of the fix that fail archetype.

This SME aims to improve their organizational culture through the teamwork as the quick-fix solution that results so many evaluations and rules to prevent the employees becoming bored. This SME should decrease their meeting for evaluation. Second, the evaluation of this SME is in every morning and also, when the work had been finished, this SME have a special meeting for every week to talk about their problems and it tends to disscuss the same problems. Last, SME should decrease the number of evaluation or remove the evaluation after they work everyday and the problems faced today by employees can be evaluated in the every morning.

In this study, we have applied the system thinking to have an understanding about feedback structures influencing the production system in one of local SME in Banda Aceh. This is can be an alternative to production management that attemps to find a weaknesses of traditional forecasting methods, that assume tomorrows's world will be much like todays.

5. Conclusion

There are 15 macroergonomics elements from 25 elements that has been chosen in this research, the most important macroergomics element that has been assessed through the SEIPS model are the organisational culture, teamwork and manufacturing technology and tools element. The CLD is developed based on the SEIPS model as preliminary CLD obtained 10 reinforcing loop and the working CLD obtained two reinforcing loop and in the final CLD there are three loops namely manufacturing technology and tools, organisational culture and sales, all of loops are reinforcing loop.

The system archetype has highlighted success to the successful archetype and fix that fail. The archetype presents the situation which can decrease the performance of work system. The action step is the manager need to spent macroergonomics investment, to give a graining toward the employees. Therefore, the SME can have an employee who has capability in manufacturing technology and tool, and also the SME should decrease the number of evaluation from twice a day to once a day and once a week for meeting. Research is needed to have better understanding about the prescriptive action as the solution. Hence, develop a dynamic model and scenarios simulation will obtain a better result.

References

[1] Hendrick, H. W. and Kleiner, B.Eds. 2002. *Macroergonomics: theory, methods, and applications*. New Jersey. CRC Press

[2] Realyvásquez-Vargas, A. Maldonado-Macías, A. A. and García-Alcaraz, J. L. 2018. *Effects of organizational macroergonomics compatibility elements over manufacturing systems'performance.* Procedia for manufacturing **3(2015)** 5715-572

[3] Guimarães L. B. d. M., Ribeiro J. L. D., Renner J. S. and De Oliveira P. A. B. 2014. *Worker evaluation of a macroergonomic intervention in a Brazilian footwear company. Applied Ergonomics.45* **(4)** 923–935

[4] Chui M. A., Pharm D., Ph, D., Mott D. A., Ph D., Maxwell L. and Ph. D. 2012. *A qualitative assessment of a community pharmacy cognitive pharmaceutical services program, using a work system approach. Research in Social and Administrative Pharmacy.* **8(3)** 206– 216

[5] Holden R. J., Or C. K. L., Alper S. J., Joy Rivera A. and Karsh, B. T. 2008. *A change management framework for macroergonomic field research. Applied Ergonomics* **39(4)** 459–474

[6] Karsh B. T. and Brown R. 2010. *Macroergonomics and patient safety: The impact of levels on theory, measurement, analysis and intervention in patient safety research. Applied Ergonomics.* **41(5)** 674–681

[7] Steege L. M. and Dykstra, J. G. 2016. *A macroergonomic perspective on fatigue and coping in the hospital nurse work system. Applied Ergonomics* **54** 19–26

[8] Jasiak A. 2015 *Ergonomic modernization in a selected automotive company.* Procedia Manufacturing. **3** 4769–4775

[9] Jasiak A and Dewicka A. 2015. *Application of selected quantitative analysis methods in the design of an ergonomic system.* Procedia Manufacturing. **3** 4776–4783

[10] Realyvásquez A., Maldonado-Macías, A. A., García-Alcaraz J. L. and Blanco-Fernández J. 2015. *Effects of Organizational Macroergonomic Compatibility Elements over Manufacturing Systems' Performance.* Procedia Manufacturing. **3** 5715–5722

[11] Sterman J.D. 2000 *Business dynamics: Systems thinking and modeling for a complex world.* Boston, MA:Irwin/McGraw-Hill

[12] Mai T.and Smith, C. 2015. *Addressing the threats to tourism sustainability using systems thinking: a case study of Cat Ba Island, Vietnam. Journal of Sustainable Tourism.* **23(10)** 15041528

IOP Conf. Series: Materials Science and Engineering **506** (2019) 012014 doi:10.1088/1757-899X/506/1/012014

Engine Modelling of a Single Cylinder Diesel Engine Fuelled by Diesel-Methanol Blend

Erdiwansyah[1,2*], A.M. Leman[3], Fitri Khoerunnisa[4], R.E Sardjono[4], Munawir[2], M.I Maulana[5], M.S.M Sani[2],
[1]Faculty of Mechanical Engineering, Universiti Malaysia Pahang, 26600 Malaysia
[2]Fakultas Teknik, Universitas Serambi Mekkah, 23249 Aceh, Indonesia
[3]Faculty of Technology, Universiti Tun Hussein Onn, Malaysia
[4]Department of Chemistry, Faculty of Mathematics and Science, Indonesia University of Education, Bandung 40522, Indonesia
[5]Faculty of Mechanical Engineering, Universitas Syiah Kuala, 23249 Aceh, Indonesia

*Corresponding author: erdi.wansyah@yahoo.co.id

Abstract. Performance on diesel engines with various combustion chamber sizes and fuel injection require effective handling so that the root combustion process in large diesel engines is timely. This study compares the experimental data with simulation through the single cylinder diesel modelling using GT-Power software. This simulation utilises parameters in diesel engines of 1000 and 2000 rpm speeds with methanol and diesel fuels. This modelling aims to develop a single cylinder diesel engine by measuring all components on the machine. The comparison between the simulation results and experimental data indicate similar collaboration with a difference about 5%–10%. Based on the results, the modelling of developed machine shows that simulation using GT-Power software is able to successfully validate all machine data parameters up to 90%. This simulation result can be expressed accurately similar to the experimental results and can be used as standardization for a single cylinder diesel engine design.

Keyword: Diesel engines, GT-Power, diesel-methanol blend

1. Introduction

Details on the diesel engine, size range, and engine performance vary greatly. Generally, the space and dissimilar combustion geometry characteristics for fuel injection need to be dealt effectively due to various design issues. Larger diesel engines need to achieve rapid airspeed from fuel combustion so that the process can be accomplished within time. Different types of geometry in the inlet port, piston shape, cylinder head, and injection pattern for the fuel were used so that the diesel engine with varied sizes can be achieved [1]. Typically, ratings on a machine indicate the strength at which the manufacturer's expected the product can provide better power, economy, robustness, and reliability that can satisfy a user's service conditions. Maximum torque and engine speed can be achieved because usually the results can be given simultaneously.

The compression of ethanol fuel has been used in the automotive with diesel and gasoline engines. Over the last decade, its application as a fuel additive for transportation has been in great demand among various cylinder developers because it can store high and light pressures as mentioned by [2–6]. Mostly, methanol fuel is used as an additive to gasoline fuel in both modified and non-modified vehicles through retrofit conversion in the market and has retained market price capability [7–11]. Vehicles with converted fuels will experience power loss and permissions for driveability, due to mounting or retrofit design. Significantly, the capabilities and strengths of the vehicles will increase so that they can be realised with retrofit kits especially on more sophisticated vehicles, or manufactured vehicles. Involvement of engine and vehicles manufacturers as original engine equipment developers is increasing. Especially in the use of diesel engine replacement applications with heavy duty, the number of these requests is still considered very limited and scattered so that large-scale manufacturing is still guaranteed. The design can come from gasoline or diesel engines to achieve that goal. While the original equipment machinery is ready and also available, however, for diesel engines has been transformed by the plant into the ignition and may continue, replacement equipment involving diesel fuel refuelling with the carburettor and also gas and addition of ignition system and spark plug on the combustion engine.

This study explores the application of using methanol fuel as a diesel fuel additive and the its impact on diesel engine performance in simulations. Diesel engines use methanol fuel additives for diesel fuel optimisation by comparing the use of methanol and diesel fuels in simulations [12,13]. GT-Power software is a very prominent simulation that can be used by industrial machinery, vehicle, and supplier is very suitable for analysis of various problems on the machine [14–16]. GT-Power also able to analyse all types of internal combustion engines and has been available for many components to model the next idea. GT-Power is based on the dynamics of one-dimensional usage, with heat transfer through piping as well as several other components in the machine's working system. The one-dimensional model simulation in GT-Power is the development of GT-Suite software or the Gamma Technologies family [14,15]. The use of four steps on diesel injection engines is manual measurement, and modelling using GT-Power to investigate machine performance will be discussed in this study.

An investigation of the importance of performance parameters on a single cylinder diesel engine is one of the properties of geometry, the terms of the efficient and other related engine parameters as discussed by [1,17–19]. Efficiency in cylindrical diesel engine mounts is as shown in thermal, relative, mechanical, volumetric, and brake heat efficiencies. Performance of the parameters associated with the engine are the pressure of the average effective, the specific power output, the engine intake valve index, the average speed of the piston, the ratio of air or air to the fuel, specific fuel consumption, and the carbon value of the fuel.

This study compares the experimental simulations by modelling a single cylinder diesel engine by using GT-Power software, as well as measuring the overall parameters of the machine with methanol and diesel fuels. The properties of diesel and methanol fuels are different in various facets and combustion systems. The differences in properties between diesel and methanol fuels are as shown in table 1. At this stage, these properties can be investigated at brake power such as brake torque, volumetric efficiency, consumption of brake specific fuel, and pressure on the cylinder.

2. Simulation and Setup

Modelling of the single cylinder diesel engine using comparable auxiliary methanol diesel fuel is discussed in this study. The specifications of the single cylinder diesel machine are as shown in table 2 while the engine used in this experiment is as shown in figure 1. This study was done by simulation using GT-Power software from the result of single cylinder injection diesel engine modelling using four steps and aims to investigate the impact of engine performance. Modelling was done based on the steps that have been determined by starting to open the component in the diesel engine based on the option to measure the parts contained in the machine. Furthermore, all machine components in the machine were inserted into the GT-Power library for matching before the next step. To create a new model in GT-Power, first select windows after that select the title in the template library contained in the list.

Table 1. Properties of methanol and diesel fuels.

Fuel Properties	Diesel	Methanol

IOP Conf. Series: Materials Science and Engineering **506** (2019) 012014 doi:10.1088/1757-899X/506/1/012014

Chemical formula	$C_{14}H_{28}$	CH_3OH
Flame speed rate (cm/s)	33	35
Boiling temperature (°C)	190-280	64.7
Denscity (g/cm^3, at 2°C)	0.84	0.79
Flash point (°C)	51	11
Auto ignition temperature (°C)	254	464
Lower heating value (MJ/kg)	42.74	20.27
Cetane number	56.5	4
Stoichiometric air/fuel ratio	14.7	6.66
Octane number	Not applicable	109
C/H ratio	0.50	0.25
Kinematic viscocity at	$2.5^{*}\ 10^{-6}$	$75^{*}\ 10^{-6}$
Carbon content (wt.%)	86	37.5
Hydrogen content (wt.%)	14	12.5
Oxygen content (wt.%)	0	50
Sulfur content (wt.%)	<50	-
Heat of vaporization (MJ/kg)	0.27	1.11

Table 2. Engine specifications.

Model	**Yanmar Model TF120M**
Stroke (mm)	96
Fuel injection type	Direct Injection
Fuel tank capacity	11 L
Displace volume (cm^3)	638
Compression ratio	17.7:1
Max power	7.7 kW @ 2400 rpm
Max torque	161 Nm @ 4500 rpm
Bore (mm)	92
Continuous rating output	10.5 HP @ 2400 rpm
Maximum rating output	12.0 HP @ 2400 rpm
Cooling system	Water-radiation
Type	Horizontal single-cylinder 4-stroke diesel engine

Figure 1. Engine experiment set up

In modelling, some templates are needed to be copied before creating a new object. Some icons that are necessary were selected for the new model creation. Among some templates and also objects have been assigned and inserted in the GT-Power template library which can't be separated, is part of the Gamma Technologies family [15]. According to [20-21] in this engine model, the damage such as tree systems, intake systems, fuel systems injection, systems exhaust, and machine cylinders may occur. While the machine selected, intake system there are components, data and different sizes. All system environments were run and terminated at the intake valve. The machine fuel system injection was focused to investigate performance machine and to support diesel fuel added with methanol from injections fuel, exhaust gases for disposal, and fresh air intake. Some machine cylinder components and injection fuel on a single cylinder diesel engine. Components such as data and size were included in the GT-Power list. Components such as cylinders, engines, and injectors. While the disposal is the last system contained in the machine. This system started from the valve to the last which was the environment. All materials that come to diesel must be connected to orifficeconn. Then, the diesel engine was modelled with GT-Power as the development of this research. The required components are fuel and diesel engine data to be able to build a new engine model. To build a model on the GT-Power list of information is needed, among other libraries. However, not all models must require all items. If the initial stage is to build the model, the optimal value for some items that are already on the list must be determined to do the simulation. The definition of an attribute should be specified as a parameter that can be executed in some cases to determine the optimal value.

The compression ratio, inline or configuration V, combustion size, V-angle (optional), two or four strokes were the data in machine characteristics. Meanwhile, the data in the geometry of the cylinder were stroke, offset pin, bore, head bowl geometry, the length of connecting rod, piston area, high clearance, and head area. All components in the geometry were also the data in the exhaust and intake systems. As for the throttle data, the throttle angle discharge coefficient in both the flow direction and the throttle location were considered. For fuel injector, the data such as the number of injector and location, injection speed, LVH, fuel type, number of nozzle hole, and nozzle diameter were considered. While the data contained in the exhaust and intake valve such as the disposal coefficient, valve diameter, valve lash, and elevator profile. Pressure, humidity and temperature were the data used at ambient conditions. This study focuses on the use of diesel fuel which was added with methanol through simulation. Modelling for development of the single-cylinder direct injection engine diesel with GT-Power is as shown in figure 2.

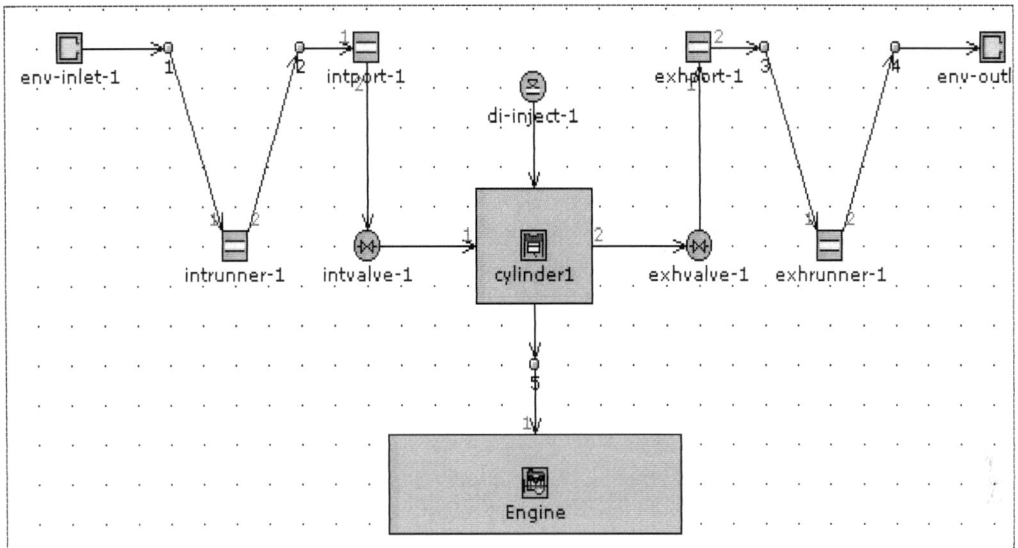

Figure 2. Engine single cylinder.

The data model included in GT-Power was the air composition of 76.7% N2 and 23.3% O_2 for the pressure of 1 bar and temperature of 26,85 °C was included into env-01 and env-02. Intrnner air filter or intake runner is an intake system for air filters, outrunner-01 and air filter-01. This diesel machine represents parts of the system intake pipes and certain parameters as shown in Table 3. Model diesel engine disposal system on GT-Power consisted of exhport-01, exhrunner-01, exhrunner-exit-01, and muffler-01. The exhaust system pipe also represented by parts contained in diesel engines with special parameters as shown in Table 4. Machine models using GT-Power such as involve-01 represents from valve and exhvalve-01 represents exhaust valve. Table 5 shows the machine diesel parameter components of the valve.

Dynamic geometric designs such as diesel machine compression ratio represents the maximum volume of cylinders or auxiliary or deleted volume (Vd) and also the volume of permit (Vc) with minimum volume division (Vc). Delivering power by a diesel engine can be absorbed by a dynamometer of angular velocity and torque product. Any machine efficiency can be determined [22]. The energy constellation (E) is the thermal efficiency (π_{ith}) is the ratio of energy in the brake power (*bp*). Mechanical efficiency (n_m) is defined as the ratio of brake power (*bp*) or transported power to the indicated power (*ip*) or power provided to the piston. It can also be defined as the ratio of the brake thermal efficiency to the indicated thermal efficiency.

The ratio of thermal efficiency to the ideal cycle is the relative efficiency or the efficiency ratio (π_{re}l), the criterion indicating the rate of development of the machine is the efficiency ratio. Lu et al. [19] To determine the performance parameters of a four-stroke engine is by volumetric efficiency (π_v), the stroke difference that the four-stroke engine possesses the breathing ability of the engine. Defining volumetric efficiency as the velocity of the air volume into the intake with the volume level transferred by the system. The normal volumetric efficiency range lies in engine speeds of about 80% to 85% for SI machines and 85% to 90% for CI machines. Determination of engine performance such as [1,19,23–26]. The mean effective pressure (*mep*) in which n_R is the sum of the crank revolutions on each stroke for cylinder power (two for four-stroke, one cycle for two-stroke) is defined according to equation (1).

$$mep = \frac{Pn_R}{V_d\, N} \tag{1}$$

Where *mep* is the mean effective pressure, Pn_R is the power number of crank revolutions per power stroke, and $V_d\,N$ is the number displaced cylinder volume.

Measurement of engine fuel conversion efficiency can be defined by Equation (2), as provided by [22]:

$$nf = \frac{W_c}{m_f\, Q_{nv}} = \frac{(Pn_R\,/N)}{(m_f\, n_R\,/N)Q_{HV}} = \frac{P}{m_f\, Q_{HV}} \tag{2}$$

Where nf is the fuel conversion efficiency, W_c is work per cycle, m_f is the mass flow rate, Q_{nv} is the fuel volumetric efficiency, n_R is the number of crank revolutions per power stroke, Q_{HV} is the fuel heating value and P represent the power.

In the engine test, the fuel consumption is measured as a flow rate mass flow unit per time (m_f). A more useful parameter is the specific fuel consumption (sfc) the fuel flow rate per unit power output. It measures the efficiency of an engine in using the fuel supplied as in equation (3).

$$SFC = \frac{m_f}{P} \tag{3}$$

Where *SFC* is the specific fuel consumption, m_f is the fuel mass flow rate and P is power.

At the time of machine testing, both the flow rate in the fuel (m_a) and the velocity of the air flow (m_f) period are measured. The ratio of the flow rate is used to define when the engine is being operated such as the ratio of fuel/air (F/A) and fuel/air ratio (A/F). Performance parameters on diesel engines can be developed. In some cases, the power (P) of engine can be measured using equations (4) and (5).

$$P = \frac{n_f\, m_a\, NQ_{HV}\,(F/A)}{n_R} \tag{4}$$

Where n_f is the fuel conversion efficiency, M_a is the air mass flow rate, Q_{HV} is the fuel heating value, n_R is the number of crank revolutions per power stroke, and F/A is the fuel-air.

$$P = \frac{n_f\, n_v\, NV_d\, Q_{HV}\, P_{a,i}\,(F/A)}{2} \tag{5}$$

Where n_f is the fuel conversion efficiency, n_v is volumetric efficiency, NV_d is the number displaced cylinder volume, and Q_{HV} is the fuel heating value.

Equation (6) was used for torque (T) measurement whole equation while equation 7) was used for mean effective pressure.

$$T = \frac{n_f\, n_v\, NV_d\, Q_{HV}\, P_{a,i}\,(F/A)}{4\pi} \tag{6}$$

$$mep = n_f\; n_v\; Q_{HV}\; P_{a,i}\;(F/A) \tag{7}$$

Where n_f is the fuel conversion efficiency, n_v is the volumetric efficiency, Q_{HV} is the fuel heating value, and *F/A* is the fuel-air.

The power specification per unit on the piston is the result of the engine design using the available piston zones at the time regardless of cylinder size. The measurement of power specifications on pistons can be obtained using equation (8).

$$\frac{P}{A_p} = \frac{n_f\, n_v\, NLQ_{HV}\, P_{a,i}\,(F/A)}{2} \tag{8}$$

Where A_p is the piston crown area, n_f: is the fuel conversion efficiency, n_v is the volumetric efficiency, NLQ_{HV} is the fuel heating value, F/A is the fuel-air. The piston velocity can be measured using equation (9).

$$\frac{P}{A_p} = \frac{n_f\, n_v\, NS_p\, Q_{HV}\, P_{a,i}\,(F/A)}{4} \tag{9}$$

Where A_p is the piston Crown area, n_f is the fuel conversion efficiency, n_v is the volumetric efficiency, NLQ_{HV} is the fuel heating value, and F/A is the fuel-air.

Pavlov et al. [1] argued that the specific strengths are proportional to the product when observed from the average piston velocity and mean pressure. It illustrates the importance of a direct link to efficient engine performance such as high fuel conversion, high-efficiency volumetric, improved engine output shifts related to the increased inlet air densities, maximum air or fuel ratios that can be used on combustion engines with an average speed of the piston.

Table 3. System intake for parameters engine

Parameter	Intrunne Air-filter	Air-fil ter-01	Intrun ner-01	Int port-01
Length (mm)	80	69.64	59.7	55.2
Dia. inlet end (mm)	44.8800	159.63	40.44	40.69
Wall temperature. (°C)	28.85	28.85	76.85	176.85
Dia. outlet end (mm)	62.13	159.63	40.1	32.78
Disc. length (mm)	34.4	34.4	34.4	34.4

Table 4. System exhaust for parameters engine

Parameter	Exhport -01	Exhrun ner-01	Muffler -01	Exhrun nerexit-01
Length (mm)	40.4	98	283.4	25.6
Wall temperat. (°C)	480	480	480	480
Dia. outlet end (mm)	29.82	27.86	138.88	34.6
Disc. length (mm)	47.3	47.3	47.3	47.3
Dia. inlet end (mm)	26.38	27.86	138.88	34.6

Table 5. Intake and exh parameters for engine

Parameter	Intvalve-01	Exhvalve-01
CAM timing angle (°CA)	462.5	214.6
Valve lash (mm)	0.125	0.125
Valve diameter (mm)	35.54	29.04

3. Result and Discussion

This chapter discusses the results of simulations using GT-Power software. These results of were available in various formats. After the GT-Suite application has been processed, the graphical interface file was processed via GT-POST and displayed, all simulated data were captured and analysed. GT-POST is a very sophisticated tool that displays graphics and output data in the form of desired excel sheet. These results represent the performance of a single cylinder machine diesel that was modelled in some standard parameters usage.

The engine model responses with parameters inputted on GT-Power were all recorded and analysed during simulation runs, with the aim of better understanding on engine reactions to the impact of diesel fuel use temperatures with methanol fuel. Different fuel temperatures were analysed with temperatures between from 300 K to 500 K before the simulation was performed. The engine speed for this test started from 1000, 2000, 3000, and 4000 rpm. Variable engine performance was consisted of heat brake efficiency, brake power, brake specific fuel consumption, brake torque, and effective brake pressure. Each burning fuel temperature has a different output trend so that it can provide different circumstances and reasons.

Figure 3 shows the simulation results by validating the experimental data for pressure differences on a single cylinder diesel engine. These results explain that the simulation results and experimental data have excellent consistency, and there is only a very small difference to some cases that has been done [27]. Overall, the maximum deviation of results between the simulations and the experimental data was less than 10%. Thus, in this case, it is considered to have excellent calibration, since it has been able to evaluate the effect on the combustion process of a single-cylinder engine diesel by using methanol and diesel fuels.

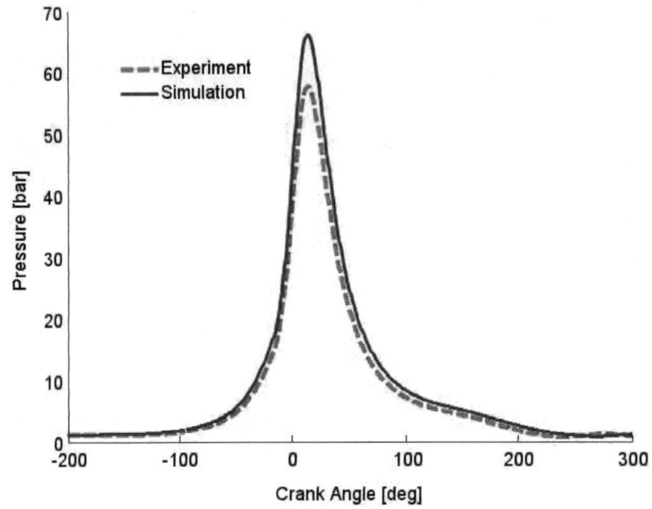

Figure 3. Comparison of simulated and experiment data

Figure 4 shows the result of P-V diagram in Cycle 1, wherein the event of intake that occurred before it was compressed. The cylinder was turned off for two cycles, and the intake valve was not deactivated so that the cylinder decreased into the sub-atmosphere during the piston movement as appropriate during the intake event. These results show a good correspondence between the simulation and the experiment during combustion, compression, and expansion in the thermodynamic cycle. The pressure on the gas exchange has shown slightly lower value than the measured result so that the difference will continue throughout the rest of the disabled stroke intake.

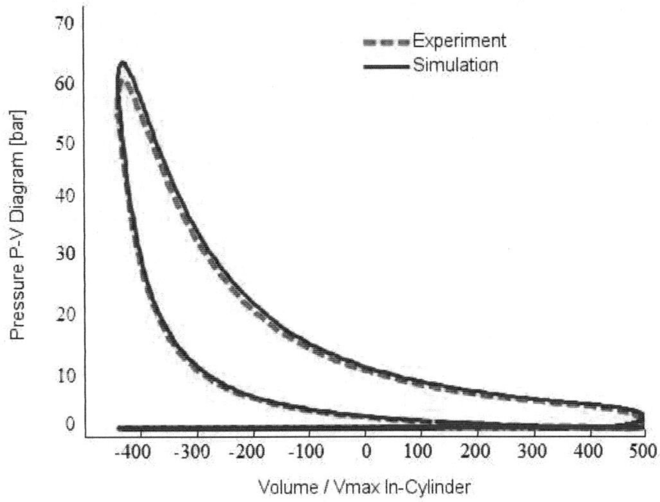

Figure 4. P-V diagram for cycle one firing event

At the time of transition, the temperature of the gas between the burning and the unburnt need to be determined. This moment can describe the beginning of the flame in the position of the sensor. To be able to determine the start of the flame, the initial result of the flux heat trace was taken and measured. This can allow us to pay more attention when the heat flux has changed rapidly due to the flame. At the time of initial trace, the heat flux has exceeded the threshold value of 1 W/cm^2 °CA extracted and selected as the beginning of the flame [28]. In this case, the fire at the crank angle has reached 358.5 °C. The ignition point went at 358 °CA, with a 900-rpm engine speed requiring of 3.8 m/s or 20.5 °C. Spark plugs with sensors about 30 mm gave an average of 7.89 m/s. The hot flux trace is shown in Figure 5. Where the data is the result of comparison between the experimental data and the simulation, as shown that this simulation can be trusted with an error difference of approximately 10%.

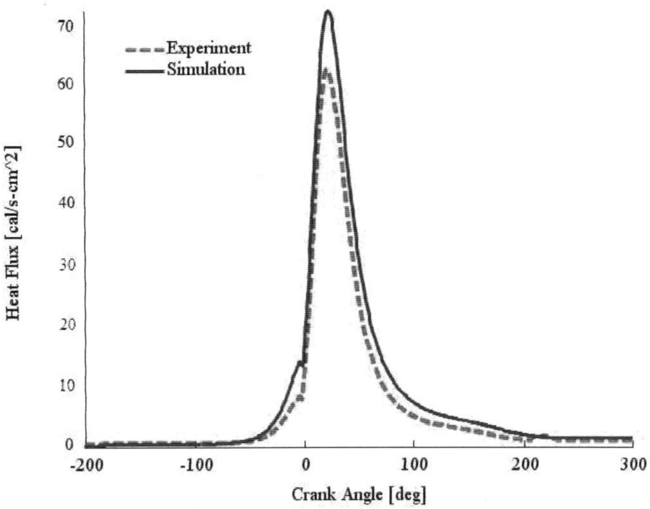

Figure 5. Heat Flux In-Cylinder Crank Angle

The simulated results with experiments on the heat transfer coefficients were predicted by the correlation formulas in the original form as shown in Figure 6. The predicted heat transfer coefficient results are very clear. Where the accuracy of the results is allowed using the original equation. This is necessary to obtain correlation results that are tailored to a specific combustion engine on a single cylinder diesel engine [29]. The purpose of this study is to build models that can predict parameters on machines similar to experiments. So that the achievement of pressure on the cylinder can be adjusted, thus the pressure of the effect as shown, the correlation adjustment is required when heat transfer occurs. Each coefficient equation will be used in the simulation as the relevant setting. The main coefficients are needed to search for the overall equation.

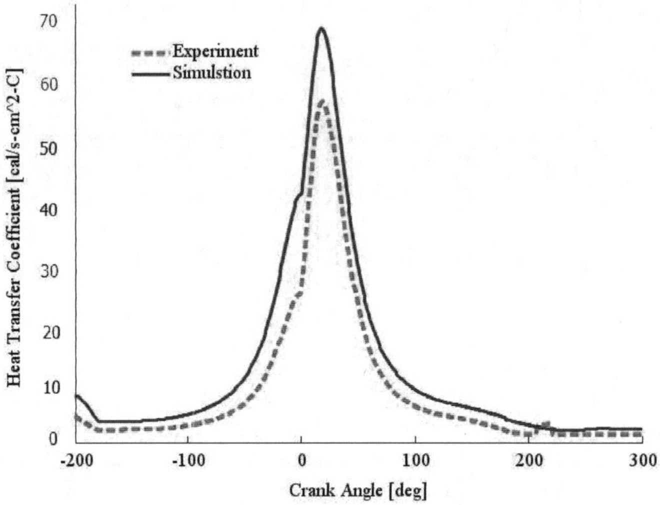

Figure 6. Heat transfer coefficient

The result of fuel variation in cylinder pressure has almost no difference between simulation and experiment as shown in Figure 7. The difference of fuel variation between the simulation and experiment modelled on GT-Power is approximately 5%. Thus, the simulation using GT-Power software can be utilised in other cases because this simulation has been able to adjust and read the parameters contained in the machine.

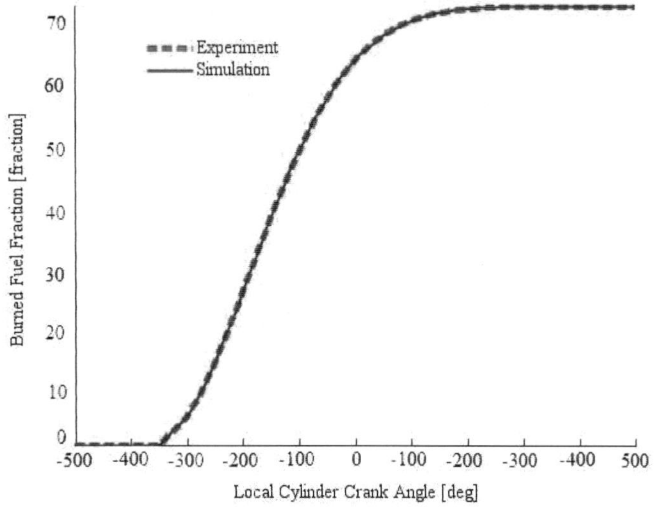

Figure 7. Burned fuel fraction in-cylinder

4. Conclusion

This study compares the simulations with experiments from diesel engines that were modelled using GT-Power software. This modelling used methanol and diesel fuel with engine speeds of 1100 and 2000 rpm. Simulation results with experimental measurements show that of all tested cases have similarities that reach between 5%–10%. This simulation aims to assess the parameters of a single cylinder diesel engine incorporated into the GT-Power software. With this result, it can be stated that this simulation has been successful and can be expressed accurately as a result of a real experiment on a diesel engine.

Acknowledgments

This research was supported by University Malaysia Pahang (UMP) Flagship Research Grant (RDU172204), PGRS160303, FRGS Research Grant (RDU130131 and UMP Grant RDU 180335). And we thank our colleagues from Automotive Engineering Center (AEC) and Advanced Automotive Liquid Laboratory (ALL), who provided insight and expertise that greatly assisted the research.

References

[1] Pavlov G I Nakoryakov P V Sukhovaya E A 2017 Development of Silencer for Low-Power Internal Combustion Engines *Procedia Engineering* vol 17 pp 1690-1695

[2] Yao C Pan W Yao 2017 A Methanol fumigation in Compression-ignition Engines: A Critical Review of Recent Academic And Technological Developments *Fuel* vol 209 pp 713–732

[3] Tang Q Fu J Liu J Zhou F Yuan Z Xu Z 2016 Performance improvement of liquefied natural gas (LNG) engine through intake air supply *Applied Thermal Engineering* vol 103 pp 1351–1361

[4] Zhang X Wang H Zheng Z Reitz R D Yao M 2016 Effects of Late Intake Valve Closing (LIVC) and Rebreathing Valve Strategies on Diesel Engine Performance And Emissions At Low Loads *Applied Thermal Engeering* vol 98 pp 310–319

[5] Khandal S V Banapurmath N R Gaitonde V N Hiremath S S 2017 Paradigm Shift from Mechanical Direct Injection Diesel Engines to Advanced Injection Strategies of Diesel Homogeneous Charge Compression Ignition (HCCI) Engines- A comprehensive review *Renewable and Sustainable Energy Reviews* vol 70 pp 369–384

[6] Zhang C Wu H 2016 Combustion characteristics and performance of a methanol fueled homogenous charge compression ignition (HCCI) engine *Jurnal of the Energy Institute* vol 89 pp 346–353

[7] Awad O I Mamat R Ali O M Sidik N A C Yusaf T Kadirgama K 2017 Alcohol and ether as alternative fuels in spark ignition engine: A review *Renewable and Sustainable Energy Reviews* vol 82 pp 2586-2605
[8] Wang J Wu J Xu Z Li M 2017 Thermodynamic Performance Analysis of A Fuel Cell Trigeneration System Integrated with Solar-Assisted Methanol Reforming *Energy Conversion Management* vol 150 pp 81–89
[9] Broekaert S Demuynck J De Cuyper T De Paepe M Verhelst S 2016 Heat transfer in premixed spark ignition engines part I: Identification of the factors influencing heat transfer *Energy* vol 116 pp 380–391
[10] Fu J Deng B Liu J Wang L Xu Z Yang J 2014 Study of SI engine fueled with methanol vapor and dissociation gas based on exhaust heat dissociating methanol *Energy Conversion and Management* vol 79 pp 213–223
[11] Ramesh Bapu B R Saravanakumar L Durga Prasad B 2017 Effects of combustion chamber geometry on combustion characteristics of a DI diesel engine fueled with calophyllum inophyllum methyl ester *Jurnal of the Energy Institute* vol 90 pp 82–100
[12] Yu M Tang X Lin Y Wang X 2018 Diesel Engine Modeling Based on Recurrent Neural Networks for a Hardware-in-the-loop Simulation System of Diesel Generator Sets *Neurocomputing* vol 283 pp 9–19
[13] Tang Y Zhang J Gan H Jia B Xia Y 2017 Development of a real-time two-stroke marine diesel engine model with in-cylinder pressure prediction capability *Applied Energy* vol 194 pp 55–70
[14] Fjallman J 2013 GT-Power Report
[15] Notes I GT-SUITE Installation Notes 2016
[16] Erdiwansyah Mamat R Sani M S M Khoerunnisa F Sardjono R E Ali O M 2018 Effects of Diesel-Biodiesel Blends in Diesel Engine Single Cylinder on the Emission Characteristic *MATEC Web of Conferences* vol 225 pp 01013
[17] Chen S C Kao Y L Yeh G T Rei M H 2017 An onboard hydrogen generator for hydrogen enhanced combustion with internal combustion engine *Internationa Journal of Hydrogen Energy* vol 42 pp 21334–21342
[18] Bae C Kim J 2017 Alternative fuels for internal combustion engines *Proceedings of the Combustion Institute* vol 36 pp 3389–34413
[19] Lu Y Paul Roskilly A Zhang Z Yu X Li Z Jiang L 2017 Conceptual Study of Scroll-type Rotary Gasoline Internal Combustion Engine *Energy Procedia* vol 142 pp 1545–1551
[20] Yum K K Taskar B Pedersen E Steen S 2017 Simulation of a two-stroke diesel engine for propulsion in waves *International Journal of Naval Architecture and Ocean Engineering* vol 9 pp 351–372
[21] Praptijanto A Muharam A Nur A Putrasari Y 2015 Effect of ethanol percentage for diesel engine performance using virtual engine simulation tool *Energy Procedia* vol 68 pp 345–354
[22] Dudareva N Y Enikeev R D Ivanov V Y 2017 Thermal Protection of Internal Combustion Engines Pistons *Procedia Engineering* vol 206 pp 1382–1387
[23] Jiaqiang E Pham M Zhao D Deng Y Le D Zuo W 2017 Effect of Different Technologies on Combustion and Emissions of the Diesel Engine Fueled with Biodiesel : A review *Renewable and Sustainable Energy Reviews* vol 80 pp 620-647
[24] Venkata Sundar Rao K Kurbet S N Kuppast V V 2018 A Review on Performance of the IC Engine Using Alternative Fuels *Materials Today: Proceedings* vol 5 pp 1989–1996
[25] Vasilyev A V Bakhracheva Y S Storojakov S Y 2016 Dynamics Simulation Model for the Internal Combustion Engine Valve Gear *Procedia Engineering* vol 150 pp 312–317
[26] Toma M Bobâlcă C 2016 Research on Drivers' Perception on the Maintenance of Air Filters for Internal Combustion Engines *Procedia Technology* vol 22 pp 961–968
[27] Shu J Fu J Liu J Zhang L Zhao Z 2018 Experimental and Computational Study on the Effects of Injection Timing on Thermodynamics, Combustion and Emission Characteristics of a Natural Gas (NG)-diesel Dual Fuel Engine at Low Speed and Low Load *Energy Conversion and*

Management vol 160 pp 426–438
[28] Cuyper T De Broekaert S Nguyen D Paepe M De Verhelst S 2017 Studying the Effect of the Flame Passage on the Convective Heat Transfer in a S . I Engine *SAE Technical Paper Series*
[29] Zak Z Emrich M Takats M Macek J 2016 In-Cylinder Heat Transfer Modelling *Journal of Middle European Construction and Design of Cars* vol 14 pp 2–10

IOP Conf. Series: Materials Science and Engineering **506** (2019) 012015 doi:10.1088/1757-899X/506/1/012015

Economic Value from the Household Environment Using EM4 Addition of Compost Solid Fertilizer in Banda Aceh

S.M. Herlina[1*], Dian Aswita[2]
[1]Education of Economic Faculty, Universitas Serambi Mekkah, Indonesia
[2]Education of Biology Faculty, Universitas Serambi Mekkah, Indonesia

*Corresponding author: herlina.smanap@gmail.com

Abstract. As population is growing in Banda Aceh city, the waste generation rate also increasing. Therefore, there is an effort to overcome this alarming increasing rate of waste generation. This study intends to find out whether the leaves around the households in the Banda Aceh region can be made into solid compost with the addition of EM4 to meet the C/N ratio. Preparation of solid compost fertilizer that comes from household environment uses raw materials of vegetable residues in the form of cassava leaves and stems, water spinach leaves and stems, spinach leaves, papaya skin, mulberry leaves, rimbang nuts, dried leaves fermented using microorganism 4 (EM4). The nutrient content of solid compost fertilizer has met the minimum technical requirements of organic solid fertilizer so that it can serve as business opportunity in Aceh. Thus, the element C/N ratio of 10 indicates that it meets the minimum standards.

Keywords: EM4, compost fertilizer, organic solid waste

1. Introduction

The growing population of Banda Aceh City over the last 5 years has apparently been in line with the increasing waste generation in Banda Aceh City [1,2]. Therefore, there is an effort to overcome the increasing waste generation. Composting is one way to reduce the waste produced by household. Law No.18 Year 2008 states that household waste is garbage derived from daily household activities, excluding stools and specific waste (waste containing toxic materials) [3]. The type of waste that can decompose is organic waste, also called wet waste, such as food and leaf blowing [4]. Composting is necessary because it is a process of degrading organic matter by microorganisms and organisms under controlled conditions into materials such as stable humus [5]. Utilization of organic waste into compost can reduce the burden of the environment. In addition, the sorting results can reduce waste generation and generate compost fertilizer which can also be sold or managed so as to increase household income [6].

The people of Banda Aceh generally reduce solid waste such as food scraps, plastics, paper, leaves, and others by burning or disposing them directly to landfills. Burning waste is considered not good because the smoke from the burning process will cause air pollution and eventually damage the human respiratory system. Meanwhile, when made compost, then the manufacturing process takes about 6 months [7].

Efforts to shorten the time of composting can be done by adding Microorganism 4 (EM4). EM4 was first discovered by prof. Terou Higa from Japan's Ryukyus University. The solution contains numerous fermented microorganisms that can work effectively in the fermentation of organic matter. However in urban areas EM4 use for waste management has not been widely used [8]. The benefits of this study are

IOP Conf. Series: Materials Science and Engineering **506** (2019) 012015 doi:10.1088/1757-899X/506/1/012015

as follows: 1) It can provide knowledge about compost fertilizer generating from household environment in Banda Aceh City by adding microorganism 4 (EM4), 2) It is an effort to support the concept of sustainable development that is environmentally sound, 3) Composted solid fertilizer produced by households can open business opportunities.

1.1. Problem Formulation

Does the leaves around the households in the Banda Aceh region meet the C/N ratio elements if liquid compost is made with the addition of EM4?

1.2. The Research Purpose

The purpose of this study is to find out if the leaves around the households in the Banda Aceh region are made of solid compost with the addition of EM4 to meet the C/N ratio or not.

2. Experimental Method

2.1. The Location to produce Compost Fertilizer

Implementation of composting that comes from the household environment in the form of organic waste was done in the yard of the resident's house on Jl. Nyak Moebin No.92, Kota Banda Aceh. Meanwhile, the execution time lasted for a month and laboratory tests in Baristand lasted for a month as well. Thus, the length of labor to get the results of laboratory tests was two months along February-March in 2018.

2.2. Components of Solid Compost Fertilizers

Preparation of solid compost fertilizer that comes from household environment uses raw materials of vegetable residues in the form of cassava leaves and stems, water spinach leaves and stems, spinach leaves, papaya skin, mulberry leaves, rimbang nuts, dried leaves fermented using microorganism 4 (EM4). EM4 is a microorganism culture that can be applied as an inoculant to increase the diversity of soil microorganisms. EM4 can also be used as a starter to accelerate the decomposition process, so that the composting process can take place more quickly [9].

The comparison between dry and wet waste comparisons is 2: 1. EM4 amounts as much as 50 ml, sugar ¼ kg, and enough water. While the required equipment is a perforated trash bin, 1 meter high, a machete to smooth the raw material, black tarps 1.5 meters, the tub in which to stir organic trash, cutting wooden board as a place to chop, and 2 meter plastic strap.

When illustrated in the flow chart diagram, then the manufacture of compost fertilizer is as follows:

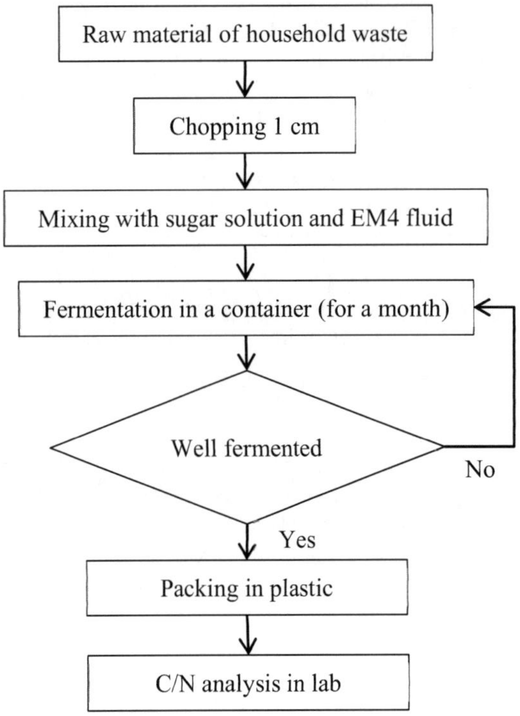

Figure 1. Flow chart of the composting process of organic solid waste

The success of composting fertilizer can be seen from its soft texture, blackish brown color, and sting smell like the humus forest aroma.

2.3 Preparation Procedure

The preparation procedure of making solid compost are as follows: 1) Cutting organic waste for 1cm on the cutting board until it becomes smooth, 2) Prepare 2 liters of water and mix with EM4, insert sugar, and then dissolve, 3) Enter the wet and dried garbage that has been chopped into the tub; then mix them with the water solution until they are well mixed, so that all the waste are exposed to the solution, 4) After being well mixed, put into the trash that has been perforated using a drill, 5) Cover tightly with tarp and wrap with plastic. The purpose of being put into a tightly closed tub is to have a good fermentation of the compost, 6) Within a month, the organic waste will be destroyed evenly indicating that the temperature or heating on the organic waste has been appropriate, 7) Sift the garbage and put it in plastic to take to the laboratory, 8) Plants need macro nutrients in the form of nutrients Nitrogen (N), Phospor (P), and Calium (K) in large quantities. Nitrogen, an important component of chlorophyll, nucleotides, nucleic acids, amino acids, acts to stimulate overall plant growth, especially in the leaves and stems, branches, and stems. The element P plays a role in stimulating sapling formation, root development, and accelerating flowering, and cooking. Meanwhile, the element K serves to process the formation of carbohydrates and proteins that can strengthen roots, stems, leaves, flowers and fruit apart from overcoming drought [10], so the eight step to conduct an analysis of C, N, P, K and C/N ratios in the Baristand laboratory, Banda Aceh.

3.0 Result And Discussion

3.1 The components of Compost Solid Fertilizer

The results of the test on solid compost fertilizer are shown on the table of compost test of organic solid compost of household waste in table 1.

Table1. Compost Solid Fertilizer

No	Components	Test Result Compost Solid (%)
1	N	3.00
2	P	0.60
3	K	1.49
4	C	25.02
5	C/N	10.00

**Resource: Lab. Baristand Aceh, 2018*

Minister of Agriculture Regulation No. 70/Permentan /sr.140/10/2011 on organic fertilizers, biological fertilizers and enhancers states that the minimum technical requirements of minimal organic fertilizer for macro nutrients (N + P + K) are at least 4 (Table 2, Minimum Technical Requirements of Organic Solid Fertilizers) [11]. Meanwhile, in Table 1, the test result shows that macro nutrient is 5.09%, so that the solid compost fertilizer that comes from household waste in this research can be said to have fulfilled the minimum requirement of organic solid fertilizer, while the organic C element is at least 15% where in Table 1 the results show a value of 25.02.

Table 2. Minimum Technical Requirements of Organic Solid Fertilizers

No	Parameter	Unit	Competency Standard			
			Granule / Pellet		Crumb / Crude	
			Pure	Enriched microbe	Pure	Enriched Microbes
1.	C – organic	%	min 15	min 15	min 15	min 15
2.	C/N ratio		15-25	15-25	15-25	15-25
3.	Accompanied Ingredients (Plastic, glass, gravel)	%	max 2	max 2	max 2	max 2
4.	Water Level	%	8-20	10-25	15-25	15-25
5.	Heavy metal:					
	As	ppm	max 10	max 10	max 10	max 10
	Hg	Ppm	max 1	max 1	max 1	max 1
	Pb	Ppm	Max 50	max 50	max 50	max 50
	Cd	ppm	max 2	max 2	max 2	max 2
6	pH	-	4-9	4-9	4-9	4-9
7	Macro Elements (N+P2O2+K2O)	%	**min 4**			

Source: Minister of Agriculture Regulation No.70/Permentan/sr.140 /10/2011

For the C/N ratio, according to the results of the discussion of the experts within Puslitbangtanak, Directorate of Fertilizers and Pesticides, IPB of the Department of Soil, Disperindag, and the Fertilizer and User Entrepreneurs Association in table 3, it is agreed that the minimum standard for C/N ratio is 10-25. C/N ratio is one of the parameters used to determine the quality of compost for the mammers that regulated by the Peraturan Menteri Pertanian (Permentan) and Standard Nasional Indonesia (SNI). It means that the value of test result of 10 meets the minimum standards.

Table 3. Minimum Technical Requirement of Organic Solid Fertilizer

No	Parameter	Composition	
		Solid	Liquid
1.	C-organic (%)	≥ 12	≥ 4.5
2.	C/N ratio	**10-25**	-
3.	Accompanied Ingredients % (Plastic, glass, gravel)	≤ 2	-
4.	Water Level (%)		
	- Granule	4-12	-
	- Crude	13-20	-
5.	Level of heavy metal		
	As (ppm)	≤ 10	≤ 10
	Hg (ppm)	≤ 1	≤ 1
	Pb (ppm)	≤ 50	≤ 50
	Cd (ppm)	≤ 10	≤ 10
6.	pH	4-8	4-8
7.	Total Level		
	- P2O5 (%)	< 5	< 5
	- K2O (%)	< 5	< 5
8.	Pathogen Microbe (Ecoli, Salmonella)	Stated	Stated
9.	Macro Elements (%)		
	Zn, Cu, Mn	Max 0.500	Max 0.2500
	Co	Max 0.002	Max 0.0005
	B	Max 0.250	Max 0.1250
	Mo	Max 0.001	Max 0.0010
	Fe	Max 0.400	Max 0.0400

Source: The results of the discussion of Puslitbangtanak experts, Directorate of Fertilizers and Pesticides, IPB Department of Land, Disperindag, and the Association of Fertilizers and Users Entrepreneurs.

3.2 Development Potential of Compost Solid Fertilizer

Based on the results of solid compost test conducted, the organic compost fertilizer can be said to have economic value. The selling price of organic solid compost in Banda Aceh City can be seen in table 4.

Table 4. Selling price of organic solid compost in Banda Aceh City

No	Stores/garden	Price 4 kg (Rp)	Price/kg (Rp)
1	Taman Sari Garden	10.000/4kg	2.500,-
2	Lilywandy Nursery Garden	13.000/4kg	3.250,-
3	Flowers Garden	15.000/4kg	3.750,-
4	Usaha Tani Store	14.000/4kg	3.500,-

Source: primary data, 2018.

The cheapest price of organic compost fertilizer is Rp.10.000,-, while the highest is Rp.15.000,- per 4 kg (per pack). The average price of organic solid compost in Banda Aceh City is Rp.13.000 per 4 kg (per pack). If the average per kilo is Rp.3.250,-. Dengan asumsi bahan pembuatan per 4 kg sebesar Rp. 1.000, maka keuntungan rata-rata dari seluruh sample adalah sebesar Rp. 2.250 (1.500 + 2.250 + 2.750 + 2.500 = 9.000/4 = 2.250).

All packaging of organic solid compost product sold in the market does not have the composition of N, P, K, C, C/N ratio. This is because the content of the fertilizer is not always the same in every production. In terms of color, the solid composts sold in Banda Aceh City are all blackish brown, indicating an indication of fertilizer that meets minimum standards. However, fertilizers that meet the minimum standards are not only seen from their color, but also from their components. Since all the fertilizers sold in the market are not listed as to how big the minimum content of N, P, K, C, C/N is, then the possibility of the fertilizer for not having met the minimum standards could have happened.

It is possible to get the cheapest price of the organic compost fertilizer in the plant sellers (garden) who also supply organic compost fertilizer to the Department of cleanliness of Banda Aceh. Meanwhile, at the place of the plant sellers (garden), the organic compost fertilizers are sold Rp. 15.000,- to the maximum because they make on their own. There are also the same fertilizers supplied from Medan and thus sold at the price of Rp. 13.000,-. The price is also affected by the location of the sale. In places located in the middle of the city and on the roadside or in a strategic place, then the selling price tends to be higher. In contrast to the location of sales outside the city area, the selling price is relatively cheaper. The raw material of the fertilizer also affects, but the composition of the raw material is not listed on the product packaging.

The comparison between raw materials and the price of organic compost fertilizer on the market can illustrate that this product has a promising economic potential. In addition, sustainable economic development with environmental insight is a good concept, because maintaining the environment and cleanliness and also doing waste management properly, one of which is to utilize compost from household waste actually has benefits received by the local community. The execution time of this liquid compost takes one month. However, since it took a laboratory test in Baristand laboratory for one month, the total labourship to get the results was two months.

4.0 Conclusion

The nutrient content of solid compost fertilizer has met the minimum technical requirements of organic solid fertilizer so that it can serve as business opportunity in Aceh. Thus the element C/N ratio of 10 indicates that it meets the minimum standards. It means that solid compost fertilizer can be used as business opportunity for household in Aceh because has economic value as much as Rp. 2.250.

References

[1] Jumlah Penduduk Banda Aceh Badan 2017 Pusat Statistik Kota Banda Aceh Available at *https://bandaacehkota.bps.go.id/*
[2] DLHK3 Banda Aceh Dinas Lingkungan Hidup Kebersihan dan Keindahan Kota Banda Aceh 2017 Available at *http://dlhk3.bandaacehkota.go.id/*
[3] Undang-undang No.18 Tahun 2008 tentang Pengelolaan Sampah; 2008.
[4] Daniel V 2009 Easy Green Living (Bandung) *Hikmah*
[5] Markel J A 1981 Managing Livestock Wastes. *Westport Conn. AVI Pub Co*
[6] Artiningsih N K A 2008 Peran serta masyarakat dalam pengelolaan sampah rumah tangga (Studi kasus di Sampangan dan Jomblang, Kota Semarang) *Master's thesis of Universitas Diponegoro.*
[7] Sundari E Sari E Rinaldo R 2012 Pembuatan Pupuk Organik Cair Menggunakan Bioaktivator Biosca dan EM4 *Prosiding SNTK TOPI 2012* pp 1907 – 0500
[8] Irianto I K 2013 Peranan Effective Microorganism (EM4) dalam pengelolaan Sampah Tinjauan dari Perspektif Pengelolaan Lingkungan secara Berkelanjutan *Artikel Mikrobiologi Lingkungan. Fakultas Pertanian Universitas Marwadewa*
[9] Diver S 2001 Nature Farming and Effective Microorganism

[10] Dobermann A and Fairhurst T Rice Nutrient Disorders & Nutrient Management. Potash & Potash Institute/Potash & Potash Intitute of Canada; 2000.
[11] Peraturan Menteri Pertanian No No:70/Permentan/sr.140/10/2011 tentang Persyaratan Teknis Minimal Pupuk Organik Padat; 2011.

IOP Conf. Series: Materials Science and Engineering **506** (2019) 012016 doi:10.1088/1757-899X/506/1/012016

Mechanical Properties of Hybrid Marble Powder and Kenaf Fibre Reinforced Polyester Resin Composites

NuzuliFitriadi[1*], Lindawati[1], IsmiArdiansyah[1]
[1]Mechanical Engineering Study Program, Politeknik Aceh Selatan

*Corresponding author: nuzuli@poltas.ac.id

Abstract. Kenaf fibre is one of the strongest natural fibres in the world today. Therefore, the purpose of this study is to create a new composite of Kenaf fibres combined with marble waste powder. The incorporation of marble powders and Kenaf fibres using a matrix of a polyester resin type has never been done. Thus, this study aims to create and test various composite material composites (mixture of matrix, marble powder and Kenaf fibre) which will produce different mechanical properties and obtain good tensile strength in test specimens with several composition variations. Preparation of test specimens was carried out with 4 (four) compositions of matrix mix and marble powder; 50:50, 60:40, 70:30 and 80:20 with the weight of Kenaf fibre is fixed 2 gram. The largest tensile test value is 63,407 MPa, obtained from 80:20 composition with the largest E value 3,03GPa. The largest compression test value is 33.49, obtained from the composition of 60:40 with the largest E value of 2,35GPa. The largest bending test value is 19.80 MPa, obtained from the composition of 80:20 with the largest E value of 36,55GPa. From the research results of tensile test, compression test, and bending test which possess good value is between 60:40 and 70:30 of composition ratio.

1. Introduction

The kenaf plant (Hibiscus cannabinus L.) is a short-lived herbaceous plant-producing fibres such as hemp, roselle, and jute. The kenaf fibres are produced from the bark of the stem. In Indonesia, kenaf planting spread in Lampung, West Java, East Java, and South Kalimantan with average productivity of 0.9-1.2 tons / ha of dry fibre [1]. Low productivity leads to high competition at the farm level in obtaining the kenaf plant [2].

Kenaf fibres have been widely used in various aspect of daily life products such as for sacking raw materials, pulp materials, paper, and polypropylene composites in the polymer industry and car dashboards. In addition to environmentally friendly, kenaf fibre is also strong. The mechanical properties of kenaf pulps derived from kenaf leather with 6% sodium hydroxide (NaOH) usage have met the NUKP slurry specifications as specified in SNI 6106-1999, ie 4,24 mm long fibre, freeness 760 mL CSF, tensile index 66,54 Nm / g, crack index 5,72 kPam2 / g and tear index 18,09 mNm2/g [3].

Various study on the feasibility of using kenaf fibre have been carried out. Ishak et al. conducted comparative study of mechanical properties of short kenaf bast and core fibre reinforced unsaturated polyester composites with varying fibre weight fraction. The result showed that the kenaf bast fibre reinforced-composites had higher mechanical properties than kenaf core fibre composites. In addition, the optimum fibre content, 20%, provided the highest tensile strength for both bast and core fibre composites [4]. Other studied the mechanical properties of kenaf/polyester composites processed through vacuum infusion method. Vacuum infusion process for the kenaf-polyester composite manufactured resulted in a higher strength, low cost alternatives, and environmental friendly [5]. Investigated the mechanical properties of kenaf/polypropylene composite with variation of fibre weight and length. The result showed higher fibre length improved the impact properties of composite. Based on these related studies, it can be seen that kenaf fibre is feasible to be used with various filler combination for bio composite material development [6].

On the other hand, South Aceh District is well known for marble stone production. The marble stone has been widely produced for various home interior need, home appliances, and as a material for carving art. The activity of processing of raw stone into a valuable product produces marble waste that lead to environmental problem. Thus, this marble waste should be reused to obtain new products that have economic value. For that purpose, this study is conducted to create a new composite of kenaf fibres combined with marble waste powder.

2. Method

2.1. Manufacturing Method

Before making specimens, special treatment of kenaf fibres is previously soaked in 1% alkaline solution (NaOH) which serves to remove some contents such as lignin, oil, and protein that can cause decay in the fibre. The composition of the specimen was determined as shown in table 1. The composition differences focused on the resin and marble powder of 20%, 30%, 40% and 50%, while the use of the kenaf fibre was with a fixed composition of 2 grams. Use of 2 grams of kenaf fibre based on the volume of the mold.The steps taken in making specimens are as follows:

i. The mashed wood mold is given evenly lubricated.
ii. A catalyst of 1% of the resin volume is mixed with resin.
iii. Kenaf fibre is inserted into the mold.
iv. Resin that has been mixed with marble powder and catalyst is poured into a mold that has been stacked with kenaf fibres.
v. The composite mixture is then dried for ± 4 hours.
vi. After drying, the composite is removed from the mold.
vii. Next will be tested.

Table 1. Composition of test specimens.

No	Specimens type	Polyester Resin %	Marble powder %	Kenaf Fibre gr
1	Specimen A	80	20	2
2	Specimen B	70	30	2
3	Specimen C	60	40	2
4	Specimen D	50	50	2

Tools and materials used in this study as in figure 1.

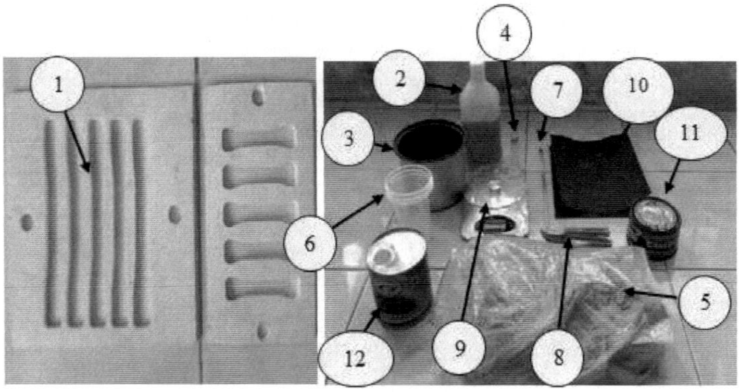

Figure 1. Materials and tools.

Image Description			
1.	Mold	7.	Brush
2.	Resin	8.	Spoon
3.	Marble Powder	9.	Digital scale
4.	Catalyst	10.	Sandpaper
5.	Kenaf fibre	11.	Wax
6.	Stirrer container	12.	Thinner

2.2. Measuring Method

After the specimen is made with the dimensions as in ASTM standard, some tests are performed as follows:

i. Tension test according to ASTM D 638-03 standard [7].
ii. Compression test according to ASTM D 695-02a standard [8].
iii. Bending test according to ASTM D790-02 standard [9].

The test equipment used in aServo Pulser test equipment located at Laboratory of Impact Research Center South Aceh Polytechnic.

3. Results and Discussion

3.1. Tensile Test

Tensile strength testing is one effort to know the material characteristics. Tensile tests were performed on 5 specimens per composition so that the average values as shown in table 2 can be obtained.

Table 2. The average yield of tensile testing.

Specimens type	Max. Stress (MPa)	Break Point Strain (%GL)	Upper Yield Stress (MPa)	Max. Load (N)	E (MPa)	E (GPa)
A	26.47	11.23	26.50	546.16	1588.12	1.59
B	33.71	13.42	48.58	513.66	2022.82	2.02
C	33.57	14.37	44.20	432.29	2014.32	2.01
D	50.53	21.86	42.80	454.62	3032.04	3.03

From the results of performed test, the maximum tensile stress is 50,53 Mpa. This number is to be found from the specimen as shown in figure 2.

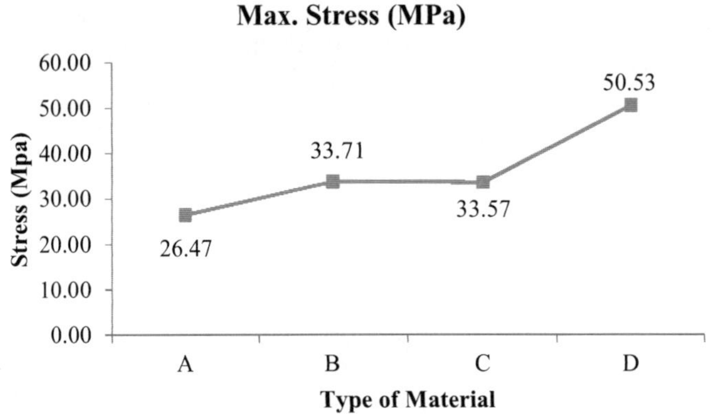

Figure 2. Maximum stress tensile test.

From the figure 2, it can be concluded that the stress is generally increased by increasing the percentage of resin content. Increasing the percentage of marble powder will reduce bonding properties of the polyester resin as a matrix to bind the kenaf fibres as reinforcement. The average of elastic modulus for this specimen is 3.03 GPa.

3.2. Compression Test

Compression strength test is one effort to know the material characteristics. Compression tests were performed on 5 specimens/composition types so that the average values as shown in table 3 can be obtained.

Table 3. The average yield of compression testing.

Specimens type	Maximum point Stress	Maximum point Load	Break point Strain	Upper yield Stress	Lower yield Stress	Elastic modulus
	MPa	N	%GL	MPa	MPa	Gpa
A	19.03	1141.698	3.624	13.070	12.585	1.61
B	23.09	1361.352	4.586	19.518	18.063	2.35
C	24.46	1467.78	4.867	21.175	20.069	2.19
D	23.88	1432.72	6.102	18.403	17.931	1.96

IOP Conf. Series: Materials Science and Engineering **506** (2019) 012016 doi:10.1088/1757-899X/506/1/012016

From the results of this test obtained the best composition on Specimens type C with kenaf fibre as much as 2 grams of 24,46 MPa, as shown in figure 3. Average of elastic modulus for this specimen is 3,03Gpa. It is known that the more marble powder will increase the compressive strength. However, it is necessary to further study by bending so that the characteristics of this composite material can be known comprehensively.

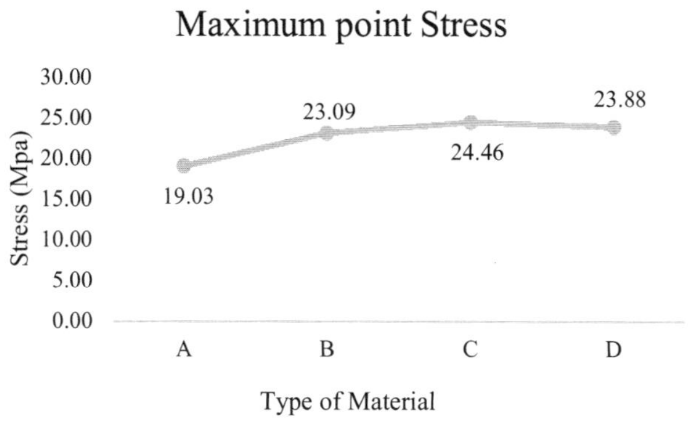

Figure 3. Maximum stress of compression test.

3.3. Bending Test

From table 4 it can be seen that there is a difference in value during bending testing.

Table 4. The average yield of bending testing.

Specimens type	Maximum point Stress	Maximum point Load	Break point Strain	Upper yield Stress	Elastic modulus
	MPa	N	%GL	MPa	Gpa
A	9.02	180.34	-4.93	9.65	49.88
B	11.15	223.02	4.66	9.20	33.52
C	15.91	318.16	12.18	14.80	43.87
D	17.05	341.02	8.00	15.47	36.55

The highest value is specimen D as shown in figure 4, stress point of 17,05 MPa and the lowest in specimen A with a stress of 9,02. The average stress of this composition is 13,28 MPa. Then obtained elastic modulus 36,55GPa. It is known that the more the marble powder will reduce the properties of polyester resin as the matrix but kenaf fibre as reinforcement able to increase the bending strength.

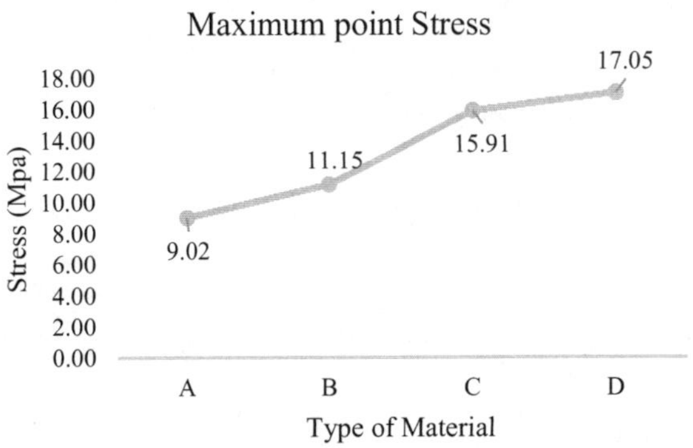

Figure 4. Maximum stress of bending test.

From the research of composite tensile test between the mixture of resin and marble powder, it is known that the highest value is specimen A with the highest value reaching 58,91 MPa, while the lowest value is specimen D with the composition kenaf fibre as much as 2 grams with value 21.35 MPa. The highest elasticity modulus is also possessed by the specimen A with a 3,03Gpa elasticity modulus value.

The results of the compressive test show that the highest value is specimen C with a value of 33,49 MPa with an elastic modulus of 2,35 GPa, whereas the lowest is specimen D with a value of 16,35 MPa and the modulus of elasticity 1,61 GPa.

The result of bending test shows that the specimen type which has the highest value is specimen A with a value of 19,80 MPa while the lowest value is specimen D with a value of 6.125 MPa.

From the result of tensile test, compressive test, and bending test all of specimen types is having ideal value will be recommended for the manufacture of the product. From the results of this study the type of specimen that has good and ideal value is specimen type A.

4. Conclusions

The conclusion of thhose 3 tests found the mechanical characteristics of this composite material are:

i. The largest tensile test value is 63,407 MPa, obtained from specimen A with the largest of elastic modulus value 3.03 GPa. The largest compression test value is 33,49, obtained from the specimen C with the largest of elastic modulus value of 2.35 GPa. The largest bending test value is 19,80 MPa, obtained from the specimen A with the largest E value of 36.55 GPa.
ii. From the research results of tensile test, compression test, and bending test which possess good value are specimen number B and C.

References

[1] Ayu Aizatul Natasa, ElzaZuhry, and Adiwirman 2016 Growth and Fibre Content of some Kenaf Varieties (Hibiscus Cannabinus L.), JOM Faperta Vol 3 No 2 pp 1-7
[2] Sudjindro 2004 Budidaya Kenaf (Hibiscus cannabinus L.) Seri Edisi Khusus Balai Penelitian Tembakau dan Tanaman Serat. Malang.
[3] Teddy Kardiansyah and Susi Sugesty 2014 Karakteristik Pulp Kimia Mekanis dari Kenaf (hibiscus cannabinus l.) Untuk Kertas Lainer *Jurnal Selulosa* Vol 4 No 1 Juni 2014 pp 37-46
[4] Ishak M R, Leman Z, Sapuan S M, Edeerozey A M M and Othman I S 2010 Mechanical properties of kenaf bast and core fibre reinforced unsaturated polyester composites. IOP Conf. Series: Materials Science and Engineering 11 of 9th National Symposium on Polymeric Materials (NSPM 2009).

[5] Moh Yuhazri, Phongsakorn Y P T, Haeryip Sihombing, Jeefferie A R, Puvanasvaran Perumal, Kamarul A M and Kannan Rassiah 2011 Mechanical Properties of Kenaf/Polyester Composites, International Journal of Engineering & Technology IJET-IJENS Vol: 11 No: 01.
[6] Loh X H, Mohd Daud M A, and Selamat M Z 2016 Mechanical properties of kenaf/polypropylene composite: An investigation, *Journal of Mechanical Engineering and Sciences* (JMES, Volume 10 Issue 2 pp 2098-2110
[7] Tension test according to ASTM D 638-03 standard.
[8] Compression test according to ASTM D 695-02a standard.
[9] Bending test according to ASTM D790-02 standard.

Development of Leadership Transformation Module for Off-Campus Students

R Hamzah[1*] , N A F Baharuddin[1] and Z Ali[1]
[1]Centre of Modern Languages and Human Sciences, Universiti Malaysia Pahang, 26600 Pekan, Pahang, Malaysia

[*]Corresponding author: rohanahamzah@ump.edu.my

Abstract. Young generation is the hope and the beneficiary for the nation's development in the future. However nowadays, the increase of participation among students in Higher Education Institutions (HIE) in various social problems are very devastating. Because of that, the need for serious efforts in helping students to improve the internal strength to form a strong personality and understanding of social responsibility as a leader is not only to themselves but also for the society. Additional focus should be given to off-campus students as they are more susceptible to negative influences such as friends and the surrounding environment. Through the preliminary findings, studies found that off-campus students are not able to understand, practice and carry out responsibility as a leader in terms of self-management; relationship with the Creator; social relationship with housemate; and environmental management leadership. Therefore, this article discusses the process of Development of Leadership Transformation Module of Off- Campus Students (TKDPLK) by looking at the module's content validity. This module is developed by integrating spiritual intelligence knowledge and giving awareness of social responsibility.

Keywords: Leadership transformation module, university students, social skills development

1. Introduction

Nowadays, social problems among young people are extremely troubling, especially among university students since they are the hope and national asset in building a developed and visionaries nation in the future. However, if this problem continues, the country will suffer, particularly in the skilled human resources. On January 19, 2017, Dato' Seri Muhammad Najib Tun Abdul Razak the sixth Prime Minister of Malaysia launched the National Transformation 2050 (TN50). It aims to build a new idea that will be carried out by young people for the realization of the agenda of Malaysia to become a thriving and developed country alongside the nation`s identity. Emphasis is given to young generations because they are the heirs who will be continuity of national development. These can be realised if starting from the student's level, established with a strong spiritual intelligence, mastery of knowledge, having an effective social responsibility and a high personality characterised as a leader in life. Human development needs to be taken seriously by university students who will help the nation in the future. It requires more attention from all parties, especially students who live outside the campus.

In the current study, the researchers focused on off-campus students since they are highly exposed to external influences. They are bound by the regulations of the University but not bound by the rules of the residential college. In fact, these students are not monitored by the management of the University as they are beyond the control of the authority of the university. Therefore, they need to take care of themselves in dealing with external and internal issues. To identify the problems, a need analysis was conducted to the off- campus students staying in Gambang, Malaysia. The preliminary results through interviews and indirect observations have been carried out and found that most of the students faced problems in self-management; relationship with the Creator; social relationships between housemates and residential management.

2. Preliminary Studies

Through interviews in the preliminary studies, it was found that students have problems in self-management such as managing their time, lacking self-control in terms of emotion and action and having less interest in learning. As for the relationship with the Creator, most of the students were not hesitate to share the problem whereas they admitted missing their solat, recite Quran, often forgetting to remember Allah. Result from these findings show that students lack of spiritual intelligence and spiritual knowledge. Once man can recognize God, worship and reach out to Him, and able keep away from doing things against with spiritual knowledge [1]. Internal requirements are vital to form superior human beings with good personality.

In terms of the students' social life with their housemate, disagreements between them that occurred until their relationship became worst. According to Miskawayh [2], human cannot get the basic necessities but only with the help from others. This shows the importance of the relationship between human beings to one another, where they need and complement each other. In addition, for residential management problems, from indirect observation have been made. The observations found that, the condition of the house was unpleasant. Their house was messy and unorganized. In fact, there were no specific places to put their books. Instead they were all over the places in the house (mattress, floor and on the dining table). In fact, it was not wrong to say that they were not bothered with hygiene matter in the house. A few of them just simply cleaned their own room and left out other places in a house such as a kitchen, living room, and toilet untidy and messy. Comfortable houses will be able to make us happy [3]. Moreover, the environment in the house plays an affective role for both atmosphere and resident's emotion.

Poor internal strengthening with lack of spiritual knowledge led the students to lose their life referrals in forming a harmonious and positive way of life. Without the proper self-management knowledge, it will cause the student get out from the right path and led to unhealthy symptoms in life. This will affect the lifestyle, attitudes, thoughts, self-management, interaction and social relationships with friends and community.

3. Theoretical Framework

It is clear that in human development process, the Western challenge to education with scientific and technological, while the Muslim challenge to education is spiritual, moral and social development. However, the West has no reliable, holistic and authentic source to know and draw the boundaries of human freedom and social life for moral and social development. Only Islam has a revealed, authentic and sound source of holistic and comprehensive knowledge of the cause and nature of human life on earth [4]. This research had referred to the Holistic Self- Development Model by Afzamiman. According to the framework, as shown in figure 1, human beings that have a good relationship with the Creator basically, they can establish a good relationship between human beings and the environment. Man has a responsibility of being a leader and has an identity as a spiritual servant to the Creator. A good relationship with God Almighty makes them faithful and devoted. Besides, a devoted human being are able to refrain themselves from negative influence. They, in fact can adapt to a good deed that God has commanded. Also, a good relationship between human and the environment will naturally occur if the human inner core is well preserved and maintained [5]. It begins with the internal spiritual construction of one`s will manifest the deeds and behaviour. Therefore, this module is developed by integrating spiritual intelligence that strengthens internal strength and promote awareness on the implementation of social responsibility. Social responsibility is from the aspect of leadership in self-transformation which is related to Creators, to themselves, housemates and the management of a residential environment.

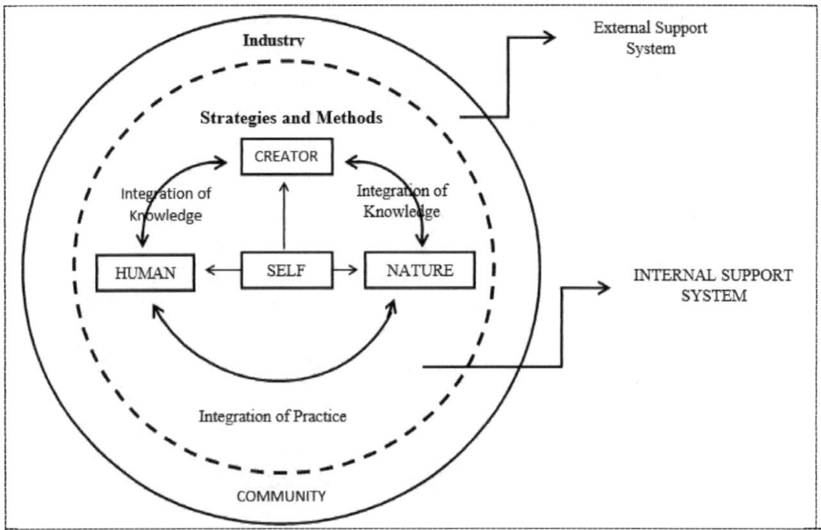

Figure 1. Holistic Self- Development Model [6].

4. Instructional Model for the TKDPLK Module

The Development of Self- Leadership Transformation Module focused on identifying the validity of the TKDPLK Module content. Various rules and procedures need to be considered in order to produce the best effects in module construction [7]. There are Sidek Mohd Noah 2003, Sharifah Alwiah Alsagoff 1981, and Rusell 1974 [7]. For the construction of TKDPLK module, the researcher use Sidek Module Development Model 2003, as a reference to develop the module as shown in Figure 2. There were two main stages in this model, the first stage was the preparation of the draft and the second stage was to try and evaluate the module.

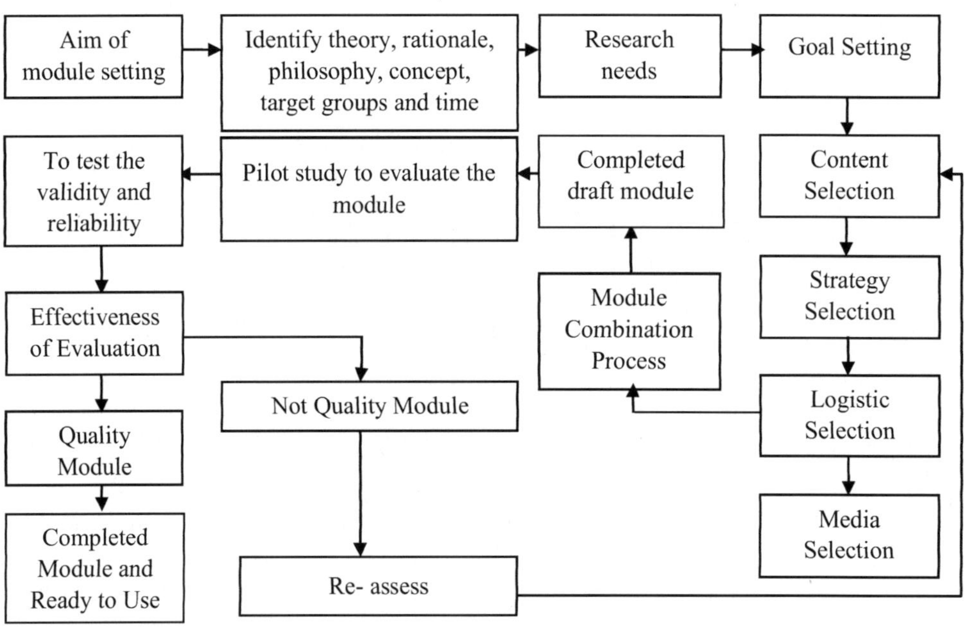

Figure 2. Sidek Module Development Model [7]

In order to identify the content of the module, preliminary studies were conducted. There were four (4) major problems that were identified. They were self-management, relationship with the Creator, relationship with the housemate and management of students' residence. Thus, these 4 components became the main content in the TKDPLK module. For each section, researcher set up activities to strengthen the knowledge of readers. In the process of arranging content of the module, the researchers began with focusing on the spiritual intelligence that is relationship with the Creator. Selection to start with strengthening the relationship with the Creator is because strengthening internal strength will help in shaping the good personality between human being and also between humans with the environment. Next is the transformation of self-leadership. Six (6) steps were identified to transform the self-leadership to become a better person. The next section is concerned with social leadership among housemate. There were three (3) steps to improve the relationship between housemates and help reader to enhance of social awareness. The final part concerns with the leadership of residential management. In this section, the researchers provide some examples of innovation that can be done by students to improve the resident environment to be better and conducive. Table 1 explains the organization process of arranging the content of the module.

Table 1. The Organization Process of the Content Module

Main Emphasis	Topic	Sub- Topic of Module	Activity	Learning Outcome
Spiritual Intelligence	4 Life Formula without Expire Date	Formula 1: Cherish Time, Fair with Creator Formula 2: Power and Balance Formula 3: Think, Lead and Manage Changes	Blessing that rarely thanked	Increase Spiritual Awareness
Transformation of self-leadership	6 Steps of Transformation of Self-leadership	Step 1: Check Yourself Step 2: Identify Negative Side and it`s Level Step 3: Self Focused Step 4: The Purpose of Changes Step 5: Build confident Level Step 6: Take Charge, Your Space and Time is Yours	Determination of Life	Give Awareness
Social Leadership	3 Steps to Transform Social Leadership	Step 1: Socialize in real life not only through Screen Step 2: Action hone Social Intelligence, You are the Role Model Step 3: Action hone Social Intelligence, Communicate Decorously	Life if not alone	To Increase Social Awareness

Resident Management Leadership	5 Steps of Transformations of Resident Management Leadership	Step1: Learn from Almighty God Step 2: Start with what we have Step 3: Innovation of Low Cost High Impact Step 4: Fair; Take and Put it back in the Place Step 5: The difference in duties need to be consolidated	To Increase Social Awareness

There a few interactive elements as in the use of texts and graphics are included. Such is necessary as the elements can attract the users to use this module as a medium of reference and provide simulations for critical thinking.

5. Research Methodology

This research focuses on identifying the validity of the module content. Five lecturers from the Centre of Modern Languages and Human Sciences, University Malaysia Pahang (UMP) were involved in the validation module content. They are from Humanitarian Sciences and Soft Skill department. The content validity is using Sidek and Jamaludin Formula. Equation 3 shows the formula of module content validity, where the total number from lecturers scores (x) will be divided by the actual score (y) and multiply by 100 [7]. The result will present in percentages.

$$\frac{Lecturer\ Score\ (y)}{Total\ Score\ (x)} \times 100\ = Content\ Validity\ Score \quad (1)$$

Next, to get the average percentage, the total score for each lecturer will be divided by the number of lecturer involved. According to [7] a percentage that exceeds 70% is a good content as shown in table 2.

Table 2 Average Percent of Content Validity Module

Lecturer (L)	**Lecturer Score Amount**	**Module Validity Level by Lecturer %**
L1	A	A/40 X 100% = V
L2	B	B/40 X 100% = W
L3	C	C/40 X 100% = X
L4	D	D/40 X 100% = Y
L5	E	E/40 X 100% = Z
Module Validity Level	=	**(V+W+X+Y+Z)/5**
	=	F % ≥ 70%

6. Results and Discussion

Validity is an effort of defining a built measurement tool to measure exactly what should be measured. In this case, the validity of content was used. [8] Content validity refers to the degree to which measurement tool can collect data including the reviewed data. A measurement tool has high content validity if it can measure all the contents that examined. Review process took about a month, together with the recommendations that were made by the lecturers involved. Researcher use formula Sidek and

IOP Conf. Series: Materials Science and Engineering **506** (2019) 012017 doi:10.1088/1757-899X/506/1/012017

Jamaludin to analyse the data validity. From the validity of the data, the average percentage of contents validity derive from the formula is 78% more than the value of minimum 70% as shown in Table 3. Therefore, the TKDPLK Module has a good content validity. This indicates the module is suitable for students to use. According to Sidek and Jamaludin formula a percentage that exceeds of 70% has good content validity [7].

Table 3. Module Validity Level

No	Statement	Score				
		L1	L2	L3	L4	L5
1	The design of cover page module is suitable	4	4	4	5	4
2	The objective of the module corresponds to the target groups	4	3	4	4	4
3	The contents module suitable with the objective of the spiritual intelligence development	4	4	4	4	3
4	The contents module suitable with the objectives of the development of student leadership transformation	4	4	4	4	4
5	The module features correspond to the target group	4	4	4	4	4
6	The human development theory is in line with objectives and target groups	4	4	4	4	4
7	The development of the module content is interesting and appropriate	3	3	4	4	4
8	The activities for each topic are suitable and interesting	3	3	4	4	4
	TOTAL	30	29	32	33	31
	Indicators					
	1-Strongly Disagree, 2-Disagree, 3-Less Agree, 4-Agree, 5 Strongly Agree					

Lecturer (L)	Lecturer Score Amount	Module Validity Level by Lecturer %
L1	30	30/40 X 100% = 75%
L2	29	29/40 X 100% = 73%
L3	32	32/40 X 100% = 80%
L4	33	33/40 X 100% = 83%
L5	31	31/40 X 100% = 78%
Module Validity Level		**= (75%+73%+80%+83%+78) / 5** = 78 % ≥ 70%

7. Conclusion

It is proven that TKDPLK module is suitable for off campus students and can be used as a guide in shaping their self-personality. In return, it can lead to improve their values and makes them become much wiser in managing themselves based on practicing religion of Islam in their daily life. Students are exposed to the knowledge of spiritual intelligence, which to develop internal strength and facing the challenges of life as off-campus student. A strong spiritual intelligence will be able to drive the development of good social depravity with friend and the managements of the student`s residential environment. It can be concluded that the contents of the Module TKDPLK is seen to be accurate and able to help off-campus student to overcome neither internal nor external problems.

Acknowledgements
The authors would like to be obliged to Universiti Malaysia Pahang for providing financial assistance under project no. RDU1603145.

References

[1] Zulkiple A G and Salimah A M 2006 Penghayatan agama sebagai asas pembangunan pelajar: Analisis terhadap beberapa pandangan al-Imam al-Ghazali

[2] Mohd N O 2010 Falsafah akhlak (Bangi) *Universiti Kebangsaan Malaysia Publisher*

[3] Syazwani A 2010 The Perception of Low Cost Urban Town Residents Against Comfort and Satisfaction Based on Sustainable Design Elements

[4] Abdul H A A S 1994 Islamization: Reforming Contemporary Knowledge *International Institute of Islamic Teaching*

[5] Baharuddin E and Zainab I 2015 Spiritual Intelligence Builds a Man of Excellence *Islamiyyat* vol 37 (2) pp 97-105

[6] Afzamiman 2012 Holistic Integration Education Approach Module In Training College of Technical and Vocational Education at University Teknologi Malaysia

[7] Sidek M N and Jamaludin A 2005 Module Development; Build the Training Module and the Academic Module *University Putra Malaysia* (Serdang)

[8] Darryl L S 2014 Design Models and Learning Theories for Adults. American Society for Training and Development

Evaluation of Accessibility for People with Disability in Public Open Space

Q Aini[1*], H Marlina[1], A Nikmatullah[1]
[1]Faculty of Engineering, University Muhammadiyah Aceh, Bathoh 23245, Banda Aceh, Indonesia

*Corresponding author: iqur_ennie@yahoo.com

Abstract. Public space should be accessible for all people regardless of certain circles. Accessibility is the availability of pathways for all people, including persons with disabilities and the elderly, measured by safety, convenience, usability and independent. This study evaluates the accessibility for PwDS in public open spaces in Banda Aceh City, Aceh Province, Indonesia. The areas used as the case study are Putroe Phang, Blang Padang and Taman Sari. The research method used is quantitative and qualitative by analyzing the willingness and suitability based on facilities and disability technical guidelines. The convenience of access in public open space measured by factors related to the form, dimensions, colors and textures of an access point. The results showed that Taman Sari has a better facilities and disability technical guidelines than the other two cases. In terms of conformity with the standardization of technical guidelines and convenience of access, Taman Sari is not yet fully functional, especially for persons with disabilities.

1. Introduction

The Public open space is an area used by various groups for both personally and community groups. The formation of public open spaces related to the parts of the natural and guided environment, where the public has the same opportunity to access these spaces freely. Public open space must be accessible to users without distinguishing physical or non-physical factors.

Generally, public open spaces created have not considered the convenience of persons with disabilities [1-3]. There is still some discrimination against persons with disabilities, with the unavailability of special facilities and access for them in the public area. In addition, the perception of the public that persons with disabilities are second-class citizens, who fulfil their needs and rights, can only be given if the first class community (normal society) has prospered. The failure access is also can be seen in pedestrian lines that do not care for children with special or normal needs, both of that have the same risk of accidents. For this reason, public open space should be accessible from a variety of ages for both those with normal physical conditions and those with physical disabilities.

In Banda Aceh, access to public spaces such as markets as public facilities is not sufficient for people with disabilities. Other research in Malaysia also explained that mall buildings could not fully facilitate the visitors from various backgrounds, both in terms of age and physical abilities. Public facilities in the external environment must pay attention to parking access, pedestrian, signs and markers, making it easier for users, especially persons with disabilities [4-5]. The impact of city growth physically and functionally measured in terms of accessibility and utilization. In urban areas, the size of accessibility viewed from aspects of proximity, distribution, road conditions and availability of transportation. While the benefits of public space in urban areas measured by comfort, quality, aesthetics and appearance. In Famagusta, Cyprus, public spaces are not useful and inaccessible due to lack of access for pedestrians, physical structures that do not support pedestrians, lack of safety and discomfort [6]. Both of public

IOP Conf. Series: Materials Science and Engineering **506** (2019) 012018 doi:10.1088/1757-899X/506/1/012018

buildings and public spaces, access for people with special needs should be comfortable, useful and easy to access. People with Disability (PwDs) also rely heavily on public transportation in urban areas.

Accessibility in the area of public transportation is still not maximal considering the needs of persons with disabilities. The case in the facilities of public transportation, often encountered with the lack of signs or facilities markers. The absence of special access for PwDS in the terminal area, physical transportation that does not support PwDs and the conditions of shelters that are not suitable for PwDs. Transportation access that is not friendly to diffable and does not facilitate the achievement of public spaces [7- 8].

Based on the mentioned theories, the access to public areas that are commonly considered is the public buildings such as markets and malls and the access to urban transportation areas. Public open space is a city facility that used as a recreational area of the city with varied users of space. In this case, a deeper evaluation study would be focus on facilities in the public open space.

In evaluating the availability of public access there are four factors that affect the level of comfort, namely the condition of public furniture and physical attractions, vegetation and the condition of public furniture, the pathway for people with disabilities, and the existence of shelter [9]. Accessible must imply that the availability of facilities will not make people with special needs feel they need help to reach out, walk in and out, use facilities and are not considered beggars. For that reason, public open space must pay attention to the needs of PwDs [10]. On the surface of the pedestrian there are four factors can interfere with comfort or satisfaction, namely form, dimensions, colour and texture [11]. These factors are uses as benchmarks in evaluating public space in this study.

The Government has regulated the Facility Technical Guidelines and Accessibility in Building and the Environment [12]. One of the regulations related to accessibility, which must provide convenience for all people, including persons with disabilities and the elderly, in order to realize equal opportunities in all aspects of life and livelihood. For public access facilities, that need consideration as a disabled channel is in the parking area, drop of area and pedestrian lane. On diffable access points must pay attention to clear and safe surface conditions, sloping roads and have safety and have interconnected lines, thus assisting in achieving a facility from city scale to micro environment [13-14].

Control of design in the park can be considered feasible when fulfilling the principles of facilities and disability, namely the principle of safety; health; convenience; convenience; security; beauty and independence [15-16]. The facilities and disability technical guidelines that must be considered in public spaces [3,6,17-18] are:

i. Public toilets should be equipped with handrails where position and height are adjusted for wheelchair users and other disabled persons. The recommended handle has right-angled shape to help move the wheelchair user.
ii. The basic size of space in three dimensions refers to the anthropometry, equipment and space functions
iii. Pedestrian should be attention to the road surface, slope, stopping point, lighting, drainage, where minimum width was 120 cm and safety along the edge of pedestrian.
iv. The guideline should have guiding block located in front of the vehicle traffic lanes, in front of the entrance, on pedestrian as the environmental and building link and guides from public facilities to public transport.
v. The parking area for PwDs has an access point located between two parks and a ramp leading to a circulation pathway that has elevation differences.
vi. The angle of the ramp no more than 6°, the maximum length of the horizontal ramp no more than 900 cm, it minimum width 95 cm without the safety edge and 120 cm with the edge of the guard. The borders on the prefix or ending of the ramp should be free and flat, where illuminated with sufficient lighting and equipped handrails.
vii. Signs and markers needed on the direction and purpose of pedestrian paths, public toilets, special parking disabilities, facility names and places, public phone and atm.

Based on the information, the existence of public space also provides accessibility in accordance with the technical principles and technical guidelines and disability. This study will evaluate the extent of the feasibility of public space in Kota Banda Aceh for persons with disabilities (PwDs).

2. Research Methodology

2.1 Type of Research

The research is a case study by evaluating the accessibility in public spaces in Banda Aceh City, Aceh Province, Indonesia. The method of discussion used is qualitative and quantitative methods.

2.2 Location of the research

The observations of the public space as its functions, gathering place and public recreation place for communities. Figure 1 shows that the public open space chosen as the research location in the center of Banda Aceh. The first location is BlangPadang which is a public open space that is often used as a sports center, culinary center and venue for events such as exhibitions and concerts. Second location is a Taman Sari also known as Bustanussalatin Park. This park is often used as a playground for children and a place for organizing cultural events. The third location is PutroePhang, which is a historical park from the time of the Sultan Iskandar Muda Kingdom. Currently it functions as a playground and educational park for students from kindergarten to students. The three locations of public open spaces are the dominant open spaces that are the choice of the people of Banda Aceh and surrounding areas to gather to do recreational activities on a daily basis with different forms of activity and age of visitors.

1. Blang Padang
2. Taman Sari
3. Taman Putroe Phang

Figure 1. Location of the research.

2.3 Research variables and operational definition

In this study, the dependent variable is accessibility, while the independent variable is the element of open space related to availability, conformity with standardization and convenience. Related to these variables, then the operational definitions are:

i. Accessibility is the availability of a path for everyone including PwDs and the elderly as measured by safety, convenience, usefulness and independence to achieve equal opportunity in all aspects of life and livelihood [9].
ii. Availability is the presence of persons with disabilities and the elderly access.
iii. Conformity is disabled facilities in public spaces that meet the technical guidelines and standardization.
iv. Convenience is disabled facilities in public space does not interfere with the user satisfaction by analyzing the form, dimension, color and texture.

2.4 Data collection techniques
The collected data is measurement data through field observation to assess the environmental situation of the three case studies related to the availability and fulfillment of standardization based on the principle of facility and accessibility. Surveys conducted with direct observation and measurement. Documentation is needed to determine the physical condition of the environment and feel the environment. Official data from related agencies in this case Dinas Tata Kota Banda Aceh and Dinas Pertamanan are also required to know the existing condition.

2.5 Data analysis
Stages of analysis performed are:

i. Identify available of accessibility for PwDs and the elderly by reviewing the facility's technical guidelines and disability.
ii. Evaluate accessibility compliance based on standardization on facility technical guidelines and disability.
iii. Evaluate the convenience of accessibility based on the study of form, dimension, color and texture.

3. Results and Discussion
3.1 Existing Conditions
3.1.1 Taman Sari. Figure 2 shows an overview of circulation and facilities that affect accessibility. The conditions of access and public facilities in this area can be seen in figure 3.

Figure 2. The existing conditions in Taman Sari.

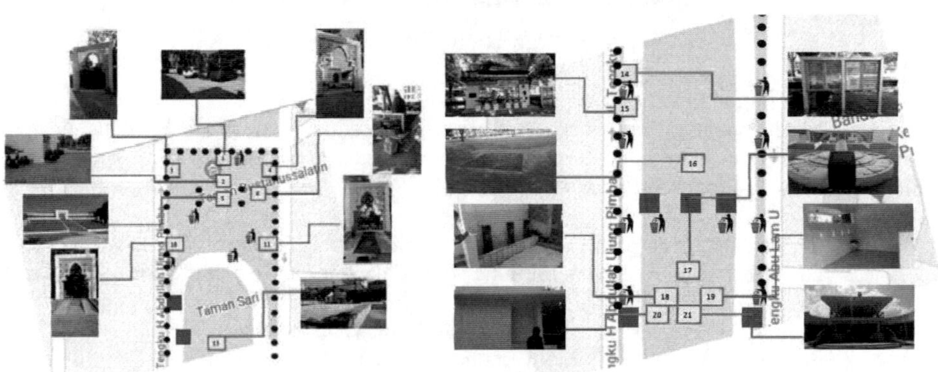

Figure 3. The existing conditions in PutroePhang.

3.1.2 Taman PutroePhang. The figure 4 and figure 5 shows the accessibility and existing conditions in Putroe Phang also the facilities.

Figure 4. The accessibility conditions in PutroePhang.

Figure 5. The existing conditions in PutroePhang.

3.1.3 Blang Padang. The figure 6 and 7 shows the accessibility and existing conditions in Putroe Phang also the facilities.

Figure 6. The accessibility conditions in Blang Padang.

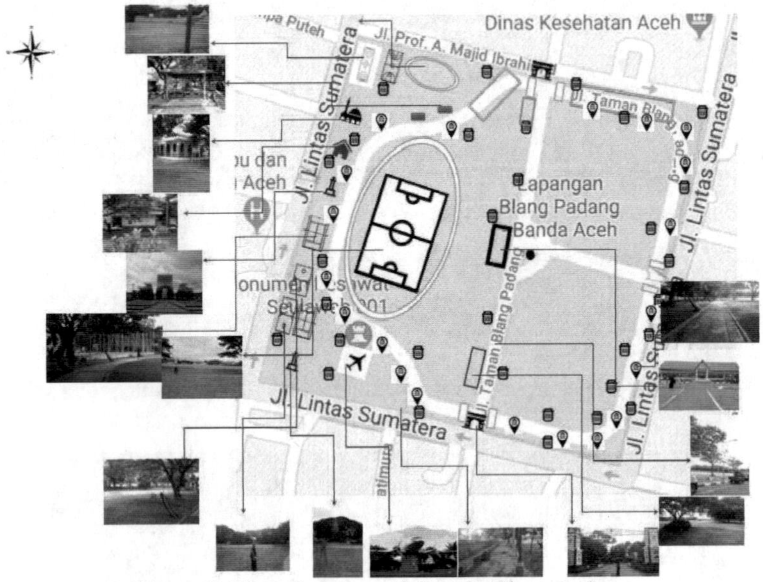

Figure 7. The existing conditions in Blang Padang.

3.2 Identification of Accessibility

Identification of facility base on the survey results shown in Figures 2, 4 and 6. Identification refers to the Facility's Technical Guidelines and Accessibility to Buildings and the Environment [17]. Available of accessibility based on technical guidance in each case study described in table 1. Table 1 and figure 8 show that Taman Sari has more complete technical availability of facilities and accessibility than the other two cases.

Table 1. The availability of Accessibility on Three public spaces.

No	Technical guidelines	Available		
		PutroePhang	Blang Padang	Taman Sari
1.	Public toilets	Available	Available	Available
2.	The basic size	Available	Available	Available
3.	Pedestrian	Available	Available	Available
4.	Guiding line	Not Available	Not Available	Available
5.	Parking Area	Available	Not Available	Available
6.	Ramp	Available	Not Available	Available
7.	Signs and Markes	Not Available	Not Available	Available

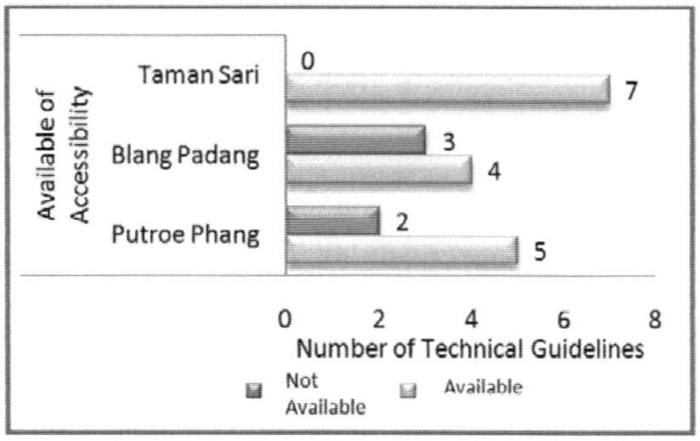

Figure 8. Available of Accessibility.

3.3 Evaluate accessibility compliance

Accessibility compliance in figure 11 evaluated based on the identification of access availability (figure 8). Conformity refers to the standardization of Facility Technical Guidelines and Accessibility on Building and environmental [2][5]. According to table 2 and figure 9, it is concluded that almost all of the accessibility in Blang Padang and Taman Sari is not up to standard.

Table 2. The accessibility compliance with technical guidelines.

No	Technical Guidelines	Accessibility compliance		
		PutroePhang	Blang Padang	Taman Sari
1.	Public toilets			
	a. Embossed signs	NS	NS	NS
	b. Space	NS	NS	NS
	c. Hight Closet 45-50 cm	NS	NS	NS
	d. Handrail	NS	NS	NS
2.	The basic size			
	a. Match the function	NS	NS	NS
	b. According to the need	NS	NS	NS
3.	Pedestrian			
	a. Stable, Strong, weatherproof	NS	S	NS
	b. Mound <1,25 cm	NS	S	NS
	c. Max slope 2°	S	S	S
	d. Stoping point	S	NS	S
	e. Lighting	S	NS	S
	f. Drainage away from pedestrian edge	S	S	S
	g. One way lane 120 cm, two-way lane 160 cm	S	S	S
	h. The safety edge is 10 cm high and 15 cm wide	NS	NS	NS
4.	Guiding line			
	a. Directional block	NS	NS	S
	b. Warning block	NS	NS	S
5.	Parking Area			
	a. Access point	NS	NS	S
	b. Ramp	NS	NS	NS
6.	Ramp			
	a. Max slope 6°	NS	NS	NS

	b. Horizontal length <900 cm	S	S	S
	c. Min width 95 cm without edge protection	NS	NS	NS
	d. Max width 120 cm edge secured	NS	NS	S
	e. Handrail	NS	NS	NS
	f. Lighting	NS	NS	NS
7.	Signs and Marks	NS	NS	S

Note:
S: Suitable
NS: Not Suitable

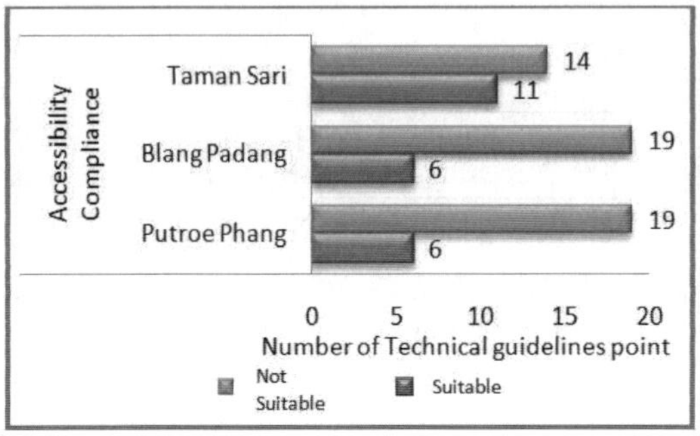

Figure 9. Accessibility compliance with technical guidelines.

Of all cases, most of the access to three cases of public space does not meet the technical guidelines. Compatibility is mostly found only in the pedestrian, but for connecting links to other public facilities are not easily accessible for persons with disabilities.

3.4 The Evaluation of the Comfort Access

Evaluation of comfort measured based on four factors that can interfere with comfort or satisfaction, i.e. form, dimension, color and texture [4]. These four factors are reviewed based on access availability reviews. The results obtained based on figure 10 and table 3 are Blang Padang and Taman Putroe Phang does not have convenient access for PwDs, where Taman Sari has considered the convenience although not yet optimal.

Table 3. Measure of Comfort Access.

No	Measure of comfort	Comfort Access		
		PutroePhang	Blang Padang	Taman Sari
1.	Public Toilet			
	a. Form	UC	UC	UC
	b. Dimension	UC	UC	UC
	c. Color	UC	UC	UC
	d. Texture	UC	UC	UC
2.	The basic size			
	a. Form	UC	UC	UC
	b. Dimension	UC	UC	UC
	c. Color	UC	UC	UC
	d. Texture	UC	UC	UC
3.	Pedestrian			

IOP Conf. Series: Materials Science and Engineering **506** (2019) 012018 doi:10.1088/1757-899X/506/1/012018

	a. Form	C	UC	UC
	b. Dimension	UC	C	C
	c. Color	UC	C	C
	d. Texture	UC	UC	UC
4.	Guiding line			
	a. Form	UC	UC	C
	b. Dimension	UC	UC	C
	c. Color	UC	UC	C
	d. Texture	UC	UC	C
5.	Parking Area			
	a. Form	UC	UC	UC
	b. Dimension	C	UC	C
	c. Color	UC	UC	UC
	d. Texture	C	UC	C
6.	Ramp			
	a. Form	UC	UC	C
	b. Dimension	UC	UC	UC
	c. Color	UC	UC	UC
	d. Texture	UC	UC	C
7.	Signs and Markes			
	a. Form	UC	UC	C
	b. Dimension	UC	UC	C
	c. Color	UC	UC	C
	d. Texture	UC	UC	UC
	Note: C: Comfort UC: Uncomfort			

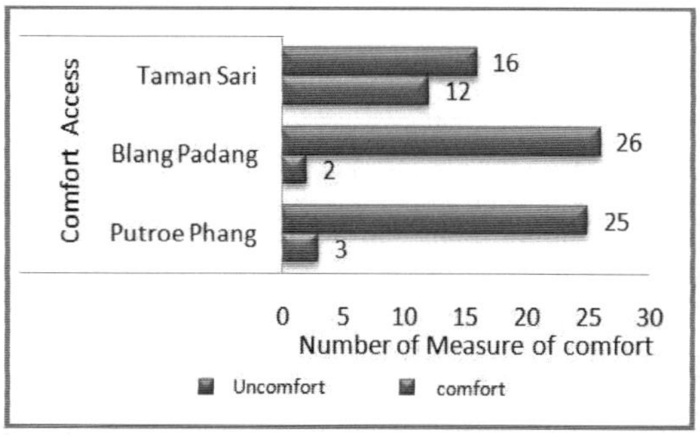

Figure 10. Comfort Access.

4. Conclusions

The conclusions of this study are:

i. Evaluation of suitability and comfort of access to public spaces is largely impartial for access for persons with disabilities (PwDs).
ii. Standard conformity is seen only on pedestrian lines but connecting to nearby public facilities is not easily accessible to PwDs (not accessible).
iii. The facilities and accessibility in the public spaces in Banda Aceh are not comprehensive to provide comfort to PwDs.
iv. Of the three cases in this study, Taman Sari has considered the facilities and accessibility of people with disabilities, but the suitability and comfort has not been considered thoroughly.

IOP Conf. Series: Materials Science and Engineering **506** (2019) 012018 doi:10.1088/1757-899X/506/1/012018

Acknowledgments
The research was supported by the Engineering Faculty University of Muhammadiyah Aceh and Institute for Research, Publishing, Community Service and Development (LP4M) University of Muhammadiyah Aceh.

References

[1] Novita, Rita. Aksesibilitas Bangunan Fasilitas Publik bagi Penyandang Disabilitas di Kota Banda Aceh, (Riset). FKM-BKA Natural Aceh, (2015)

[2] Thohari.S, Pandangan Disabilitas dan Aksesibilitas Fasilitas Publik bagi Penyandang Disabilitas di Kota Malang, Indonesian Journal of Disability Studies, Volume 1 Issue 1, (Juni 2014), 27-37

[3] Huiyun Xiang, Motao Zhub, Sara A. Sinclair , Lorann Stallones , J.R. Wilkins III , Gary A. Smith, Risk of vehicle–pedestrian and vehicle–bicyclist collisions among children with disabilities, Elsevier, Accident Analysis and Prevention 38 1064–1070, (2006)

[4] Irfan, Izziah, Anggraini.R, Kajian Aksesibilitas Kaum Difable Pada Gedung Pasar Aceh Berdasarkan Persepsi Masyarakat, Lansia dan Penyandang Cacat, Jurnal Teknik Sipil Universitas Syiah Kuala, Volume 1 Special Issue, Nomor 2, (2017), 533-542.

[5] Alaa Bashiti1, Asiah Abdul Rahim, A Study On The Accessibility In Shopping Malls For People With Disabilities (Pwds) In Malaysia, International Journal of Natural Sciences Research *Vol. 3, No. 1,* (2015), *pp. 9-20*

[6] Nil Pasaogullari, Naciye Doratli, Measuring accessibility and utilization of public spaces in Famagusta, Elsevier, Cities, Vol. 21, No. 3, p. 225–232, 2004.

[7] Seyed Hassan Khalifeh Soltani, Mashita Sham, Mohamad Awangb & Rostam Yaman, Accessibility for Disabled in Public Transportation Terminal, Elsevier, Procedia - Social and Behavioral Sciences 35 (2012) 89 – 96,

[8] Endang Sri Wahyuni, Bhisma Murti, Hermanu Joebagio, Public Transport Accessibility for People with Disabilities, Journal of Health Policy andManagement(2016), 1(1): 1-7.

[9] Pratiwi. A.D, dan Ernawati.J, Tingkat Kenyamanan Fungsional Alun-Alun Batu sebagai Ruang Publik, Architecture Student Journal (2018)

[10] Harry Kurniawan, Familiarity in Designing a Mosque: A Practice of Universal Design, Programme Work Programme-Architecture for All Regions I, II, III & IV-Oral Presentations Tokyo, 26th September 2011

[11] Prijadi.R, Sangkertadi, Tarore.R.C, Pengaruh Permukaan Jalur Pedestrian Terhadap Kepuasan dan Kenyamanan Pejalan Kaki di Pusat Kota Manado, Media Matrasain, Volume 11, No.1, (Mei 2014), 43-54

[12] Pedoman Teknis Fasilitas dan Aksesibilitas Pada Bangunan Gedung dan Lingkungan. Nomor: 30/PRT/M/2006.

[13] Cullen, M. Improving Transport Accessibility for All; Guide to Good Practice, France: OECD Publications, (2006).

[14] Henry, T. Policy On Persons with Disabilities, Office of the Prime Minister, Social Services Delivery, (2009).

[15] Widyawati.K, dan Laksmitasari.R, Penilaian Ruang Bermain Anak di Kota Depok Sebagai Salah Satu Indikator Tercapainya Kota Layak Anak, Faktor Exacta 8 (3), (2015), 195-2017

[16] Persyaratan Teknis Fasilitas Publik bagi Penyandang Disabilitas. *Handicap International*. 2004.

[17] Bahri, Samsul. Upaya dan implementasi pemerintah kota terhadap peraturan menteri pekerjaan umum No: 30/prt/m/2006 tentang Pedoman teknis fasilitas dan aksesibilitas pada bangunan gedung dan lingkungan di Kota Banda Aceh. (Seminar). Seminar Terwujudnya Aksesibilitas Pada Bangunan Fasilitas Publik. Nopember 2015.

[18] C Venter, T Savill, T Rickert, H Bogopane, A Venkatesh, J Camba, N Mulikita, C Khaula, J Stone and D Maunder, Enhanced Accessibility for People with Disabilities Living in Urban Areas, (2002)

Development of Candlenut Shell Carbonization Tool Using a Vertical Multi Chambers

A Amri[1*], D Maulina[1] and Muhibban[2]

[1] Industrial Engineering Department, South Aceh Polytechnic, Aceh, Indonesia
[2] Mechanical Engineering Department, South Aceh Polytechnic, Aceh, Indonesia

*Corresponding author: asbahrul.arl@gmail.com

Abstract. Candlenut shell, a solid waste from the process of taking candlenut seeds, is potentially able to improve the quality of the life of rural community by converting it to charcoals. It seems that the local people has difficulty to produce the charcoals from the candlenut shell. The aims of the research are to produce a carbonization tool using energy derived from the candlenut shells and to determine the tool's capacity. The tool was designed by utilizing Autodesk Inventor 2018 software. The result shows that the tool has three vertical Ø60cm x 30cm chambers that facilitate the carbonization process. The main components are a metal drum, a reservoir, two chamber barriers, and an exhaust pipe. It is also found that the best result of the tool is to process 24kg of the shell with 7.75 hours of processing time. Thus, the tool can carbonize the candlenut shell without any additional energy and the production of the charcoals is raised three times in one day.

Keywords: Carbonization tool, Candlenut shell, Charcoal, Autodesk inventor

1. Introduction

Candlenut (*Aleurites moluccana*) is one type of industrial plant from the family Euphorbiaceae [1,2,5]. Candlenut is classified as a medium-sized tree with a wide header that can reach height up to 20 m and has chest height diameter up to 90 cm [2]. Candlenut seeds are an economic value commodity and are the mainstay of farmers in various regions in Indonesia.

In Indonesia, candlenut plants are spread throughout the archipelago and always experience an increasing land area and production. Smallholder plantation statistics show that in 2004 candlenut production reached 94,000 tons with 206,300 Ha of land area and in 2010 increased to 100,600 tons with 215,000 Ha of area. The latest data shows that candlenut production increased significantly in 2013 to 107,200 with 215,410 Ha of land area. Based on data from the Ministry of Agriculture, Aceh Province has an average candlenut production in the last five years (2010-2014) is 11,621 tons per year. In this case, South Aceh District produced 92 tons of candlenut in 2014.

Most of the candlenut farmers in South Aceh Regency only rely on candlenut seeds as a worth selling commodity. However, the candlenut shell as a waste of the candlenut fruit [3] is not used optimally. It is approximately 60% of candlenut shell become a solid waste [4]. Since the shells are rigid waste, it remains in the environment and tend to give a negative impact on the environment. If it is utilized properly, the candlenut shell may give a significant impact toward the farmers' livelihood. Candlenut seeds and shells can be seen in figure 1 below.

Figure 1. Seeds and candlenut shell.

Candlenut shell has hard properties with a calorific value of 4164 cal/gram, so it can be used as fuel in drying candlenut seeds [5]. It can also be processed to produce activated charcoal that is useful for various fields, such as industry, health, environment and agriculture [6]. Activated carbon is made through a process of carbonization and activation of all carbon-containing materials [7,8]. Carbonization is a process to convert organic material to charcoal [9]. The carbonization process is an incomplete combustion process, so that the material is only carbonized and not oxidized. It is expected that the process will produce good quality charcoal.

Based on the observation, candlenut shells are not being utilized as charcoal due to the unavailability of technology to be adopted and used by candlenut farmers. The most possible technology to use is a palm shell pyrolysis drum reactor [10] and coconut shell burning air controlled drum [11], but it still uses other energy to carry out the carbonization process. Therefore, it requires efforts to help candlenut farmers to increase the economic value of candlenut seeds through the use of candlenut shells into activated charcoal. One of the efforts is to design and make a candlenut carbonization tool that is suitable for the needs of candlenut farmers, both in terms of process and cost.

2. Method

2.1. Tools and Material

The candlenut carbonization tool using a vertical multi chambers will be made using materials that are easy to find and familiar with candlenut farmers. The materials are a second-hand 200 litres metal drum which expected to be able to reduce the production costs, L-Shape angle bar, 2 mm iron plate, 1 mm aluminium plate, Ø 2 inch galvanized-pipe, 2 cm iron strip and Ø10 mm iron bar. The tool is designed by utilizing Autodesk Inventor 2018 software as an ambitious 3D parametric modeller [12]. Welding, grinding, and drilling machines and others additional tools such as hammers, chisels and scales are used in manufacturing the carbonization tool.

2.2. The Design

Considering the tool's user, traditional farmers, the tool is designed as simply as possible in order to provide an easy operational process of carbonization. The tool utilizes human power and fully uses the energy from the candlenut itself in carrying out carbonization process. The candlenut is not only is the raw material to be proceed for charcoal, but also as fuel in the burning process. The process offered may lead to a possible low operating cost. Low operating costs will have a better impact on the candlenut farmers by increasing the value of the products they produce. Therefore, the tool is expected to be affordable by the community in terms of cost and process.

The carbonization tool is designed using 3 vertical carbonization chambers, carbonization chamber I, II, and III, which aims to provide space for adequate air supply. It will attain a perfect burning process that generates enough heat to facilitate other chambers, carbonization chamber II and III, having carbonized. The carbonization chambers are shown in figure 2.

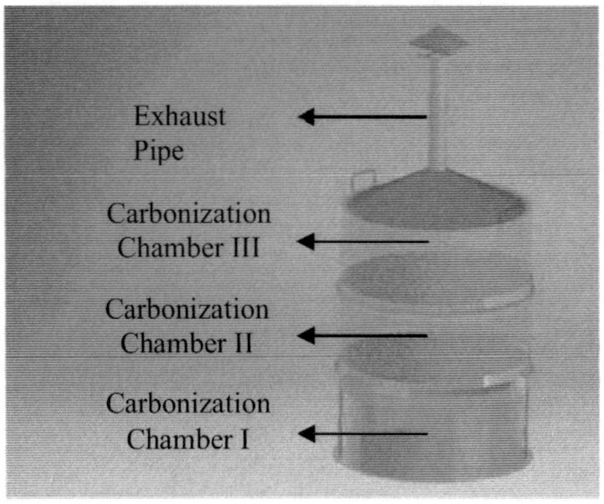

Figure 2. Candlenut shell carbonization tool using a vertical multi chambers.

The three carbonization chambers mentioned above require 4 main components, namely drum as the main frame to constrain the outside and inside environment, reservoir as a reservoir for carbonization product and a place of initial burn, chamber barrier and chimney as a cover and also carries the gas out of the tool. figure 3 below shows the detail of the 4 main components.

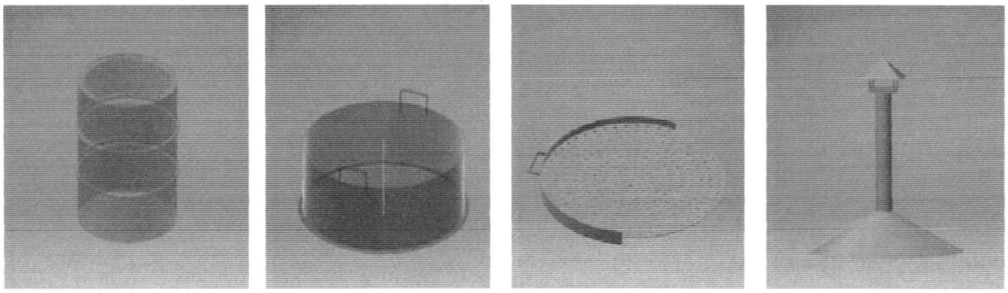

(a) (b) (c) (d)

Figure 3. The main components of the tool. a) Drum, b) Product reservoir, c) Chamber barrier, and d) Exhaust pipe.

2.3. *The Experiment Design*

To determine the capacity of the tool and processing time of the carbonization process, 3 times experiment will be carried with varying the amount of candlenut shells that will be carbonized. Variations in the amount of shells in each experiment can be seen in table 1. Indicators that determine the capacity of the tool are the number of processes that can be done in 24 hours and maximum volume per process.

Table 1. Variation of the amount of candlenut shells in each experiment.

Test Number	The amount of Candlenut (kg)		
	Chamber I	Chamber II	Chamber III
I	10	10	10
II	8	8	8
III	6	6	6

2.4. Flowchart of Manufacturing and Designing Tool
The following is a flow diagram in designing and manufacturing of candlenut shell carbonization tool using a vertical multi chambers as shown in figure 4.

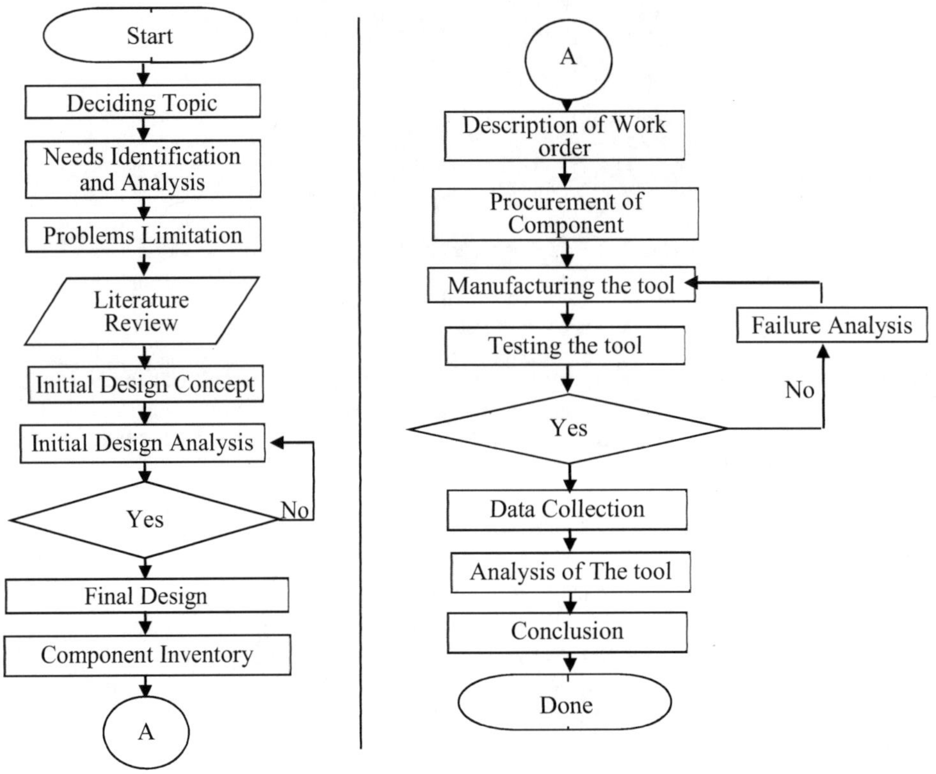

Figure 4. Flowchart of tool design and manufacture.

3. Results and Discussion

3.1. Candlenut Shell Carbonization Tool Using a Vertical Multi Chambers
The process of manufacturing of the tool followed the design shown in Figure 2 and 3. Through the process of manufacturing, the main parts of the tool which were produced and the complete one shown in figure 5 and 6 respectively.

The carbonization process was carried out through predetermined stages. The stages of the carbonization process by using this tool were:

i. Burning Stages. The Carbonization Chamber I (CC I) was filled with some candlenut shells that were burnt. When burnt, a chamber barrier was put to close the CC I providing a base as an area to put new shells called the CC II. This step was repeated for the CC III. Finally, an exhaust pipe was put as a lid of the tool.
ii. Cooling Stages. The burning stage completed, which was indicated by the little amount of and bluish smoke [10], then the cooling stage began. At this stage, all of the air flows, including the exhaust pipe was blocked to prevent the oxidation process taking place in the cooling process. The cooling stage completed if the chamber reaches room temperature [13].

IOP Conf. Series: Materials Science and Engineering **506** (2019) 012019 doi:10.1088/1757-899X/506/1/012019

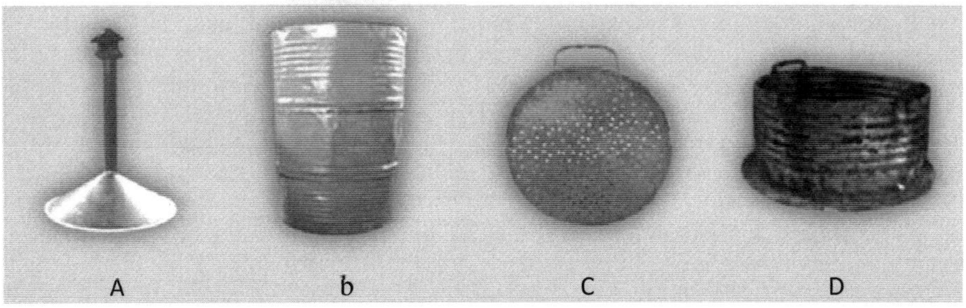

Figure 5. Manufacturing results of the main parts; a) exhaust pipe, b) drum, c) chamber barrier, d) product reservoir.

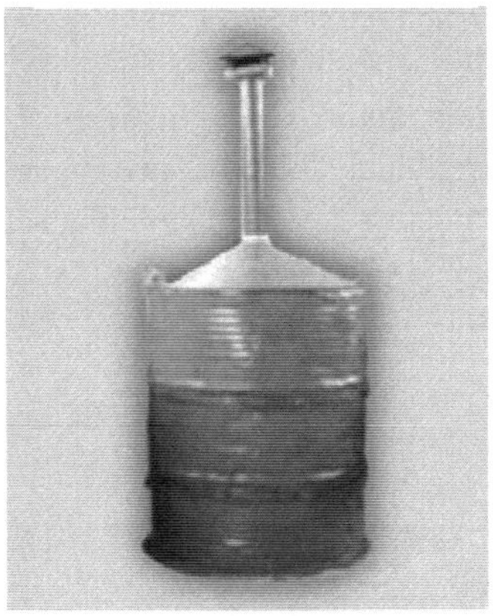

Figure 6. Candlenut shell carbonization tool using a vertical multi chambers after assembly.

3.2. Result of the Experiment

The carbonization process consists of 2 stages; burning and cooling. The carbonization process is counted from the combustion process begins until the cooling process is completed. The burning process starts when the firts burn is carried out in the CC I and it is completed when all candlenut shell in all CCs are burnt. The burning temperature data were measured in all CCs every 15 minutes using thermocouple. The data are shown in table 2.

Table 2. Burning process's temperature of candlenut shell.

Time (Minutes)	Experiment I				Experiment II				Experiment III			
	Temperature (°C)				Temperature (°C)				Temperature (°C)			
	CC I	CC II	CC III	Avrg.	CC I	CC II	CC III	Avrg.	CC I	CC II	CC III	Avrg.
0	32	32	32	32.0	32	32	32	32.0	32	32	32	32.0
15	240	130	110	160.0	210	92	77	126.3	229	106	89	153.8
30	355	232	164	250.3	311	133	94	179.3	340	190	94	236.4
45	330	275	190	265.0	330	265	183	259.3	296	249	172	268.1
60	315	285	214	271.3	297	344	258	299.7	283	258	193	280.2
75	224	323	252	266.3	230	269	209	236.0				
90	219	298	195	237.3								

The cooling process begins when the burning process is stopped. It is done when the temperature of the CCs outer wall is the same as the ambient temperature which is 32°C. The temperature is measured every half an hour using a thermocouple. Table 3 shows the temperature data during the cooling process.

Table 3. Cooling process's temperature of candlenut shell.

Time (hour)	Experiment I Temperature (°C)				Experiment II Temperature (°C)				Experiment III Temperature (°C)			
	CC I	CC II	CC III	Avrg.	CC I	CC II	CC III	Avrg.	CC I	CC II	CC III	Avrg.
0.5	80	73	64	72.3	82	77	67	75.3	85	87	70	80.7
1	55	47	44	48.7	58	55	50	54.3	60	62	55	59.0
1.5	49	45	42	45.3	47	43	41	43.7	42	43	40	41.7
2	46	43	40	43.0	42	41	37	40.0	38	39	40	39.0
2.5	44	43	40	42.3	37	40	36	37.7	37	37	36	36.7
3	43	42	40	41.7	35	40	35	36.7	35	37	36	36.0
3.5	40	40	38	39.3	33	38	35	35.3	33	36	35	34.7
4	39	40	37	38.7	33	35	34	34.0	33	36	35	34.7
4.5	39	39	37	38.3	32	35	34	33.7	32	34	34	33.3
5	38	38	37	37.7		34	34	34.0		34	34	34.0
5.5	37	38	37	37.3		33	33	33.0		32	33	32.5
6	36	37	36	36.3		32	33	32.5			32	32.0
6.5	34	35	35	34.7			32	32.0				
7	33	34	35	34.0								
7.5	32	33	34	33.0								
8		32	33	32.5			Ambient Temperature = 32°C					
8.5			32	32.0								

The maximum temperature in the burning process reached 355°C in experiment I in CC I, 344°C in experiment II in CC II and 340°C in experiment III in CC I. While the time needed for burning process is 1.5 hours at experiment I, 1.25 hours in experiment II and 1 hour in Experiment III. While the cooling process of experiment I took 8.5 hours, experiment II took 6.5 hours and experiment III took 6 hours. table 4 shows the amount of processed candlenut shells, the carbonization time and the number of repetitions of the carbonization process in one day.

Table 4. The measurement of the capacity of candlenut carbonization tool.

Description	Experiment I	Experiment II	Experiment III
Σ one time processed candlenut shell (kg)	30.00	24.00	18.00
Single Processsed Time (hour)	10.00	7.75	7.00
Σ Carbonization Process in one day (time)	2.40	3.10	3.43
The amount of candlenut shell that can be processed in one day (kg)	72.00	74.32	61.71

It can be concluded that experiment II has a maximum capacity in processing candlenut shell into charcoal which is 74.32 kg in one day. Hence, the candlenut carbonization tool using a vertical multi chambers maximum capacity is at 24 kg per process and able to do 3 times carbonization process in a day.

3.3. Discussion

Our results confirm that candlenut shell can be carbonized using its own energy by utilizing the candlenut carbonization tool using a vertical multi chambers. The chambers provide enough space for oxygen to be involved in the burning process. The adequate heat from the CC I will generate burning process in the last two chambers. Other tools may not applicable for candlenut shells, such as coconut and palm shell carbonization tool. Coconut shell is quiet easy to be carbonized because of its shape which provide enough space for oxygen. Palm shell has much similarity with candlenut shell in term of size. Palm shell pyrolysis reactor is still required additional energy to perform burning process. Thus the research proves that there is another way to carbonize the candlenut shell without any other additional energy.

4. Conclusion

The experiments revealed that the tool can carbonize the candlenut shell without any additional energy and the Experiment II gave the maximum production capacity. It can process 24 kg candlenut shell in 7.75 hours which indicates the process can be performed 3.1 times in a day. In consequence, Experiment II can carbonize candlenut shell 74.32kg per day, while Experiment I and III can only carbonize 72kg and 61.71 of candlenut shell.

Therefore, it can be concluded that the candlenut carbonization tool using a vertical multi chambers has 24kg in capacity with 7.75 hours in each process and able to perform carbonization process 3.1 times in a day. Thus, the tool is able to process 74.32kg candlenut shell into charcoals in a single day.

References

[1] Rosman R and Djauhariya E 2005 *Status dan Teknologi Budidaya Kemiri* (Bogor: Balai Penelitian Obat dan aromatika)
[2] Krisnawati H, Kalio M and Kanninen M 2011 *Aleurites Moluccana (L). Wild.* (Bogor: Center for International Forestry Research)
[3] Tambunan B H, Saptoadi H and Syamsiro M 2014 A Preliminary Study on Use of Candlenut Shell as a Renewable Source of Energy, Min Indonesia *Journal of Ocean, Mechanical and Aerospace – Science and Engineering* **9**
[4] Yovial, Marthiana W, Duskiardi and Habibi 2017 Pemanfaatan Cangkang Kemiri Dengan Ukuran Serbuk d <250 μm Sebagai Bahan Penguat Pada Komposit Resin Epoksi *Jurnal Agroindustri* **7** 1 56 - 62
[5] Estrada F, Gusmaol R, Mudjijati and Indraswati N 2007 Pengambilan minyak kemiri dengan cara pengepresan dan dilanjutkan ekstraksi cake oil *Widya Teknik* **6** 2 121-130
[6] Setiawan Y 1992 Penganekaragaman Produk Olahan Kemiri *Laporan Penelitian Tahun 1992* (Bogor: Balai Besar Penelitian dan Pengembangan Industri Hasil Pertanian)
[7] Lempang M 2014 Pembuatan dan Kegunaan Arang Aktif *Info Teknis Eboni* **11** 2 65-80
[8] Sunarsih S, Hastutiningrum S & Nisa T D 2016 Activated Carbon from Jackfruit Peel Waste as Decolouring Agent of Screen Printing Waste Water *Prosiding Seminar Nasional Teknik Kimia "Kejuangan" Pengembangan Teknologi Kimia untuk Pengolahan Sumber Daya Alam Indonesia, Yogyakarta, Maret 2016*
[9] Jamilatun S and Setyawan M 2014 Pembuatan Arang Aktif dari Tempurung Kelapa dan Aplikasinya untuk Penjernihan Asap Cair *Spektrum Industri* **12** 1 73-83
[10] Habiati 2014 Kajian Potensi Produk Pirolisis Limbah Padat Kelapa Sawit *Jurnal Biologi Edukasi*
[11] Hadi R 2011 Sosialisasi Teknik Pembuatan Arang Tempurung Kelapa dengan Pembakaran sistem

[12] Suplai Udara Terkendali *Buletin Teknik Pertanian* **16** 2 77-80
[13] Waguespack C 2014 *Mastering Autodesk® Inventor® 2015 and Autodesk® Inventor LT™ 2015* (Indiana: Autodesk Official Press)
[14] Merzari F, Lucian M, Volpe M, Andreottola G and Fiori L 2018 Hydrothermal Carbonization of Biomass: Design of a BenchScale Reactor for Evaluating the Heat of Reaction *Chemical Engineering Transactions* **65**

Pattern recognition of electromyography (EMG) signal for wrist movement using learning vector quantization (LVQ)

D S Putra[1], Y U W Weru[2] and Fitriady[3]

[1,2] Informatics Engineering, Politeknik Aceh Selatan, Tapaktuan, 23711, Indonesia
[3] Industrial Electronics Engineering, Politeknik Aceh, Banda Aceh, 23119, Indonesia

Corresponding author: darma.poltas@gmail.com

Abstract. EMG is an electric signal in the human muscle layer. This signal is caused by muscle contraction activity. The main purpose of this study was to explore the pattern of electromyography signal for wrist movement in open finger extensor. The EMG can be recorded using a device called electromyography (EMG). It can be acquired by attaching an electrode to the surface of the skin and the electrode was capturing the raw of EMG signal. Volunteers involved in this study were six people where each individual have 10 datasets the EMG signals. The electrodes are installed in the lower arm muscles. The EMG raw signal was processed by normalizing zero-mean. After pre-processing, the EMG signal has been done a feature extraction process to get the EMG data which was be an input vector in Learning Vector Quantization (LVQ). The feature extraction method was mean absolute value (MAV), root mean square (RMS), minimum value (Min), maximum value (Max), variance (Var), standard deviation (STD), and length of data (LoD). This study indicates that the classification accuracy for training and testing data of the EMG signal for wrist movement in open finger extensor (OFE) and grasping finger extensor (GFE) was 70.83% and 83.33% respectively. Therefore, the EMG signal can be used for identifying muscle disorder, artificial hand control and biometric identity.

Keywords: EMG signal, LVQ, human muscle layer, muscle contraction activity

1. Introduction

The development of science and technology that has occurred currently had a significant impact on the biomechanics [1-8] and biomedical fields (biomechanics and biomedical engineering) [9-11]. Biomechanics is a study of human body movements while biomedical is a science that uses technical concepts applied in the medical and health fields.

Research on the signal electromyography (EMG) [12-19] is one of the existing studies in the biomedical field. Research in this field has developed rapidly which is shown by the solution to the problems in human rehabilitation and treatment.

Every human has a unique EMG signal. This EMG signal can help humans to identify abnormalities that occurs in muscles for the rehabilitation of patients and artificial hand control [20-24. Some relevant research that have been published discussed about basic components in artificial hands that use EMG signals are used to control the artificial hand. Two electrodes are attached to the lower arm muscles used to record EMG signals. The EMG signal has been obtained then it has been extracted and classified to get a different types of classification [12, 25-26]. Flexion movement in a single finger can be identified with a single channel surface EMG device. This surface EMG device has recorded signals from the muscles during flexion in different fingers [21-22, 27]. The use of multi-channel EMG electrode surface is used to classify the combination and single finger movements for artificial hand control. Feature extraction method is using time-domain-auto regression. Feature reduction method uses orthogonal fuzzy neighborhood discriminant analysis and classification method using linear discriminant analysis [28-29]. Pattern recognition in finger movements using artificial neural networks is extracting EMG signals using the time domain feature to classify finger movements. The network training process using the Lavenberg-Marquardt algorithm and performance measurement uses a mean square error (MSE).

The researches stated above explain that the classification of EMG signals in the lower arms is only in flexion and extension of the fingers while the flexion and extension of the wrist are ignored. On the other hand, flexion and extension of the wrist was affected EMG signals in the lower arm muscles.

Hence the present study aims to investigate the finger recognition for opening or grasping combined with flexion and wrist extension by classifying EMG signals in each of these movements.

2. Methods

The research method in this study based on the flow diagram shown in figure 1.

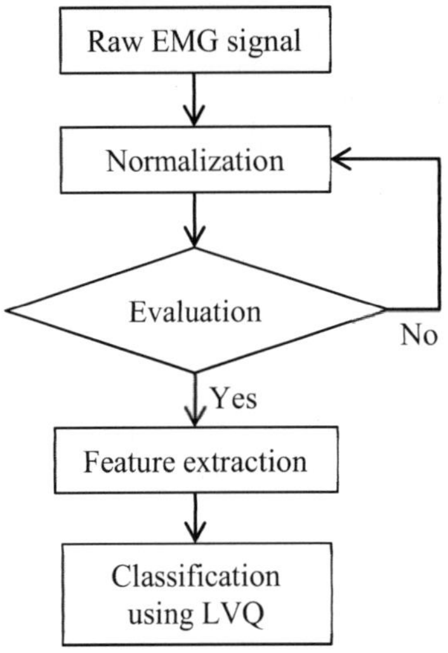

Figure 1. Research method flow diagram.

2.1. Object and research location

Six male participants have been chosen for this research. The average of participant's weight and height is 60.5kg and 162.5cm respectively. The research was held at Politeknik Aceh Selatan's Multimedia and Programming Laboratory.

2.2. Data collection

The EMG signal data collection techniques used in this study includes preparation, collecting data, and processing data stage.

2.2.1. Preparation stage. The preparation stage was the initial stage before the data collection stage was carried out. In this stage, the needs of the equipment needed for EMG signal data are determined. The tools and equipment needed are the surface electrode, the EMG myoware sensor, arduino uno minicomputer, cable, and laptop. The physical of the surface electrode and the EMG sensor used as shown in figure 2 and figure 3. All devices and materials required are installed into a single unit.

Figure 2. Surface electrode

Figure 3. EMG myoware sensor

2.2.2. Data recording stage. The data recording stage is carried out to retrieve the data for each subject. Before data collection, the subject must first fill out the inform consent to become the participants. The participants were normal subjects who had no abnormalities and injuries in the muscles. Subjects were attached to three electrodes in the lower right arm muscles, namely the wrist extensors and the wrist flexors. The figure of the attached electrodes as shown in figure 4.

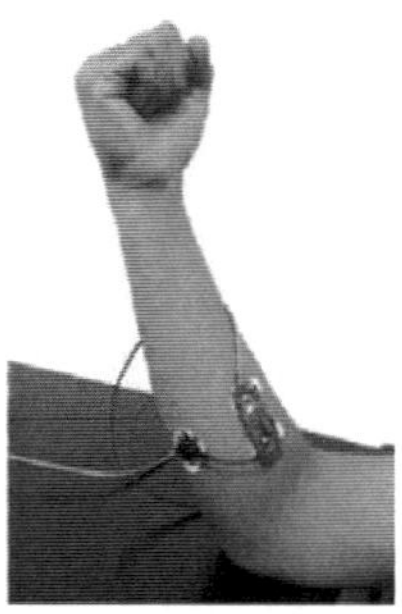

Figure 4. Electrode placement in the lower arm muscle

The two types of movement for recording EMG data for each individual are open finger extensor (OFE) and grasping finger extensor (GFE) as shown in figure 5 and figure 6.

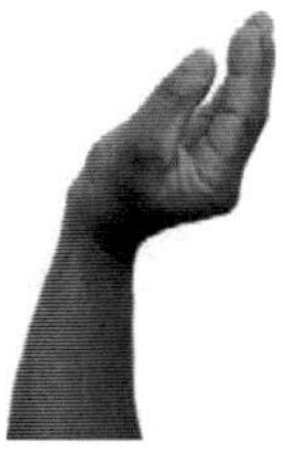

Figure 5. Open finger extensor movement

Figure 6. Grasping finger extensor movement

The process of recording EMG signals was using the CoolTerm software. Every movement starts on the 5th second and ends on the 15th second. The 1st to 5th seconds was called the initial stages of the EMG signal while the 16th to 20th seconds are called the end stages of the EMG signal. The recording of EMG signals on one of participant for OFE and GFE movements was shown in Figure 7.

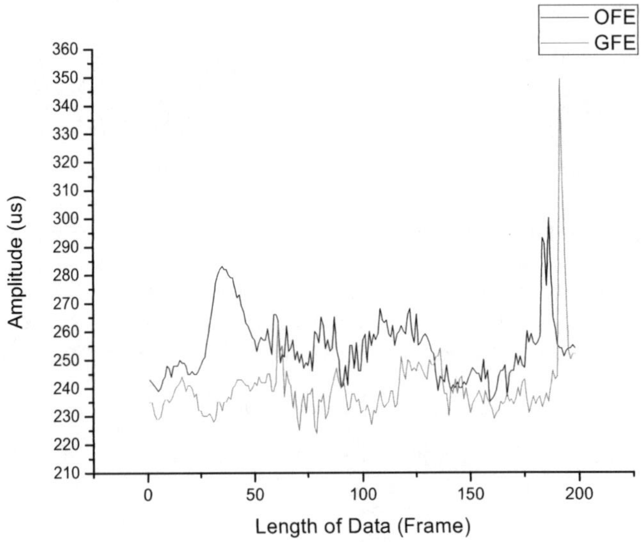

Figure 7. The shape of EMG signal for OFE and GFE movement

2.2.3. Data processing stage. In this stage, the raw EMG signal that has been recorded from the subject will be processed by the EMG signal. The EMG signal processing was to ensure that the EMG signal recording results do not contain interference. All EMG raw signals has normalized so that the raw EMG signal was within the same standard. Normalization method uses the zero mean methods as shown in equation (1).

$$zero\ mean = a - x \tag{1}$$

Where $\boldsymbol{a}$ is EMG signal data and $\boldsymbol{x}$ is the average value of $\boldsymbol{a}$.

After normalizing the EMG raw signal, the EMG signal from the normalized feature extraction was used as input data in the classification process. This EMG signal feature was used to get the accuracy of the classification results. In this study, the feature extraction method used was time domain features. The reason for choosing time domain feature is due to the fact that the feature does not require mathematical transformation.

2.2.4. Mean absolute value (MAV). This method was one of feature that popularly used in EMG signal analysis. The MAV feature was the average absolute value of the EMG signal amplitude. MAV equation is as shown by equation (2).

$$MAV = \frac{\sum_{i=1}^{N} |x_i|}{N} \tag{2}$$

Where $|x_i|$ is the absolute value of the EMG signal and N is the amount of data from the EMG signal.

2.2.5. Root mean square (RMS). Root mean square is the root of the average value of the squared EMG signal. This aims to get the effective value of the EMG signal. The equation of the RMS is as shown in equation (3).

$$RMS = \sqrt{\frac{\sum_{i=1}^{N} x_i^2}{N}} \tag{3}$$

Where x_i^2 is the square of the EMG signal amplitude value and N is the amount of data from the EMG signal.

2.2.6. Minimum value (Min). The minimum value is the minimum amplitude value of the EMG signal.

2.2.7. Maximum value (Max). The maximum value is the maximum amplitude value of the EMG signal.

2.2.8. Variance (Var). The variance is the average of the two-degree value of deviation variables. Variance equation is as shown in equation (4).

$$Var = \frac{\sum_{i=1}^{N} x_i^2}{N-1} \tag{4}$$

Where x_i^2 is the square of the EMG signal amplitude value and N is the amount of data from the EMG signal.

2.2.9. Standard Deviation (STD). The standard deviation is the distribution of the value of the EMG signal and how closed each EMG value is to the average value line of the EMG signal. The greater the standard deviation value, the greater the distance of each data with the average value. The standard deviation equation is shown in equation (5).

$$s = \sqrt{\frac{\sum_{i=1}^{n} \left(x_i - \bar{x}\right)^2}{n-1}} \tag{5}$$

Where s is standard deviation,x_i is the EMG signal data, $\bar{x}$ is the average value of EMG signal, and n is the amount of data.

2.2.10. Length of Data (LoD). Length of data is the length of EMG signal data. LoD equation as shown in equation (6).

$$LoD = length(N) \tag{6}$$

Where N is the data of EMG signal.

2.2.11. Stage of EMG signal pattern recognition. In this stage, the pattern recognition method is used to classify the EMG signal. This classification is to determine the level of differences in EMG signal patterns in each participant. The classification method used in this study is Learning Vector Quantization (LVQ). This method works determining the initialization vector or reference vector taken in each class. This reference vector will be updated by calculating the difference between the input vector and the reference vector to obtain an input vector in a stable (convergent) state. In the LVQ algorithm, the input vector x calculates the shortest distance to the reference vector w_c using the Euclidean distance as shown in equation (7). If the reference vector w_c and the input vector x are in the same class, the reference vector w_c will be updated using equation (8) so that the distance w_c will be closer to x. If the reference vector w_c and the input vector x are in different classes, the reference vector w_c will be updated using equation (9) so that the distance w_c will be further away from x.

$$d_c = \sum_{j=1}^{n} \left(x^j - w_c^j\right)^2 \tag{7}$$

$$w_c(t+1) = w_c(t) + \alpha(x(t) - w_c(t)) \tag{8}$$

$$w_c(t+1) = w_c(t) - \alpha(x(t) - w_c(t)) \tag{9}$$

Where t is a time and α adalah learning rate where $0 < \alpha < 1$. The LVQ network simulation uses Weka software[30]. Ten EMG data for each individual was selected. This data was divided into two parts; six data as learning data and four data as testing data. The method for learning and testing uses supplying testing sets, where test data was separated by learning data. The number of features involved in the input vector is seven units. The LVQ type used is the LVQ1 network. The total codebook vector in the competitive layer is six units where one codebook vector states one class. The network performance is measured using accuracy, precision, and recall as shown in equations (10), (11) and (12).

$$accuracy = \frac{TP + TN}{TP + TN + FP + FN} \tag{10}$$

$$precission = \frac{TP}{TP + FP} \tag{11}$$

$$recall = sensitivity = TPR = \frac{TP}{TP + FN} \tag{12}$$

3. Results and Discussion

From the research methodology that has been explained above, the results can be obtained as shown in the confusion matrix in table 1 and table 2. To measure the performance of the LVQ classifier network as shown in table 3 and table 4.

Table 1. Result of learning and testing LVQ network to OFE movement

		Prediction (%)					
		Subject1	Subject2	Subject3	Subject4	Subject5	Subject6
Actual (%)	Subject1	100	0	0	0	0	0
	Subject2	0	100	0	0	0	0
	Subject3	0	33.33	66.67	0	0	0
	Subject4	0	0	0	66.67	0	33.33
	Subject5	0	0	66.67	0	0	33.33
	Subject6	0	33.33	0	0	0	66.67

Table 2. Result of learning and testing LVQ network to GFE movement

		Prediction (%)					
		Subject1	Subject2	Subject3	Subject4	Subject5	Subject6
Act	Subject1	66.67	0	0	0	0	33.33

	Subject2	0	50	0	25	0	25
	Subject3	0	0	100	0	0	0
	Subject4	33.33	0	0	66.67	0	0
	Subject5	0	0	0	0	100	0
	Subject6	0	0	0	0	0	100

Table 3. Performance of testing LVQ network for OFE movement

Subject	TP (%)	FP (%)	FN (%)	Precision (%)	Recall (%)
1	100	0	0	100	100
2	100	66.66	0	60	100
3	66.67	66.67	33.33	50	66.67
4	66.67	0	33.33	100	66.67
5	0	0	100	0	0
6	66.67	66.66	33.33	50	66.67

Table 4. Performance of testing LVQ network for GFE movement

Subject	TP (%)	FP (%)	FN (%)	Precision (%)	Recall (%)
1	66.67	33.33	33.33	66.67	66.67
2	50	0	50	100	50
3	100	0	0	100	100
4	66.67	25	33.33	72.72	66.67
5	100	0	0	100	100
6	100	58.33	0	63.15	100

By using equation (10), the result of accuracy classification average is 70.83% for OFE movement and 83.33% for GFE movement.

From table 1, it show that the LVQ network classify the EMG signal for OFE movement with true positive (TP) for subject 1 and subject 2. It indicates that the EMG signal in open finger extensor combining wrist movement is unique. Meanwhile, the LVQ network classifying the EMG signal with false negative (FN) for subject 3, subject 4, subject 5 and subject 6 indicates that the EMG signal for this subject is not unique. Table 2 shows that the LVQ network classifies the EMG signal for GFE movement with true positive (TP) for subject 3, subject 5 and subject 6. It shows that the EMG signal in grasping finger extensor combining with the wrist movement is unique. Meanwhile, the LVQ network classify the EMG signal with false negative (FN) for subject 1, subject 2 and subject 4 then the EMG signal for this subject as not unique.

Table 3 explains that subject 1 has the best precission and recall value. It indicates that the EMG signal for this subject has the best performance of testing network criteria. The subject 2, subject 3, subject 4 and subject 6 have a good performance. Meanwhile, the subject 5 has a poor performance. Table 4 shows that the subject 3 and subject 5 have the best precission and recall value. It indicates that the EMG signal for this subjects have the best performance of testing network criteria. Meanwhile subject 1, subject 2, subject 4 and subject6 have a good performance of testing network criteria.

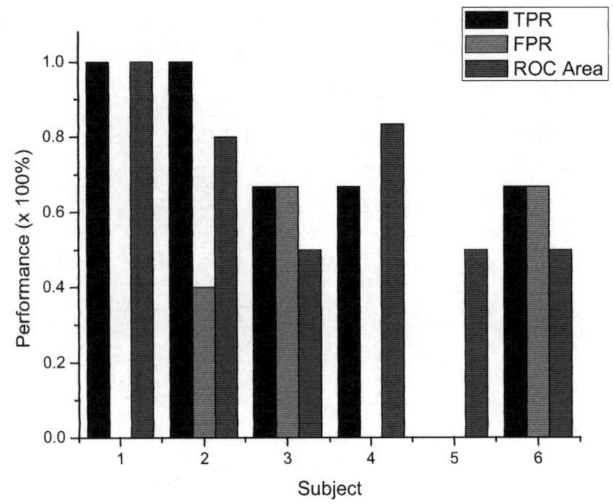

Figure 8. LVQ network performance evaluation for OFE movement

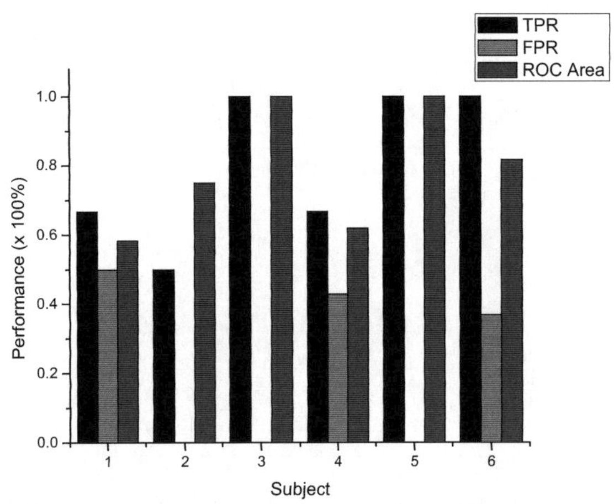

Figure 9. LVQ network performance evaluation for GFE movement

Figure 8 and 9 shows ROC Area to TPR for LVQ network performance. Total area under ROC curve (AUC) is a single index to measure the performance a test. The larger the AUC, the better the overall performance of LVQ network test is.

4. Conclusion

The results of the present of the study also suggest that the EMG signal in the lower arm muscle can be used to identify person's wrist movement. The EMG signal in lower arm muscle can also be used to determine the level of movement pattern variation from one to another. The accuracy of the classification process is 70.83% for OFE movement and 83.33% for GFE movement. This study has shown that the feature used in this research and the LVQ network classifier was sophisticated method in studying the EMG signal pattern. Further research might investigate the EMG signal in lower arm muscle involving more participants so that the classification process can achieve better accuracy.

Acknowledgement
This work was supported by Republic of Indonesia Ministry of Research, Technology and Higher Education (No. 117/SP2H/LT/DRPM/2018).

References

[1] Garza-Rodriguez A Sanchez-Fernandez L Sanchez-Perez L A Ornelas-Vencez C and Ehrenberg-Ihzunza M 2018 Pronation and supination analysis based on biomechanical signals from Parkinson's disease patients *Artificial Intelligence in Medicine* vol 84 pp 7-22

[2] Ferber R Osis S T Hicks J L and Delp S L 2016 Gait biomechanics in the era of data science *Journal of Biomechanics* vol 49 pp 3759-3761

[3] Chian-Her L Ching-Chi H and Po-Yuang H 2017 Biomechanical study of different fixation techniques for the treatment of sacroiliac joint injuries using finite element analyses and biomechanical tests *Computer in Biology and Medicine* vol 87 pp 250-257

[4] Hindalgo I Vilda P G and Garayzabal E 2018 Biomechanical Description of Phonation in Children Affected by Williams Syndrome *Journal of Voice* vol 32 pp 515.e15-515.e28

[5] Antwi-Afari M F and Lie H 2018 Fall risk assessment of construction workers based on biomechanical gait stability parameters using wearable insole pressure system *Advanced Engineering Informatics* vol 38 pp 683-694

[6] Hunt M A and Hatfield G L 2017 Ankle and knee biomechanics during normal walking following ankle plantarflexor fatigue *Journal of Electromyography and Kinesiology* vol 35 pp 24-29

[7] Kim E and Freivalds A 2018 Two-dimensional biomechanical thumb model for pipetting *International Journal of Industrial Ergonomics* vol 68 pp 165-175

[8] Razavian S R Greenberg S and McPhee J 2019 Biomechanics Imaging and Analysis *Oxford-Elsevier* pp 488-500

[9] Krishnan S and Athavale Y 2018 Trends in biomedical signal feature extraction *Biomedical Signal Processing and Control* vol 43 pp 41-63

[10] Rajeswari J and Jagannath M 2017 Advances in biomedical signal and image processing – A systematic review *Informatics in Medicine Unlocked* vol 8 pp 13-19

[11] Waris A Niazi I K Jamil M Gilani O Englehart K Jensen W Shafique M and Kamavuako E N 2018 The effect of time on EMG classification of hand motions in able-bodied and transradial amputees *Journal of Electromyography and Kinesiology* vol 40 pp 72-80

[12] Samuel O W Zhou H Li X Wang H Zhang H Sangaiah A K and Li G 2018 Pattern recognition of electromyography signals based on novel time domain features for amputees *Computer & Electrical Engineering* vol 67 pp 646-655

[13] Heales L J Tucker K Vicenzino B Hodges P W and MacDonald D A 2018 A comparison of fine wire insertion techniques for deep finger flexor muscle electromyography *Journal of Electromyography and Kinesiology* vol 41 pp 77-81

[14] Lipinski C L Donovan L McLoughlin T J Armstrong C W and Norte G E 2018 Surface electromyography based muscle fatigue detection using high-resolution time-frequency methods and machine learning algorithms *Physical Therapy in Sport* vol 154 pp 45-56

[15] Garikayi T Van den Heever D and Matope S 2018 Analysis of surface electromyography signal features on osteomyoplastic transtibial amputees for pattern recognition control architectures *Biomedical Signal Processing and Control* vol 40 pp 10-22

[16] Campos D P Abatti P J Bertotti F L Hill J A G and da Silveira A L F 2018 Surface electromyography segmentation and feature extraction for ingestive behavior recognition in

ruminants *Computer and Electronics in Agriculture* vol 153 pp 325-333

[17] Lipinski C L Donovan L McLoughlin T J Armstrong C W and Norte G E 2018 Surface electromyography of the forearm musculature during an overhead throwing rehabilitation progression program *Physical Therapy in Sport* vol 33 pp 109-116

[18] Duan N Liu L Z Yu X J Li Q and Yeh S C 2018 Classification of multichannel surface-electromyography signals based on convolutional neural networks *Journal of Industrial Information Integration*

[19] Putra D S Wibawa A D and Purnomo M H 2016 Classification of EMG during Walking using Principal Component Analysis and Learning Vector Quantization *International Seminar on Intelligent Technology and Its Application* pp 145-150

[20] Noce E Dellacasa Bellingegni A Ciancio A L Sacchetti R Davalli A Guglielmelli E and Zollo L 2019 EMG and ENG-envelope pattern recognition for prosthetic hand control *Journal of Neuroscience Methods* vol 311 pp 38-46

[21] Shi W T Lyu Z J Tang S T Chia T L Yang C Y 2018 A bionic hand controlled by hand gesture recognition based on surface EMG signals: A preliminary study *Biocybernetics and Biomedical Engineering* vol 38 pp 126-135

[22] Karabulut D Ortes F Arslan Y Z and Adli M A 2017 Comparative evaluation of EMG signal features for myoelectric controlled human arm prosthetics *Biocybernetics and Biomedical Engineering* vol 37 pp 326-335

[23] Ganesan Y Gobee S and Durairajah V 2015 Development of an Upper Limb Exoskeleton for Rehabilitation with Feedback from EMG and IMU Sensor *Procedia Computer Science* vol 76 pp 53-59

[24] Shakti D Mathew L Kumar N and Kataria C 2018 Effectiveness of robo-assisted lower limb rehabilitation for spastic patients: A systematic review *Biosensor and Bioelectronics* vol 117 pp 403-415

[25] Powar O S Chemmangat K and Figarado S 2018 A novel pre-processing procedure for enhanced feature extraction and characterization of electromyogram signals *Biomedical Signal Processing and Control* vol 42 pp 277-286

[26] Khushaba R N, Kodagoda S, Takruri M, and Dissanayake G 2012 Toward improved control of prosthetic fingers using surface electromyogram (EMG) signals *Expert System with Applications* vol 39 pp 0731-10738

[27] Kumar D K Arjunan S P and Singh V P 2013 Towards identification of finger flexions using single channel surface electromyography – able bodied and amputee subjects *Journal of Neuroengineering and Rehabilitation* pp 1-7

[28] van Beek N Stegeman D F van den Noort J C Veeger D and Maas H 2018 Activity patterns of extrinsic finger flexors and extensors during movements of instructed and non-instructed fingers *Biocybernetics and Biomedical Engineering* vol 38 pp 187-196

[29] Al-Timemy A H Bugmann G Escudero J and Outram N 2013 Classification of Finger Movements for the Dexterous Hand Prosthesis Control with Surface Electromyography *Journal of Biomedical and Health Informatics* vol 17 pp 608-618

[30] KentUniversity 2011 Weka Classification

Forecast Analysis of Instant Noodle Demand using Support Vector Regression (SVR)

E Fradinata[1*] , Z M Kesuma[2] , S Rusdiana[3], and N Zaman[1]
[1]Industrial Engineering Department, Faculty of Engineering, Syiah Kuala University, Banda Aceh, Indonesia.
[2]Statistics Department, Faculty of Mathematics and Natural Science, Syiah Kuala University, Banda Aceh, Indonesia.
[3]Mathematics Department, Faculty of Mathematics and Natural Science, Syiah Kuala University, Banda Aceh, Indonesia.

*Corresponding author: edinata69@gmail.com

Abstract. Support Vector Regression (SVR) is a part of Data Mining (DM) techniques where it can be used for forecasting the instant noodle. The cycle of the product demand is hard to predict. It will influence the resistant of the product quality where the product be expired easily and the other thing is the market demand. The objective of this research is approaching the predictive models with their performance measured with Mean Square Error (MSE) of SVR The data was collected from the determinant of instant noodle demand dataset. The random normal generated data was explored to get the amount of specific data. Then, it used SVR to forecast the demand. The result of this study the MSE of standard is 1.612 and the SVR is 1.436, means it increases around 11% better the performance than the original dataset. Since, we conclude that the SVR method would be promising to be one of a forecast demand method.

1. Introduction

Nowadays, the demand of instant noodle might be hard to anticipate due to those self-destructive considerations, it was caused of the behavior of a customer and the groups, for example, the society and the government regulation. The technique of support vector machine (SVR) will turn into established to utilize to solve in this area. The precision of demanding from this practice will be promising to be a better approach.

The learning machine theory would minimize the structural risk where it is construct with the boundaries upper bound and lower bound before it is generalized error. The gap is prepared after the link to the predicting. SVM obtains to be an elective also a chance to be those capable strategies on unraveling those relapse issues eventually utilizing the loss function, risk minimization, and this called SVR. An additional point of interest of SVR might make discovered [1-2].

The SVM could be working on the risk minimization (SRM) concept; its idea will be on minimizing the error and reduce the issue for overfitting. This technic could utilize in both proposes. They are in classification and regression cases in many areas. A portion investigation on which utilization this technique connected to forecasting modeling. Some studies in which use this method applied for forecasting modeling. The concept of support vector regression is finding the function that can be the best approximation the output of the vector *y* with a tolerance of error, to make it as flat as possible [3-5].

SVR may be a strategy on include the preparation focuses on the little subset of getting a yield result with the computer. The loss function of SVR is to pretend the optimization in a global minimum area. SVR can improve various interesting features and produce a better performance [6-8].

The contribution of this study might have invested the SVR method to forecast the points in the future for instant noodle industry. The indication of this matter, the SVR proves the small mean square error in prediction. This is the condition where the SVR to be one of the data mining method to predict the certain points in the future when it is used to be the predicting method.

2. Background

2.1. Support vector regression

SVR is a machine learning where it is called support vector machine (SVM) to use the useful computational load [9]. The dots results would constantly utilize to figure the variables in the unique space with portion work chosen by the SVM with gatherings made the mapping of the problem [10]. SVR will develop degeneration strategy that formed from SVM, the place might have been taken those hyperplanes closest to a measure of information the place it turns into a purpose. The little standard hyperplane needs to be gathered when the separation starting with information focuses on the hyperplane should keep minimizing [11]. It is demonstrated in figure 1.

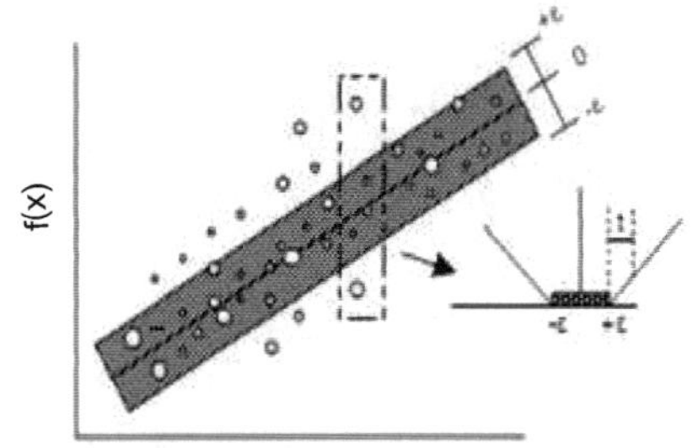

Figure 1. ε-sensitive loss function for linear of SVR [12].

The optimization to an e-insensitive loss function can be expressed in matrix notation as SVR finds the role of *f(x)* where has the biggest deviation $\mathcal{E}$ from the original target y_t, for all the training data with SVR, we got the something like a trench as in figure 1. Moreover, the formula is as follow:

$$f(\mathrm{x}) = \mathrm{wT}.\varphi(\mathrm{x}) + \mathrm{b} \quad (1)$$

Where φ(x) shows the dot in the feature space *f* from the map of *x* in the input space, coefficient, *w,* and *b* is estimated the minimization of risk function, this is defined in the formula of regression risk as follow:

$$\mathrm{Min} = 1/2 \ \|\mathrm{w}\| 2 + \mathrm{C}\ 1/\ell \Sigma \mathrm{i} = 1\mathrm{L}\epsilon(\mathrm{yi}, f(\mathrm{xi}) \quad (2)$$

Constraints to:

$$\mathrm{Yi} - \mathrm{w}\,\varphi(\mathrm{xi}) - \mathrm{b} \ \leq \epsilon, \text{'i} =, \dots, \ell$$
$$\mathrm{w}\varphi(\mathrm{xi}) - \mathrm{yi} + \mathrm{b} \ \leq \epsilon \ \text{'i} =, \dots, \ell$$

The e-insensitive loss function is usually as a cost function, where the function as given by (3).

$$L_{\epsilon}(y_i, f(x_i)) = \begin{cases} |y - f(x_i)| - \epsilon \ 1 y_i - f(x_i)] \geq, \\ 0, \ otherwise \end{cases} \tag{3}$$

The factor of $\|w\|^2$ is named regularization. Minimize the $\|w\|^2$ will make the function flat so that it can control the function of capacity. The second factor in the objective function is an error calculation measured with the e-insensitive loss function, [13]. In the process, it should be minimized *w* value. It has the aim of to find the better generalization for the regression function *f*. So it must be solved the optimization problem at (4), or another way to solve with a Matlab program is as follow:

$$min \ \tfrac{1}{2} \, x\mathrm{T} \ \mathrm{H}x + cTx \tag{4}$$

Where:

$$H = \begin{bmatrix} XX^T & -XX^T \\ -XX^T & XX^T \end{bmatrix}, c = \begin{bmatrix} \epsilon + & Y \\ \epsilon - & Y \end{bmatrix}, x = \begin{bmatrix} \alpha \\ \alpha^* \end{bmatrix}$$

Subject to:

$$x.(1, \dots, 1, -1, \dots, -1) = 0, \alpha i \,, \alpha i * \geq 0, i = 1, \dots, \ell$$

Where:

$$\mathrm{X} = \begin{bmatrix} x_1 \\ \cdot \\ \cdot \\ x_l \end{bmatrix}, \quad \mathrm{Y} = \begin{bmatrix} y_1 \\ \cdot \\ \cdot \\ y_l \end{bmatrix},$$

The ϵ-insensitive can be described as a tube in the generation of precision on the individual situation of data preparation. SVR regression was used to estimate the capacity may be claimed where the information is located on the axis (xi, yi)*n* where x_i, y_i, and *n* need the support to input, output, then the others can focus on data, separately. Separate works under the direct jobs are also a linear and nonlinear function [14].

The specific essential ideas from SVR are giving the solution from the small subset of training points to contribute the huge computational advantages. The process uses the epsilon intensive loss function to maintain the global minimum and the same time produces the generalization bound for optimization. The better settings of parameters are *C*, Ɛ, and the kernel parameters. The parameter will be determined by the degree of the larger deviation of ε and *C* parameter. Some very common of kernel functions are shown in table 1.

Table 1. Kernel functions.

Kernel	Function
Linear	x*y
Polynomial	$[(x*x_y)+1]^d$
Radial Basis Function (RBF)	$\mathrm{Exp}(-\gamma\,\|x-x_i\|^2)$

2.2. Mean Square Error (MSE)

Mean square error (MSE) is the estimator of measure the average of squares of the error. It is a risk function where to measure the quality of an estimator and the value closer to zero are better. The MSE is a measurement of the estimator quality with incorporates both mean and square of error with the formula is:

$$MSE = \sum_{i=1}^{n} \frac{(Y_t - F_t)^2}{n} \tag{5}$$

Where:
Y_t = the actual value.
F_t = the forecast value a period
n = the number of periods.

3. Methodology

The method of this study was used quantitative methods to process and analysis of data. The data were collected from the secondary data then created them with random normal generating. The instant noodle variables are price and demand were normalized. The method of this study was used quantitative methods to process and analysis of data. The data collected from the secondary data. This study runs with the Steve Gunn's algorithm where it is used the parameter of kernel function is linear, the upper bond C (not- separable case) is one, the loss function is e-insensitive [15]. SVR worked with 100 data, and it was separated into two-part data. The 90% was used for training and the rest 10% was used for testing, figure 2.

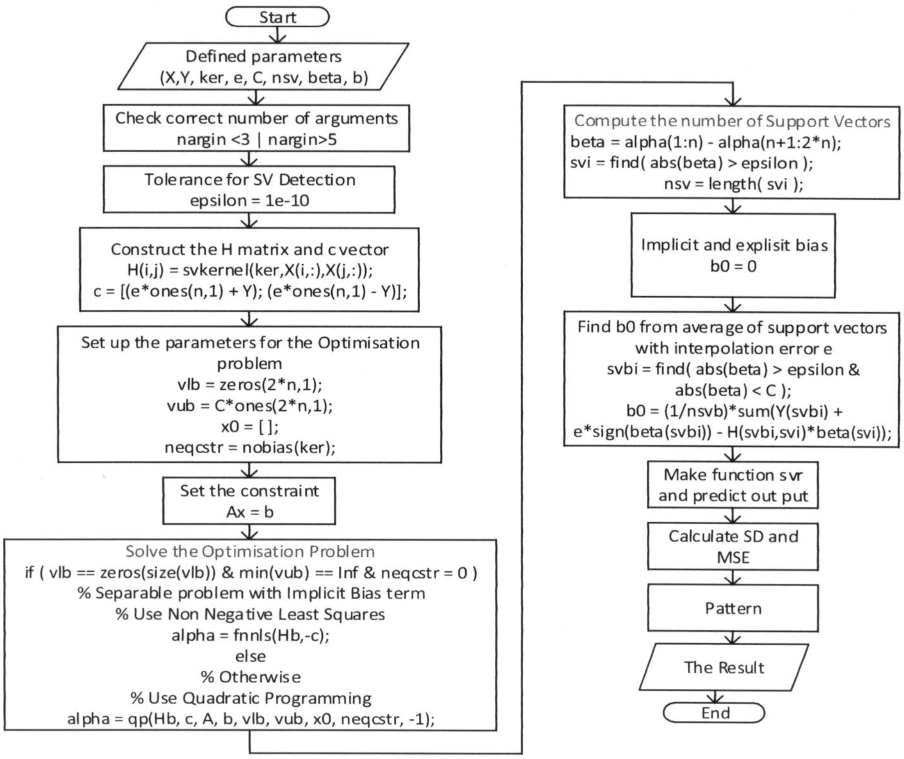

Figure 2. Block diagram of SVR algorithm.

The SVR would present the solution by means of small subset of training points where it would contribute the massive computational advantages. The global minimum would be produce between the area points to determine the optimal condition of reliable prediction. The algorithm explanation of SVR is as followed:

i. Define the parameters that for sales (x) and demand (y) , kernel, insensitive, upper bond, Number of support vector, Lagrange multiplayer, and the bias.
ii. Check the correct number of arguments.
iii. Tolerance for Support Vector Detection.
iv. Construct the Kernel matrix.
v. Set up the parameters for the optimization problem.
vi. Set the bounds: alphas ≥ 0 and $\leq C$.
vii. Configure the number of equality constraints (1 or 0).
viii. Set the constraint Ax = b.
ix. Add a small amount of zero order regularization to avoid problems when Hessian is badly. Rank is always less than or equal to n.
x. Solve the optimization problem.
xi. Compute the number of Support Vectors.
xii. Find bias from an average of support vectors with interpolation error e ($0 < alpha < C$).
xiii. Calculate the SVR for number support, beta and bias and svr output as follow:
xiv. [nsv beta bias] = svr (x,y,Metker, C,lossf, insen).
xv. tstData = svroutput (x, y,Metker, beta, bias).

xvi. SVR output
xvii. SD and MSE
xviii. Forecast the demand for the pattern from the characteristic of data
xix. Plot the comparative pattern

4. Results and Discussion

From this study, we prepared the data to check the cleaning data. In this research used two variables. They are price and demand variable of the Instant noodle industry, (D_Y). The parameters used in this study with kernel function were linear, loss-function was insensitive, the number of insensitive was 1, and upper-bound (non-separable case) was 1.

The SVR value worked under its loss function was Ɛ-insensitive, C was 1 and kernel function was linear. Those parameters affected by the amount of SV to build the regression function. The generated dataset from the characteristic of dataset after testing step to generalized the dataset in the area of upper bond and lower bond, it should be produced the forecast dataset in the certain point for predicting. The data produced the statistical characteristic of dataset where it can be compared the performance through the MSE for both running. Since, the resulted of MSE show that the actual data of standard was 1.612 and the predicting method with SVR was 1.436. Furthermore, the SVR decreased the MSE from the standard value to predict the certain point in the future, it is shown in table 2 as follow:

Table 2. Compared actual and forecasting of SVR for instant noodle.

Time (Moths)	Actual Data (Box/period)	Predicted (Box/Period)
91	343,141,145	342,089,212
92	339,503,098	358,153,780
93	362,888,381	358,153,780
94	339,800,508	348,113,425
95	326,827,872	361,768,308
96	343,468,963	347,711,811
97	359,389,685	350,121,496
98	353,704,355	366,989,293
99	347,794,147	345,302,126
100	366,989,293	326,827,872
Standard Deviation	0.0554	0.0271
Mean	0.3722	0.3619
MSE	1.612	1.436

SVR was used Steve Gunn's algorithm code and then it was modified to calculate the output of SVR function in the program code [15]. The 90% of data was used for training and the rest of 10% as testing. The testing resulted from SVR-output and then calculates the statistical output to get the characteristic of data. It was drawn in Figure 3.

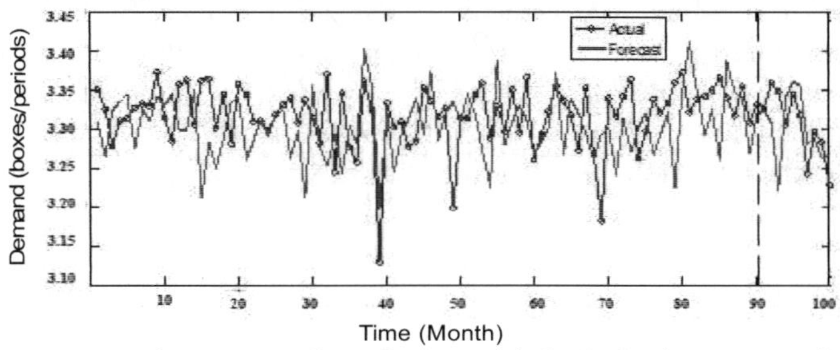

Figure 3. The output of prediction with SVR for instant noodle.

Figure 3 shows the result of SVR output for predicting point compare between the actual dataset to the forecast with the SVR.

5. Conclusions

This paper proposes the methods of SVR as a promising technique of forecasting. It is controlled by the combine generalisation control and other technique to obtain the curse dimensionality. This condition produces the result in the global quadratic optimisation with using the kernel function to view the model. The optimization points drop into the border of e-insensitive between positive and negative. After it is generalized compute the support vector and calculate the SVR for number support, beta, bias and SVR output, then this study obtain the value to compare between the standard and the SVR method. The result of the MSE is 1.436. from 1.612. It means that the performance of SVR method is better 11% than the standard. The accuracy of SVR would be able to be an alternative for the factory of instant noodle to predict the condition of customer demand. This method also can be used to predict the length of time (expired date) for instant noodle is consumed by the customer.

References

[[1] Tay F E and Cao L 2001 Application of support vector machines in financial time series forecasting *Omega* **29** 309-17

[2] Drucker H, Burges C J, Kaufman L, Smola A J and Vapnik V 1997 Support vector regression machines. In: *Advances in neural information processing systems,* pp 155-61

[3] Yu P-S, Chen S-T and Chang I-F 2006 Support vector regression for real-time flood stage forecasting *Journal of Hydrology* **328** 704-16

[4] Santosa B 2007 Data Mining (Teori dan Aplikasi) *Graha Ilmu, Yogyakarta*

[5] Jiang H, Zhang Y, Muljadi E, Zhang J J and Gao D W 2018 A short-term and high-resolution distribution system load forecasting approach using support vector regression with hybrid parameters optimization *IEEE Transactions on Smart Grid* **9** 3341-50

[6] Drucker H, Wu D and Vapnik V N 1999 Support vector machines for spam categorization *IEEE Transactions on Neural Networks* **10** 1048-54

[7] Basak D, Pal S and Patranabis D C 2007 Support vector regression *Neural Information Processing-Letters and Reviews* **11** 203-24

[8] Santamaría-Bonfil G, Reyes-Ballesteros A and Gershenson C 2016 Wind speed forecasting for wind farms: A method based on support vector regression *Renewable Energy* **85** 790-809

[9] Burges C J 1998 A tutorial on support vector machines for pattern recognition *Data mining and knowledge discovery* **2** 121-67

IOP Conf. Series: Materials Science and Engineering **506** (2019) 012021 doi:10.1088/1757-899X/506/1/012021

[10] Ghaedi M, reza Rahimi M, Ghaedi A, Tyagi I, Agarwal S and Gupta V K 2016 Application of least squares support vector regression and linear multiple regression for modeling removal of methyl orange onto tin oxide nanoparticles loaded on activated carbon and activated carbon prepared from Pistacia atlantica wood *Journal of colloid and interface science* **461** 425-34

[11] Hearst M A, Dumais S T, Osuna E, Platt J and Scholkopf B 1998 Support vector machines *IEEE Intelligent Systems and their applications* **13** 18-28

[12] Santosa B 2010 Tutorial Support Vector Machine *Teknik Idustri, ITS.[Online].*

[13] Cortes C and Vapnik V 1995 Support-vector networks *Machine learning* **20** 273-97

[14] Smola A J and Schölkopf B 2004 A tutorial on support vector regression *Statistics and computing* **14** 199-222

[15] Gunn S R 1998 Support vector machines for classification and regression *ISIS technical report* **14** 5-16

IOP Conf. Series: Materials Science and Engineering **506** (2019) 012022 doi:10.1088/1757-899X/506/1/012022

Description of *Chromolaena odorata* L. R.M King and H. Robinson as medicinal plant: A Review

Meutia Zahara[1*]
[1]Department of Biology, Islamic Faculty, University Muhammadiyah Aceh. Banda Aceh 23245. Indonesia

[*]Corresponding author: teeya_razali@yahoo.co.id

Abstract. *Chromolaena odorata* L. belongs to *Asteraceae* family (sunflower family) is an important and serious perennial herb in the world, while this weed also acts as medicinal plant. Several parts of this plant widely used to treat wound, burns, skin infections as well as to possess anticancer, antidiabetic, anti-hepatotoxic, anti-inflammatory, antimicrobial, and antioxidant properties. Native to North America and has been introduced to Asia, West Africa and Australia. Siam weed is one of common name of *Chromolaena odorata* L. grown as medical herbs and ornamental plant. The medicinal values of *Chromolaena odorata* L. lie in their phytochemicals component, the dried leaf of *Chromolaena odorata* contained flavonoid aglycones (flavanones, flavonols, flavones) including acacetin, chalcones, eupatilin, luteolin, naringenin, kaempferol, quercetin, quercetagetin, and sinensetin, terpenes and terpenoids, essential oils, and other phenolic compounds, which produce define physiological action in our body. Therefore, this review study provides detail information on general description of *Chromolaena odorata* L as medicinal plant.

1. Introduction

Chromolaena odorata is one of herbs belongs to *Asteraceae,* sunflower Family, synonyms to *Eupatorium odoratum* act as a traditional medicinal plant. Several parts of this plant widely used to treat wound, burns, skin infections as well as to possess anticancer, antidiabetic, anti-hepatotoxic, anti-inflammatory, antimicrobial, and antioxidant properties [1]. This flowering shrub native to North America, from Florida and Texas to Mexico and the Caribbean [2], has been introduced to Asia, West Africa and Australia. Siam weed is one of common name of *Chromolaena odorata* L. grown as medical herbs and ornamental plant. In Indonesia, this plant is one of the important traditional medicines, crushing the young leaves and resulting liquid used to treat skin wounds [3]. *Chromolaena odorata* L. also considered as a serious weed in plantation crops in the world such as palm oil plant, coconut, rubber and citrus. It is consisted high allelophatic and suppresses neighbouring vegetation. During the dry season the stem dry and readily burn, but the stumps remain alive and grow very rapidly and cover in the succeeding rainy season [4,5]. This weed affects particularly small farms, production in agricultural sectors and as well as natural ecosystems. It can quickly invade cleared lands and crops resulted in low yield and income [5]. *Chromolaena odorata*, one of the wound healing plants that investigated for its diverse health benefits [1].

Chromolaena odorarata L. found throughout the world especially in the pacific region under different names; Siam weed, devil weed, French weed, hagonoy, co hoy, in Indonesia known as Ki Rinyuh and si koko [8,4]. The mentioned herb is an important weed plant that extend its territory from America to Asian countries like Indonesia, India, China, Bangladesh, Thailand and other [14,15]. *Chromolaena odorata* L is being use traditionally as medicinal properties, especially for external uses as in wound skin, skin infections, and inflammation [8]. In Indonesia, most of the people used this plant to cure stomachs problem or dyspepsia, reducing the cholesterol and hypertension, vertigo, diet and mainly for external uses as well.

2. General Information

Chromolaena odorata L. belongs to the Asteraceae family. Common names of Chromolaena odorata; Eupatorium affine Hook & Arn., Eupatorium brachiatum Wikstrom, Eupatorium clematitis DC., Eupatorium conyzoides M. Vahl., Eupatorium divergens Less., Eupatorium floribundum Kunth, Eupatorium graciliiflorum DC., Eupatorium odoratum L., Eupatorium sabeanum Buckley., Eupatorium stigmatosum Meyen & Walp., Osmia conyzoides (Vahl) Sch.-Bip., Osmia divergens (Less) Schults-Bip., Osmia floribunda (Kunth) Schultz-Bip., Osmia graciliflora (DC) Sch.Bip., Osmia odorata (L) Schultz-Bip. [3]

i. Kingdom : *Plantae*
ii. Subkingdom : *Tracheobionta*
iii. Superdivision : *Spermatophyta*
iv. Division : *Magnoliophyta*
v. Class : *Magnoliopsida*
vi. Subclass : *Asteridae*
vii. Order : *Asterales*
viii. Family : *Asteraceae*
ix. Genus : *Chromolaena*
x. Spesies : *Chromolaena odorata* (L) [3]

Chromolaena odorata L. has a short life cycle approximately ten years maximum. It is an herbaceous perennial forms dense tangled bush up to 2 min in height. It occasionally reaches its maximum height of 6 m (as climber on other vegetation). The paired of branches grow freely through the main stem. The older stem in the base are brown and woody, while the tips are soft and green. The root formation is fibrous and does not penetrate beyond 20-30 cm in most soil. The flowers are white or pale bluish-lilac, and form covering the whole surface (figure 1) [3,6,7]. The leaves are arrowhead-shaped, 50–120 mm in length and 30–70 mm in wide. They grow in opposite pairs along the stems and branches (figure 1). As the species name '*odorata*' suggests, the leaves emit a pungent odour when crushed [6]. The seeds of Siam weed are small (3-5mm long, ~1mm wide, and weigh about 2.5mg seed-^{1}) [3].

Siam weed is a perennial species, as it lives for more than one year (table 1.) and very well adapted to the wet–dry tropical climate [6]. *Chromolaena odorata* L is a perennial scrambling shrub with straight, pithy, brittle stems which readily branch, bear three-veined, ovate-triangular leaves placed oppositely and with a shallow (figure 1), fibrous root system [16]. Within the native range, C. *odorata* shows the variability of morphology in term of flower color, leaf shape and hairiness, smell of the crushed leaves and its architecture. In some part of regions, its shows several forms and intermediate, while in others appears homogenous and this variability presently remains unexplained [17].

Figure 1. Leave and flowers morphology of *Chromolaena odorata* L.

Siam weed grows on a wide range of soils and in a range of vegetation types, such as in forests (rainfall 1500mm per year), grassland and arid bushveld (rainfall less than 500mm per year) [3]. The range of relative humidity should be around 60-70% for good growth of this herbs, at values higher than 80% the growth performance was bad [13]. A research showed that *Chromolaena odorata* L. seedling grew very well at 30°C and much better on mulched soil at 25°C, whereas in a heavy shade this medicinal plant will not seed. This phenomenon has a negative relationship with tree canopy cover and appears to be the most abundant on the edge of the forested area [14, 3].

Table 1. General growth pattern of *Chromolaena odorata* L. [6]

Growth	Jan	Feb	Mar	Apr	May	June	Jul	Aug	Sept	Oct	Nov	Dec
Flowering						■	■					
Seed fation ormation								■	■			
Seed drop									■	■		
Dieback					■	■	■	■	■	■		
Regrowth	■	■	■								■	■
Germination	■	■	■	■							■	■

The leaves are in oval form with 6-10 cm long and 3-6 cm width, the bottom part is wider and more pointed in the tip. The edge of the leaves is jagged and facing the base and leaves position are face to face to others. The wreath is located at the end branch (terminal), consists of 20-30 flowers with bluish color for young flower and brown for mature flower (figure 1) [4]. The flower is blooming during the dry season, inflorescence simultaneously happens within 3-4 weeks. The seed ripe then the plants will dry out; seed break out and it carried by wind. During early rainy season, the stem pieces, branches and the base of the stem sprout again, the falling seed starting to germinate and next two months sprout and shoots dominates the area. The observation that have been done by Yadav and Tripathi showed that plant density reaches 36 adults per m^2 and no less than 1300 sprouts, even though plant already mature, it's still have the potential to produces bud [4].

Siam weed is growing and breeding very fast, so can form a tight community and get in the way growth of other plants through the competition. C. *odorata* can grow up to 1000-2800 m above sea level [18] in Indonesia mostly found in the lowlands (0-500 m above sea level) as rubber and coconut plantations [4]. The plant height of C. *odorata* can reach up to 5 m and even more [19]. The young stem shows green color with soft structure and becomes brown with hard structure (woody) when it is old. The branches system is tight, reduces sunlight to the bottom part, and inhibits the growth of other plant species including grass. This weed grows very fast and dominates the area very quick [4].

3. Phytochemical content and health benefits

The medicinal values of *Chromolaena odorata* lie in their phytochemicals component (table 2.) such as tannins, alkaloid, flavonoids and other phenolic compounds, which produce a physiological action in our body [20]. The dried leaf of *Chromolaena odorata* contained ash (11%), crude fat (11%), fibre (15%), moisture (15%), crude protein (18%), and carbohydrate (31%). The active phytochemical substances are: (1) flavonoid aglycones (flavanones, flavonols, flavones) including acacetin, chalcones, eupatilin, luteolin, naringenin, kaempferol, quercetin, quercetagetin, and sinensetin, (2) terpenes and terpenoids, (3) essential oils, (4) alkaloids including pyrrolizidine, (5) saponins and tannins, (6) phenolic acids including ferulic acid, protocatechuic acid, (7) phytoprostane compound including chromomeric acid [1]. While the fresh *Chromolaena odorata* contained moisture (59.50%), total ash (2.50%), crude protein (6.56%), crude lipid (0.10%), carbohydrate (20.58%), fibre (10.76%) and total metabolize energy (109.46 kcal/100g) [12].

Table 2. Phytochemical component of *Chromolaena odorata* L. [12]

Phytochemical	Status
Alkaloids	+
Cyanogenic glycosides	+
Flavonoids	
Z Aurone	+
Z Chalcone	+
Z Flavone	+
Z Flavonol	+
Phytates	++
Saponins	+++
Tannins	++

Key: + = slightly present; ++ = moderately present; +++ = highly present

Chromolaena odorata is popularly uses for traditional wound healing; moreover, the leave aqueous extract have been used for the treatment of soft-tissue burns or skin infections [1].The efficiency of healing wounds come from the antioxidant property of the drug or plant which enhances conserving the fibroblast and keratinocyte proliferation on those wounds. Also used for a variety of ailments in many tropical countries for a long time, especially to stop bleeding. Numerous studies have demonstrated that Siam weed extract (SWE) accelerates haemostasis and wound healing [1,9].

For malaria treatment, the leaves of *Chromolaena odorata* L ingredient with lemon grass and guava leave. Other traditional medicinal uses include include anti-diarrheal, astringent, antispasmodic, antihypertensive, anti-inflammatory, diuretic, tonic, antipyretic and heart tonic and also cough remedy [1,10, 11]. The fresh leaves and extract of *C. odorata* are a traditional herbal treatment in some developing countries for burns, soft tissue wounds and skin infections. A formulation prepared from the aqueous extract of the leaves has been licensed for clinical use in Vietnam [10].

In Ghana and Benin, the infusion from fresh *Chromolaena odorata* L leaves is used to treat malaria and internal haemorrhage. In Thailand, the leaves are used to cataplasm to stem external haemorrhage. In Guatemala, microbiological investigations carried out in alcoholic extracts of *Chromolaena odorata* L extracts indicated the antimicrobial activities [21].

4. Conclusions

Chromolaena odorata L.is native to North America and now spread to all over the world. It can reach up to 6 m in height. Synonym to *Eupatorium odoratum*, and some others common name; Siam weed, devil weed, French weed, hagonoy, co hoy, in Indonesia known as Ki Rinyuh and si koko. C. *odorata* is being used traditionally as medicinal properties, especially for external uses as in wound skin, skin infections, inflammation, a therapeutic agent for a variety of diseases, such as wound healing, anti-inflammatory, analgesic, antipyretic, diuretic, and antimicrobial, anti-mycobacterial and many more. More researches needed to find a formula to make a marketable medicine from *Chromolaena odorata* L. for easier to use without crashing the leaves, especially for external uses.

Acknowledgements
The author would like to thank the University Muhammadiyah Aceh for providing the financial support for the conferences. The author also would like to acknowledge all the family members during the writing process of the manuscript.

References

[1] Sirinthipaporn A and Jiraungkoorskul W 2017 Wound healing property review of Siam Weed, *Chromolaena odorata L. Pharmacogn* Rev. 11(21) pp 35-38
[2] "*Chromolaena odorata*" Flora of North America.
[3] Cakraborty A K, Rambhade S and Patil U K 2011 *Chromolaena odorata (L.)* : An Overview. *Journal of Pharmacy Research* 4 (3) pp 573-576
[4] Prawiradiputra B R 2007 Ki Rinyuh *(Chromolaena odorata (L) R M King* dan *H Robinson:* Gulma padang rumput yang merugikan *Wartazoa* Vol. 17 No 1
[5] Muniappan R and Bamba J 2000 Biological control of *Chromolaena odorata:* success and failures *Proceeding of the X international Symposium on Biological Control of Weed* pp 81-85 USA
[6] CRC Weed Management 2003 Siam Weed or *Chromolaena-Chromolaena odorata L.*
[7] Cruttwell R, McFadyen and Skarratt B 1996 Potential distribution of *Chromolaena odorata* (siam weed) in Australia, Africa and Oceania Agriculture *Ecosystems and Environment* 59 (1-2) pp 89-96)
[8] Vaisakh M N and Pandey A 2012 Pharmacognostic study of leaves of *Chromolaena odorata UNM International Journal of Pharmaceutical Research and Development (IJPRD)*
[9] Akah P A 1990 Mechanism of hemostatic activity of Eupatorium odoratum *International Journal of Crude Drug Research* vol. 28 no. 4 pp 253–256
[10] Vaisakh M N and Pandey A 2012 The invasive weed with healing properties: A review *on* Chromolaena odorata *International Journal Pharmaceutical Science and Research* 3:80–3.
[11] Vital P G and Windell L R 2009 Antimicrobial activity and cytotoxicity of *Chromolaena odorata (L. f.*) King and Robinson and *Uncaria perrottetii* (A. Rich) Merr. Extracts *Journal of Medicinal Plants Research* 3(7): pp. 511-518
[12] Ngozi I G, Jude I C and Catherine I C 2009 Chemical profile of *Chromolaena odorata* L. (King and Robinson) leaves *Pakistan Journal of Nutrition* 8 (5) pp 521-524
[13] Ambica S R and Jayachandra The problem of Chromolaena weed *Newsletter on Biocontrol and Management of Chromolaena odorata L.*
[14] Luwun P 2002 Control of invasive *Chromolaena odorata*; An evaluation in some land use types in Kwazulu Natal-South Africa, Thesis submitted to *International Institute for Geo-information Science and Earth Observation.*
[15] Warea O 2004 *Chromolaena* (Siam weed) Pest Advisory *leaflet No. 43/Secretariat of the Pacific Community*
[16] Zachariades C, Day Munịappan M, Reddy R G V P 2009 *Chromolaena odorata* L. King and Robinson (*Asteraceae*) DOI: 10.1017/CBO9780511576348.008 · Source: OAI
[17] Zachariades C, Von Senger I, Barker N P 2004 Evidence for a northern Caribbean origin for the southern African biotype of *Chromolaena odorata In Proceedings of the Sixth International*

Workshop on Biological Control and Management of Chromolaena, ed. M. D. Day and R. E. McFadyen. ACIAR Technical Reports 55 Canberra Australia: ACIAR, pp. 25–27.
[18] FAO 2006 Alien invasive species; Impacts of forest and forestry-A review
[19] Department of Natural Resources, Mines and Water 2006 Siam weed declared no. 1 *Natural resources mined and water Pesr series* Queensland Australia
[20] Akinmoladun A C, Ibukun E O, Dan-Ologe I A 2007 Phytochemicals constituents and antioxidant properties of *Chromolane odorata Scientific Research and Essay* Vol 2(6) pp. 191-194
[21] Tonzibo Z F, Wognin E, Chalchat J C, and Guessan Y T N Chemical investigations of *Chromolaena odorata* L. King Robinson from Ivory Coast *Journal of essential oils –Bearing Plants* 10 (2) pp. 94-100

Determination of Mineral Contents in Meukek Marble Using XRD and SEM-EDS Analysis

L Lindawati [1] M Mursal[2] and A Afdhal[3]
[1]Department of Mechanical Engineering, Polytechnic of South Aceh, Indonesia
[2]Department of Physics Department, Syiah Kuala University, Indonesia
[3]Department of Industrial Engineering Department, Polytechnic of South Aceh, Indonesia

Email: lindawati203@gmail.com

Abstract. Marble is a metamorphic rock that has been widely used in many applications. In general, it is composed of recrystallized carbonate minerals. The aim of this study was to investigate the minerals present in the marble sample from Meukek area, South Aceh. The mineral identification was conducted by using the X-Ray Diffractometer (XRD). The morphological and elementary analysis was carried out by using Scanning Electron Microscopy supported with Energy Dispersed Spectroscopy (SEM-EDS). Based on the XRD pattern data analysis, it was revealed that the marble sample from Meukek area in South Aceh mainly consists of calcium carbonate ($CaCO_3$). The SEM-EDS analysis indicated the presence of 55.479% Oxygen (O), 0.249% Magnesium (Mg), 1.197% Aluminium (Al), 0.716% Silicon (Si), 0.202% Sulphur (S), 1.388% Potassium (K) and 40.769% Calcium (Ca) elements in Meukek marble. The highest element contents were shown by Oxygen (O) and Calcium (Ca). It confirmed that calcite is the major mineral phase composed in Meukek marble.

Keyword : *meukek marble, x-ray diffraction, scanning electron microscopy and energy dispersed spectroscopy*

1. Introduction

Meukek is one of areas in South Aceh that is well known for its marble deposits. The marble was primarily used for sea-wave barrier before. The abundance of marble in the area offers the potential for economic growth for the community [1]. The marble has been produced into many valuable products. The properties of marble were highly influenced by the mineral composition [2-3]. Principally, marbles are commonly composed of mineral calcite and dolomite.

However, the information related to mineral content of marble from Meukek was still questioned. Consequently, mineral content of marble from Meukek should be determined. In recent years, numerous studies have been conducted in order to identify mineral composition of marble from an area [4-12]. Marble from various areas were analysed by using X-ray diffraction (XRD). X-Ray diffraction allowed the identification of both major and minor mineralogical components [13-14]. These previous researches reported that marbles from various areas are commonly consist of calcite and dolomite.

Hence, the objective of this study is to determine mineral composed in the marble from Meukek, South Aceh. The mineral content was determined by using X-Ray Diffraction (XRD) analysis. In order to confirm the mineral content, the elements presence over the marble should be determine using Scanning Electron Microscopy and Energy Dispersed Spectroscopy (SEM-EDS) analysis. The results of this research will be useful as reference in subsequent studies.

IOP Conf. Series: Materials Science and Engineering **506** (2019) 012023 doi:10.1088/1757-899X/506/1/012023

2. Materials and methods

2.1 Sample Preparation

The marble from Meukek area in South Aceh were selected as raw materials. The samples of marbles from Meukek were collected from Marble Production Unit in Tapaktuan, South Aceh, Indonesia. The samples used for XRD and SEM-EDS analysis were in marble powder form. Marble stones were cut into small pieces using a marble stone cutter. Then, they were crushed and mashed using mortar until became powder that is ready to be analysed. Marble stone from Meukek has a grey colour as shown in Figure 1a and Figure 1b.

Figure 1a. Marble Stone **Figure 1b**. Marble Powder

2.2 Mineral Phase Identification

In order to determine the mineral content, the sample of marble should be characterized by X-ray diffraction (XRD). Mineral characterization with X-Ray diffraction (XRD) was carried out at the Materials Laboratory. X-ray Diffraction analyses was conducted by using powdered sample the Shimadzu XDR-6100. The sample was radiated with Cu radiation at the wavelength of 1.54060 Å. Besides mineral composition, XRD allowed to determine the crystal structure. The particle size can be calculated by using the following Scherrer equation [10].

$$\tau = \frac{K\lambda}{\beta cos\theta} \tag{1}$$

Where τ is the average crystal size (nm), K is Scherrer constant (0.9), λ is X-ray (Å) wavelength, θ is diffraction angle (°), and β (FWHM) is maximum half width (radians).

2.3 Morphological Surface

Scanning Electron Microscopy (SEM) testing was carried out in order to observe the surface morphology of the marble sample. The elemental content in the sample was detected by using Energy Dispersive X-Ray Spectroscopy (EDS). The result showed in term of certain peaks that represent a number of elements contained in marble sample. The SEM-EDS analysis was performed at the Material Engineering Laboratory.

3. Results and Discussion

3.1 Mineral Phase

Basically, X-ray diffraction (XRD) has been used to detect the mineral and structure crystal contained in the material. The parameters are resulted in form of diffraction angle (*2θ*), distance between two fields (*d*), full width at half maximum value (FWHM) and intensity (*I*). The mineral contents in marble were determined based on the mineral diffraction peaks resulted from the relationship between diffraction angle (*2θ*) and intensity (*I*). Table 1 shows the data obtain from X-Ray Diffraction analyses on marble sample from Meukek, South Aceh. The table presents the three highest X-Ray Diffraction peaks found at *2θ* = 29.4789 °; 26,5781°; and 48,597. The highest peak appeared at *2θ* = 29.4789° with Intensity 1928. According to mineral database equipped in X-Ray Diffractometer, the mineral content identified at that diffraction angle are calcite ($CaCO_3$). Hence, the major mineral phase composed in marble stone from Meukek, South Aceh are Calcite. This finding is in line with the theory that Calcite is the main minerals that form limestone or marble. Marble with large number of calcite content tends to appear in white colour. The presence of some impurities such as oxide, silicate, quartz and graphite may affect the texture and colour of marble in such area. Marble which composes of graphite will pose grey colour. Regarding marble shows a grey colour with some orange and red fibres inside, it was assumed that the Meukek marble consists of calcite and graphite minerals.

Table 1. XRD data

No	*2θ* (°)	d (Å)	FWHM (°)	Intensity (a.u.)
1.	29.4789	3.02762	0.15680	1928
2.	26.5781	3.35136	0.13790	501
3.	48.5972	1.87197	0.17650	318

3.2 Crystalline Size

The size of the crystal of mineral is determined by widening full width at half maximum (FWHM) of the X-ray diffraction peaks. The crystalline sizes, which were calculated by using the Scherrer equation (1), are tabulated in Table 2. The crystalline size counted from each mineral is 10.19, 11.48 and 9. 58 nm. It can be seen that the peak width is inversely proportional to the size of the crystal. The larger the size of the crystal, the smaller the width of the diffraction peak appeared and vice versa.

Table 2. Crystalline size of Composed Mineral in Marble from Meukek

No	FWHM (°)	θ (°)	Crystalline Size (nm)
1.	0.15680	14.73	10.19
2.	0.13790	13.29	11.48
3.	0.17650	24.29	9.58

3.3 Morphological Surface

The presences of certain elements over marble surface were then confirmed by SEM-EDS analysis. The morphological surface of marble from Meukek area are shown in Figure 2a and Figure 2b. The morphology of the surface was obtained by capturing and processing of secondary electrons emitted by the test sample. The picture shows surface morphology resulted from 500 times and 1000 times magnification in each marble stone sample from Meukek. The greater the magnification given, the clearer the crystal formed seen. It can be seen that there are beam-shaped and angled but irregular particles. This is due to impurity factors that allow new phases to emerge. SEM analysis was carried out in line with Energy Dispersive X-Ray Spectroscopy (EDS). Characterization with EDS was conducted to determine the elemental content of a tested sample.

IOP Conf. Series: Materials Science and Engineering **506** (2019) 012023 doi:10.1088/1757-899X/506/1/012023

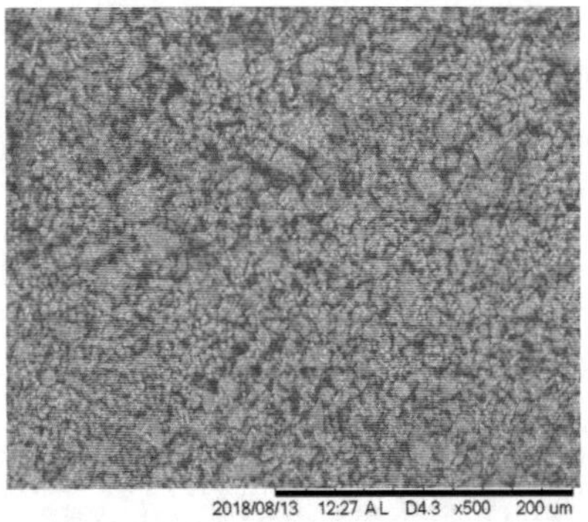

Figure 2a. SEM image of marble with 500x magnification

Figure 2b. SEM image of marble with 1000x magnification

Figure 3 shows the results of Energy Dispersive X-Ray Spectroscopy (EDS) characterization for marble stones from Meukek. The spectrum of EDS analysis indicates the presence of some elements in marble from Meukek area. The elements present are Oxygen (O), Magnesium (Mg), Aluminium (Al), Silicon (Si), Sulphur (S), Potassium (K), and Calcium (Ca) on Meukek marble. The percentage of mass of each element is shown in table. The highest element contents were shown by Oxygen which is equal to 55.479 % and then followed by Calcium which is equal to 40.769 %.

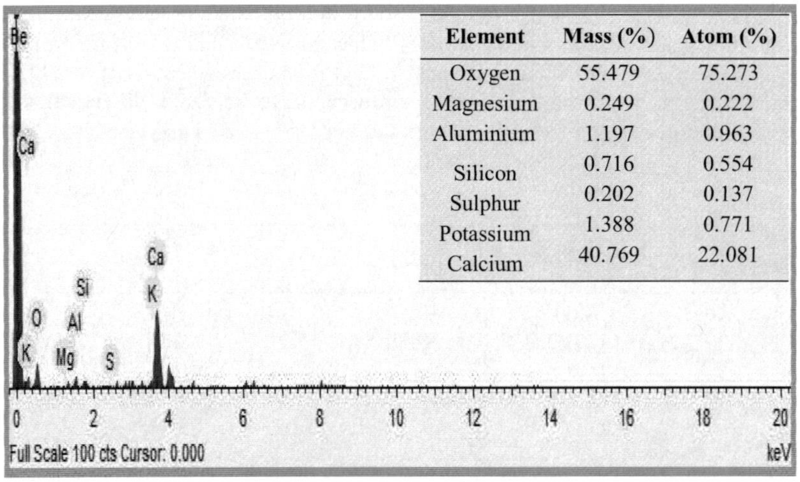

Element	Mass (%)	Atom (%)
Oxygen	55.479	75.273
Magnesium	0.249	0.222
Aluminium	1.197	0.963
Silicon	0.716	0.554
Sulphur	0.202	0.137
Potassium	1.388	0.771
Calcium	40.769	22.081

Figure 3. EDS Spectrum for Marble from Meukek

4. Conclusion

In this study, the mineral composition of South Aceh marble was investigated. XRD analysis reveals that marble stone from Meukek has dominantly consisted of calcite ($CaCO3$). A grey appearance related to the presence of some impurities such as graphite. The SEM-EDS analysis indicated the presence of 55.479% Oxygen (O), 0.249% Magnesium (Mg), 1.197% Aluminium (Al), 0.716% Silicon (Si), 0.202% Sulphur (S), 1.388% Potassium (K) and 40.769% Calcium (Ca) elements in Meukek marble. The highest element contents were shown by Oxygen (O) and Calcium (Ca). It confirmed that calcite is the major mineral phase composed in Meukek marble.

Acknowledgments
Authors thanks Indonesian Minister of Higher Education through General Director of Research and development for finding this research.

References

[1] Fauzan dan Maman Surahman 2016 Rancangan Penambangan Marmer Desa Kebutuhjurang, Kecamatan Banjar Negara, Kabupaten Banjar Negara, Jawa Tengah. *Jurnal Bahan Galian Industri* Vol 10 No 27 pp 1-9

[2] Lindawati Nuzuli Fitriadi Afdhal 2018 Analysis of Noise Level Generated by Stone Cutter Machine A Case Study in Marble Production Unit, South Aceh *Jurnal Inotera* Vol 3 No 1

[3] Lindawati and Mursal 2018 Mineral Identification of Marble Stones from Gunung Kerambil, South Aceh Using X-ray Diffraction, *Journal of Aceh Physics Society* vol 7 (3) pp 152-156.

[4] Abdul Manan dan Yaseen Iqbal 2007 Phase Microstructure and Mechanical Properties of Marble in North-Western Part of Pakistan: Preliminary Findings. *J Pak Mater Soc.* Vol. 1 No.2.

[5] M Fahad Y Iqbal M Riaz R Ubic M Abrar 2016 Geo-mechanical Properties of Marble Deposits from the Nikani Ghar and Nowshera Formations of the Lesser Himalaya, Northern Pakistan A Review, *Himalayan Geology*, Vol. 37 (1) pp. 17-27

[6] Octavian G Duliu Maria Nicoleta Grecu Corina Cristea 2009 EPR and X-Ray Diffraction Investigation of Some Greek Marbles and Limestones *Romanian Reports in Physics* Vol 61 No. 3 pp 487–499

[7] S Khrissi M Haddad L Bejjit S Ait Lyazidi M El Amraoui and C. Falguères 2017 Raman and XRD Characterization of Moroccan Marbles. XII Maghreb Days of Material Sciences *IOP Conf. Series: Materials Science and Engineering* vol 186

[8] S Khrissi L Bejjit M Haddad C Falguères S Ait Lyazidi M El Amraoui 2018 Study of marbles from Middle Atlas (Morocco): Elemental, Mineralogical and Structural Analysis Sustainable Buildings and Cities *IOP Conf. Series: Materials Science and Engineering* vol 353

[9] Leontakianakos G Baziotis I Profitis E Chatzitheodoridis and Tsimas S 2013 Assessment of the Quality of Calcination of Marbles from Thassos Island Using Raman Spectroscopy and X– Ray Diffraction Bulletin of the Geological Society of Greece vol XLVII *Proceedings of the 13th International Congress* Chania

[10] Onimisi M Obaje N G and Daniel A 2013 Geochemical and Petrogenetic Characteristics of the Marble Deposit in Itobe Area Kogi State Central Nigeria *Advances in Applied Science Research* vol 4(5) pp 44-57

[11] Ali Moumouni Nathaniel Goeter Goki Mohammed Suleiman Chaanda 2016 Geological Exploration of Marble Deposits in Toto Area, Nasarawa State, Nigeria, *Natural Resources*, vol 7 pp 83-92

[12] M Brilli M P Lapuente Mercadal F Giustini H Royo Plumed 2018 Petrography and Mineralogy of the White Marble and Black Stone of Göktepe (Muğla, Turkey) used in antiquity: New data for provenance determination *Journal of Archeological Science* Reports Volume 19 pp 625-642

[13] JD Lynn 2011 A New Method to Determine Marble Provenance with XRD Accessory Mineralogy Society of Petroleum Engineers SPE/DGS Saudi Arabia Section Technical Symposium and Exhibition Al-Khobar, Saudi Arabia

[14] H Aguiar S Chiussi M López-Álvarez P González J Serra 2018 Structural Characterization of Bio Ceramics and Mineralized Tissues Based on Raman and XRD Techniques *Ceramic International* Volume 44 Issue 1 pp 495-504

Mathematics Modeling of Diabetes Mellitus Type SEII$_T$ by Considering Treatment and Genetics Factors

Asmaidi[1*] and Eka Dodi Suryanto[2]

[1]Polytechnic of South Aceh, Merdeka Street, Komplek Reklamasi Pantai,Tapaktuan city, 23751, Indonesia
[2]Prima Indonesia University, Belanga Street, Simp. Ayahanda, Medan City, 061-4578890, Indonesia

[*]Corresponding author: asmedmat@gmail.com

Abstract SEII$_T$ stands for susceptible (S), exposed (E), infected population without treatment (I) and infected population with treatment (I$_T$). For infected population is grouped into infected without treatment and infected with treatment by insulin injection. Susceptible group migrate to exposed by genetics factors. The aims of this study are such as composing mathematics models for diabetes mellitus tipe SEII$_T$, composing mathematics models for diabetes mellitus, determining fixed point and basic reproduction number, stability analysis for fixed point, and fixed point stability. The result of the study is mathematical modeling or compartment diagram for diabetes mellitus. Compartment diagram was analyzed analytically and numerically. Analyses result gained two fixed points that are, fixed point without disease and fixed point with disease. Each fixed point was analyzed based on basic reproduction number in order to obtained data analyzed both analytically and numerically which the fixed point without disease was stabilized when $R_0 < 1$, while its counterpart stabilized at $R_0 > 1$. Human behavior at $R_0 < 1$ is when susceptible population proportion (s) was increased from initial value then stabilized about $s = 0.9999$. For exposed (e) was diminished in the beginning then rested around $e = 0$. For infected without treatment(i)was lowered first, then stabilized around $i = 0$. For infected population with treatment (i_T) were increased from the beginning then lowered and stabilzed at around $i_T = 0$. Human behavior at $R_0 > 1$ shown as the susceptible (s) population increased from initial point to fluctuated and then rested around $s = 0.54711$. Exposed (e)group lowered at first, then stopped around $e = 0.05655$. Infected population without treatment (i) diminished at first then rested at around $i = 0.00393$. Infected population with treatment (i_T) went up first then fluctuated and finally rested at around $i_T = 0.39241$.

1.Introduction

1.1 Background

Ulfah develop mathematical models for diabetes mellitus disease type SEII$_T$ which considering treatment factors. SEII$_T$ stands for susceptible (S), exposed (E) and infected (I) [1, 3-4]. Infected population is divided into two groups; infected population without treatment (I) and infected population with treatment (I$_T$). Next, mathematical model type SEII$_T$ was developed by considering genetics factors. This assumption was obtained from studies conducted by [2, 5]. In their study, susceptible population shifted to exposed population engineered by genetic factors.

IOP Conf. Series: Materials Science and Engineering **506** (2019) 012024 doi:10.1088/1757-899X/506/1/012024

1.2 Problem Formulations

How the effect of infective contact rate to the spread of diabetes mellitus type $SEII_T$ by considering treatment and genetic factors in population? This study aims to see the effect of changes in the rate of infective contact to population behavior in mathematical modeling of disease spread of diabetes mellitus type $SEII_T$ by considering treatment and genetic factors.

2. Methods

2.1 Mathematical Modelling Type $SEII_T$

Mathematical modeling for diabetes mellitus desease in this research is designed as figure 1 below:

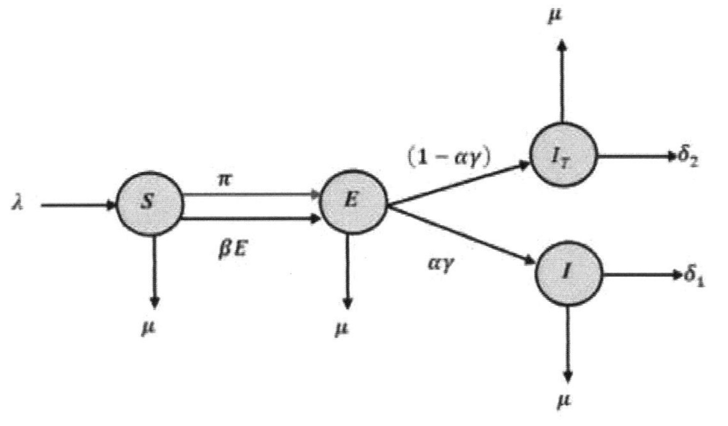

Figure 1. Mathematical Model Type $SEII_T$ with treatment factors and genetics factors

2.2 Research Conduct

The stages of the research are as follows;

i. Firstly is the modeling stage.
ii. Secondly, fixed point determination stage.
iii. Thirdly, the stage of determining the basic reproduction number.
iv. Fourthly, the fixed point stability analysis phase.
v. Next is the simulation phase of fixed point stability.
vi. The simulation phase of the effect of contact rate infective for the spread of disease in the population.

2.3 Data and Resources

Data in this research is secondary which gained from references in the form of journals and literatures.

3. Result and Discussion

3.1 Differential Equation

Differential equation for compartment diagram as follows;

$$\frac{dS}{dt} = \lambda - (\pi + \mu + \beta E)S$$

$$\frac{dE}{dt} = (\pi + \beta E)S - (\mu + 1)E$$

$$\frac{dI}{dt} = \alpha\gamma E - (\mu + \delta_1)I \quad (1)$$
$$\frac{dI_T}{dt} = (1 - \alpha\gamma)E - (\mu + \delta_2)I_T$$
$$N = S + E + I + I_T$$

by N is total population. To solve equation (1), the simplification is done by making the proportion of each population to the total population.

$$s = \frac{S}{N}, e = \frac{E}{N}; i = \frac{I}{N}; i_T = \frac{I_T}{N}$$

Hence, equation (1) can be written as follow and from that can get the equation (2):

$$\frac{ds}{dt} = \frac{\lambda}{N}(1 - s) - (\pi + \beta eN - \delta_1 i - \delta_2 i_T)s$$
$$\frac{de}{dt} = (\pi + \beta eN)s - \left(1 + \frac{\lambda}{N} - \delta_1 i - \delta_2 i_T\right)e$$
$$\frac{di}{dt} = \alpha\gamma e - \left(\delta_1 + \frac{\lambda}{N} - \delta_1 i - \delta_2 i_T\right)i \quad (2)$$
$$\frac{di_T}{dt} = (1 - \alpha\gamma)e - \left(\delta_2 + \frac{\lambda}{N} - \delta_1 i - \delta_2 i_T\right)i_T$$

with $s + e + i + i_T = 1$. The parameters contained in $SEII_T$ type mathematical modeling can be seen in table 1.

Table 1. Parameter in mathematical modeling Type $SEII_T$.

Parameter	Annotation	Units
λ	Birth rate	time^{-1}
μ	Mortal rate by nature	time^{-1}
δ_1	Mortal rate by disease towards population infected without treatment	time^{-1}
δ_2	Mortal rate by disease towards infected with treatment	time^{-1}
π	Population shift rate from susceptible to exposed by gene factors	time^{-1}
β	Population shift rate from susceptible to exposed by infective contacts inter-population	time^{-1}
α	Population shift rate from exposed to infected without treatment	time^{-1}

3.2 Fixed Point

3.2.1 Fixed Point without disease

$$T_0 = (s, e, i, i_T)\ ; s = \frac{\lambda}{\lambda + \pi N}; \quad e = 0; \quad i = 0; \quad i_T = 0.$$

3.2.2 Fixed Point with disease

$$T_1 = (s^*, e^*, i^*, i_T^*)$$

$$s^* = \frac{\lambda}{\lambda+N(\beta eN-\delta_1 i-\delta_2 i_T+\pi)}$$

$$e^* = -\frac{\pi Ns}{N(N1s\beta+\delta_1 i+\delta_2 i_T-1)-\lambda}$$

$$i_1^* = \frac{\mathrm{N}(\delta_1-\delta_2 i_T)+\lambda-\sqrt{(\delta_1 N-\delta_2 N i_T+\lambda)^2-4\alpha\gamma\delta_1 e\mathrm{N}^2}}{2\delta_1\mathrm{N}}$$

$$i_{T1}^* = \frac{-\delta_1 Ni+\delta_2 N+\lambda-\sqrt{4(\alpha\gamma-1)\delta_2 N^2 e+(-\delta_1 Ni+\delta_2 N+\lambda)^2}}{2\delta_2\mathrm{N}}$$

3.3 Basic Reproduction Number Equation (3) is determine the basic of reproduction number:

$$\begin{aligned}\frac{de}{dt} &= (\pi+\beta eN)s-\left(1+\frac{\lambda}{N}-\delta_1 i-\delta_2 i_T\right)e\\ \frac{di}{dt} &= \alpha\gamma e-\left(\delta_1+\frac{\lambda}{N}-\delta_1 i-\delta_2 i_T\right)i \\ \frac{di_T}{dt} &= (1-\alpha\gamma)e-\left(\delta_2+\frac{\lambda}{N}-\delta_1 i-\delta_2 i_T\right)i_T\end{aligned} \quad (3)$$

Basic reproduction number gained by following equation (4)

$$R_0 = \frac{\beta\lambda^{\ 2}}{(\lambda+N)(\lambda+\pi N)} \quad (4)$$

3.4 Population shift rate from susceptible to exposed by infective contacts inter-population simulation
Population shift rate from susceptible to exposed by infective contacts inter-population simulation ($\boldsymbol{\beta}$) is done to show the dynamics of disease in its population. Besides, it will show the trend of diminishing value due to the shift of susceptible to exposed by infective contact that lower the score of basic reproduction number R_0. There are four values ($\boldsymbol{\beta}$) taken from interval [0.0903: 0.0003] with this step $\mathbf{3\times 10^{-2}}$, while other parameter fixed as shown in table 2.

Table 2. Parameter value for model $R_0 > 1$.

Parameter	Value	Sources
λ	2.000	[5]
μ	0.132	-
δ_1	0.139	-
δ_2	0.134	-
β	0.900	-
α	0.004	[5]
γ	1.000	[5]

R_0 value at the population shift rate susceptible to exposed by genetics factor becomes lower as shown in table 3.

Table 3. R_0 value with parameter (β).

β	Value of R_0
0.9030	0.820
0.0603	0.550
0.0303	0.280
0.0003	0.003

Based on table 3, it can be seen that value of $R_0 < 1$. The diminishing of parameter β causes R_0 lower, so it helps to suppress or to control the disease speed rate in the population. Based on these parameter values, the result of population shifts rate simulation from susceptible to exposed by infective contact as shown in figure 2a-2d.

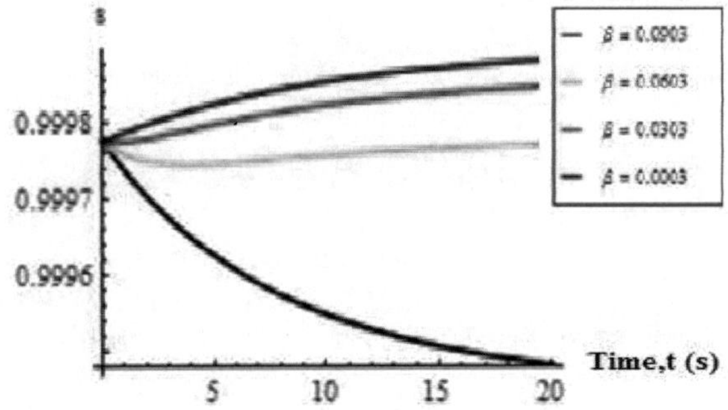

a) Susceptible

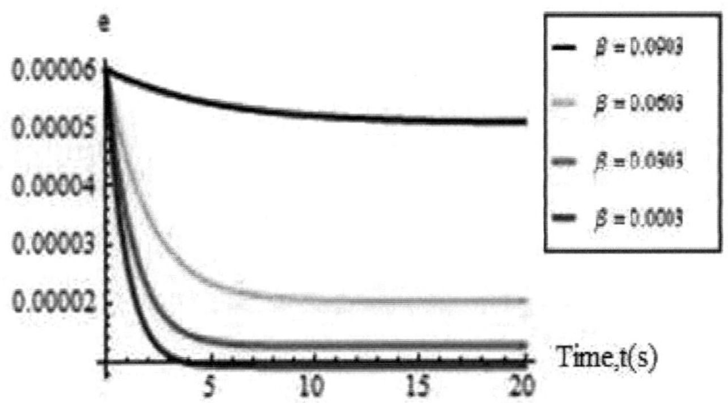

b) Exposed

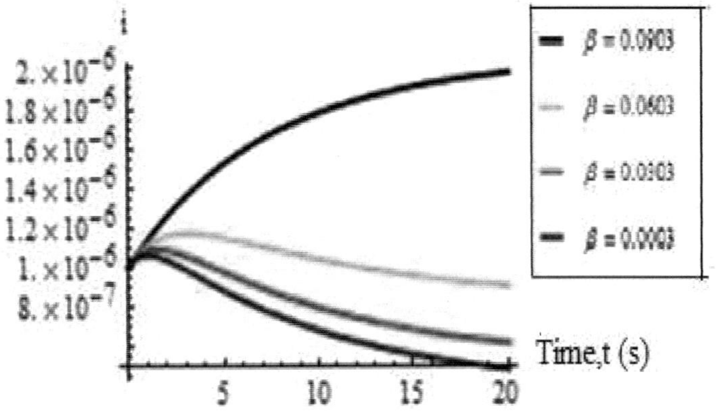

c) Infected population without treatment

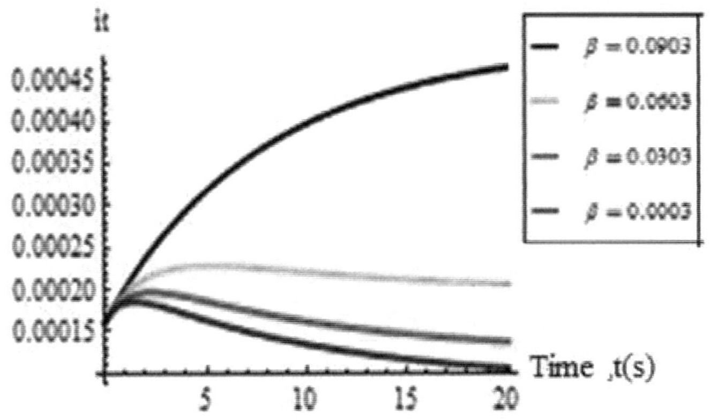

d) Infected population with treatment

Figure 2a-2d. Human Population with different coefficient

Figure 2a-2d shows that the shift of human population when parameter value for infective contact (β) is differentiated while other parameter is fixed. Susceptible population with lesser β increase from initial start and at the end of the simulation has the largest population. Exposed population with lesser β lower than at the beginning and at the end of simulation has the lesser population. Infected population without treatment with lesserr β lower at the first place and at the end of simulation has the lesser population. Infected population with treatment with lesser β lower at the beginning and at the end of simulation has the lesser population.

Based on the simulation, it can be concluded that by decreasing infective contact or lowering the communication between susceptible individual and exposed one trend to increase susceptible population and lowering the rest groups. This infective contact rate served as controlling parameter in decreasing the number of disease in population.

4. Conclusion

Based on the analysis and discussions section, as what has be done to mathematical modeling type SEII$_T$, the writers are able to draw some conclusions;

i. Susceptible population with lesser β went up at initial stage and at the end of simulation has the largest population.
ii. Exposed population with lesser β diminished at first and at the end of simulation has smaller population.
iii. Infected population without treatment with lesser β lowered at the beginning and at the end of simulation has lesser population.
iv. Infected population with treatment with lesser β went down from start and at the end of simulation has smaller population.

References

[1]. Ulfah J., Kharis M., Chotim M. 2014. *Model Matematika Untuk Penyakit Diabetes Mellitus Tanpa Faktor Genetik Dengan Perawatan*. Unnes Journal of Mathematics 3(1).

[2]. Abraham dan San. R. 2015. *Analisis Model Matematika Model Penyebaran Penyakit Diabetes Dengan Faktor Genetik. SAINS*. **15(1)** 31-37

[3]. Resmawan. 2013. *Model Matematika Tipe SEIRS-SEI Untuk Transmisi Penyakit Malaria. Tesis*. FMIPA Bogor

[4]. Edelstein dan Keshet L. 2005. *Mathematical Models in Biology*. Edisi 7. Random House. New York-USA.

[5]. Jones. 2007. *Note on* $\mathcal{R}_0$. Tesis. Department of Anthropological Sciences Stanford University, California.

IOP Conf. Series: Materials Science and Engineering **506** (2019) 012025 doi:10.1088/1757-899X/506/1/012025

Improvement of Mathematical Problem Solving Ability Through Problem Based Learning Model In Applied Mathematics Course II In Industrial Engineering

Resky Rusnanda[1*] and Safwandi[1]
[1]Polytechnic of South Aceh, Merdeka Street, Komplek Reklamasi Pantai,Tapaktuan city, 23751, Indonesia

[*]Corresponding author: reskyrusnanda@gmail.com

Abstract. Problem based learning model is applicable in Applied Mathematics II course in order to obtain better students' problem-solving skill in mathematic. In this research, one can learn the differences of student performance to solve problem by problem-based learning compared to traditional learning. Besides, observation has been done to student activities during class session. Data analysis shows significant of $\alpha = 0.05$ gained $t_{count} = 2.16$ and $t_{Table} = 1.72$ in other hand $t_{count} > t_{table}$, one can conclude that problem-based learning increase student performance in solving problem compared to those in control group in Applied Mathematics II course, Study Program of Industrial Engineering Polytechnic of South Aceh. The student activity was rate at 69.64% average.

Keyword *: problem solving ability, problem based learning, applied mathematics*

1. Introduction

The problem is the gap between desire or hope and reality. According to [1] a problem is the state of cause or event that must be altered in certain ways in order to gain desired result. In problem-based learning, small group of students are created in order to share ideas while discussing theme, plotting hypothesis, designing tests, conducting experiments, collecting data, interpreting data and make a conclusion. Then, students could prepare presentation, present their findings, have small discussion and finally submit their reports.

According to [2] problem-based learning is a learning where students work on authentic problems in order to gather their own knowledge, develop inquiry, improve learning skills, develop independence and self-confidence. As the interest grew, a research on the application of problem-based learning in Applied Mathematics II course aiming the better students' performance in solving math problem in Industrial Engineering Study Program of Polytechnic of South Aceh has being conducted.

The objective of this research is to emulate the success of application problem-based learning model in Applied Mathematics II course and to know the average percentage of student activity during the learning process by using problem-based learning model in the Applied Mathematics II course in Industrial Engineering Study Program in Polytechnic of South Aceh. Besides, writers are able to comprehend student activities during the class session.

2. Methods

2.1 Types and research Approach

This is a quantitative research by experiment approach. Quantitative research can be seen in the use of numbers at the time of data collection, interpretation of data and the results [3].

IOP Conf. Series: Materials Science and Engineering **506** (2019) 012025 doi:10.1088/1757-899X/506/1/012025

2.2 Research Design

Research design that engaged in this study; random, pre-test, post-test. Traditionally, random, pre-test, post-test design models were shown below in figure 1.

E O_1 X O_2

R

K O_3 O_4

Figure Error! No text of specified style in document.**1**. Design models for the research

where the E is the experiment group, K is the control group, R is the random, O_1 is the initial stage of experiment group, O_2 is the final stage of experiment group, O_3 is the initial stage of control group and O_4 is the final stage of control group.

2.3 Population and Sample

Population is the number of all registered students in Industrial Engineering Study Program of Polytechnic of South Aceh. From This population, taken as sample as much as two group of students in the second terms of their first year namely group A and group B.

2.4 Data Collection

Data in this study were collected using test instruments both zero treatment (pre-test) and after treatment (post-test). In addition, data collection techniques used are observation sheets of student learning activities.

2.5 Data Analysis

Comparing the pre-test and post-test scores to look for gains that occur after learning in each group calculated by the normalized gain formula [4], namely:

$$g = \frac{S_{post} - S_{pre}}{S_{max} - S_{pre}} \quad (1)$$

where the g is the normalized gain, S_{post} the post-test score, S_{pre} is the pre-test score, and the S_{max} is the maximum score.

The data to be analyzed is the data from the pre-test and post-test results in the form of test data on problem solving abilities. Data of the test results were analyzed using the t-test.

$$t = \frac{\bar{x}_1 - \bar{x}_2}{S\sqrt{\frac{1}{n_1} + \frac{1}{n_2}}} \quad (2)$$

with combined variance ($S^2_{combined}$), as [4] can be measured as:

$$S^2{}_{gab} = \frac{(n_1-1)S_1{}^2 + (n_2-1)S_2{}^2}{n_1+n_2-2} \quad (3)$$

Hypothesis testing uses right-sided t-test at a significant level $\alpha = 0.05$. The testing average $\mu_1$and μ_2, correlative hypothesis H_0 and its counter part H_1:

H_0: $\mu_1 = \mu_2$ There is no difference in the improvement of students mathematical problem solving abilities between using problem-based learning models and increasing students' mathematical problem solving abilities using conventional learning.

H_1: $\mu_1 > \mu_2$ Improving students 'mathematical problem solving skills using problem-based learning models is better than improving students' mathematical problem solving abilities using traditional learning.

Because the test used is the right-hand test, the applicable testing criteria rejects H_0 if $t > t_{1-\alpha(n_1+n_2-2)}$ and accepts H_0 if t has other prices, with degrees of freedom $dk = n_1 + n_2 - 2$ and chance $(1 - \alpha)$. If the data is not normally distributed, the hypothesis will be tested by non-parametric statistical tests. The test used is the Mann-Whitney test. According to [5] the Mann Whitney test is used to examine the average differences of two groups of samples mutually independent if one or both groups of samples are not normally distributed.

3. Result and Discussion

3.1 Pre-test and Post-Test

Table 1 and 2 below will illustrates the score on mathematical problem-solving abilities pre-test and post-test in the experiment class (A) and the control class (B).

Table 1. Experiment Class (A) Score on mathematical problem-solving abilities pre-test and post-test

Students	Pre-Test	Post-Test
Student 1	55	95
Student 2	60	75
Student 3	52	90
Student 4	65	85
Student 5	75	80
Student 6	45	75
Student 7	70	100
Student 8	55	85
Student 9	60	70
Student 10	50	80
Student 11	65	90
Student 12	52	80
Student 13	55	70

Table 2. Control Class (B) Score on mathematical problem-solving abilities pre-test and post-test

Students	Pre-Test	Post-Test
Student 1	50	75
Student 2	70	80
Student 3	52	70
Student 4	50	60
Student 5	45	60
Student 6	60	80
Student 7	55	68
Student 8	50	65
Student 9	62	70
Student 10	45	90

3.2 Result Data Analysis

3.2.1 Variance and Standard Deviation

Variance (s_1^2) and standard deviation (s_1) from pre-test in experiment class in order are 72.423 and 8.51. Variance (s_2^2) and standard deviation (s_2) from control group are 99.266 and 9.96 in sequence. Based on data analysis carried out by using the Lilliefors Test for both classes, it was found that the distribution of the experimental class and the test class initial test data was normally distributed. In addition, using the variance homogeneity test obtained that both classes are homogeneous for data pre-test.

3.2.2 N-Gain from pre-test and post-test

Table 3. N-Gain from pre-test and post-test

Students in Experiment Group	*N-Gain* Experiment Group	Students in Control Group	*N-Gain* Control Group
Student 1	0.89	Student 1	0.50
Student 2	0.38	Student 2	0.33
Student 3	0.79	Student 3	0.38
Student 4	0.57	Student 4	0.20
Student 5	0.20	Student 5	0.27
Student 6	0.55	Student 6	0.50
Student 7	1.00	Student 7	0.29
Student 8	0.67	Student 8	0.30
Student 9	0.25	Student 9	0.21
Student 10	0.60	Student 10	0.82
Student 11	0.71		
Student 12	0.58		
Student 13	0.33		
Sum	7.53		3.80
Mean	0.58		0.38

From table 3, the average value($\bar{x}$) N-Gain the score of mathematical problem-solving ability for each class is obtained by the average N-Gain of 0.58 experimental class into the "medium" category, while the control class 0.38 is also in the "medium" category. Then, calculated the variance value and standard deviation of N-Gain Score of Mathematical problem-solving performance. The calculation results obtained by the experimental class variance value = 0.056 and the standard deviation value of the experimental class = 0.24. Control class variance value = 0.035 and control class standard deviation value = 0.19.

3.2.3 N-Gain Data Distribution Normality Test
Based on the N-Gain normality test scores of mathematical problem solving abilities using the Lilliefors Test, the N-Gain score data for both classes were normally distributed. In addition, the N-Gain data variance for both variance classes is homogeneous.

3.2.4 Hypothesis Test of N-Gain Data
The mean value and N-Gain variance obtained are:

$$\bar{x}_1 = 0.58,\ S_1^2 = 0.056,\ S_1 = 0.24,\ n_1 = 13$$

$$\bar{x}_2 = 0.38,\ S_2^2 = 0.035,\ S_2 = 0.19,\ n_2 = 10$$

Combining equation (2) gained combined variances (s^2), std dev combined (s) consecutively 0.047 and 0.22. Then using equation (1) obtained t value of 2.16. At a significant level $\alpha = 0.05$ and degrees of freedom $dk = n_1 + n_2 - 2 = 21$,so $t_{table} = t_{(1-\alpha)(n_1+n_2-2)} = t_{(0.95)(21)} = 1.72$. The testing criteria is to reject Ho if $t_{count} > t_{Table}$, for other t-value, Ho is accepted. Based on the results of the above calculations obtained $t_{count} = 2.16$ and $t_{Table} = 1.72$ or $t_{count} > t_{Table}$. Thus, Ho is rejected at a significant level of $\alpha = 0.05$. It can be concluded that the improvement of students mathematical problem-solving skills taught with problem-based learning models is better than students taught with traditional learning in Applied Mathematics II courses in the Industrial Engineering Study Program Polytechnic of South Aceh.

3.3 Students Activities
The average student activity during the first meeting until the third meeting was 69.64%, so it can be concluded that student activities during the learning process use problem-based learning models in

IOP Conf. Series: Materials Science and Engineering **506** (2019) 012025 doi:10.1088/1757-899X/506/1/012025

applied Mathematics II courses in Industrial Engineering Study Program Polytechnic of South Aceh in the active category. This data can be seen in Table 4 below.

Table 4. Data on Average Student Activity

Meeting	1	2	3	Mean
Percentage	66.07%	69.64%	73.21%	69.64%
Criteria	Active	Active	More Active	Active

4. Conclusion

This work can be conclude by the improvement of students mathematical problem solving skills taught with problem based learning models which is better than students taught with conventional or traditional learning in applied Mathematics II courses and other than that, the average percentage of student activity during the learning process by using problem-based learning models in applied Mathematics II courses in the Industrial Engineering Study Program Polytechnic of South Aceh experienced an increase with the active category.

References

[1] Yusna Decy Pramita Sari 2015. Penerapan Model Problem Based Learning (PBL) dalam Materi Relasi dan Fungsi bagi Siswa Kelas X MAN Model Banda Aceh. Skripsi: Unsyiah.

[2] Hosnan M. 2014 Pendekatan Saintifik dan Kontekstual Dalam Pembelajaran Abad 21. Jakarta: Ghalia Indonesia.

[3] Arikunto Suharsimi 2006. Prosedur Penelitian Suatu Pendekatan Praktik. Jakarta: Rineka Cipta

[4] Syahputra Edi 2012 Pengaruh Penggunaan Model Pembelajaran Archored Instruction terhadap Peningkatan Kemampuan Komunikasi Matematis dan Self-Concept siswa. Tesis Bandung: Universitas Pendidikan Indonesia.

[5] Sundayana Rostina 2010 Statistika Penelitian Pendidikan Garut: *STKIP Garut Press*

The Analysis of Willingness to Pay for Operations and Financing Road Maintenance at Tengku Hasan Street Aceh Besar District

A Fatimah [1*], Jurisman[1], W Soviana[1]
[1]Civil Engineering Department, Faculty of Engineering, Universitas Muhammadiyah Aceh, 23245 Banda Aceh, Indonesia

[*]Corresponding author: aldinafatimahcivil@gmail.com

Abstract. Infrastructure components are generally divided into two major parts, namely cost-back and no-cost return. Infrastructure financing is not returned by involving the community financially by using a system that is often referred as connection fees are charges imposed by the company or government services to individuals. This study was carried out to investigate the willingness to pay (WTP) by civilian for operations and financing road maintenance in Tengku Hasan Street, Aceh Besar. The study was conducted by distributing questionnaires to 140 respondents who are villagers Bakoy and employees of Tirta Mountala Company. The value of the average WTP community is IDR 1.800 per month. With the amount of WTP values of IDR 2.4642 million per month, or approximately IDR 29.5704 million per year. The values indicate that a very small amount of people is willing to pay for the operations and maintenances in Tengku Hasan street but it can be concluded that there is awareness of the public to get a decent quality infrastructure despite having to pay a fee. The majority of respondents chose to agree to be involved financially in the operation and maintenance of Tengku Hasan street Bakoy village, Aceh Besar district with citations per year, as many as 41 people or 56.9%.

Keyword : *infrastructure components, willingness to pay, descriptive analysis*

1. Introduction

Infrastructure can be defined as a physical element (urban) to support community activities systems. The main functions of the infrastructure are to serve the needs of the community in the activities of both economic and social [1]. Generally, Infrastructure financing components are divided into two major parts, namely cost and the cost is not returned back.

Financing infrastructure is not returned by involving the community financially by using a system that is often referred to as connection fees are fees charged by the company or government services to individuals. Associated with this system, the researcher conducted research about society participation toward operations financing and maintenances of roads on Tengku Hasan street Bakoy village which is a local road in the district of Aceh Besar.

Infrastructure of transportation that include all parts of the road, including complementary buildings and equipment intended for traffic, which is at ground level, above ground, below ground or water, as well as above the water surface, except railwaysand lorries roads. The road has so many roles not only as a transportation strip, serves as a support in various mayors such as social, economic, cultural, and environmental as well as supporting the growth and distribution region [2].

In general, the role of the road be seen in two main dimensionsof the road as providers of acces to land parcels (micro scale) and road as service providers the mivement of people and goods. In a broader perspektive the road is able to deliver a significant contribution in encouraging the growth of the region both physically as well as economically [3].

Limited sources of financing to support the operation and maintenance of roads is a classic problem faced by almost all developing countries. This is reflected in the amount of funds allocated for road maintenance needs of only about less than 20% of the funding requirements for operations and maintenance of roads that in fairly good condition. This matter arises because of the tendency to spend the budget for the construction and repair [4].

There are several factors that affect people's participation in development. These factors can be grouped into two categories, internal factors and external factors. Internal factors are factors of the affected communities in the public participation program. While external factors are factors that come from outside the community [5]. Hence, this study was conducted to determine the how far these factors can build the consciousness in the civilian by investigate the WTP for operations and financing road maintenance in Tengku Hasan Street, Aceh Besar.

2. Method

2.1 Willingness to Pay

The willingness to pay is defined as the amount of money willing to pay by individuals to obtain a good or service [6]. Conception of willingness to pay (WTP) of society closely related to aspect public welfare. [7] that the concept of WTP is closely related to the concept of welfare economy (welfare economics). Welfare Economics uses a cost benefit analysis approach as an attempt to explore whether there is a potential change to usability an item due to changes in economic variables, such as price changes.

According [8] WTP is a method for measuring how much the price of loss resulting from damage to the road but we do not directly know the market price. Several approaches are used in the calculation WTP to calculate raising or deterioration of environmental conditions are:

i. Calculate the costs incurred by an individual who is willing to reduce the negative impact on the environment due to development activity
ii. Calculate the reduction in the value or price of the goods due to the declining quality of the environment.
iii. Through a survey to determine the level of people's willingness to pay in order to reduce negative impacts on the environment better.

2.2 Data Collection

Tengku Hasan Street, Village Bakoy connect between Lambaro and Blang Bintang area with Limpok and Ulee Kareng. Where, ± 3.3 km road length and width ± 5 m. The required data in this study of primary and secondary data. The samples were taken approximately 10% of the total population [1] Total population is 1369 and this study has 140 respondents. Flowchart of experimental design for this study were shown in the Figure 1.

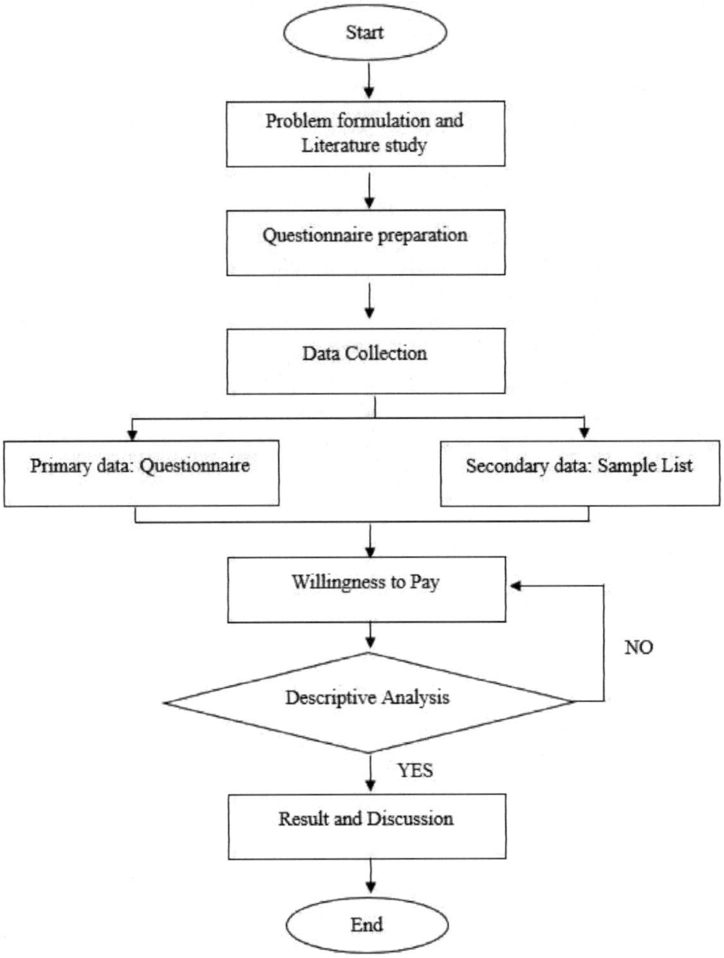

Figure 1. Experimental design flowchart

3. Results and Discussion

3.1 Respondent Characteristic

The characteristic of the gender shows that most of the respondent are male which are 75 respondents (53.6%),and for female are 65 respondents (46.4%). The distribution of the respondents based on the gender were shown in the Figure 2

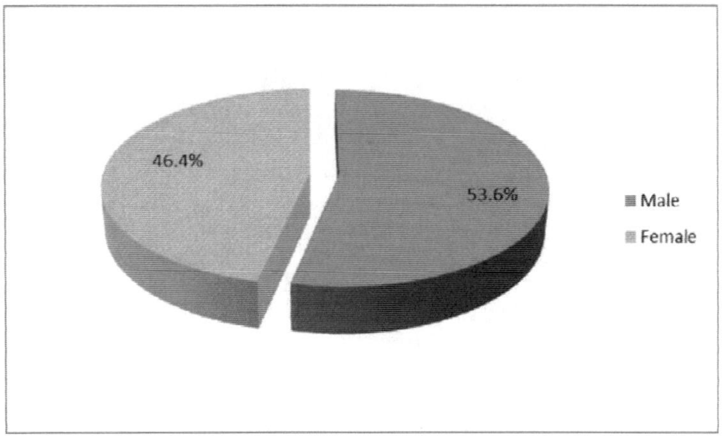

Figure 2. Characteristic of the gender

The characteristic of the age shows that most of the respondents are 30-40 years old which are 55 respondents (39.3%), some are < 30 years old which are 45 respondents (32.1%). The respondent are 41-50 years old which are 30 respondents (21.4%). Only few respondents are > 50 years old which are 10 respondents (7.1%). The distribution of the respondents based on the age can be seen in the Figure 3.

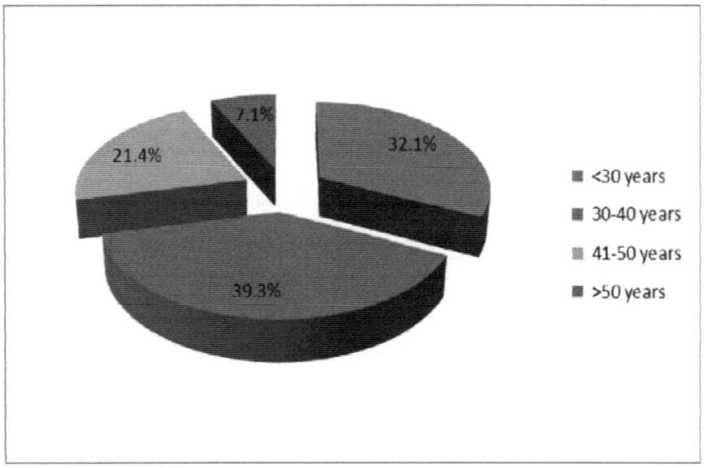

Figure 3. Characteristic of the age

The characteristics of the educational background shows that almost the respondents are bachelor degrees which are 81 respondents (57.9%), some are senior high school which are 44 respondents (31.4%). The respondents are junior high school which are 11 respondents (7.9%). Only few have master degrees which are 3 respondents (2.1%) and the rest ones are doctor degrees which is 1 respondent (0.7%). The distribution of the respondents based on the educational background were shown in the figure 4.

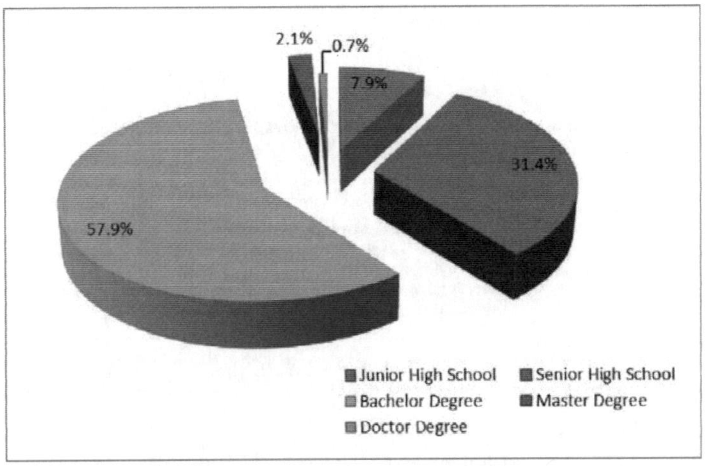

Figure 4. Characteristic of the educational background

The characteristics of the monthly income shows that almost the respondents are IDR 2,100,000 – IDR 3,000,000 which are 38 respondents (27.2%), some are IDR 1,100,000 - IDR 2,000,000 which are 37 respondents (26.4%). The respondent with IDR 3,100,000 - IDR 4,000,000 are 35 respondents (25.0%). Only few have IDR < 1,000,000 which are 16 respondent (11.4%) and the rest ones are IDR > 4,000,000 which are 14 respondents (10.0%). The distribution of the respondents based on the monthly were depicted in the figure 5.

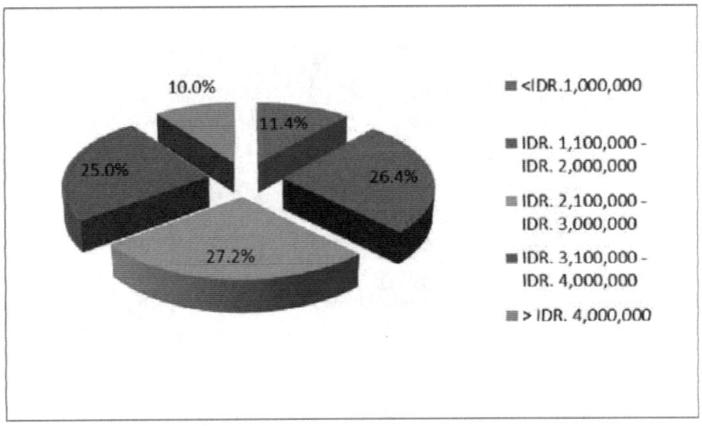

Figure 5. Characteristic of the monthly income

The characteristic of the type vehicle ownership shows that most of the respondent have motorcycle which are 107 respondents (76.4%), some are having a car which are 20 respondents (14.3%). Only few have car and motorcycle which are 11 respondents (7.9%) and the rest have truck which are 2 respondents (1.4%). The distribution of the respondents based on the type vehicle ownership were illustrated in the figure 6 below.

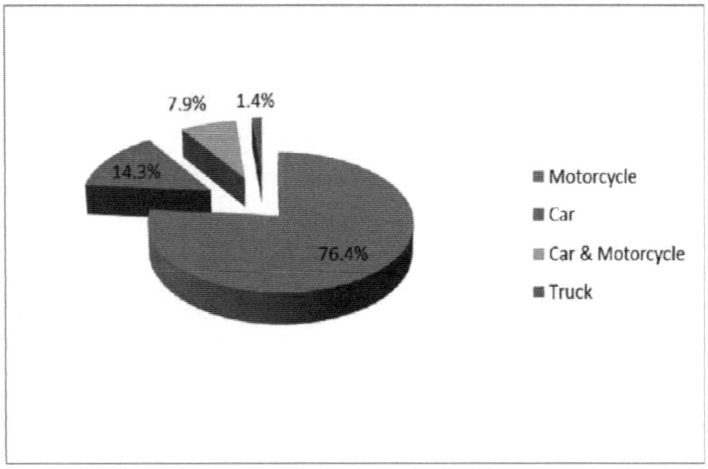

Figure 6. Characteristic of the the type vehicle ownership

3.2 Analysis Willingness to Pay for Operations and Financing Road Maintenance

Based on the results of 140 respondents who responded to questionnaires distributed, it turns out the respondents who expressed a willingness to pay for operations and maintenance funding Tengku Hasan street Bakoy village, Aceh Besar district is as much as 72 people or 51.4%.

In this study, the respondents are divided into 4 groups. Group 1 are the respondents who refuse to be involved financially in the operation and maintenance in Tengku Hasan street which amount up to 68 people or equal to 48.6%. Group 2 are the respondents who are willing to be financially involved in operations and maintenance in Tengku Hasan street paid IDR 1,000 – IDR 2,000 per month is 57 people or 40.7%. While, Group 3 are the respondents who are willing to be financially involved in operations and maintenance in Tengku Hasan street paid IDR 2,100 - IDR 3,000 per month is 11 people or 7.9% and Group 4 are the respondents who are willing to be financially involved in operations and maintenance in Tengku Hasan street paid IDR 3,100 – IDR 4,000 per month is 1 person or 0.7%. And the last group, Group 5 are the respondents who are willing to be financially involved in operations and maintenance Tengku Hasan streeet who paid for > IDR 4,000 per month is 3 people or 2.1%. From all respondents who agreed to be financially involved in the operation and maintenance of the road, the average value of the WTP is IDR 1797.22 or rounded up to IDR 1,800 per month. With the magnitude of the average value of WTP IDR. 1,800 monthly x 1369 (number of local community) = IDR 2.4642 million monthly, or approximately IDR 29.5704 million per year. This value is very small to fund operations and maintenance Tengku Hasan street but it can be concluded that there is awareness of the public to get a decent quality infrastructure despite having to pay a fee. Distribution WTP on user respondents Tengku Hasan street can be seen in Table 1.

Table 1. Percentage of willingness to pay on the User Respondents Tengku Hasan Street.

No	Group	Number of Respondents (People)	%
1	Group 1	68	48.6
2	Group 2	57	40.7
3	Group 3	11	7.9
4	Group 4	1	0.7
5	Group 5	3	2.1
Total		140	100

In this section will describes about the response of respondents to the operations and maintenance funding mechanism Tengku Hasan street Bakoy village, Aceh Besar district. Based on the table above, the majority of respondents chose to agree to be involved financially in the operation and maintenance of Tengku Hasan street Bakoy village, Aceh Besar district with citations per year, as many as 41 people or 56.9%. The distribution of the operations and maintenance funding mechanism Tengku Hasan street Bakoy village, Aceh Besar district can be referred in the Figure 7.

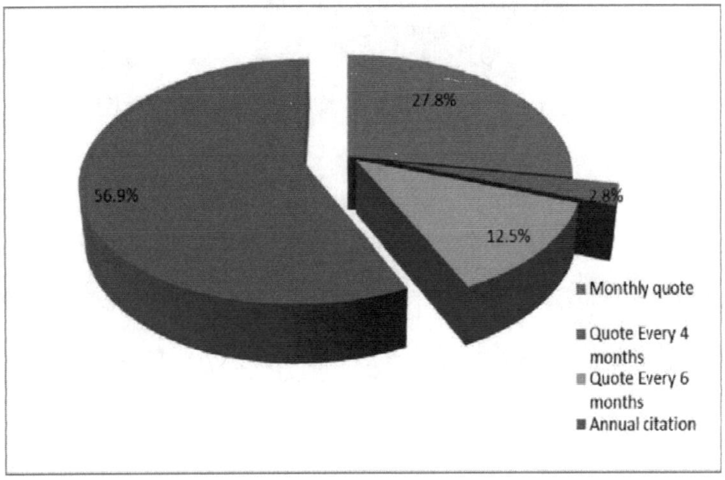

Figure 7. Method of Payment

4. Conclusion

Based on the results of 140 respondents who responded to questionnaires distributed, it has been revealed that the respondents who expressed a willingness to pay for operations and maintenance funding in Tengku Hasan Street Bakoy village, Aceh Besar district is as much as 72 people or 51.4% of respondents and those who are not willing as many as 68 people or 48.6%. The average value of WTP is IDR 1793.22 or rounded IDR 1800 monthly. With the amount of WTP values of IDR 2.4642 million per month, or approximately IDR 29.5704 million per year, this value was considered to be very small to pay the operations and maintenances in Tengku Hasan street but it can be concluded that there is still some awareness from the people to get a decent quality infrastructure despite having to pay a fee. The majority of respondents chose to agree to be involved financially in the operation and maintenance of Tengku Hasan street Bakoy village, Aceh Besar district with citations per year, as many as 41 people or 56.9%.

Acknowledgements

The author would like to thank to Dr. Hj. Hafnidar A.Rani, ST, MM (Dean of Engineering Faculty,Universitas Muhammadiyah Aceh) and Universitas Muhammadiyah Aceh for gives a good support and providing an excellent facilities in order to accomplish this work.

References

[1] Arikunto S 2006 Prosedur Penelitian (Suatu Pendekatan Praktik) *Rineka Cipta* Jakarta.

[2] Pemerintah Republik Indonesia 2002 Undang – Undang Republik Indonesia Nomor 29 *Tentang Jalan,* Jakarta.

[3] Kristianto A W 2008 Peningkatan Partisipasi Masyarakat dalam Pembangunan Infrastruktur Jalan (Studi Kasus Pelaksanaan Program Pembangunan Infrastruktur Pedesaan (PPIP) Di Desa Campurejo Kecamatan Pancang Kabupaten Gresik) *Jurnal Teknik Sipil* Institut Teknologi Sepuluh Nopember

[4] Widjonarko 2007 Menuju Pembiayaan Prasarana Kota Berbiaya Tak Kembali (Studi Kasus Jalan Lokal Di Semarang), Tesis Mahasiswa Jurusan Teknik Sipil, Universitas Diponegoro, Semarang.
[5] Levinson David 2005 Paying The Fixed Cost Of Roads, Manuscript, Department of Civil Engineering University of Minnesota.
[6] Wechel Tamara V and Kimberly Vachal 2004 Investment in Rural Roads: Willingness to Pay for Improved Gravel Road Service in Freight Transportation, MPC Report No. 04168, Mountain-Plains Consortium, North Dakota State University, Fargo, ND.
[7] Carson Richard T Nicholas E Flores and Norman F. Meade 2000 Contingent Valuation: Controversies and Evidence JEL: Q26 D61
[8] Mandasari, L dan Ciptomulyono U 2014, Valuasi Ekonomi Kali Surabaya Dengan Pendekatan Willingness to Pay Dan Fuzzy MCDM, *Jurnal Teknik Sipil* Institut Teknologi Sepuluh Nopember

Determination of the Best Location for Freshwater Fish Ponds in South Aceh Districts by Using Multi Criteria Analysis

Meraty Ramadhini[1*] and Arnold Rikardo Sihombing[1]

[1]Industrial Engineering, Politeknik Aceh Selatan, Aceh, 23711, Indonesia

*Corresponding author: meraty.ramadhini@yahoo.com

Abstract. South Aceh is one of the areas in the development of freshwater cultivation. The purpose of this study was to determine the values and factors for freshwater fish ponds using multi criteria analysis using the Analytical Hierarchy Process (AHP) method in South Aceh District. Criteria consist of ecological and socio-economic, assessment is carried out on the level of scoring and weighting factors of each parameter. The results of the analysis of ecological answer factor indicate that slope factors, soil types and rainfall must be anticipated in choosing a location for freshwater cultivation, because these three factors get the same weight and highest of the other factor is 0.2900. The results of the factor analysis of socio-economic criteria indicate that the distance factor from the road gets the highest weight of 0.4100, and this needs attention because it relates to economic principles, such as the distance of the marketing location to the location of cultivation.

Keywords: Multi criteria analysis, Freshwater cultivation, Soil type, Slope factor

1. Introduction

The sea waters of South Aceh are part of the Indian Ocean with a coastline reaching 140 km which has very large and abundant fish resource potential. In addition, the waters of South Aceh also have good potential for fish farming, the potential area for aquaculture in South Aceh Regency is also quite large both in fish cultivation in ponds, public waters and in fish farming in the Minapadi system. Overall the potential of freshwater fish farming is estimated to be 20,156 Ha. However, the utilization has reached 48.67 Ha with a production of 265.97 tons / year [1].

The potential for developing brackish water cultivation (ponds) reaches 4,000 Ha in 9 (nine) districts, namely Trumon District, Bakongan, North Kluet, South Kluet, Pasie Raja, East Labuhan Haji, Labuhan Haji, and Samadua. However, the new community has cultivated 17.50 Ha of the potential of the existing land with the level of application of simple technology with a production of only 5.75 tons, this is a huge potential to advance fisheries and marine affairs.

One of the factors to achieve a successful pond culture business, in addition to the investment costs, quality, and specific character of cultivated biota, operator discipline, cultivation methods with applied technologies such as design, layout and construction, as well as production levels, must also be consider the biophysical characteristics of locations such as biology, hydrology, meteorology, quality of soil, and water that are in accordance with the environmental carrying capacity of the region [2].

One method for selecting the location of the cultivation of freshwater fish ponds is through multi-criteria analysis. The purpose of this study was to determine the weight and location factors for freshwater fish ponds in South Aceh Regency. The results of this study are expected to be the basis for further planning in fisheries development in South Aceh Regency.

2. Material

This research was carried out in the district of Southern Aceh, Aceh Province. South Aceh Regency has an area of 384200 ha with 18 districts. The data used in this study is primary data, primary data was obtained through interviews with five respondents and filling out questionnaires for three farmers per district.

Analytical Hierarchy Process (AHP) is applied to set goals and criteria for this research. This research used two criteria, namely ecological criteria and socio-economic criteria. Ecological criteria consists of four factors; slope, soil texture, rainfall and distance from the river while socio-economic criteria consists of three factors, namely the distance from road, distance from settlements and current land use.

Site selection is a very important step in determining the success of aquaculture business since its ecological and socio-economic should be taken into considerations. Technically, if the location of the fish pond is good it will influences the construction of the fish pond to be built and the operational costs of its maintenance. In this case, the calculation of important values in the AHP analysis will produce a weighted value for each criterion, where the weight value produced is then used as a priority sequence that affects the final focus set in the form of potential fisheries for freshwater fish ponds. Table 1 shows the suitable criteria for ponds.

Table 1. Land suitability criteria for ponds.

Criteria	Factor	Classification of Land 1	2	3	4
Ecology	Slope (%)	0-3	>3-8	>3-8	>8
	Type of soil	loamy/sandyclay		Sandy clay	Sand
	Rainfall (mm thn^{-1})	<1,000	>1,000-2,000	>2,000-2,500	>2,500
	Distance from river (m)	<500	>500-1,000	>1,000-1,500	>1,500
Socio-	Distance from road (m)	<500	>500-1,000	>1,000-1,500	>1,500
	Distance from the settlement	<400	>400-800	>800-1,200	>1,200
	Land Use	Pond	Field	Garden	Shrubs

Source : Cahyaningrum, dkk (2014)

2.1. Analytical Hierarchy Process (AHP)

AHP is a decision support model developed by Thomas L. Saaty. This decision support model will describe complex multi-factor or multi-criteria problems into a hierarchy, according to [3], hierarchy is defined as a representation of a complex problem in a multi-level structure where the first level is the goal, followed by factor level, criteria, sub criteria, and so on down to the last level of alternatives.

According to previous study by Saaty, for various information, a scale of 1 to 9 is the best scale in interviews [4]. Each paired comparison is evaluated in the author scale 1-9 as follows;

	Most Important				Neutral				Most Important	
Element A	9	7	5	3	1	3	5	7	9	Element B

The value and definition of qualitative opinions from the Saaty comparison scale can be seen in the following table 2.

Table 2. Pairwise comparative assessment scale.

Interest Intensity	Information
1	Both elements are equally important (equal Importance)
3	One element is slightly more important than the other (Slightly more Importance)
5	One element is more important than the other (Materially more Importance)
7	One element is clearly more important than other elements (Significantly more Importance)
9	One element is absolutely important than other elements (Compromise values)
2,4,6,8	Values between two values of consideration are close together (Compromise values)

Source: Saaty, T. Lorie. 1990

2.2. Test Index Consistency and Ratio

Determination of the consistency of the matrix is based on the maximum eigen value, obtained by the following equation (1):

$$\text{Consistency Index (CI)} = \text{CI} = \frac{\lambda\,\text{max} - \text{n}}{\text{n} - 1} \qquad (1)$$

Where CI is deviation rasio consistency index, λmax is eigen value maximum from an ordered matrix n, n is matrix order n.

If the CI value is zero, then the matrix pairwise comparison that consistent. Limitation of consistency (inconsistency) determined by Thomas L. Saaty determined using the Consistency Ratio (CR), namely comparison consistency index with Random Index (RI) value. Consistency Ratio can be formulated in equation (2) as follows:

$$\text{Consistency Ratio (CR)} = \frac{\text{CI}}{\text{RI}} \qquad (2)$$

where RI is a Random Consistency Index, If the Consistency Ratio (CR) ≤ 0.1, then the results of the data assessment can be justified / consistent and if not then the assessment needs to be repeated. The results of data calculations can be justified/consistent. List of RI can be seen in table 3.

Table 3. Random index.

n	1	2	3	4	5	6	7	8	9	10
RI	0.00	0.00	0.58	0.90	1.12	1.24	1.32	1.41	1.45	1.48

Source: Saaty, T. Lorie. 1993

3. Results and Discussion

3.1. Weghting Criteria with AHP Method

Criteria weighting is done to determine the highest value and the lowest value of ecological criteria and socio-economic criteria from several factors in increasing productivity and quality of freshwater fish ponds. The results of the weighting criteria analysis obtained can be seen in table 4 below.

Table 4. Results of Criteria Weighting Analysis.

Ecological Criteria (0.39996)	Factor	Weight	Socio-economic Criteria (0.29900)	Factor	Weight
	Slope	0.2900		Land use	0.3300
	Type of soil	0.2900		Distance from road	0.4100
	Rainfall	0.2900		Distance from the settlement	0.2600
	Distance from the river	0.1400			
	Amount	1.0000		Amount	1.0000

Analysis of ecological criteria indicates that slope, soil type and rainfall have the same and highest weight of 0.2900. Slope is the height of the place / location of the pool to the surface of the sea, the location of the pond is good if the location is located between the highest tide and the lowest tide. Soil acts to hold water during the fish cultivation process, this type of soil greatly determines the success factor of freshwater cultivation, soil types that are good for freshwater cultivation are types of clay. Rainfall that tends to be high is a problem for farmers, rainwater has a level of acidity that can harm fish cultivators, the cultivator has not taken anticipatory measures to free the pond from the acidity of rainwater.

Socio-economic factor analysis shows that the distance factor from the road currently obtains the highest weight of 0.4100. This is related to economic principles, such as far from the marketing location with the location of fish farming.

3.2. Consistency Test Result

What is measured in AHP is Consistency Ratio by looking at the Consistency Index. The expected consistency is near perfect to produce decisions that are close to valid. Although it is difficult to achieve the perfect, the expected consistency ratio is less than or equal to 10%.

Table 5. Test consistency.

Criteria	λmax Value	CI Value	CR Value
Ecological	4.0009	0.0003	0.00033
Socio-economic	3.0537	0.0270	0.0465

Based on table 5. It can be concluded that Consistency Ratio (CR) ≤ 0.1, then the results of data calculation can be justified/consistent.

4. Conclusion

Based on the data shown above, it can be concluded that:

i. Ecological criteria have a greater weight than socio-economic criteria, this can improve the quality of freshwater fish ponds better, because it is related to the distance of water sources. The farther away with the water source, the more the cost of procuring water for cultivation.

ii. Factor analysis including ecological criteria shows that slope, soil type and rainfall have the highest weight, this indicates slope related to the height of the place / location of the pool to the surface of the sea, the type of soil that cannot store water and high rainfall.

iii. Analysis of factors including socio-economic criteria shows that the distance from the road has the highest weight, this factor can increase productivity and quality of fishery products, because it is associated with easier / affordable accessibility.

iv. Ecological criteria can be set as the top priority for ponds development, because in this criterion there are still many inhibiting factors in the coastal areas of South Aceh Regency to be used as ponds land, related to slope, type of soil and rainfall.

References

[1] Rahmat. 2015. Media Publikasi dan Informasi Pemerintah Kabupaten Aceh Selatan.

[2] Radiarta, I.N, A. Saputra, O. Johan. 2005. Pemetaan Kelayakan Lahan untuk Usaha Budidaya Laut dengan Aplikasi Inderaja dan Sistem Informasi Geografi di Perairan Lemito, Provinsi Gorontalo. *Jurnal Penelitian Perikanan Indonesia*, 11(1):1-14.

[3] Saaty, T. Lorie. 1993 Pengambilan Keputusan Bagi Para Pemimpin, Proses Hirarki Analitik untuk Pengambilan Keputusan dalam Situasi yang Kompleks. Pustaka Binama Pressindo.

[4] Saaty, T. L. 1990. The Analytical Hierarchy Process:Planning, Priority Setting, Resource Allocation. Pittsburgh University Pers. P. 97.

IOP Conf. Series: Materials Science and Engineering **506** (2019) 012028 doi:10.1088/1757-899X/506/1/012028

Effect of Moderate Physical Activity to Muscle Fatigue on Untrained People

T D Santi
Public Health Faculty, University Muhammadiyah Aceh, Banda Aceh

Corresponding author: taharads@yahoo.co.id

Abstract. Muscle fatigue is a condition of discomfort, decreased efficiency due to excessive activity, loses energy / stamina, a common non-specific symptom experienced by many people and is associated with many health conditions which leading to achievement and performance to decrease. This study aimed to determine the effect of moderate physical activity (AFS) to muscle fatigue on untrained people. This investigation compared levels of lactic acid and the Borg scale before and after AFS. Fifteen man ages from 15 to 20 years volunteered to participate in the study. The subjects of the study were given moderate physical activity in the form of stepping up and down test until they reach the targeted maximum heart rate which is 70-79% for 15-20 minutes, 3 times a week for 4 weeks. Muscle fatigue was obtained by looking at the levels of lactic acid and the Borg scale before and after AFS. The results show the decrement after the AFS for the both lactic acid and the Borg scale with the mean levels of 9.02 ± 2.55 before AFS and 6.00 ± 2.76 after the AFS for the lactic acid and 3.70 ± 0.94 before AFS and 1.80 ± 1.0 after AFS for the the Borg scale tightness. It was found that AFS on untrained people can reduce muscle fatigue and the physical ability can be improved by exercise with appropriate intensity, duration and frequency.

Keyword : *muscle fatigue, lactic acid, borg scale, moderate physical activity*

1. Introduction

Physical activity is the movement of the limbs that cause the expenditure of energy (activity). Any movement of the body produced by skeletal muscle requires energy expenditure. Less physical activity is an independent risk factor for chronic disease, and overall is thought to cause death globally. Data of Basic Health Research in 2010 revealed that the prevalence of Indonesian population lack of physical activity was 72.9% and the residents that do not do the physical activity regularly was 42.8% [1-2].

Moderate physical activity is being classify with stepping up and down test at least 15-20 minutes for 3 times a week during 4 weeks. Several factors that influence physical activity are age, gender, calorie intake and nutrient. The benefits of physical activity when done in a healthy state regularly, with mild to moderate intensity will improve health and fitness. The data for the study were obtained through levels of lactic acid and the_Borg scale. A physical activity performed in accordance with its basic principles can improve physical quality. Improvement of biological quality parameters as a result of proper physical activity, among others, accumulation of lactic acid is reduced so as to prevent fatigue. [3-5].

Lactic acid is the final product produced from pyruvic acid during anaerobic glycolysis. The build up of lactic acid will inhibit glycolysis, resulting in muscle fatigue. High levels of lactic acid and lack of energy reserves the problems such as cramps, nausea, weakness, and feeling exhausted that are often faced by all people including professional athletes and amateurs because this condition will cause fatigue during the move so that it can lead to a decline in performance and achievement. Efforts to overcome these problems can be done by setting the right exercise program, giving nutrition, emotional and physical environment [6-7].

In a study it was found that regular physical activity with exercise frequency of three to five times per week, exercise intensity 60-80% of maximal heart rate, and duration of exercise 20-60 minutes resulted the smooth flow of blood and speed up the removal of metabolic waste substances so that recovery takes place quickly, and can slow the occurrence of fatigue after exercise. Even the American College of Sports Medicine (ACSM) explains the intensity of physical activity must reach the target zone of 60-90% of the maximum heart rate or Maximum Heart Rate (MHR), this area range is commonly referred to as Training Zone or training area. The intensity of the exercise is said to be mild if it reaches 60-69% of MHR, while it reaches 70-79% of MHR and high if it reaches 80-89% of MHR. The intensity of the exercise can be increased by increasing the burden of exercise by jumping-bob or by speeding up the exercise movement [8-9].

The Borg scale is a simple method to measure the intensity in training. Perceived exertion of exercise intensity by assessing physical signs such as heart rate, breathing rate and perspiration/sweating. Borg scale is based on percentage of maximum pulse. The Borg scale is an excellent tool to use for guiding training intensity. Heart rate and lactate give secondary feedback. If the effort and speed were increased, more lactate will be produced and the body must compensate with increased breathing and raised heart rate. Measuring pulse and lactate level gives excellent feedback and helps us to describe and learn the different intensity levels. Lactic acid measurements are usually taken from blood samples from the tip of the finger. This study is an exercise program designed to introduce students to regular physical activity and intentionally to reduce the muscle fatigue.

2. Method

This study is a quasi experimental with pretest-posttest group design. The study included selection of subjects according to the criteria and signed informed consent form, lactic acid and Borg scale assessment before moderate physical activity (AFS), medium physical activity training with stepping up and down test to reach the target pulse of 70-79 % maximal heart rate. The subject study is 15 high school students, aged 15–20 year. Subjects underwent 4 weeks of medium physical activity training, 3 days/week, for 15 min. Before stepping up and down test, subjects were introduced the research tool (desk high as 40 cm) and how to do the exercise. A pulse rate meter is attached to the arm the subject to determine the pulse target 70-79 % maximal heart rate. After that the stop watch is executed, for start counting the time needed for medium physical activity training with the metronome rhythm is slow to beat pulse reaches the target already determined. Once achieved, keep it for 2 to 3 minutes then the result of the pulse the target of the pulse. The study was conducted to compare the levels of lactic acid peripheral to fatigue and the Borg scale before and after the exercise program. Research data is processed using SPPS. The mean values obtained were analyzed descriptively, followed by T-test.

3. Results and Discussion

Characteristics of subjects in this study were presented in Table 1. Based on this table, it can be seen that the variable weight, height, body mass index, age and heart ratemax in the study subjects have met the study criteria.

Table 1. Mean ± SD Characteristics of Research Subjects

No	Variable	Mean±SD
1	Weight (kg)	55.90±4.17
2	Height (cm)	168.90±4.45
3	Body Mass Index (kg / m2)	19.70±0.90
4	Age	16.40±0.51
5	Heart Rate$_{max}$ (x / min)	203.60±0.51

n= 15 (man)

Table 2 shown that there is a decrease in lactic acid levels before (pre-test) and after (post-test) the implementation of physical activity exercise is in untrained people

Table 2. Mean ± SD Lactic Acid Variable

Variable	Untrained People with Medium Physical Activity	
	Pre-Test	Post-Test
Lactid Acid (mmol/L)	9.02±2.55	6.00±2.76

From table 3 it can be seen that there are differences of the Borg scale before (pre-test) and after (post test) of physical exercise activity in untrained pople.

Table 3. Mean ± SD Borg Scale Variable

Variable	Untrained People with Medium Physical Activity	
	Pre-Test	Post-Test
Borg Scale	3.70±0.94	1.80±1.03

From table 4 normality test results obtained on the pre-test and post-test has a significance p value>0.05 which means the data is normally distributed. In bivariate analysis using paired t test statistic.

Table 4. Normality Test

No	Variable	Untrained People with Medium Physical Activity	
		df	Sig
1	Pre-Test	10	0.418
2	Post-Test	10	0.943

Table 5. below shows T paired test for lactic acid content and the Borg scale with significance level p = <0,05 resulted significant difference between pretest and posttes.

Table 5. T test Lactic Acid Levels dan Borg Scale

No	Variable	Pretest		Posttest		t	P
		Mean	± SD	Mean	± SD	Value	
1	Lactic Acid	9.02	2.55	6	2.76	2.32	0.04
2	Borg Scale	3.70	0.94	1.80	1.03	8.14	0.000

P = < 0,05

The body composition should be proportionate between muscle mass and fat. By exercising moderate physical activity in addition to forming an ideal body can also improve performance, fitness and performance. Good physical fitness causes one to not get tired [10-11]. Table 1 shows that the body mass index of untrained persons is included in the normal range of 18.5-25. This is consistent with the normal limit of IMT of Europe according to WHO is 18.5 - 24.9, Asians according to International Obesity Task Force (IOTF) and WHO are 18.5 - 22.9 and Indonesian according to MOH is 18.5 - 25.0 [12-14].

Based on tables 2, 3 and 5 it can be seen that the mean decreases in lactic acid level of pre-test and post-test is 3.02 mmol / L. For the Borg scale, the shortness of pre-test and post-test decreased. The paired t-test results showed significant differences in pre-test and post-test of untrained people given moderate physical activity (AFS). Exercise done regularly based on the accuracy of duration, frequency and intensity impact on the decrease in lactic acid levels so that it can slow fatigue. The buildup of lactic acid will inhibit glycolysis, resulting in muscle fatigue. High levels of lactic acid will cause acidosis in and around the muscle cells, impede coordination, increase the risk of injury, inhibit the energy system from creatine phosphate. High levels of lactic acid in athletes will have a negative impact on athlete performance [8,15].

Physical activity according to ACSM (American College of Sports Medicine) will be more meaningful if performed between 70% and 80% maximum heart rate, regularly three times a week with intensity that increases heart rate. This concept is inline with the results of the study that a significant reduction in lactic acid levels can be seen in untrained people who get moderate physical activity with a 70-79% HRmax workload performed three times a week for four weeks [6,16].

Decreased scaling of Borg scale can be seen from pre-test and post-test. The finding of this study was consistent with other studies that showing significant differences in Borg scale at week 4 and 5 ($p < 0.05$) through t-test to see the effect of walking practice on Borg scale score and six minutes walking distance test. Scale Borg rate fatigue stage, the greater the value means the higher the level of fatigue. This scale increases slowly during the first hour during exercise. This study illustrates that during pre-test scoring, the score of Borg scale were short around 3.70 and after obtaining moderate physical activity for four weeks scores, the Borg scale decreased to 1.80 and according to statistic t-test both showed significant differences [17-19].

4. Conclusion

The conclusion that can be drawn from this research is based on the result of paired T test for lactic acid concentration and Borg congested scale that has significance level $P = <0,05$ between pretest and posttes. This shows moderate physical activity with HRmax 70-79% having an effect on slowing fatigue in untrained people. Physical activity is being done in accordance with its basic principles can improve the physical quality, performance, achievement, and performance of a person.

5. References

[1] McArdle W D Katch F I and Katch V L 2001 Exercise Physiology: Energy, Nutrition, and Human Performance. Philadelphia: Lippincott Williams and Wilkins.
[2] Departemen Kesehatan Indonesia. Profil Kesehatan Indonesia 2010.
[3] Singh R 2005 Nutritional Requirements of Athletes Exercising in a Hot Environment *Malaysian Journal of Nutrition* vol.11(2) pp 189-198.
[4] Warburton D Nicol Chrystal W Bredin Shannon 2006 Health Benefits of Phisycal Activity: The Evidence *Canadian Medical Association Journal* vol 174(6).
[5] Octa L 2012 Pengaruh Latihan Aerobik Terhadap Pembentukan ATP–Mitokondria (suatu tinjauan intramolekuler) Universitas Tunas Pembangunan Surakarta: Surakarta.
[6] Sherwood Lauralee 2014 Fisiologi Manusia dari Sel ke Sistem Edisi ke 8 EGC
[7] Laursen PB 2005 Models to Explain Fatigue during Prolonged Endurance Cycling *Sports Med* vol 35 (10) pp 865-898.

[8] Hernawati 2010 Produksi Asam Laktat pada Exercise Aerobik dan Anaerobik FMIPA, Universitas Pendidikan Indonesia Bandung.

[9] Ratmawati Y 2013 Latihan Aerobik Intensitas Sedang Dengan Diet Rendah Kolesterol Lebih Baik Dalam Memperbaiki Kognitif Daripada Intensitas Ringan Pada Penderita Sindroma Metabolik Dalam Pollock M.L & Wilmore J H Exercise in health and disease Evaluation and Prescription for Prevention a Rehabilitation 2nd Ed Saunders

[10] Irawan AM 2007 Nutrisi Energi & Performa Olahraga *Polton Sports Science & Performance Lab* vol 01 No 04

[11] Stolen T Chamri K Costagna C Wisloff U 2005 Physiology of soccer an update *Sport Med* vol 35 (6) pp 501

[12] Chrisly M Djon W Shane H 2015 Manfaat Latihan Olahraga Aerobik terhadap Kebugaran Fisik Manusia *Jurnal E-Biomedik (Ebm)* vol 3 No 1

[13] Susilowati 2010 Burnout Pada Atlet.Tersedia dalam: http://bgmpsikologi. blogspot.com

[14] Heryudarini H Yekti W Sri M 2005 Penggunaan Berbagai Cut-Off Indeks Massa Tubuh Sebagai Indikator Obesitas Terkait Penyakit Degeneratif Di Indonesia Pusat Penelitian dan Pengembangan Gizi dan Makanan, Badan Penelitian dan Pengembangan Kesehatan Depkes. Gizi Indon pp 31

[15] Guntara P 2014 Pengaruh Recovery Aktif Dengan Recovery Pasif terhadap Penurunan Kadar Asam Laktat Universitas Pendidikan Indonesia Bandung

[16] Guyton A C and John E Hall 2008 Buku Ajar Fisiologi Kedokteran Edisi 11 Editor Irawati Setiawan *Jakarta Penerbit Buku Kedokteran EGC*

[17] Gusbakti R 2006 Pengaruh Pemberian Minuman Berkarbohidrat Berelektrolit dapat Memperlambat Kelelahan Selama Berolahraga Bagian Ilmu Faal Fakultas Kedokteran Universitas Islam Sumetera Utara. *Majalah Kedokteran Nusantara* vol 39 no.1

[18] Powers S Edwar H 2007 Exercise Physiology Theory and Application to Fitness and Performance (Sixth Edition) McGraw-Hill Companies Inc Newyork

[19] Adams Scott and S Delorey Darren and Davenport Margie and S Fairey Adrian and North Scott and S Courneya Kerry 2018 Effects of high-intensity interval training on fatigue and quality of life in testicular cancer survivors *British Journal of Cancer* 118.doi 10.1038/s41416-018-0044-7

Performance of A Small Diesel Engine by Experimental and Numerical Simulation Analysis

Erdiwansyah[1,2*], A.M Leman[3], A. M. Norkhizan[1], L. Mulyadi[3], Fitri Khoerunnisa[4], R.E Sardjono[4], M.I Maulana[5] and M.S.M Sani[2]
[1]Faculty of Mechanical Engineering, Universiti Malaysia Pahang, 26600 Malaysia
[2]Fakultas Teknik, Universitas Serambi Mekkah, 23249 Aceh, Indonesia
[3]Institut Teknologi Nasional, Malang, Indonesia
[4]Department of Chemistry, Faculty of Mathematics and Science, Indonesia University of Education, Bandung 40522, Indonesia
[5]Faculty of Mechanical Engineering, Universitas Syiah Kuala, 23249 Aceh, Indonesia

*Corresponding author: erdi.wansyah@yahoo.co.id

Abstract. A single cylinder diesel engine is an internal combustion engine; this machine is typically a compression ignition engine, fuel ignition through a high one made for compression of fuel and air mixtures. This study conducts a modelling with one dimension on a single cylinder four-stroke diesel engine and analyses the simulation results with experimental data. The simulation model covers the entire engine cycle consisting of compression, power, intake, and disposal for different engine speeds such as 1000, 2000, 3000, 4000, 5000, and 6000 rpm. Simulation modelling was done for the development of computation using software GT-POWER. The result of this modelling measures the accuracy level of machine parameter data from the modelled simulation. Based on the simulation results with modelling the machine, the data accuracy rate reached 90% in various cases tested with different engine speeds. Furthermore, the results also show that the modelling provides accurate data and can be used in the industry as a tool for development of different machines in the future.

Keyword : *single cylinder direct injection, in-cylinder heat transfer, performance*

1. Introduction

A single cylinder diesel engine is an internal combustion engine; this machine is typically a compression ignition engine, which fuel ignition through a high one made for compression of fuel and air mixtures [1-6]. The engine operating system uses a cycle on a diesel engine. Unlike the petrol machine, the incoming air does not throttle so the engine will experience over speed if not immediately done. The engine wheel system will encourage older injection [7-8].

The modelling of direct injection diesel engines in this study was measured using four distinctive steps. The modelling of the diesel engine with the GT-POWER simulation shown for this four-cylinder engine. The GT-POWER liner device is a very prominent simulation tool that can be used for the manufacture of vehicle engines and suitable to analyse various problems found on the machine [9]. Designing models for steady state and transient simulations using GT-POWER can be used as an analysis tool on engine control and powertrain. These are all for internal combustion engines, and many components are provided to users as advanced modelling concepts. GT-POWER software is based on the dynamics of gas for one dimension, beginning with heat transfer and flow into piping in other components found on the system machine. GT-POWER software is a typical one-dimensional tool of the GT-SUITE family [10-14].

The details contained in the diesel engine design have many significant variations in the size range and performance on the machine. In particular, the geometry of the combustion chamber is very different from the required fuel injection characteristics, which can greatly enhance possibility to solve the problem of large diesel engine designs that can achieve rapid fuel blending speed in the process of available time completion for fuel combustion in the engine [1], [15-19]. Types of geometry contained in inlet port, piston shape, cylinder head, and pattern on fuel injection that can be used over a diesel size range.

A high increase in engine strength can be shown where their products can be expected by the manufacturer to provide strength, reliability, resilience, and economics that can be satisfactory below service situations. Generally, the maximum torque velocity at which it should reach is also provided. According to [20-22], machine performance parameters are the most important sala one in geometric properties, engine performance parameters, and of the engine efficiency. Engine efficiency includes thermal, mechanical, relative, brake heat, and volumetric efficiencies. Parameters related to engine performance are average effective pressure, specific power output, engine intake valve index, average piston speed, fuel-air, air-fuel ratio, specific fuel consumption, and the calorific value contained in the fuel.

Internal combustion [22-24] investigating the importance of performance parameters on diesel engines such as efficient terms, machine performance parameters, and other related geometric properties. Thermal efficiency can show machine, mechanical, brake heat, relative, and volumetric efficiencies. The parameters contained in the performance of other related machines are the effects of average pressure, specific power consumption for fuel output, the average speed of piston, fuel, intake valve match index, and air at fuel output or air-fuel ratio. Designs built on the geometry of a four-cylinder diesel engine, the compression ratio of a diesel engine is the maximum volume of the cylinder as well as the displaced volume cylinder (V_d) and cylinder volume demand (V_c) as in the division for lowest cylinder volume (V_c). The measurement of the compression ratio contained in a four-cylinder diesel engine can be done using Equation (1).

$$r_c = \frac{V_d + V_c}{V_c} \qquad (1)$$

The power delivered by the diesel engine and absorbed by the dynamometer is the torque product and the angular velocity. The power of a diesel engine can be defined by Equation (2).

$$P = 2\pi NT \qquad (2)$$

Every engine efficiency is as predetermined by [20]. The energy ratio (E) is the measured indicated thermal efficiency (π_{ith}) of the indicated power (*ip*) with the energy at the fuel input. The ratio of energy in brake power (*bp*) is the efficiency of the brake thermal efficiency (π_{bth}), the ratio of braking power (*bp*) or power delivery to indicated power as shown by the indicated power or also the power supplied to the piston is defined by mechanical efficiency (π_m) which also can be defined as the efficiency ratio for brake heat and thermal efficiency. The thermal efficiency ratio of the actual cycle has an ideal cycle which is the relative efficiency or efficiency ratio (π_{rel}). The efficiency ratio is very necessary to be used in the criteria for the level of engine development used. One of the most important parameters in determining the performance of the four-stroke engine is volumetric efficiency (π_v). This is because in the four-stroke machine, there is a difference of suction stroke and therefore volumetric efficiency can show the machine respiration ability. The volume rate into the shared intake system is the volumetric efficiency at a rate at which the system removes volume. The normal volumetric efficiency range is about 85%–90% when at full speed for the CI machine and 80%–85% for the SI machine [24], [25].

$$\Pi_{ith} = \frac{ip}{E} \qquad (3)$$

$$\Pi_{ith} = \frac{bp}{E} \tag{4}$$

$$\Pi_m = \frac{bp}{ip} \tag{5}$$

$$\Pi_v = \frac{m_a}{P_n V_{dip}\, N/2} \tag{6}$$

$$\Pi_{rel} = \frac{Actual\ therm\quad efficiency}{Air\text{-}standar\ efficiency} \tag{7}$$

The research related to machine performance has been done by [20], [26-27]. Mean effective pressure (*mep*) is where number of crank revolutions per power stroke (n_R) is the sum of all crank turns for each power on the cylinder (one for two-stroke and two for four-stroke per cycles) as shown in Equation (8).

$$mep = \frac{Pn_R}{V_d N} \tag{8}$$

The efficiency of a given fuel conversion is an efficient measure of the engine as mentioned by [1]:

$$nf = \frac{W_c}{m_f Q_{HV}} = \frac{(Pn_{R/N})}{(m_f n_R/N) Q_{HV}} = \frac{P}{m_f Q_{HV}} \tag{9}$$

Specific fuel consumption as:

$$sfc = \frac{m_f}{P} \tag{10}$$

In testing this modelled engine, both the fuel mass flow rate (m_f) and the mass flow rate (m_a) were measured. The use of the flow rate ratio to define the conditions in engine operation is the fuel or air ratio (A/F) and also the water ratio (F/A). Development of parameters of machine performance can be correlated using the following Equations (11) and (12).

For power P:

$$P = \frac{n_f m_a N Q_{HV}\,(F/A)}{n_R} \tag{11}$$

$$P = \frac{n_f n_v N V_d\, Q_{HV} P_{a,i}\,(F/A)}{2} \tag{12}$$

The torque (T) measurement used Equation (13). Where power per piston crown area is fuel conversion efficiency (*nf*), volumetric efficiency (*nv*), (N_d) fuel heating value (Q_{HV}), ($P_{a,i}$), and (F/A) divided by 4π.

$$P = \frac{n_f n_v N_d\, Q_{HV} P_{a,i}\,(F/A)}{4\,\pi} \tag{13}$$

For mean effective pressure measurement, Equation (14) was used.

$$mep = n_f n_v Q_{HV} P_{a,i}(F/A) \tag{14}$$

The specific strength of each piston unit is a success in measuring and designing the engine used in the existing piston area regardless of the available cylinder size. Specific power can be measured using Equation (15). Where power per piston crown area is fuel conversion efficiency (*nf*), volumetric efficiency (*nv*), fuel heating value (NLQ_{HV}) ($P_{a,i}$), and (F/A) divided by two.

$$\frac{P}{A_p} = \frac{n_f n_v NLQ_{HV} P_{a,i}(F/A)}{2} \tag{15}$$

Meanwhile, the piston velocity can be measured using Equation (16). Where power per piston crown area is fuel conversion efficiency (*nf*), volumetric efficiency (*nv*), fuel heating value (NS_p), (Q_{HV}), ($P_{a,i}$), and (F/A) divided by four.

$$\frac{P}{A_p} = \frac{n_f n_v NS_p Q_{HV} P_{a,i}(F/A)}{4} \tag{16}$$

Thus, the specific power is related to the product such as average piston velocity and average effective pressure [1]. This relationship has illustrated the importance of engine efficiency performance in high fuel conversion, high volumetric efficiency, can increase engine output at the time of displacement by growing the incoming air density, the maximum fuel ratio that can be used for the machine and the average speed pistons.

2. Experiment Setup

The modelling of diesel engine direct-injection (DI) four single cylinder strokes developed for the simulation is presented in this paper. The diesel engine specifications were modelled as shown in Table 1. Figure 1 shows the engine used in the experiment.

Table 1. Specifications diesel engine single cylinder

Model	**Yanmar Model TF120M**
Ratio Compression	17.7:1
Continuous output	10.5 HP @ 2400 rpm
Fuel tank capacity	11 L
Displace volume (cm^3)	638
Fuel injection type	Direct Injection
Bore (mm)	92
Maximum output	12.0 HP @ 2400 rpm
Type	Horizontal single-cylinder four-stroke engine diesel
Max power	7.7 kW @ 2400 rpm
Max torque	161 Nm @ 4500 rpm
Cooling system	Water-radiation
Stroke (mm)	96

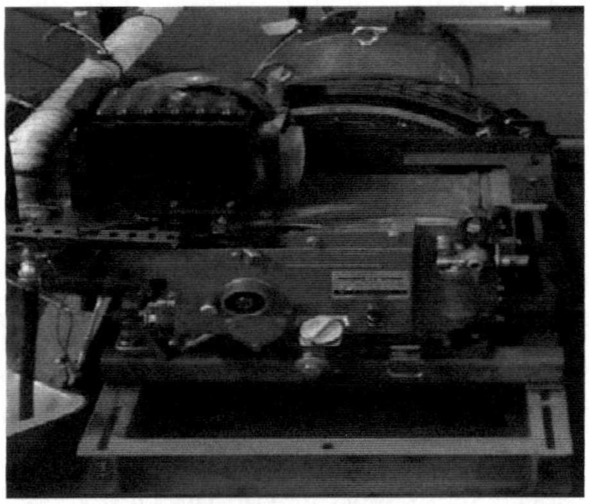

Figure 1. Engine test

In the development of engine modelling in GT-POWER, the first step was to investigate and open the diesel engine to measure all components contained in the machine to be incorporated into the GT-POWER library. Then, for modelling machines in the GT-POWER, the window was selected at first and the title template was contained in the library menu. The GT-POWER in template library was placed on the left side of the screen. All templates were pre-loaded, and everything available on the menu were used for the machine modelling. To begin the modelling purposes, the icons listed in the template library were clicked and dragged into the project library. Among them was the required template and also some contained objects which were included in the GT-POWER library. Fuel injection systems and engine cylinders were focused on the engine cylinder performance with fuel support from fuel injection systems, exhaust gases, and systems on fresh air intake up to the exhaust system. In cylinders and fuel injection systems on diesel engines have several components, but all diesel engines have the same components. Components, data, and size were recorded and inserted into the GT-POWER library. All components of the system on fuel injection and engine cylinders were injectors, engines and cylinders. The exhaust system was the last system on the machine. This system started from the exhaust valve and ended in the environment. Exhaust system found on GT-POWER are exhaust port, muffler, exhaust valve, exhaust runner exit, exhaust runner, and environment.

Once the diesel engine system modelling was completed, the development of diesel engine model was done using GT-POWER software. GT-POWER computing was used for modelling diesel engines, engine cylinders, intake systems, and fuel injection systems as well as exhaust systems. Fuel injection systems, intake systems, and engine cylinder systems were connected to the involved in the engine and intake cylinder systems as well as the fuel injection system. Fuel injection systems and engine cylinders were connected into the exhaust system inside the cylinder and fuel injection system were located on exhvalve in the exhaust. Orifficeconn connected all components contained in the diesel engine. Once the work was completed, diesel engine modelling was extended and developed using GT-POWER software. Modelling diesel engines were fired by GT-POWER as shown in Figure 2.

The data information was used to build diesel engine models included in the GT-POWER library. This modelling does not require all items, and only additional information may be required. However, this list is an excellent starting point. If the initial stage of model development is being done, the optimum value must be determined for some items on the list to which the simulation ends. In this case, the determination of the attribute must first be defined as a parameter to run a series of cases in determining the optimal value. The data from machine characteristics were compression ratio, inline or V configuration, combustion sequence, V-angle (optional), two or four strokes. While the data contained in the geometry of the cylinder were a stroke, pin offset, bore, high clearance TDC piston, length connecting rod, head area, piston area, and head of bowl geometry. All components in geometry were data in intake and exhaust systems. While the data on the throttle was a throttle location and coefficient for discharge with throttle angles in a two-way flow. Meanwhile, the data in fuel injector were some injector and location, nozzle diameter and a number of the nozzle hole, LVH, injection rate, and type of fuel. Data on the exhaust and intake valves were profiles of elevators, valve lashes, valve diameters, and discharge coefficients. For data on ambient conditions were temperature, humidity, and pressure. While for performance data is very used when setting up the model after the development has been completed.

These preparations such as reviewing the finished model, setting the case, making arrangements, plot settings, and plot requests need to be done. The parameters contained in the created model were listed automatically in each case, and case settings must be defined for the initial case of the simulation. Computational time can often be reduced to steady-state simulation systems in the planned sequence and take advantage of the initialisation state to be able to run the settings. Time plots and also cycles can be requested by selecting some appropriate plots to be used from the folders in the plot options for each section. Requests from all plots contained in individual sections will be automatically saved, the result of this plot regardless of the request selected.

When diesel engine modelling is ready, the simulation with GT-POWER can be started and run. Simulation progress was displayed in a window with a scrolling text form. After all, the data input was successfully read, the simulation was run, and further developments in the form of reports were produced.

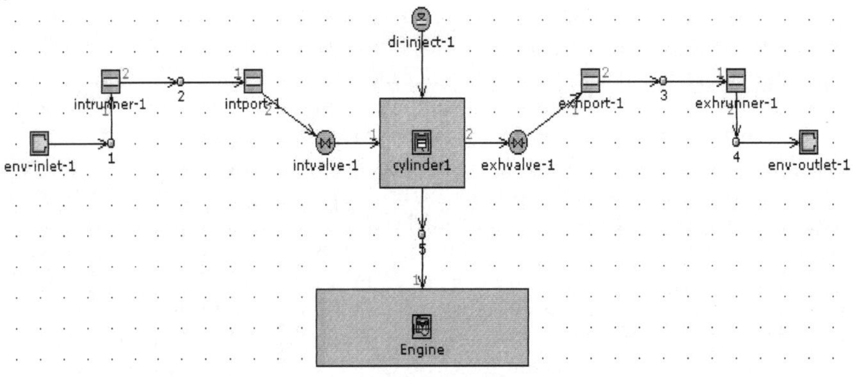

Figure 2. Single cylinder direct injection diesel engine on GT-POWER [28]

3. RESULTS AND DISCUSSION

Whenever a simulation is run, some output files will be generated from the GT-SUITE process which contains the report results in graphs and tables in various formats. The output is mostly available on GT-POST post-processing applications. This tool is very sophisticated as it can be used to display the animation and output of the analysis as it has been selected during the setup before simulation [2].

The results of all running simulations were all data from different engine performance and speed (rpm). The modelling of this machine was run at different engine speeds such as 1000, 2000, 3000, 4000, 5000, and 6000 rpm. Simulation results in this paper to produce engine performance such as brake specific CO, brake specific CO_2, brake power (kW), brake power (HP), and brake torque (bar).

The experimental results data needs to be verified in determining the accuracy of the data in the simulated model. The accuracy of the verified values in the modelling were compared with the data from the experimental results. Verification at the initial stage that the simulation results there are some differences in the results data from the experiment. Thus, this modelling requires calibration so that the simulation result will be close to the experiment result. Calibrated models include lifting at a crank angle with a value of 0.83 for the duration of a single cylinder diesel engine to obtain similarity to the engine specifications. Furthermore, the maximum coefficient number was 5 mm and advanced to 0.95. Hence, this model has a new array to be able to perform validation on the experimental data.

Figure 3 shows the results after tuning and calibration of the data. The difference between the simulation results and the experimental data was 7%–10%. Thus, the newly completed model can be optimised and used in the next simulation phase. The average percentage of simulation difference with the experiments of all cases tested was about 7%. Hence, the modelling with GT-POWER software has been successful. This result is a development of real research experimental data.

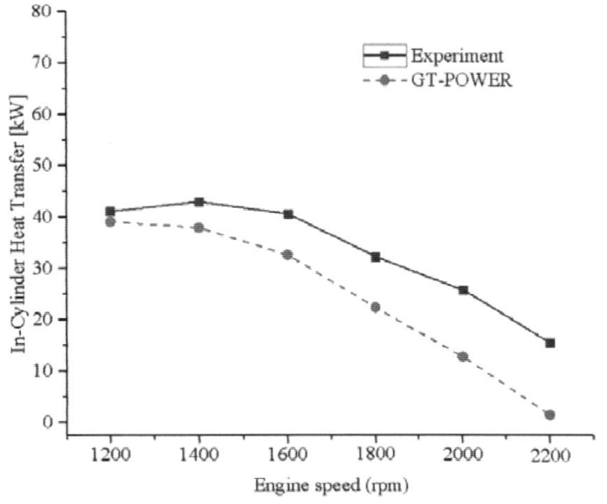

Figure 3. In-Cylinder Heat Transfer for different speed

Variations of brake efficiency with different engine speeds and loads are as shown in Figure 4. Different ratios of results between experimental data developed and simulations of models made using GT-Power with different engine speeds and loads was only about 7%. Thus, the level of accuracy of simulation data with built models can be trusted and can be used for engine development in automotive industry [29].

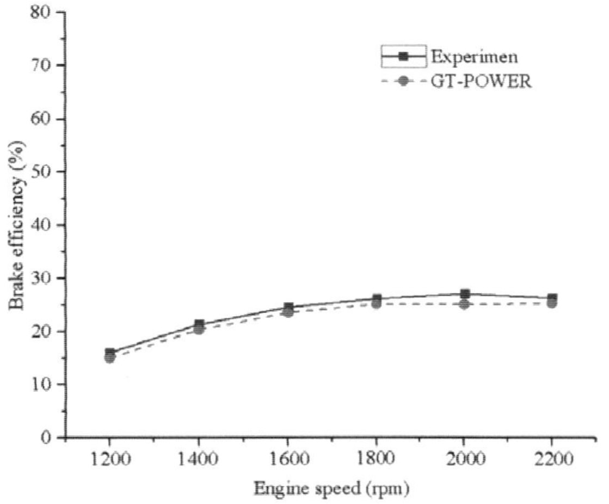

Figure 4. Brake specific for different speed

A measurement of the accuracy of the engine modelling built to verify the experimental data is required. The model was built, and its accuracy was verified by comparing experimental engine performance with modelling. Of the two results, the value was correlated based on the trend. The initial stage of verification shows a very significant difference when compared with the experimental results. Thus, this model needs to be calibrated so that the results will be able to resemble experimental data. In order to calibrate the model, some parameters must be adjusted such as the angle multiplier is set so that the duration of the simulation is the same as that on the machine. The forward coefficient was also adjusted, and the maximum number was made to 5 mm to generate new array numbers which can be used in the valve model. The result of experimental validation data with model simulation is as shown in Figure 5.

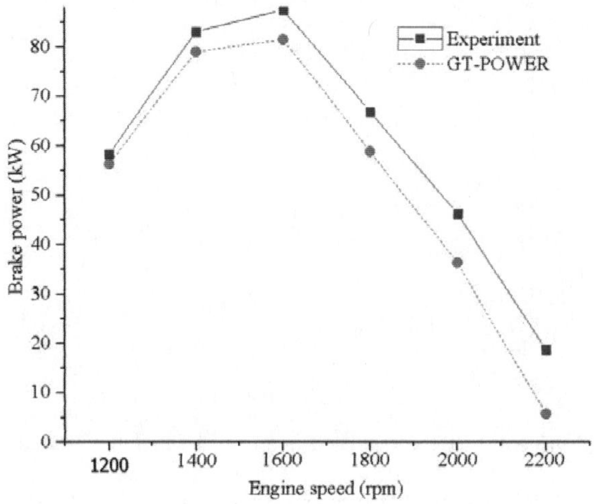

Figure 5. Brake power (kW) for different speed

The simulated results were validated based on the results data from experiments that have been performed to serve as practical data [30]. The graph of the comparative results modelled between the simulation results and the results of the experimental studies is as illustrated in Figure 6. The graph illustrates the brake power curves with full engine loads at different speeds. The percentage of the maximum and minimum deviations investigated that deviation was 5%–10%, respectively. From this result, it states that the level of accuracy of the simulation has achieved reliable results.

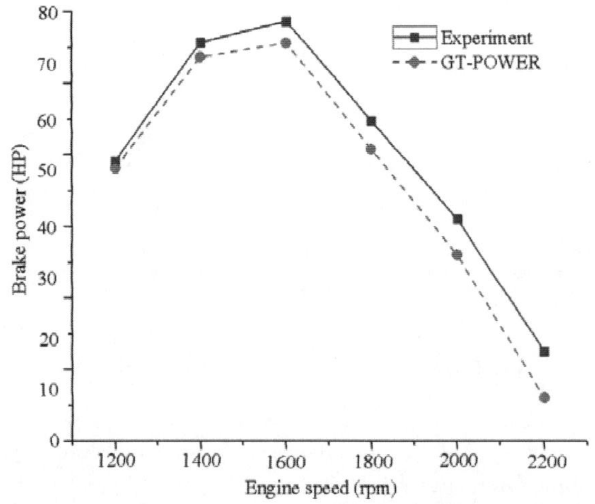

Figure 6. Brake power (HP) for different speed

The modelling machines built with GT-POWER software were evaluated with experimental data to calculate the accuracy of the data, and the results of experimental data were compared with the simulation results using GT-Power to be adjusted, as shown in Figure 7. The simulation results and experimental data got the deal which was excellent. This suitability can be seen at experimental and simulated experimental error rate at only 3%, which was within the allowed distance range. Therefore, the results of calibration with GT-POWER modelling have a level of performance accuracy in the analysis and simulation modelled was relatively reliable.

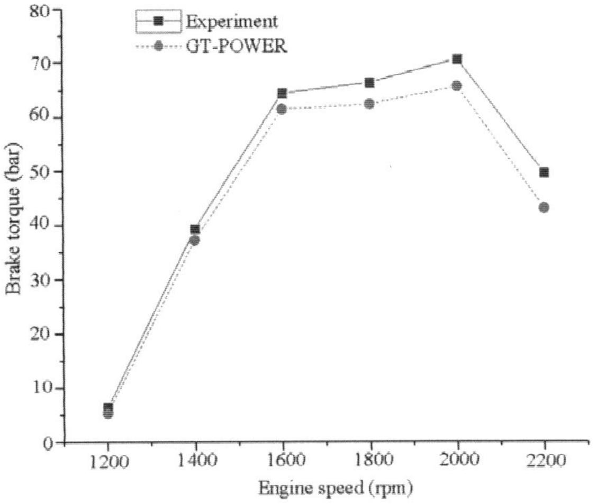

Figure 7. Brake torque (bar) for different speed

4. Conclusion

In this paper, the experimental data were compared with the simulation using GT-POWER software. The purpose of this simulation is to measure machine parameters with accuracy in the model simulation. Simulations were performed with different machine speeds (1000, 2000, 3000, 4000, 5000, and 6000 rpm). This simulation result can be used as a standard for machine in the future. Testing at different engine speeds indicates a level of data efficiency similar to that of experimental data. From the simulation results, it shows that the accuracy of the data in the simulation with experimental reached 90% for the entire tested case. Thus, the simulation of diesel engine modelling with GT-POWER software can be used for industry in the development of vehicle engines and other engines.

Acknowledgement

This research was supported by University Malaysia Pahang (UMP) Flagship Research Grant (RDU172204), PGRS160303, FRGS Research Grant (RDU130131 and UMP Grant RDU 180335). And we thank our colleagues from Automotive Engineering Center (AEC) and Advanced Automotive Liquid Laboratory (ALL), who provided insight and expertise that greatly assisted the research.

References

[1] Z Yang F Zhang L Wang K Wang and D Zhang 2018 Effects of injection mode on the mixture formation and combustion performance of the hydrogen internal combustion engine *Energy* vol 147 pp 715–728

[2] D C Gosai and H J Nagarsheth 2016 Diesel Engine Cycle Analysis of Two Different TBC Combustion Chamber *Procedia Technol* vol 23 pp 504–512.

[3] E. Mattarelli G Cantore C A Rinaldini and T Savioli 2017 Combustion System Development of an Opposed Piston 2-Stroke Diesel Engine *Energy Procedia* vol 126 pp 1003–1010

[4] N Y Dudareva R D Enikeev and V Y Ivanov 2017 Thermal Protection of Internal Combustion Engines Pistons *Procedia Eng.* vol 206 pp 1382–1387

[5] L V Plotnikov and B P Zhilkin 2017 The gas-dynamic unsteadiness effects on heat transfer in the intake and exhaust systems of piston internal combustion engines *Int. J. Heat Mass Transf.* vol 115 pp 1182–1191

[6] C Bae and J Kim 2017 Alternative fuels for internal combustion engines *Proc. Combust. Inst.* vol 36 no. 3 pp 3389–3413

[7] E Jiaqiang M Pham D Zhao Y Deng D Le and W Zuo 2017 Effect of different technologies on combustion and emissions of the diesel engine fueled with biodiesel : A review vol 80 no May, pp 620–647 2017.

[8] N Zamel and X Li 2006 Life cycle comparison of fuel cell vehicles and internal combustion engine vehicles for Canada and the United States *J Power Sources* vol 162 no 2 SPEC ISS pp 1241–1253

[9] I Notes 2016 GT-SUITE Installation Notes

[10] A Praptijanto A Muharam A Nur and Y Putrasari 2015 Effect of ethanol percentage for diesel engine performance using virtual engine simulation tool *Energy Procedia* vol 68 pp 345–354

[11] C Guo et al. 2018 Effect of fuel injection characteristics on the performance of a free-piston diesel engine linear generator: CFD simulation and experimental results *Energy Convers. Manag.* vol 160 no February pp 302–312

[12] M Božić A Vučetić M Sjerić D Kozarac and Z Lulić 2018 Experimental study on knock sources in spark ignition engine with exhaust gas recirculation *Energy Convers. Manag.* vol 165 pp 35–44

[13] M Sjerić I Taritaš R Tomić M Blažić D Kozarac and Z Lulić 2016 Efficiency improvement of a spark-ignition engine at full load conditions using exhaust gas recirculation and variable geometry turbocharger Numerical study *Energy Convers. Manag.* vol 125 pp 26–39

[14] Erdiwansyah et al. 2018 Effects of Diesel-Biodiesel Blends in Diesel Engine Single Cylinder on the Emission Characteristic *MATEC Web Conf.* vol 225

[15] L Tartakovsky and M Sheintuch 2018 Fuel reforming in internal combustion engines *Prog. Energy Combust. Sci.*, vol 67 pp 88–114

[16] A Nourbakhsh M Bayareh A Mohammadi and S Jahantighi 2018 Effect analysis on boiling heat transfer performance of an internal combustion engine at the shutdown time *Int. J. Therm. Sci.* vol 129 no December 2017 pp 365–374

[17] F. Li *et al.* 2017 Experimental study on spray characteristics of long-chain alcohol-diesel fuels in a constant volume chamber *J. Energy Inst.* vol c pp 1–14

[18] T Badawy M A Attar H Xu and A Ghafourian 2018 Assessment of gasoline direct injector fouling effects on fuel injection, engine performance and emissions *Appl. Energy* vol 220 no September 2017 pp 351–374

[19] A Broatch R Novella J García-Tíscar and J Gomez-Soriano 2018 Potential of dual spray injectors for optimising the noise emission of gasoline partially premixed combustion in a 2-stroke HSDI CI engine *Appl. Therm. Eng.* vol 134 pp 369–378

[20] A V Vasilyev Y S Bakhracheva and S Y Storojakov 2016 Dynamics Simulation Model for the Internal Combustion Engine Valve Gear *Procedia Eng.* vol 150 pp 312–317

[21] C Caligiuri and M Renzi 2017 Combustion modelling of a dual fuel diesel-producer gas compression ignition engine *Energy Procedia* vol 142 pp 1395–1400

[22] M Toma and C Bobâlcă 2016 Research on Drivers' Perception on the Maintenance of Air Filters for Internal Combustion Engines *Procedia Technol.* vol 22 no October 2015 pp 961–968

[23] G Szabados and Á Bereczky 2018 Experimental investigation of physicochemical properties of diesel, biodiesel and TBK-biodiesel fuels and combustion and emission analysis in CI internal combustion engine *Renew. Energy* 2018.

[24] Y. Xu H Kang J Gong S Zhang and X Li 2018 A study on the combustion strategy of gasoline/diesel dual-fuel engine *Fuel* vol 225 no March pp 426–435

[25] S C Chen Y L Kao G T Yeh and M H Rei 2017 An onboard hydrogen generator for hydrogen enhanced combustion with internal combustion engine *Int. J. Hydrogen Energy* vol 42 no. 33 pp 21334–21342

[26] Y Chen and R Raine 2016 The effects of hydrogen supplementation on idle performance and emissions of an SI engine *Fuel* vol 176 pp 190–199

[27] Y Kurtgoz M Karagoz and E Deniz 2018 Biogas engine performance estimation using ANN *Eng. Sci. Technol. an Int. J.*

[28] R Mamat and N R Abdullah 2010 Emissions of a Single Cylinder Diesel Engine Operating with Ethanol Emissions of a Single Cylinder Diesel Engine Operating with Ethanol no November pp 3–4

[29] B Menacer 2015 Simulation and Modelling of a Turbocharged Compression Ignition Engine *Int. J. Energy Power Eng.* vol 4 no 3 pp 129

[30] A R Mahmoudi I Khazaee and M Ghazikhani 2017 Simulating the effects of turbocharging on the emission levels of a gasoline engine *Alexandria Eng. J.* vol 56 no. 4 pp 737–748

IOP Conf. Series: Materials Science and Engineering **506** (2019) 012030 doi:10.1088/1757-899X/506/1/012030

The Study of Area Infiltration Rate by Using Biopori Holes as an Effort to Reduce Water Level and Groundwater Conservation

Yulia[1*] and Effendi Nurzal[2]

[1]Civil Engineering, Faculty of Engineering, Muhammadiyah University, 23245 Banda Aceh, Indonesia
[2]Architecture, Architecture Study Program, Faculty of Engineering, Muhammadiyah University, 23245 Banda Aceh, Aceh, Indonesia

[*]Corresponding author: jeng_yuliaoke@yaho.co.id

Abstract. The expansion of land development for settlement and other infrastructure has impact for lessen of green open space that serves to absorb rainwater. Accordingly to that, there is a need of effort to increase the absorbtion rate of rainwater into the ground; one of the methods is using biopori holes. The area that uses as a sample to observe the infiltration rate using biopori holes is Kopelma Darussalam, Banda Aceh. The aim of this study is to measure the infiltration rate before and after the application of biopori holes, find out the amount of holes that required, discover the differences of ground water level every 2, 5, and 10 years return period and find out the amount of water that absorbed to ground water. The data which uses to calculate the water level is the annual maximum daily rainfall, land area under forest cover and texture, and infiltration rate. Meanwhile, the data which uses in order to calculate the volume of water infiltrated into the soil in each month is monthly rainfall data, land area which classified by the cover and texture and infiltration rate. Inundation height at annual maximum rainfalll plan for 2, 5, and 10 years return period with using 100, 160 and 400 holes for each 100 m^2 is lower from 9.01% to 77.43% compared to not using biopori holes. In addition, the amount of water that infiltrated after using biopori holes is increasing up to 4120 m^3 for each month.

1. Introduction

The ongoing development of life requires a transformation of land-use that continues to increase. The expansion of land development for settlement and other infrastructure has impact for lessen of green open space that serves to absorb rainwater. These conditions has resulted intoinundation affected by rain become higher and the amount of rainwater that infiltrated reduce [1-2].

Due to the growing water demand and the main source of water comes from rainfall; efforts are required to effectively absorbp rainwater into the ground. Some technologies that increase water infiltration into the soil such as infiltration basin, infiltration trench, and french drain have long been introduced to society. However, the water absorption technology is not yet wide spread applicability for various reasons. Among others is the needed of a relatively large place, a relatively long time, and the cost is not economical. Thus, a further technology is still needs to be developed to be more appropriate, maintained at more economical cost, and environmentally friendly. One of the method is using biopori holes [3].

The aim of this study are to measure the infiltration rate before and after the application of biopori holes, determine differences for every annual maximum rainfall plan for 2, 5 and 10 years return period. Other than that, to know the amount of water that infiltrated into the soil. Biopori hole is cylindrical holes made vertically into the ground with a diameter of 10 cm and about 80-100 cm depth or in the case of shallow groundwater, not exceed the depth of groundwater level. The space between the biopori holes ranged between 50 cm to 100 cm. Holes filled with organic waste. Surface run off will go into the hole and seep into all directions around the hole through biopori [3].

Kopelma Darussalam is the center of academicals activity in Banda Aceh. Besides, the population growth in this area is increasing year by year. That is why to know the infiltration rate with and without using biopori holes, so that the differences of infiltration rate in both conditions can be measured [2]. Before doing field measurement for using or not using biopori holes condition, first to do this is to analysis the geospatial information for the area so that the area to apply biopori holes can be determined for each variation of soil textures. The soil textures for Kopelma Darussalam area according to soil textures map is divided in two types which is sand and silt. Whereas the land coverage that investigated is grass-covered, weed-covered and uncovered soil. After each type of land condition has classified, and then the field measurement for each land condition can be processed [3].

Infiltration is the entry process of the water from the surface in to the ground. Inside the soil water flow in the lateral direction known as interflow, flowing to the lakes and rivers springs, or vertically, known as percolation, flowing to the groundwater [4]. The calculations is done based on the amount of rainfall statistical formula as follows $R_T = R_{bar} + K.\ S_d$ [2, 5].

Calculate the volume of rain flow (direct runoff) that is Volume = Direct flow x Time. In this case, volume is expressed in cubic meters, direct flowing m^3/sec and the time in seconds. Furthermore, the flow depth is calculated, that is the total volume divided by the area of watershed, depth (m) = Volume (m^3)/Area (m^2) [4]. In estimating flow caused by heavy rains, evaporation and evapotranspiration that occurs within a short time periodic small and can be ignored, so that the formula becomes, flow = (presipitation-infiltration). The Geographic Information System (GIS) is a series of systems that utilize digital technology to perform spatial analysis[2].

2. Experimental apparatus and procedure

The tools used for this study includes drill biopori; steel ruler 20.0 cm and 30.0 cm; gauge; modified pipes with a diameter of ± 10 cm height, 25 cm for size measurements without biopori hole sand , 100.0 cm for the measurement biopori holes; hollow cans with ±15.5 cm diameter and 16.0 cm high; stopwatch; and stationery. Material that used in this study is composition of the fertilizer plant with a mixture of 1:1:1 (soil: rice husk: cow dung) with the fertilizer plan is using as the biopori filler and sufficient water.

The maps that used in this study is satellite image for Kopelma Darussalam from 2011, which sourced from Regional Development Agency Banda Aceh; Administration Map; Soil Texture Map for Kopelma Darussalam; annual maximum rainfall plan from BMKG Blang Bintang, Montly Rainfall Data from BMKG Blang Bintang, Aceh Besar. Spatial analysis of the study area is done by using the Geographical Information System. Input data including data from satellite imagery. Restrictions on land units are doing through the interpretation of satellite imagery. Interpretation of satellite image classification focused on the preparation stage to restrict land units that have the same physical characteristics.

For the implementation of infiltration measurements using modified pipes and hollow cans instead of ring infiltrometer. The first step is clearing the land-area of study; and prepare the tool at the location of measurement; the hollow can is pressed into the soil surfaces into ±5cm depth; modified pipes measuring 25.0 cm high and 10.0 cm in diameter is pressed in to the soil surface in to ±5cm depth; the remain high above the surface is 20.0 cm; pipe already laid ±5cm into the soil surface, vertically inside the pipe is placed a 20.0 cm steel ruler and water is poured into the pipe until full. Once the water is full, the stopwatch is started and every 2 minute interval water degradation seen in the ruler is noted; when the water runs out, but the decrease in water level has not showing the constant infiltration rate, the water can be added with observe or record the position of the clock in the delivery of addition water into the pipe. Experiments' performed in three types of land cover which is uncovered, grass-covered and weed-covered, on sand and silt soil texture.

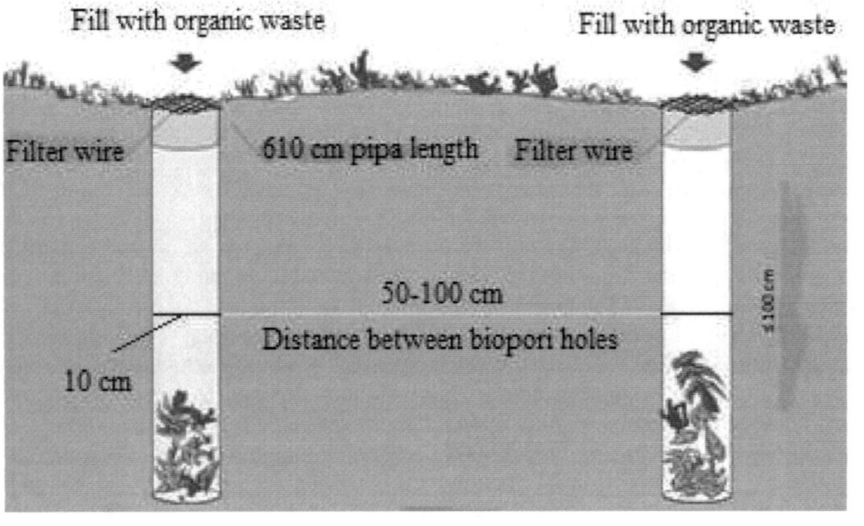

Figure 1. Biopori holes

First step is determine the land-area of study and prepare the tools on the location of measurement; create cylindrical holes using a drill biopori vertically in to the ground with a diameter of ±10.0 cm; and ±70.0 cm depth or lower if ground water level is shallow; press the modified pipes measuring 100 cm in height and diameter ±10.0 cm in to the existing hole. The height of the pipe left above the surface is 30.0 cm; fill the hole with fertilizer plants until its level with soil surfacee; affix a 30.0 cm ruler whose given double-sided tape to the inside of the pipe vertically. Water is poured into the pipe until full; once full of water, stopwatch is turned on and each time interval 30 sec the degradation of water level visible in the liner noted; when the water runs out, but the decrease of water level has not demonstrated a constant infiltration rate, more water can be added when observing or recording the position of the clock at the delivery of addition water into the pipe.

Rainfall data were obtained from a rainfall station daily rainfall data. Rainfall data were obtained at 53 years. Rainfall data that does not follow a normal distribution and lognormal, then expected to follow the Gumbel distribution. This estimate proved by Kolmogorov-Smirnov test.

To calculate the inundation level in the region of Kopelma Darussalam, the area is divided into several parts, where the parts are separated based on soil texture and land cover. In calculating the volume and inundation level using biopori holes, the biopori holes are divided into 3 different patterns. For the distance between holes is 50.0 cm the holes per 100.0 m^2 is 400 unit, distance between holes 75.0 cm the holes per 100.0 m^2 is 160 unit and distance between holes 100.0 cm then the holes per 100.0 m^2 is 100 unit.

Step calculations to determine the inundation level is recounted annual maximum rainfall plan for 2, 5, and 10 years return period; calculate the rate of infiltration without using biopori holes (uncovered, grass-covered and weed covered soil) with two types of soil texture; calculate the area of each land cover with two types of soil texture. In this case, to facilitate the calculation,each soil texture is divided into several sections or segments of land cover (egs and withgrass-covered, and so on); types of land cover and soil texture that has already grouped, calculated by finding the difference between the total area with rainfall and the rate of infiltration of the area, with return period of 2, 5, 10 years; then the volume of inundationon the uncovered, grass-covered, and weed-covered soil summed together with the volume of inundation from the rain that fell on top of the building. So the indundation pool of total volume obtained for each segment after the total volume obtained calculate the water level for each segment by dividing the volume of a inundation with total area of each segment. Do this until the volume and level of inundation for all of the area is calculated in various return period of rainfall.

Step calculations to determine the inundation level is recounted annual maximum rainfall plan for 2, 5, and 10 years return period; calculate the rate of infiltration after the aplication of biopori holes (uncovered, grass-covered and weed covered soil) with two types of soil texture and infiltration rate without biopori holes. First calculate the area of the land that using biopori hole (uncovered, grass-covered and weed covered soil) with two ground textures. Then calculated according to the distributions segment. First, for 80 biopori holes every 100.0 m^2, count the numbers of holes that will fill these lands according to existing land area in the segment. After the number of holes on the land is acquired, it will be known how much land in the segment that uses biopori holes in various types of land cover and soil texture. And the land that does not use biopori holes calculated by subtracting the extent of land that has been given biopori holes; at the calculation of the volume of the inundation with biopori holes, the type of land cover and soil texture that already grouped, calculated by finding the difference between the total area with rainfall and the rate of infiltration of the area, with return period of 2, 5, 10 years. Then the volume of inundationon the uncovered, grass-covered, and weed-covered soil that not use and using biopori holes summed together with the volume of inundation from the rain that fell on top of the building. So the pool of total volume obtained for each segment. Later, the total volumes obtained and calculate the water level for each segment by dividing the volume of a inundation with total area of each segment. Do this until the volume and level of inundation for all of the area is calculated in various return period of rainfall. Then, the water level for the whole of the land has been acquired by repeated periods of 2, 5, 10 years.

The steps taken in calculating the volume of water that goes into the ground (infiltration) with and without biopori holes are as follows, calculate monthly rainfall IR90; calculate the infiltration rate for each soil texture and variety of land cover; classify land area before using the biopori holes based on soil texture and land cover; classify land area after the aplication of biopori holes based on soil texture and land cover; area that using biopori holes calculated based on the distance between holes which ranged between 50.0 cm to 100.0 cm; to determine the volume of water that enters the soil before using biopori holes. Every month is calculated area of the land that has been classified earlier with each infiltration rate or monthly rainfall, then calculate the overall volume in a region Kopelma Darussalam, Banda Aceh; to determine the volume of water that enters the soil after use biopori holes (with the number of biopori holes variated per 100.0 m^2 with a distance between holes 50.0 cm-100.0 cm), then in each month is calculated area of land that has been classified as partially before the infiltration rate of each individual or monthly precipitation, including land which using the biopori holes. Instead of that, calculate the overall volume in the region Kopelma Darussalam, Banda Aceh and calculation of the volume of water that enters the soil before and after use biopori holes put in the chart for every month.

3. Result and Discussion

Rate of infiltration area without and by using biopori holes. Infiltration rate measurements that performed on Kopelma region Darussalam, Banda Aceh shows that the infiltration rate for sandy soil texture is faster than the silt-textured soil, with a variety of land cover. Land that is not covered absorbs water more slowly than land which has a cover such as grass and weed. In addition, by using biopori hole the ability of soil to absorb water is much faster than not using biopori. This is because the hole that has been filled biopori fertilizer plant has more cavities that allow water to be rapidly absorbed into the ground in lessen time. This result can be shown in table 1, 2, 3 and 4 as follows;

Table 1: Rate of infiltration at the various land cover and soil.

Variation of Land Cover	Infiltration rate (mm/min)

	Sand	Soil
Ground without cover	0.07	0.025
Grass	0.09	0.033
Weeds	0.100	0.042
Biopori holes	8.000	1.000

Table 2: The water level at return period of 2, 5 and 10 years.

Return period (years)	Rainfall (mm)	Biopori holes (mm)			
		0	100	160	400
2	114.86	83.11	67.03	57.37	18.76
5	172.34	140.60	124.51	108.25	76.24
10	210.40	178.66	162.57	152.92	117.31

Table 3: Water volume per month at variety number of biopori holes.

Variation in the number of biopori holes	The volume of water entering the soil per month					
	January	February	March	April	May	June
0	58253	53737	58221	58253	55667	17769
100	59122	53878	58483	58738	58613	17835
160	59643	53963	58640	59029	58830	17851
400	61729	54301	59267	60194	59695	17879

Table 4: Water volume per month at variated number of biopori holes

Variation in the number of biopori holes	The volume of water entering the soil per month					
	July	August	September	October	November	December
0	13384	54012	58037	35313	58253	58253
100	13363	54158	58259	35456	59283	58987
160	13351	54246	58393	35459	59901	59428
400	13302	54597	58928	35467	62373	61191

4. Conclusion

Infiltration rate without biopori holes: Uncovered soil with sand textured is 0.07 mm/min and silt textured is 0.025 mm/min. Grass-covered soil with sand textured is 0.09mm/min and silt textured is 0.033mm/min. Weed-covered with sand-textured is 0.10 mm/min and silt textured is 0.0416 mm/min. Infiltration rate using biopori holes for sand-textured soils is 8.0 mm/min and silt-textured is 1 mm/min. Inundation height at annual maximum rainfall plan for 2, 5, and 10 years return period with using 100, 160 and 400 holes per 100.0 m^2 is lower from 9.01% to 77.43% compared to not using biopori holes. In addition, the amount of water that infiltrated after using biopori holes is increasing up to 4120 m^3 for each month, compared to not using biopori holes.

Acknowledgments

The author would like to be very grateful for the support of the final assignment supervisor who has provided biopore drilling tools and a place to conduct drilling.

References

[1] Ahmad S, 2010 *Analisis Ketersediaan Lahan dan Kesesuaian Tanaman Untuk Reboisasi, Program* Studi Magister Konservasi Sumberdaya Lahan , Syiah Kuala University **vol 1** pp 4-5
[2] Syahrizal. 2011, Penggunaan *Citra SPOT dan SIG Untuk Menentukan Lahan Kritis di DAS Krueng Tripa Bagian Hulu Kabupaten Gayo Lues*, Program Studi Magister Konservasi Sumberdaya Lahan, Syiah Kuala University pp 14-15
[3] Brata R., Kamir 2008 *Lubang Resapan Biopori Penebar Swadaya* **vol 2** pp 25-28
[4] Triatmodjo B., 2009, *Hidrologi Terapan, Beta Offset,* **vol 3** pp 34-37
[5] Suripin, 2004, *Sistem Drainase Perkotaan yang Berkelanjutan*, *Andi Offset*, **vol 1** pp 20-22

IOP Conf. Series: Materials Science and Engineering **506** (2019) 012031 doi:10.1088/1757-899X/506/1/012031

Implementation the Trilogy Juran in SMEs Business Case Study in Indonesia

Ayi Tejaningrum[1*]

[1]Faculty of Management, STIE Ekuitas, Bandung, Indonesia.

[*]Corresponding author: ayi.tejaningrum@ekuitas.ac.id

Abstract. The study was conducted on 12 Small and Medium Enterprises (SMEs) in 4 cities in West Java Province using qualitative research methods in-depth interviews became the main source for obtaining data. SMEs have major weaknesses with regard to low product quality. The purpose of this study is to obtain information and data analysis regarding how the Juran Trilogy is applied in quality management. Grounded theory is used for research methods with the Snowball Sampling approach to determine the number of samples. Based on field data shows that Indonesian SMEs have weaknesses in the planning phase especially for the transformation of the quality dimension into the production process. This condition is caused by weak access to technology and raw materials. In the process of quality control, it is generally already well done by SMEs but it has not been supported by the process of recording a new reactive to the problem, so the high inconsistency in the quality dimension. The government must be involved in assisting SMEs to access mastery of technology and raw materials and improve the skills of SME owners to be able to carry out planning, control and quality development (The Trilogy Juran) transformation for all their employees.

1. Introduction

Joseph M. Juran is one of the world leaders who focus on quality development in the company. Juran is famous for its quality management Trilogy concept. In addition to Juran, there are other prominent figures such as W. Edward Deming who is famous for Deming's 14 point, Philip B Crosby with his Quality is free, and A.V. Feigenbaum who spawned the three steps to quality concept of Walter A Shewart, Kaoru Ishikawa and Genichi Taguchi. Joseph was born on December 24, 1904 in Braila-Moldova, Dr. Joseph M. Juran from a father named Jacob who is engaged in the shoe business. History notes how Juran suffers from anxiety when he sees his father's business suffering from an inability to manage product quality. Coming from this concern, Juran did the analysis and gave birth to the concept of Trilogy Juran's. In this model introduced three stages in the process of quality management that is quality planning, quality control and quality improvement.

SMEs in Indonesia today become the backbone of the nation's economy because it is a dominant business population that reaches 99.6%. Grows 59.08% of GDP and contributes 14.05% of exports volume (Business Profile of Micro, Small and Medium Business SME Bank Indonesia, bi.go.id). On the other hand, Indonesian SMEs experience some problems especially limitations in market access, access to financial institutions, access to raw materials as well as access to limited natural resource ownership. Many factors affect the inability of SMEs to access the market including the limited knowledge and skills of SMEs in formulating the wishes and expectations of consumers into the quality of technical dimensions; the courage to bear the low risk, the mastery of technology is still limited in terms of upstream industry downstream macro.

At the paper OECD Ministerial Conference on Small and Medium-sized Enterprises, taking place on 22-23 February 2018 in Mexico [1] mention SMEs are lagging behind in the digital transition and are disproportionately affected by trade barriers deficient intellectual property protection and quality of infrastructure and institutions [1]. SMEs managers must be able to improve understanding of quality and apply it to the ability to compete in a turbulent business environment and ensure the adoption of quality culture in all aspects of the organization [2]. Total quality management is positively related to the performance of SMEs in the manufacturing sector [3]. Shows that the low capability of SME products is due to a weak quality culture [4]. Problems relating to the ability to meet the expectations and desires of consumers or meet the quality of products that consumers expect becomes a major thing and important to be analyzed further. In previous research conducted by the authors show the quality of SME products is still far from the value expected by consumers especially with respect to the consistency of quality dimensions [5]. As mentioned earlier, Dr. Joseph M. Juran Engineering who is very interested in the field of product quality development gave birth to Trilogy Juran models are: quality planning, quality control and quality improvement [6]. This paper attempts to explain how Indonesian SMEs are implementing the Trilogy Juran, to further analyze why the quality of SME products has not been as expected.

2. Theoretical review

2.1 The Juran's model

The Juran Trilogy can summarize in three primary managerial functions. Juran's views on these functions are explained in the following section [7]. Hence, figure 1 shows The Juran Trilogy.

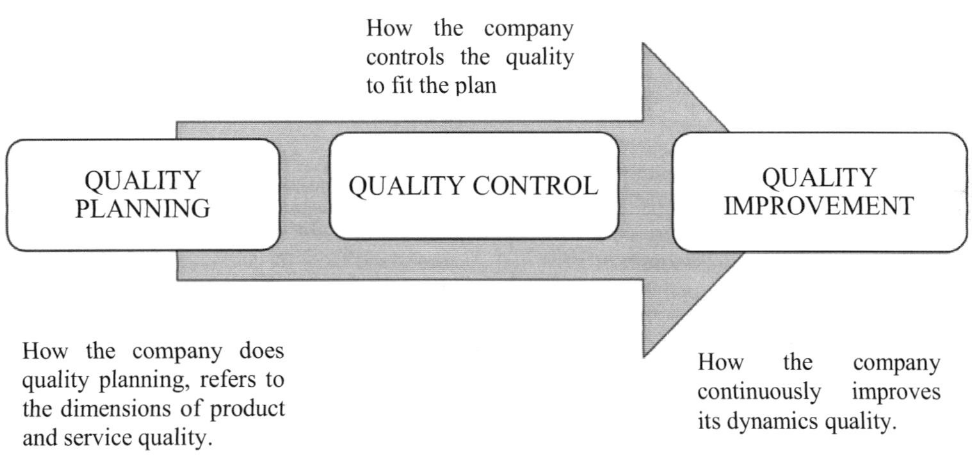

Figure 1. The Juran Models

From the picture above, the quality must first be planned, what kind of quality will be offered to consumers? Companies must have careful planning, which comes from the value of expectations and consumer desires. This refers to the opinion of Shigeru Mizuno [8], Crosby [9]. The second stage is quality control, where the company must plan well, so that the activity from the input process to the output is able to provide the quality dimension in accordance with expectations. Next, is the development stage, the company must be able to make continuous improvement of the current and future quality dimensions this is in line with the values and expectations of consumers are constantly changing [10]

2.2 Quality planning concepts
The quality planning is an activity that involves the development of products, systems and processes needed to meet or exceed customer expectations. Some of the steps needed in the first quality planning process relate to determining who the customer is [11]. The company cannot work on all customer segments because each segment has different values and wishes. SMEs must first determine who consumers will be served by being distinguished by age, income, gender, culture, place of residence and so on. The second stage of quality planning is determining what consumers need or want. SMEs must be able to identify consumer needs that have been determined in the first stage. For this stage, UKM can refer to eight dimensions of product quality namely: performance, features, reliability, and conformity with specifications, durability, service, aesthetics, and perceived quality. The third stage is to develop products with features that respond to customer needs: at this stage SMEs must be able to develop products according to the features needed by customers. The fourth stage is to develop systems and processes that enable organizations to produce these features: SMEs must think and plan all components of the process that determine the quality features or dimensions needed by consumers. The final stage is to spread the plan to the operational level

2.3 Quality control
The Quality control involves the following process:

i. Firstly is assess actual quality performance in the first stage with regard to quality control is to assess the product quality directly in the field, set the sample analysis, record and process it into information.
ii. Secondly, compare performance with goal. The next stage is to compare with the criteria set in the planning stage to analyze whether there are deviations.
iii. Thirdly, act on differences between performance and goal: the last stage is to take action on the deviation that occurred [11].

2.4 Quality improvement
The improvement of quality should be ongoing and continual;

i. Develop the infrastructure necessary to make annual quality improvement: this stage SMEs have to invest with regard to infrastructure either in the form of machines or tools needed for product quality development.
ii. Identifying the specific areas need improvement and implement projects: Of the many dimensions of quality, SMEs must determine what priority dimensions should be developed to assign key performance indicators.
iii. Establish a project team with responsibility for completing each improvement projects
iv. Provide teams with what they need to be able to diagnose problems to determine root causes, develop solutions and establish control that will maintain gains made [11].

3. Methodology and sample

3.1 Methodology
In this research use qualitative research method with grounded theory approach and case study as its base. Grounded Theory by Creswell [12] is qualitative strategy in which the researcher derives general abstract theory of a process, action or interaction grounded in the views of participation in a study. Here it is seen that grounded theory is one of the qualitative methods of research where researchers can draw generalizations from an abstract theory of process action or interaction based on the views of the participants. Creswell describes the understanding of case studies namely the stages of the process to explore in depth programs, events, activities, processes on one or more individuals must be limited by

time and activity and researchers must gather detailed information using various data collection procedures during the period continuous time [12].

3.2 Sample research
The research was conducted on creative industry-based SMEs located in 4 cities in West Java namely Bandung, Cimahi, Sumedang and Tasikmalaya. Research is conducted by direct interviews with business owners. Research is conducted by direct in-depth interviews with business owners. For data collection using Snowball Sampling where the sample will be selected directly by the researcher, the analysis is done with grounded research approach, to be studied in more depth and the sample will continue to be added until the data is in relatively stable condition. In the initial stage one sample is taken for each city, analyzed followed by the second sample then analyzed and will stop the number of samples when getting homogeneous data and information.

4. Results and discussion

4.1. Description of sample SMEs
As stated earlier that the initial stage in the implementation of Trilogy Juran is quality planning related to determining who consumers will be served. There are 12 SMEs analyzed which are shown in table 1:

Table 1. List of sample SMEs

No	SMEs	Product	Description product
1	Opak Odang	Traditional Food	This food is made from glutinous rice flour which is chopped with grated young coconut and fried in palm oil. This product is traditional food of the Sumedang people, tastes salty and crunchy with a thin and round shape .
2	Krupuk Agung	Traditional Food	This "Krupuk" is made from cassava flour mixed with fish flavourings. Traditional Indonesian food at a price that is cheap and popular with most people was eaten with rice.
3	Noga Mawar	Traditional Food	This food is made from palm sugar mixed with peanuts. Usually eat as a snack with hot tea.
4	Paprici	Agriculture	Engaged in agricultural business with the main product being cultivation and distributor of paprika. This business has been running for 7 years and absorbs 4 workers and 50 farmers.
5	Qukyta Shoes	Craft shoes	This company produces special types of handmade shoes with the female target market. Established since 2016, the owner of the company is based on his hobby which is adventure to see opportunities for fashion shoes for women who have these hobbies to look fashionable.
6	Amenity	Design Interior	Engaged in interior design business specifically for household needs with a workforce of 14 and established since 2010.
7	Kulit Rancabali	Tanning Of Goat And Sheep Skin	This business is the tanning of raw goat skin and sheep to be used as material for making leather jackets. This business has been operating from the time of Japanese occupation or around the 1940s. Currently it has a workforce of 30 people.
8	Harum Sari	Candied fruit	Producing processed food with the main raw material of fresh fruit mixed with cane sugar. Fresh fruit prepared is native

			Indonesian fruit called kedondong. This business has been running since 1992, the funds currently have 10 workers
9	CV Putra Galuh	Coconut Essence	This food is processed as a result of coconut water extraction. Founded in 2008 with a current workforce of 8 people.
10	Mulayana Fasyen	Office Uniform	Office uniforms are the main business of Mulyana Fasyeh. Uniforms that are commonly produced orders from government offices such as schools, government, military, etc. This business has been running since 2007 and currently has 19 workers
11	Krupuk Apip	Traditional Food	This "Krupuk" is made from cassava flour mixed with fish flavourings. Traditional Indonesian food at a price that is cheap and popular with most people. I was eaten with rice
12	Kebaya	Embroidery	Kebaya is traditional Indonesian clothing. This business moves in embroidery for Kebaya clothing. Founded in 1974 with 30 employees

From table 1, the types of businesses and names of SMEs are sampled in 4 cities with 12 samples. Some samples are engaged in the food processing industry and a small part in the fashions and craft industries. In the analysis process explained descriptively regarding the stages of Trilogy Juran. The first concerns quality planning, both quality control and third quality development.

4.2 Quality planning implementation in SME's

How to carry out quality planning related to which consumers will be targeted by their business. There are five stages in the process of implementing quality planning namely:

i Determining Customers.
ii Identify customer's needs.
iii Product development with features that meet customer needs.
iv Implement the plan to the operational level.
v Implement the plan to the operational level.

4.2.1. Determining customers. Generally, SMEs have defined the segmentation of the customers to be served. They realize the importance of segments because it will determine the price and quality offered. The segmentation used by the SME sample is age and income. The culinary industry is 100 percent based on revenue. While for agriculture and garment industry is age and income. Price segmentation in food SMEs is a lower segment with a price range of 30% to 40% cheaper than similar products issued by large businesses. For example, in the fruit processing industry, the price offered is IDR 15,000 / kg; the price of a large business is IDR 28,000. In interior design, shoes and fashion, the segmentation used is age and income. For shoes, take the lower middle segment with the main target of women. In the target interior design is the lower middle class for new households (age) while the border takes the segment of young people with lower middle income.

4.2.2 Identify customer's needs. Of the 12 samples, all have the same understanding of the quality requirements needed by the target audience. In food SMEs, they understand the dimensions of quality needed namely about good taste, does not contain excessive oily, clean, healthy, does not contain harmful substances and preservatives. While for the design and clothing industry, the dimensions needed regarding quality according to UKM are designs that keep abreast of the times are durable, do not contain defects and are pleasing to the eye.

4.2.3 Product development with features that meet customer needs. How SMEs develop the quality dimensions needed by consumers in the previous stage. In general, they understand and consistently make various efforts to achieve this. For example, in the interior design industry, SMEs are trying to see the development of the interior design industry through the internet and visiting developers. In the food industry, SMEs continue to seek technology to be able to meet the needs of these consumers. In the leather tanning industry, in addition to trying to improve the existing process also conduct a comparative study of tanning SMEs in other regions.

4.2.4 Develop system and process. This stage relates to the implementation planning process of the quality dimension of the process. How SMEs do this stage, in general the main problem is here, namely how to transform the dimensions of product quality in the process of input to output. SMEs generally only use felling and periodic trials. The results achieved are not a maximum, so the quality dimensions in consistencies continue to occur defective products are still being sold because they avoid losses.

4.2.5 Implement the plan to the operational level. Most SMEs have limited access to technology to meet the dimensions of product quality that consumers expect. In the food industry there is no technology that makes "opak" and crackers not contain excessive oil, so that the coconut oil will remain in the wrapper. This condition is not liked by consumers because it will reduce the level of crispy and high cholesterol. In leather tanning SMEs, the technology they have not been able to access is how to reduce fat levels on the skin which makes the skin rot faster and contain bacteria. Also related to technology to smooth the skin is still traditional so it is not evenly distributed on all products. In the paprika farming industry, it is related to the handling process that will make paprika durable and not easily damaged.

4.3. Quality control implementation in SME's
The next stage of the Juran Trilogy is how SMEs control quality. The following is a description of the condition. There are 4 parts of the quality control process, namely first with regard to as follows;

- i Assessing the quality of actual performance.
- ii Compare performance with objectives.
- iii Act on differences between performance and objectives.
- iv Take action against deviations that occur.

4.3.1. Assessing the quality of actual performance. Overall (100%) sample SMEs inspect the quality of products well analysed since the raw material comes into the final product. The weakness is only based on felling SME owners without measurable standard and not recording events. This condition results in a variety of product variations that make in consistencies in the quality of the products produced. In addition to this, felling in the quality control process is less distributed to employees, so that if the owner experiences obstacles, the tendency of the actual performance quality assessment process will be neglected. This condition occurs in 100% of SMEs or all SMEs.

4.3.2. Compare performance with objectives. The process of comparison between the expected dimensions and events generally has been carried out by SMEs although only by using instincts and explanations, it has not been scientifically recorded and analysed. But SMEs understand what dimensions contribute to the biggest disability. The weakness at this stage is that the comparison is only qualitative such as comparison of colours, comparison of crispness, comparison of oil content, comparison of performance that is difficult to quantify due to weak access to technology.

4.3.3. Act on differences between performance and objectives. Some measures have been taken by SMEs to maintain product quality performance including:

- i Firstly is involving human resources to control quality and working with the action plan in quality maintenance.
- ii Secondly is using good equipment, temperature regulation and storage.

iii Then, in order to keep the product from moisture, once again the process of distributing knowledge regarding quality control is not fully owned by all company employees.

4.3.4. Take action against deviations that occur. Some actions often deviate from SME quality planning, especially when prices of raw materials and auxiliary materials rise. SMEs have control over the use of raw materials; they understand the very close relationship between the use of raw materials and product quality. In processing SMEs try to fulfil the raw material in quality one. The problem arises when the price of raw materials is high, there are two actions taken by SMEs, namely raising prices or decreasing product quality by using two quality raw materials. Based on the results of interviews with 12 SMEs, samples of 9 SMEs 75% did the first act, namely replacing raw materials with the second quality, and as many as 25% or 3 SMEs will increase prices by first communicating them to consumers.

4.4. Quality Improvement Implementation in SME's

There are 4 actions that must be carried out in the quality development phase namely:

i Developing the infrastructure needed to carry out annual quality improvements.
ii Identifying certain areas that need improvement and implementing improvement projects improvement projects.
iii Provide the team with what they need to be able to diagnose problems to determine the root cause develop solutions.
iv Establish controls that will maintain the results made.

4.4.1. Developing the infrastructure needed to carry out annual quality improvement. In general, the infrastructure used by SMEs is still simple both from buildings, road access, warehouses, production sites, and machinery and production equipment. All SMEs combine buildings for businesses with homes. From field research and interviews, SMEs generally set aside a portion of the profits to repair business buildings, buy equipment and machines that are needed. With regard to quality improvement in infrastructure development it is not specifically intended for quality improvement.

4.4.2. Identifying certain areas that need improvement and implementing improvement projects. There is no specific area for quality improvement; the area used is combined with the location of the process. The business owner is the main initiator in the process of repairing and developing quality. SMEs consistently have quick reactions to quality improvements, but are incidental when problems arise.

4.4.3. Establish the project team with responsibility for completing each improvement project. Have not formed a committee or project that is specifically served to make quality improvements. Repairs are carried out by all employees and business owners [13]. There is no specific section for quality development.

4.4.4 Provide the team with what they need to be able to diagnose the problem to determine the root cause, develop a solution, and establish controls that will maintain the results made. Conditions that have not been scientifically conducted are based on written data and facts. UKM only conducts it based on the initiative of the owner and also felling which has been carried out for generations. The stage of diagnosing the problem has been done well and determining the control that must be done has also been done well but the quality development decision is always hampered by the costs of production and access to the technology that has not been obtained.

Acknowledgments

The author highly appreciates the Director General of Higher Education of the Republic of Indonesia who has provided financial support for the implementation of this activity.

IOP Conf. Series: Materials Science and Engineering **506** (2019) 012031 doi:10.1088/1757-899X/506/1/012031

References

[1.] OECD 2018 *Strengthening SMEs and entrepreneurship for productivity and inclusive growth* . SME Ministerial Conference 22 - 23 February 2018 Mexico City
[2.] Saumyaranjan .S Y 2018 *Total Quality Management in Indian Manufacturing* SMEs. Procedia Manufacturing **Volume 21** Pages 541-548
[3.] Edwin A, Evawere A 2016 *Impact of Product Quality Management on the Growth of Small and Medium Sized Enterprises*. *European Journal of Social Sciences,* **Vol. 52**, No. 2 2016 Page16
[4.] Tejaningrum 2016. *Quality Culture and Capability Process Supply Chain of SMEs. International Journal of Innovation and Organization*. **Vol.9.2** Oct 2016: 214-225
[5.] Tejaningrum . A. 2014 ..*Inconsistency Quality Products (on Six Sigma Programs) Effect on Customer Loyality: Case Study of Industri "Tahu". Journal "Management Study" ISSN* 2328-2185, January 2014. **Vol 2 No 1**. David Publishing USA
[6.] Juran 2015. *Juran's Quality Management and Analysis*. Mc Graw Hill International Edition. Singapore.5-19
[7.] Juran J. M.1995 *Quality designing, creating the new definition of quality into goods and services*. Jakarta: Library Binaman Pressindo
[8.] Shigeru M 1994 Wide Total Quality Control. Jakarta. Management Institute
[9.] Crosby P.B 1986 *Quality is Free, The Art Of making Quality Certain* New York, McGraw-Hill Book Co
[10.] Juran M. Joseph 1989 *Juran on Quality By Design*, New York, McMillan Company
[11.] Goetsch L. D., Davis B. S.2014. *Quality management for production, processing and services*. New Jersey: Prentice-Hall
[12.] Cresswell Jhon W 2009 *Reseach Design, Qualitative, Quantitative, and Mixed Methods Approaches* Sage, LA
[13.] Tejaningrum. A. 2016. T*rasformational Knowledge and Skills Realted to The Establisgment of quality culture and effect on products quality. Proceeding The Fifth InternationalConference on Entrepreneurship and Business Management* (ICEBM 2016)

The Influence of Water Level Fluctuation Reservoir Stability of the Earth Dam

Maimunah[1], M.Yeni[1]and D.Kumala[1]

[1]Civil Engineering Department, University Muhammadiyah of Aceh

[*]Corresponding author: maimunah@unmuha.ac.id

Abstract. A large dam construction failure which caused in reservoir by ground water level fluctuations very extreme. Construction of earth dam failures can be avoided if the conditions of the dam slopes are in stable condition. This research discusses the influence of reservoir water level fluctuations on the stability of the earth dam models in the laboratory by analyzing the stress-strain and deformation using PLAXIS program version 8.2. Implementation of the research done in stages which include testing properties of the building blocks of the model earth dam and physical modelling in laboratory experiments with variations in soil types, namely: a mixture of 90% sand-silt 10%, and 80% sand-silt 20% and 100% soil. The slope of the model on the upstream 1: 1 and 1: 1.5, while in the downstream dam slope was maintained constant 1:1. Instrumentation used includes 6 soil moisture sensors with three accuracy of 2%, 1 deformation pressure sensor with detection range, 3 dial gauges, capacity level 1 meter with accuracy 2%. Furthermore, analysis on the stability of the earth dam modelling in laboratory tests using PLAXIS version 8.2. The results of this analysis are total displacement Plaids, total stress, total strain, pour water pressure and the safe rate (safety factor). Based on the review of the three nodal, at upstream slope of the dam are same, the effective value of the largest stresses occur in the model 10% soil 90% sand with the upstream slope 1:1, that is equal to -16.52 kN /m^2 lowering phase. Total value of largest stresses occur in the model 20% soil 80% sand with the upstream slope 1:1, that is equal to -12.89kN/m^2 phase rising 3 phase displacement (vertical displacements, the total displacements, the total incremental displacements and incremental strains) , the largest value occurs in the model 100% of the land, the vertical displacements 8.12 x10^{-6} m, the total large displacements 3.92 x 10^{-5} m, 1.43 x 10^{-5} m total incremental displacements and incremental strains for-1.75 x 10^{-2}. Lowest safety factor (critical condition) due to raising and lowering of water level in the upstream occurred at a slope 1:1 at 1.4710.

1. Introduction

Dam is a construction that serves water reservoir and retaining water on a large scale. Earth dam serves as a lifting surface of water in reservoir and downstream towards retaining water seepage and water buffer tank. Many of the dam construction failures caused by ground water level fluctuations in the reservoir. Dams land prone to overtopping. This study is continuation of several previous studies that aim to analyse the influence of rising reservoir water level (rising of water level) [1-5] and a decrease in water level (rapid drawdown) to the structural integrity of the earth dams and modelled in 2D modelling [6-7]. In this study, earth dam is physically modelled in laboratory experiments with 3D modelling. The purposes of this study are;

i. Analyzing deformation and land slope stability weir reservoir water level fluctuations due to the physical way modelling in laboratory models and numerical calculations using Plaxis software version 8.2.
ii. Knowing the mechanisms and causes of avalanches on land weir body.
iii. Knowing earth dam deformation patterns visually in the laboratory and compared with the deformation generated using Plaxis program.
iv. Providing solutions precaution against the possibility of instability slope and collapse due to the soil body dam reservoir water level fluctuations.

2. Research Methods

2.1 Location Research
The experiment had been done at the Hydrology Laboratory of the Faculty of Civil Engineering, Gadjah Mada University, and Yogyakarta. Testing value of index properties of soil and sand carried out in the Laboratory of Soil Mechanics Faculty of Civil Engineering, Gadjah Mada University, Yogyakarta.

2.2 Organisations
Materials used for the study are as follows:

i. Mix of soil and sand through sieve No. 4 (4.75 mm) and retained at No. 200 (0.075 mm).
ii. Sand through sieve No. 10 (2.00 mm) sieve and retained on the No. 200 (0.075 mm).
iii. Water.
iv. Preview research reports.
v. Text books, proceedings, journals, referent, final report and thesis statements.
vi. Tool which is used in this research is computer, digital camera, handy cam, six soil moisture sensors with an accuracy of 2%,1 deformation pressure sensor with detection range, 3 dial gauge, capacity level 1 meter with accuracy 2%, moisture sensor, ruler, measure cup, malls and stationery,.

Figure 1. Material and Tool

2.3 Stages of research

Research conducted consists of several stages as follows;

i. Study literature.
ii. Preparation of equipment and materials.
iii. Sampling.
iv. Preparation of mixture of soil and sand properties testing.
v. Testing properties for mix soil.
vi. Determination theoretical compaction of water content.
vii. Making a model laboratory test tube.
viii. Value testing of soil density and water content for model.
ix. Simulation (simulation model dam in the laboratory and simulation with finite element models using PLAXIS version 8.2).
x. Analysis the results and discussions.

3. Results and Discussions

3.1 Result and analysis

The results of laboratory experiment is seepage discharge data (q), coefficient of permeability (k), amount of deformation and landslide visual on earth dam model. The results can be seen in the table as follows:

The finding in table 1 shows the results of material properties of the models.

Table 1: Results of material properties of the model.

Soil Type	Specific Gravity	Direct Shear Test		k (cm/s)
	Gs	c (kg/cm^2)	Φ(°)	
Sand 100%	2.80	0.01	42.44	0.560579
Soil : Sand (10%: 90%)	2.81	0.01	39.91	0.011496
Soil :Sand (20%:80%)	2.75	0.12	43.71	0.009561
Soil 100%	2.62	0.54	38.86	0.000235

Next is the laboratory experiments result that can tabulate as table 2 below;

Table 2: Results of laboratory experiments.

No	Data	Electronic	Manual
1	Water raising and rapid drawdown	○	○
2	Soil Moisture	○	✖
3	Soil Stress	○	✖
4	Deformation	✖	○
5	Seepage	✖	○

Other than that, there are figures that can be explained in details for this paper. The details are as figures shown. The figure 2 shows the example of plot vertical displacement due to gravity load. Next, figure 3 is the example of plot total displacement on the rising 1 phase. Meanwhile, the figure 4 visualized the incremental strains on rising 2 phase while figure 5 show the effective stresses on rising 2 phase After that, for figure 6 is the example of plot total incremental displacements on rising 3 phase Hence, all this figures can presents as follow;

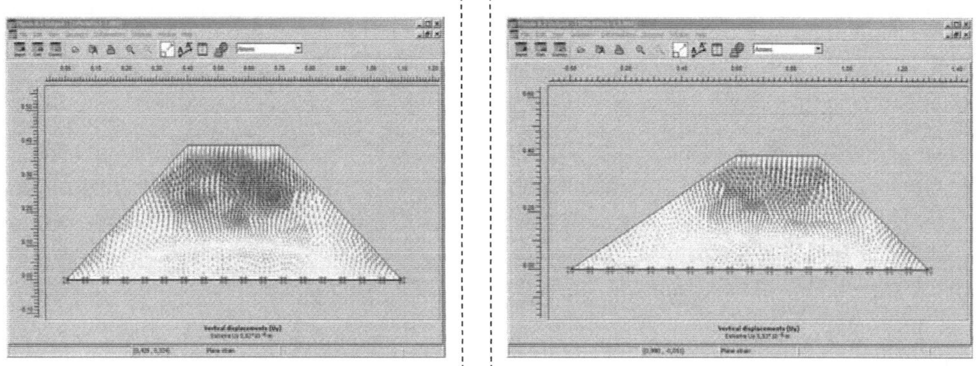

a) 10%:90% upstream 1:1

b) 10%90%, upstream 1:1.5.

Figure 2: Example of plot vertical displacement due gravity load

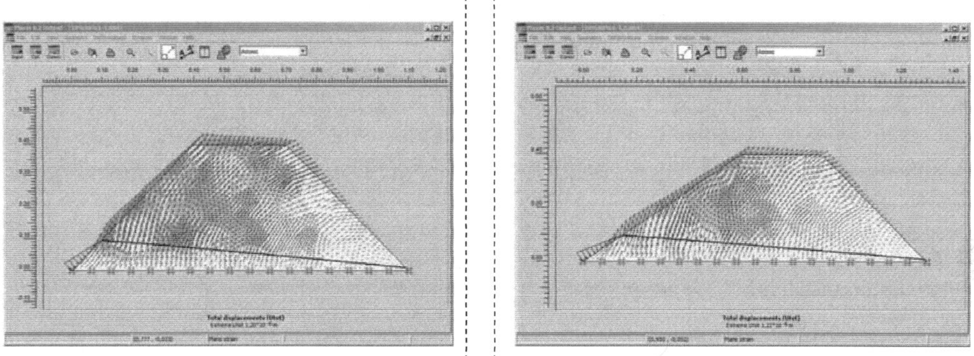

a) 10%:90% upstream 1:1

b) 10%90%, upstream 1:1.5.

Figure 3: Example of plot total displacement on Rising 1 phase

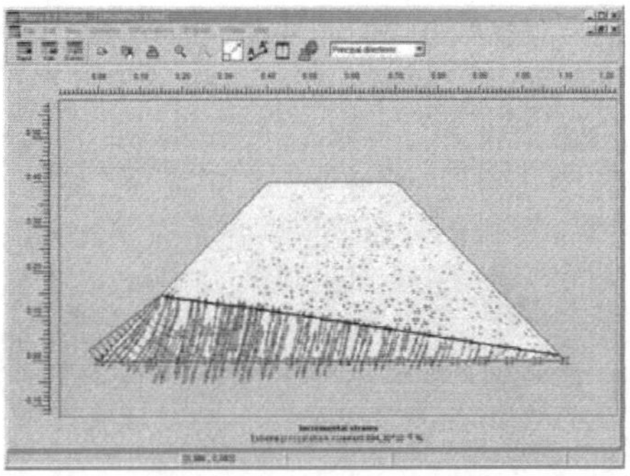
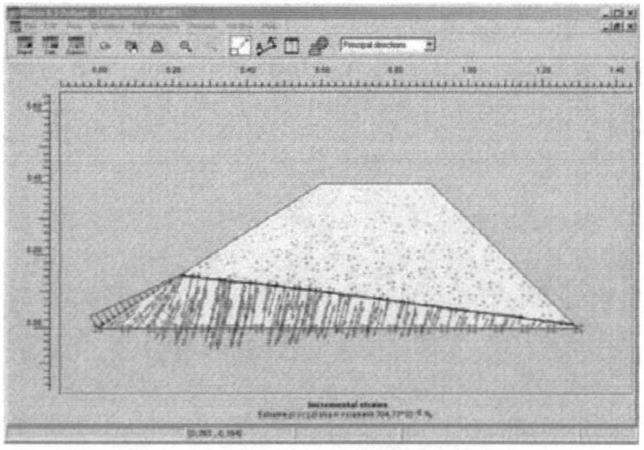

a) 10%:90% upstream 1:1 b) 10%90%, upstream 1:1.5.

Figure 4: Incremental strains on Rising 2 phase

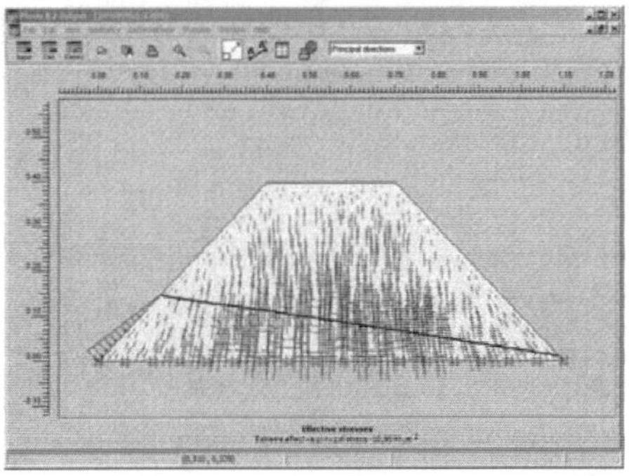
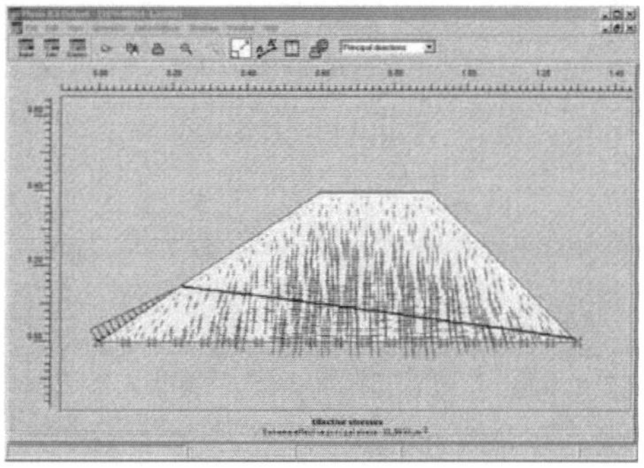

a) 10%:90% upstream 1:1 b) 10%90%, upstream 1:1.5.

Figure 5: Effective stresses on Rising 2 phase

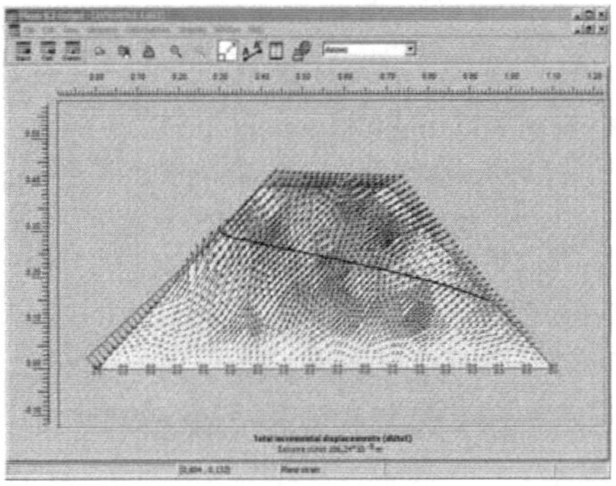
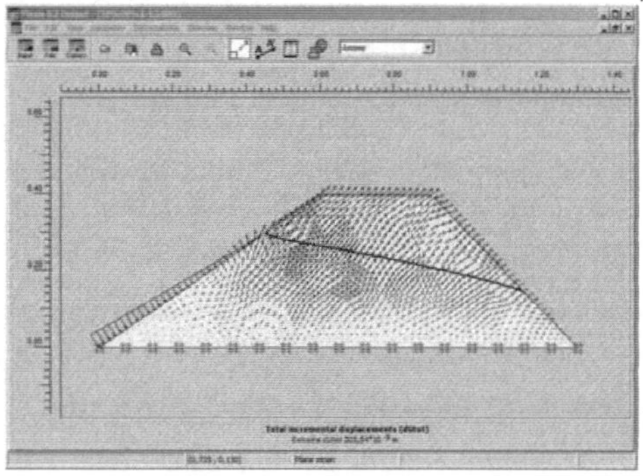

a) 10%:90% upstream 1:1 b) 10%90%, upstream 1:1.5.

Figure 6: Example of plot total incremental displacements on Rising 3 phase

3.2 Analysis of safety factor control
The safety factor control result can be tabulate as table 3 below;

Table 3. The result of safety factor.

Calculation	Phase	Soil: Sand							
		Sand				Soil			
		100%		10%:90%		100%		20%:80%	
		1:1	1:1.5	1:1	1:1.5	1:1	1:1.5	1:1	1:1.5
Safety Factor	Rising	1,7083	1,6966	5,5353	5,5684	> 10	> 10	9,1915	9,2225
	Lowering	2,1967	2,1935	6,0641	6,1921	> 10	> 10	9,4379	9,7137

4. Comparison between observation and result analysis laboratory plaids
In this study, a comparison between the observations in laboratory with PLAXIS analysis results can be only qualitative. This is because of;

i. Value land properties in laboratory models and models PLAXIS not same.
ii. Raising treatment and a decrease in water level in the laboratory and in Plaids are not same.
iii. Less accurate measuring devices (dialgauge) used in the measurements in the laboratory, especially in terms of installation

Based on observations in the laboratory and the results analysis of the analysis in PLAXIS, obtained qualitatively similar results. In test on laboratory models, models with more soil composition having a larger displacement due to the fewer sand composition, the smaller the value of the modulus of electricity.

From visual observation during the experiment and analysis results with PLAXIS Version 8.2 indicates that the model dam sandy soil mixed with material that more silt FK has a higher value compared with that mixture with silt is less. Safety factor (SF) increased in all models earth dam which increased the water level quickly in the upstream.

5. Conclusion
Based on the results model experiments in laboratory soil earth dam and the results of numerical simulations using PLAXIS Software Version 8.2 can be concluded:

i. Displacement (vertical displacements, the total displacements, the total incremental displacements and incremental strains), the largest value occurs in the model 100% of the land, while in the vertical displacements of 8.12×10^{-6} m, the total displacements amounting 3.92×10^{-5} m, the total incremental higher 1.43×10^{-5} m displacements and the incremental strains at-1.75×10^{-2}%.
ii. For models with the same composition of the constituent materials, the model with a gentler upstream slope has a smaller displacement value.
iii. For models with the same upstream slope, the model with less sand content a great value displacement which is inversely proportional to the value of E (modulus of elasticity). This is consistent with the simulation results in the laboratory. The model having the greatest displacement is 100% ground models.
iv. The main factor to determine the degree of deformation that occurs in the model has a height water level in the earth dam. Higher water level, greater deformation. Deformation values stated direction and magnitude caused the movement.

IOP Conf. Series: Materials Science and Engineering **506** (2019) 012032 doi:10.1088/1757-899X/506/1/012032

References

[1]. Craig. R. F. 1987. *Mekanika Tanah*. Edisi keempat. Jakarta: Erlangga.

[2]. Craig. R. F. 1994. *Mekanika Tanah Jilid II*. Edisi keempat. Jakarta: Erlangga.

[3]. Dunn I. S., Anderson L. R. and Kiefer, F. W. 1980. *Dasar-Dasar Analisis Geoteknik* Semarang: IKIP Semarang Press.

[4]. Dwijaka A. 2006. *Pengaruh Variasi Kadar Air Terhadap Deformasi Lereng Di Kilometre 15,9* Saluran Irigasi Kalibawang Kulon Progo. Tesis. Yogyakarta: Jurusan Teknik Sipil Dan Lingkungan Fakultas Teknik Universitas Gadjah Mada

[5]. Fathani T. F. and Nakamura H. 2005. *A New Method for Estimating the Shear Strength Parameter at Critical Slip Surface*. Journal of the Japan Landslide Society. Japan.

[6]. Fathani T. F. and Legono D.2010. *Pengaruh Fluktuasi Muka Air Reservoir Terhadap Stabilitas Bendungan Tanah Uji Model di Laboratorium*. Laporan Penelitian DPP/SPP. Yogy FakultasTeknik Universitas Gadjah Mada.

[7]. Fathani T. F.2011. *Geotechnical Analysis of Earth Dam Failure*. Prosiding Pertemuan Ilmiah Tahunan XIV HATTI. Yogyakarta.

Fast and Non-Destructive Prediction of Moisture Content and Chologenic Acid of Intact Coffee Beans Using Near Infrared Reflectance Spectroscopy

Yusmanizar[1*], Imas Siti Setiasih[2], Sarifah Nurjanah[2], Mimin Muhaeimin[2], Bambang Nurhadi[2], Santi Rosniawaty[2], Agus Arip Munawar[1]

[1] Department of Agricultural Engineering, Syiah Kuala University, Banda Aceh–Indonesia
[2] Faculty of Agriculture, Padjadjaran University, Bandung - Indonesia.

*Corresponding author: yusmanizar.unsyiah@gmail.com

Abstract. The main objective of this present research is to apply the near infrared reflectance spectroscopy (NIRS) as a fast and non-destructive method in predicting moisture content (MC) and chlorogenic acid (CGA) of intact roasted coffee beans. Diffuse reflectance spectrum were acquired for bulk coffee beans samples (Arabica and Robusta) in wavelength range from 1000 to 2500 nm. Spectra data were corrected and enhanced using standard normal variate (SNV). Prediction models, used to predict MC and CGA of intact coffee beans, were developed and performed using combination of principal component analysis (PCA) and multiple linear regression (PCA+MLR). The results showed that NIRS can be used to predict MC and CGA content of intact coffee beans simultaneously and rapidly with maximum coefficient correlation (r) were 0.92 for MC and 0.93 for CGA, whereas residual predictive deviation (RPD) indexed were 3.67 and 3.87 for MC and CGA content respectively. Based on the obtained results, it may conclude that near infrared spectroscopy can be applied in coffee quality evaluation especially to predict moisture content and chlorogenic acid. NIRS can be used and an alternative fast and non-destructive method bypassing standard laboratory procedures.

Keywords: NIRS; coffee, CGA, Quality; non-destructive

1. Introduction

Coffee is known as famous beverages for people around the worlds due to its taste, flavour and some nutrient content. Interest in coffee quality assessment is impelled by the need to supply the consumer with a consistently high quality product at an affordable price. Indeed, quality is a major aspect for the modern coffee industry because a high quality product is the basis for success in today's particularly competitive market [1]. Moreover, coffee is also one of the most important raw materials within the international trade for which quality is quintessential. Among the several species of the genus coffee identified so far, two of these varieties are economically and commercially important: Coffee Arabica and coffee Robusta. Both coffee varieties differ mutually from a botanical perspective and in terms of quality features [2].

Arabica and Robusta are globally well known and the most traded among those species. Generally, Arabica has gained greater popularity than Robusta. A greater quantity of Arabica is traded and fetches a higher price than Robusta. Differences in prices are also determined by geographical origin. Among the several species of the genus coffee identified so far, two of these varieties are economically and commercially important [2,3].

Food quality encompasses sensory properties, nutritive values, mechanical properties, functional properties and presence of defects. The majority of traditional techniques used for quality assurance are costly, labour intensive and time consuming, an example being sensory panel evaluation that, to this date, is the ultimate tool to assess coffee quality. Besides being costly and time consuming, sensory panels are inadequate for employment in routine analysis in food processing facilities. In this scenario, infrared spectroscopy is gaining attention since it is capable of solving some of the problems presented by traditional techniques [4].

Moreover, chemical compositions in green coffee beans were used to differentiate between species. Sucrose and trigonelline in Arabica have higher levels than Robusta. Conversely, caffeine and chlorogenic acid (CGA) in Arabica are lower than in Robusta [1,4,5].

Coffee quality encompasses sensory properties, nutritive values, mechanical properties, functional properties and presence of defects. The majority of traditional techniques used for quality assurance are costly, labour intensive and time consuming, an example being sensory panel evaluation that, to this date, is the ultimate tool to assess coffee quality. Besides being costly and time consuming, sensory panels are inadequate for employment in routine analysis in coffee processing facilities. In this scenario, near infrared spectroscopy is gaining attention since it has been demonstrated capable of solving some of the problems presented by traditional techniques [6,7].

These chemical compositions; caffeine, chlorogenic acid, sucrose, trigoneline are commonly measured by the high-performance liquid chromatography (HPLC). This method has the advantage of being accurate, but, on the other hand, it is time-consuming and needs extensive sample preparation. An alternative to the HPLC is therefore required to provide simplified measurement procedures, specifically for caffeine and chlorogenic acid contents in green coffee beans [8].

In particular, researchers focusing more on Fourier transform infrared (FTIR) and NIR spectroscopy. The FTIR spectrum detects fundamental vibrations in the mid-infrared region (4000– 400 cm^{-1}) whereas the NIR spectrum (2500–800 nm) arises from the molecular absorptions of overtones and combinations of fundamental vibrational bands in the mid-infrared region. These techniques are rapid, non-destructive and require minimum sample preparation [9,10].

During last few decades, NIRS has been widely employed as an effective tool for the analysis of soil properties. Compared with traditional wet chemistry analysis, IR analysis is rapid, cost effective, non-destructive, requires minimal sample preparation and can be used in situ. More importantly, it permits a quantitative assessment of several properties from a single measurement. This technique mainly measures overtones and combinations of fundamental vibrational bands for O-H, N-H and C-H bonds from the near to mid infrared region [10, 11].

The feasibility of using infrared spectroscopy in combination with multivariate statistics has been examined with notable success in coffee quality evaluation, with applications including discrimination and quantification of Robusta and Arabica blend, coffee adulteration, and prediction coffee sensory properties [11, 12].

Therefore, the main purpose of this present study is to use the near infrared spectroscopy (NIRS) for coffee quality evaluation in form moisture content (MC) and chlorogenic acid (CGA) content respectively. We develop calibration and prediction models used to predict these both quality parameters simultaneously. We use a combination method of principal component analysis (PCA) and multiple linear regression (MLR) to establish those prediction models.

2. Materials and Methods

2.1. Samples

Green Arabica and Robusta coffee bean samples were collected mainly from Gayo, Aceh Province and some of them were collected from different geographical origins around Indonesia. Coffee samples were taken and stored for three days to equilibrate before spectra acquisition and further chemical analysis were done.

2.2. Diffuse reflectance spectra acquisition

In this study, infrared spectra data of all coffee bean samples were acquired in form of diffuse reflectance spectral data. Background spectra correction was performed every hour automatically. Diffuse reflectance spectra in wavelength range of 1000 – 2500 nm with the increment of 0.2 nm resolution were acquired 64 scans, averaged and recorded in SPA and CSV extension files format [9].

2.3. Moisture content (MC) and Chlorogenic acid (CGA) measurements

Once spectra acquisitions were completed, MC and CGA content for all coffee bean samples were measured using standard laboratory method: Gravimetry for MC measurement and HPLC for CGA content measurement. Both of these actual data were measured in triplicate and averaged.

2.4. Spectra data correction

To enhance and improve prediction performance, spectra correction was employed to all spectra data of coffee bean samples. Standard normal variate (SNV) was chosen as spectra data correction method [9].

2.5. Prediction model performance

Diffuse reflectance spectra data of all coffee bean samples were used to predict MC and CGA content using a combination method of principal component analysis (PCA) and multiple linear regression (MLR). To validate the prediction result, a leverage validation method was applied during prediction model development. Model performance was quantified using statistical indicators: coefficient determination (R^2), coefficient correlation (r), the root mean square error (RMSE) and residual predictive deviation (RPD) index. It is obvious that good model should have higher R^2 and r coefficient, lower RMSE and RPD above 1.5 [9-12].

3. Result and discussion

3.1. Spectra features of coffee bean samples

Typical Diffuse reflectance spectrum of coffee bean samples in NIR wavelength region range of 1000-2500 nm is shown in figure 1. It corresponds to hydrogen bonds (C-H, N-H, O-H and S-H) and can be used to analyze the chemical composition of organic matter. Additionally, NIR spectra detect the presence of chemical contents that contain hydrogen bonds.

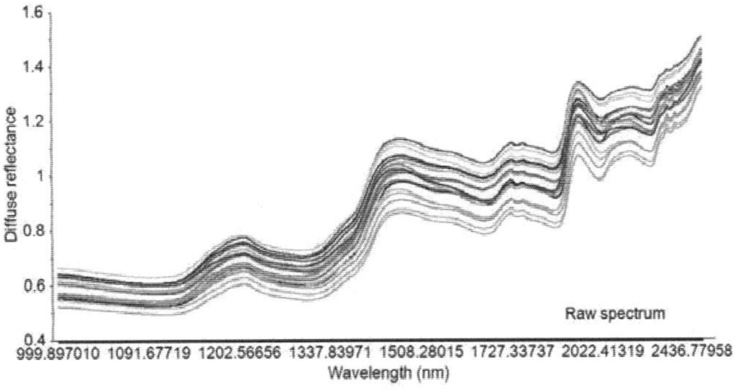

Figure 1. Typical diffuse reflectance spectra data of soil samples before correction.

Infrared spectra data obtained from NIRS instrument, sometimes generally consists irrelevant information and noises due to light scattering. These problems may cause to interfere and affect desired relevant coffee quality attributes such as moisture content and chlorogenic acid content. These noises need to be removed or minimized in order to obtain accurate, robust and stable calibration and prediction models. Therefore, we employ the SNV spectra correction method before further analysis. Spectra data for all coffee beans samples were corrected and enhanced as presented in figure 2.

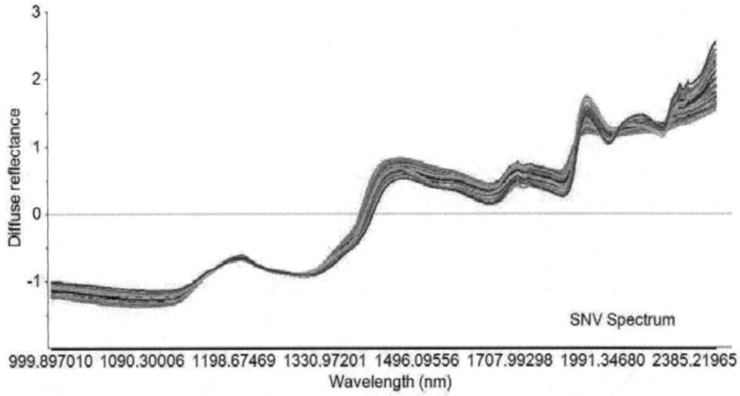

Figure 2. typical diffuse reflectance spectra data of soil samples after SNV correction.

As presented in figure 2, we may see the differences between raw and SNV corrected spectra. Some noises are reduced and we may argue that enhanced spectra may provide and achieve better and accurate prediction results compared to raw un-treated spectra.

After spectra correction was completed, then we attempted to predict MC and CGA content using a combination PCA and LMR regression method with co added of leverage validation. Raw untreated spectra data was firstly employed to predict both MC and CGA content simultaneously. Obtained result shows that MC and CGA content can be predicted better and more accurate using SNV corrected spectra data. Prediction results of MC and CGA in intact coffee bean samples were presented in table 1 and table 2.

Table 1. MC prediction results using raw and SNV spectra data.

Accuracy indicators	Raw spectra	SNV spectra
R^2	0.916	0.975
r	0.938	0.986
RMSE	0.015	0.006
RPD	3.462	4.409

MC: moisture content, R2: coefficient of determination, r: correlation coefficient, RMSE: root mean square error, RPD: residual predictive deviation. SNV: standard normal variate.

Table 2. CGA content prediction results using raw and SNV spectra data.

Accuracy indicators	Raw spectra	SNV spectra
R^2	0.512	0.557
r	0.711	0.754
RMSE	1.205	0.936
RPD	1.662	1.960

CGA: chlorogenic acid, R2: coefficient of determination, r: correlation coefficient, RMSE: root mean square error, RPD: residual predictive deviation. SNV: standard normal variate.

Maximum correlation coefficient for MC prediction was 0.986 with RPD index of 4.409, which indicated as an excellent prediction model. On the other hand, CGA prediction results was categorized as coarse sufficient model since the maximum correlation coefficient between predicted and reference

CGA data was 0.754 and RPD index was 1.960. Scatter plot derived from reference and predicted quality parameters (MC and CGA) are shown in figure 3 and figure 4.

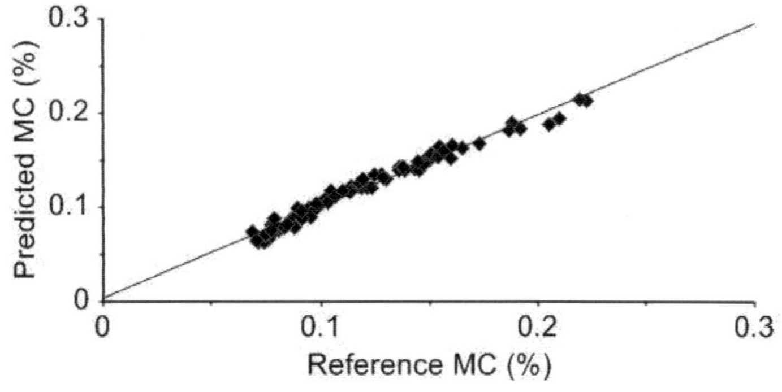

Figure 3. Scatter plot of reference versus predicted moisture content (MC) based on SNV spectra.

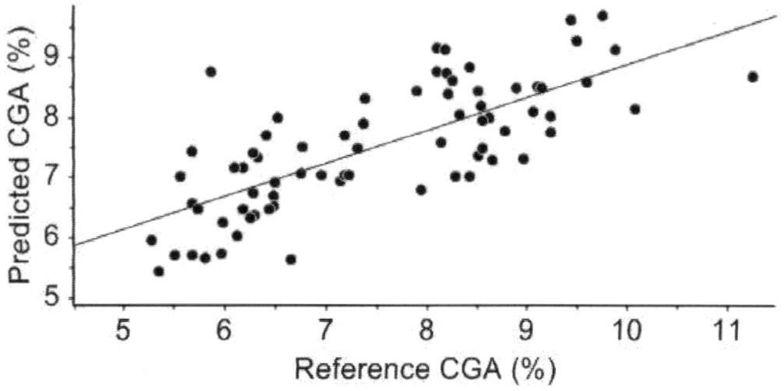

Figure 4. Scatter plot of reference versus predicted chlorogenic acid (CGA) based on SNV spectra.

Furthermore, we found on this study that MC and CGA of intact coffee bean samples can be predicted better when SNV spectra was used. Thus, we may argue that spectra data need to be enhanced and improved before further analysis. This also in agreement with some other researchers [13, 14, 15]. However, further enhancement must be conducted to increase and improve prediction accuracy. This enhancement can either adding the number of samples to be involved, or applying combination spectra enhancement method such as a combination of SNV correction method and other correction such as multiplicative scatter correction (MSC) and baseline shift correction (BSC).

4. Conclusion

Based on achieved results, we may conclude that near infrared spectroscopy (NIRS) can be used to predict moisture content (MC) and chlorogenic acid (CGA) content of intact coffee bean samples. Achieved results, shows that MC can be predicted with maximum correlation coefficient is 0.98 and residual predictive deviation (RPD) index is 4.41 with categorized as excellent result. However, CGA can be predicted sufficiently using enhanced spectra with maximum correlation coefficient of 0.75 and RPD index of 1.96. Further spectra data analysis and non-linier regression approach may be applied to enhance accuracy and prediction robustness.

References

[1] D. Perrone, Donangelo, C. M., & Farah, A., 2008. Fast simultaneous analysis of caffeine, trigonelline, nicotinic acid and sucrose in coffee by liquid chromatography–mass spectrometry. *Food Chemistry*, 110, 1030–1035.

[2] A Myles, A. J., Zimmerman, T. A., & Brown, S. D., 2006. Transfer of multivariate classification models between laboratory and process near-infrared spectrometers for the discrimination of green Arabica and Robusta coffee beans. *Applied Spectroscopy*, 60(10), 246A–272A (1097–1216. 1198-1203(6)).

[3] B. M. Nicolai, Beullens, K., Bobelyn, E., Peirs, A., Saeys, W., Theron, K. I., & Lamertyn, J. (2007). Nondestructive measurement of fruit and vegetable quality by means of NIR spectroscopy: a review. *Postharvest Biology and Technology*, 46, 99-118.

[4] H.W Siesler, Ozaki, Y., Kawata, S., & Heise, H. M. (2002). Near Infrared Reflectance Spectroscopy: Principles, Instrument and Application. Wiley VHC Verlag, GmbH, Weinheim.

[5] A.M. Feria-Morales, 2002. Examining the case of green coffee to illustrate the limitations of grading systems/expert tasters in sensory evaluation for quality control. *Food Qual*. Prefer. 13, 355–367.

[6] M. Jing, Cai, W., & Shao, X., 2010. Quantitative determination of the components in corn and tobacco samples by using near-infrared spectroscopy and multiblock partial least squares. *Analytical Letters*, 43(12), 1910–1921.

[7] M.Y, Choi, W., Park, J.H., Lim, J., Kwon, S.W., 2010. Determination of coffee origins by integrated metabolomic approach of combining multiple analytical data. *Food Chem*. 121, 1260–1268.

[8] P. Parras, Martineztome, M., Jimenez, A., Murcia, M., 2007. Antioxidant capacity of coffees of several origins brewed following three different procedures. *Food Chem*. 102, 582–592.

[9] A.A. Munawar, Hoersten, D.v., Pawelzik, E., Wegener, J.K & Moerlein, D. 2016. Rapid and non-destructive prediction of inner quality attributes of intact mango using Fourier-transform infrared and chemometrics. *J. Engineering in Agriculture, Environment and Food*, 9 (1): 208 – 215.

[10] W. S. Sepúlveda, Chekmam, L., Maza, M.T., Mancilla, N.O., 2016. Consumers' preference for the origin and quality attributes associated with production of specialty coffees: Results from a cross-cultural study. *Food Res. Int*. 89, 997–1003.

[11] D, F. Barbin, Felicio, A.L. de S.M., Sun, D.-W., Nixdorf, S.L., Hirooka, E.Y., 2014. Application of infrared spectral techniques on quality and compositional attributes of coffee: An overview. *Food Res. Int*. 61, 23–32.

[12] R. M. El-Abassy, Donfack, P., Materny, A., 2011. Discrimination between Arabica and Robusta green coffee using visible micro Raman spectroscopy and chemometric analysis. *Food Chem*. 126, 1443–1448.

[13] M. Y. Choi, Choi, W., Park, J.H., Lim, J., Kwon, S.W., 2010. Determination of coffee origins by integrated metabolomic approach of combining multiple analytical data. *Food Chem*. 121, 1260–1268.

[14] B. Clements, Hall, R., Smyth, H.R., Cottam, M.A., 2009. Thrusting of a volcanic arc: a new structural model for Java. *Pet. Geosci*. 15, 159–174.

[15] W. S. Sepúlveda, Chekmam, L., Maza, M.T., Mancilla, N.O., 2016. Consumers' preference for the origin and quality attributes associated with production of specialty coffees: Results from a cross-cultural study. *Food Res. Int*. 89, 997–1003.

Analysis of Joint Behaviour of Column Beams against Cyclic Load with Added Rebar in Joint Accordance to SK SNI T-15-1991-03

Zardan Araby[1*], Abdullah[1] and Affifudin[1]

[1]Civil Engineering Faculty, Syiah Kuala University, Banda Aceh, Indonesia

*Corresponding author: zardan_araby@yahoo.com

Abstract. Joint beam-columns are the main structural components that function to support other structural loads. The beam-column relationship area is a critical area that needs to be designed to be truly accurate so that it can dissipate energy properly well in the event of an earthquake happen. As a structural component with the role and function, column joint beams occupy important positions in the building. The ability of column joint beams to deform cyclic loads in the inelastic region provides a good structure, so as to minimize damage caused by earthquake shaking. The failure of column joint beams directly affects other structural components associated with it. The purpose of this study was to study the ability of building structures on concrete beam-column joint elements to resist cyclic load with reinforcement in accordance with SK SNI T-15-1991. Test specimens amounted to 1 (one) with a concrete quality of 19.17 MPa. The beam measuring 120 x 30 x 40 cm, column measuring 30 x 30 x 200 cm using reinforcement 8, 13.4 mm with a melting voltage (fy) 310,03 MPa and stirrup reinforcement Ø9.8-100 mm with melting stress (fy) 374,59 MPa. The test was carried out by giving a cyclic load to the end of the beam with 0.75 mm, 1.5 mm, 3 mm, 6 mm, 12 mm, 24 mm and monotonic loading, until the specimen was destroyed. The results achieved in this study are the use of reinforcement reinforcement according to SK SNI T-15-1991-03 the capacity of the cyclic load produced is higher, namely 6,97 tf (press) and 5.09 tf (tensile).

Keywords: Beam-column joint, Cyclic Load, Joint SK SNI T-15-1991-03

1. Introduction

Beam-column joint is the main structural component that functions to bind other structural loads which are cyclic loads to the column joint. As a structural component with these roles and functions beam-column joints occupy important positions in the building. Failure in beam-column joints will have a direct effect on other structural components associated with it. The planning of beam-column joint structures must be carried out carefully in order to provide higher strength reserves for other structural components, so that if a building is damaged, it is expected that the column can still support other structural components. Schodek mentions that joint stiffness giving a lot of stability to lateral forces [1]. Loads that work laterally such as wind and earthquake forces can cause structures to collapse laterally if the connecting point is not rigid enough.

In general, the behavior of destruction of beam-column joints that occur due to cyclic loads at minimum to maximum load conditions begins with shear cracks in the joint area. If the cyclic load continues to increase in the beam-column joints and the main reinforcement is not enough to bind the joint column beam and maintain it against local buckling, then immediate destruction occurs . To prevent this, it is necessary to plan a beam-column joint that is stronger in resisting cyclic loads by considering the effects of restraints. One way that can be done is to add Sengkang reinforcement to the joint in the hope that it can increase the effect of restraints on the beam-column joints so that the risk of destruction can be minimized [2].

Given the magnitude of the role of beam-column joints in preventing the total collapse of building structures, the purpose of this study is to study the ability of building structures on reinforced concrete column joint elements in resisting cyclic loads with reinforcement configuration in accordance with SK SNI T-15-1991 and in this study termed SNI91 [3]. The reason for using the old regulation is to get the results compared and applied according to the conditions in the field so that old buildings that use SK SNI T-15-1991 regulations that have suffered damage to the joints can be found the best solution to repair without having to destroy buildings.

The addition of stirrup reinforcement in the joint area and distribution of reinforcement to the column is an important variable in determining the level of capacitance of beam-column joints in reinforced concrete construction. The specimens used in this study were reinforced concrete column beams, beams measuring 30 x 40 x 120 cm3 columns measuring 30 x 30 x 200 cm3 using 8Ø14 mm reinforcement with yield stress (fy) 310.03 MPa and stirrup reinforcement Ø10-100 mm with melting stress (fy) 374.59 MPa [4]. The specimen is mounted on top of the frame with a horizontal column vertical beam configuration and the two ends of the column are tied using a steel elbow plate with a thickness of 15 mm.

When casting reinforced concrete column test specimens, cylindrical control specimens are made 15 cm in diameter and 30 cm in height as many as 3 (three) pieces as concrete quality controllers. Testing of reinforced concrete column beam objects is carried out after the concrete age reaches 28 days, the tested column balom test object is rigidly installed on the steel frame (frame) available at the Civil Engineering Department of Construction and Building Materials Laboratory. In reinforced concrete column test specimens, Linear Variable Displacement Tranduser (LVDT) is used to measure deflection of reinforced concrete column beam, Omega Tranduser measuring device which serves to measure the width of concrete cracks that occur in the beam-column joint area. All of these measuring devices are connected and monitored through the Portable Data Logger TDS 302.

Reinforced concrete column beam test is carried out by loading cyclic loads with varying pulling cycles, namely diplacement of 0.75, 1.5, 3, 6, 12, 24 mm and monotonic so that the column beam specimens are destroyed [4]. The cyclic load capacity obtained by beam-column joints referring to SK SNI T-15-1991 (SNI91) is 6.97 reject tons and 5.09 tons of tensile. The amount of displacement at maximum load is 12 mm. The ductility of the specimen is 4.22 and the largest displacement that occurs in the LVDT reading after monotonic loading is 50.98 mm with a load of only 5.63 tons.

The crack pattern that occurs in reinforced concrete column beam specimens predominantly forms a horizontal crack pattern in the beam support area and is followed by shear cracks and flexible cracks in the area joint. Larger cracks occur at the beam connection point with the column. This is due to the column beam and joint shear capacity being reinforced with stirrups so that the weak point is flexural capacity. The initial cracks of reinforced concrete column test specimens occurred when the compressive load was 2.80 tons and the tensile load was 1.70 tons [5].

2. Literature Review

2.1. A Column

Column is a vertical press rod from the frame structure that bears the burden of the beam. Column is a compressive structural element that plays an important role of a building, so that a collapse in a column is a critical location that can cause the collapse of the relevant floor and also total collapse of the entire structure [6]. SK SNI T-15-1991-03 defines columns as components of building structures whose main task is to support vertical press axial loads with high parts that are not supported at least three times the smallest lateral dimensions.

According to [7] in reinforced concrete structure books, there are three types of reinforced concrete columns, namely:

i. The column uses a lateral stirring binder. This reinforcement serves to hold the main reinforcement to keep it firmly in place.
ii. The column uses a spiral binder. The shape is the same as the first except that as a binder the longitudinal staple is a rounded spiral loop forming a continuous helix along the column.
iii. Composite column structure is a compressed structural component that is reinforced in the longitudinal direction with a profile steel girder or pipe, with or without the main reinforcement bars

2.2. Beam
Beam is a structural part of a building that is rigid and is designed to bear and transfer loads to the supporting column elements. In addition, the beam ring also functions as a binder of the columns so that if there is a movement of the columns it remains united to maintain the original shape and position. Ring beams are made of the same material as the column so that the ring beam relationship with the column is rigid, not easily deformed. Non-uniform force patterns can result in curved or deflected beams which must be held by the internal strength of the material [8].

2.3. Concrete
Concrete is a composite building material made of aggregate and cement bonding. The most common form of concrete is portland cement concrete, which consists of mineral, cement and water aggregates. The small cement water factor will cause porosity in the concrete. The porosity of the concrete is affected by less than optimal compaction of concrete due to the low cement water factor value. The greater the FAS value (Cement Water Factor), the greater the porosity and vice versa. To get a strong concrete as planned, it must use the appropriate FAS value, while the use of a low FAS value will make it difficult to work on the concrete, so that the compaction in the concrete is not optimal and results in porous concrete which will reduce the compressive strength [9].

2.4. The compressive strength of concrete
The compressive strength of concrete is the quality produced from a specimen structure. The higher the level of strength of the desired structure, the higher the quality of the concrete produced. Factors that influence the compressive strength of concrete include the proportion of constituent materials, methods of implementation, maintenance, and circumstances at the time of casting [10]. Testing of concrete compressive strength was obtained through standard testing using a compressive testing machine by giving a multilevel load at a certain load increasing speed on the concrete cylinder test specimen to disintegration. The test procedure commonly used is the standard of the American Society for Testing.

3. Method of Research
The materials used in this study were portland cement, water, reinforced iron, sand and split. The cement used is type I Portland Cement produced by PT. Semen Andalas Indonesia. The water used for concrete mixture and its treatment comes from clean water from the Regional Water Company (PDAM) Tirta Daroy Banda Aceh. The main reinforcement iron used is plain steel 8Ø14 mm, while for stirrup reinforcement plain steel Ø10-100 mm is used which is obtained from stores around Darussalam. The aggregates and sand used for concrete mixtures are the result of natural disintegration of the rocks, aggregates and sand obtained from the Krueng Aceh river.

3.1 Test of Cylinder Concrete Press Strength
Concrete cylinder compressive strength testing is done to determine the desired compressive strength of the concrete. Compressive strength testing was carried out when the specimen was 28 days old. Concrete cylinder compressive strength testing is used to represent the casting of the beam as a control specimen. Testing the compressive strength of cylindrical specimens can be seen in figure 1.

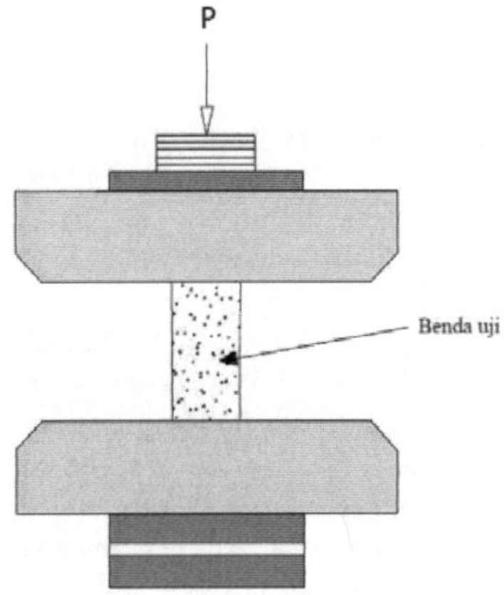

Figure 1 Cylindrical test object strength testing.

3.2 Steel Tensile Testing
The reinforcing steel used for beam columns is plain reinforcement of 8 × 14 mm and plain steel Ø10-100 mm for dash. The quality of steel is determined by the steel tensile test according to ASTM E.8-81 procedure. At the pull of this steel, the load is provided continuously until the steel breaks. Stretching readings are performed every 100 kg load by transducer dial readings. The result of this test is illustrated in a steel strain tension relationship with strain so that it can be obtained by tensile strength of steel and stretch of lengge.

3.3 Mix Proportion Planning
The concrete mix design is planned based on the American Concrete Institute (ACI 211.1-91) method for normal concrete. Based on this method, the weight of each material used will be obtained, namely the amount of aggregate, cement, and water. The concrete used in this specimen is normal concrete with a maximum size of the aggregate diameter of 19.1 mm with the quality obtained at f'c = 19.17 MPa.

3.4 Details of Joint Beam Column Test Objects
The planned specimen is a reinforced concrete beam element with a cylinder compressive strength of 19.17 MPa with columns measuring 30 x 30 x 200 cm and 30 x 40 x 120 cm for beams. Test specimens were made based on research needs, namely test specimens for testing cyclic load capacity in reinforced concrete column joints. The planned specimen consists of normal reinforced concrete with fine aggregates, coarse aggregates, reinforcing iron and water. The main reinforcement used is 8Ø14 with a melting tensile strength (fy) of 310.03 MPa. Sengkang Reinforcement uses steel Ø10-100 mm with a strong melting pull (fy) of 374.59 MPa. Clearer the shape and details of column joint beam test objects can be seen in figure 2 below.

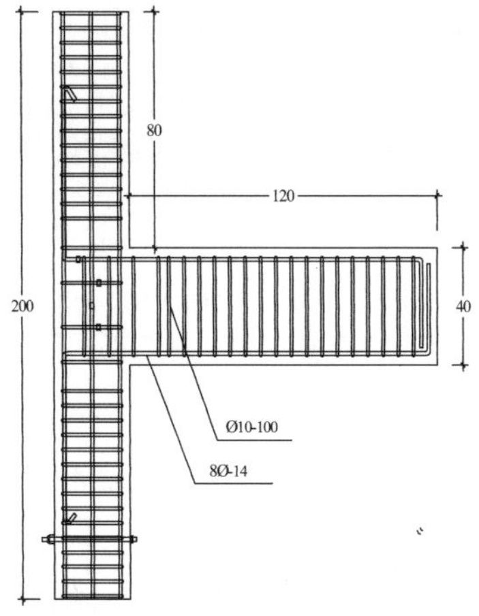

Figure 2. Form of cross-section testing.

3.5 Procedure for Testing Joint Column Test Objects

Tests are carried out when the test object is 28 days old. Testing of column beam test objects is carried out together with cylindrical specimens. The cyclic load capacity that will be taken into account is the load compressive strength resulting from the testing of the compressive strength of the cylindrical specimen. Before testing, the surface of the test object is painted in white and the grid is drawn first, to make it easier to draw the resulting crack pattern. The series of test kits and the installation of specimens in steel frames (frames) in detail are presented in the following figure 3. The test object is rigidly mounted on the frame beam. At the end surface of the beam field a two-way steel plate is attached, which is connected with a bolt, the plate is tied to the load cell to be given the desired cyclic load. Load is given horizontally against the end of the beam. Horizontal load is given by the hydraulic jack connected to the load cell and forwarded to the column beam test object. The given load is controlled by reading the dial in the data logger. Loading is carried out in a cycle and each cycle is read load and illustrated the crack pattern that arises. Load is given until the test object is destroyed. On the side of the beam a transducer is installed to read the deflection in the lateral direction.

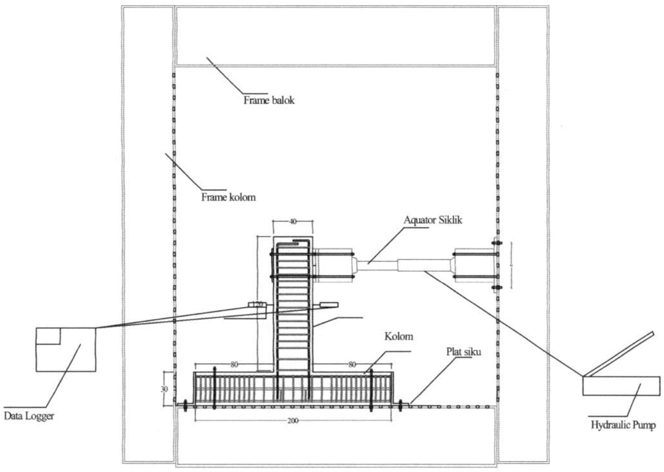

Figure 3. Beam-column joint speciment test set up.

Strains that occur in beam-column joints are read by a Portable Data Logger that has been connected to a strain gauge. Loads is stopped when the load no longer increases due to the test object being unable to accept the load so that the specimen is cracked and destroyed. The pattern of crack development is monitored at all times by making a crack image that occurs in the column according to the amount of load given.

4. Results and Discussion

The discussion carried out based on the results of testing the cyclic load capacity of the joint reinforced concrete column beam includes the cracking pattern, ductility, stiffness and energy dissipation that occurs in the joint.

4.1 Joint Beam Column Crack Pattern

Testing of beam-column joint test objects shows a crack pattern that is relatively horizontal in shape on the pedestal, as shown in figure 4.

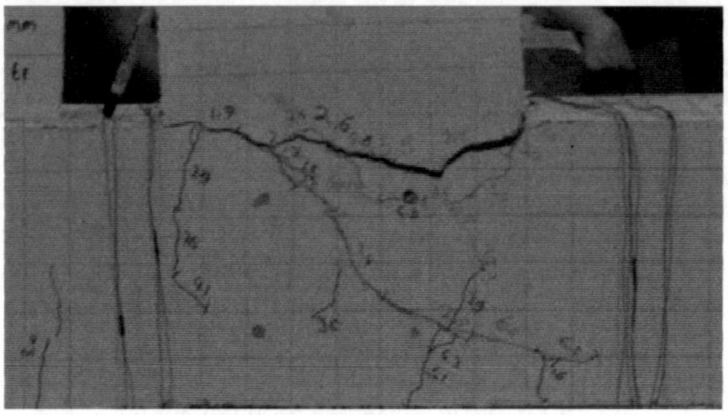

Figure 4. Crack load pattern (-), 24 mm.

The initial crack occurs in the beam-column joint test object is in the cycle (1) 3 mm tensile load where the load that occurs is equal to 2.80 tf, while the crack that occurs lies in the shear angle. Furthermore, in the same cycle of cycle (1) 3 mm, which is tensile load, cracks also occur again at 1.70 tf load, with the shape and location of cracks in the upper left corner of the joint area stretching down the column for about 16 cm and after being given flexion full ie 3 mm (+) produces a load of 3.60 tons. The results are shown in figure 5 and 6 below.

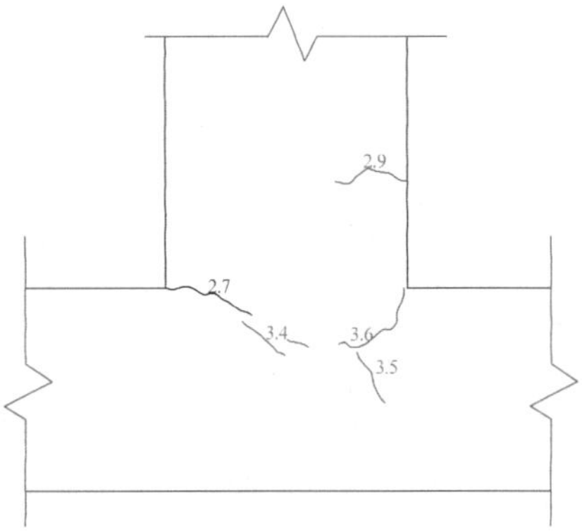

Figure 5. Loads Cracked Pattern 3 mm.

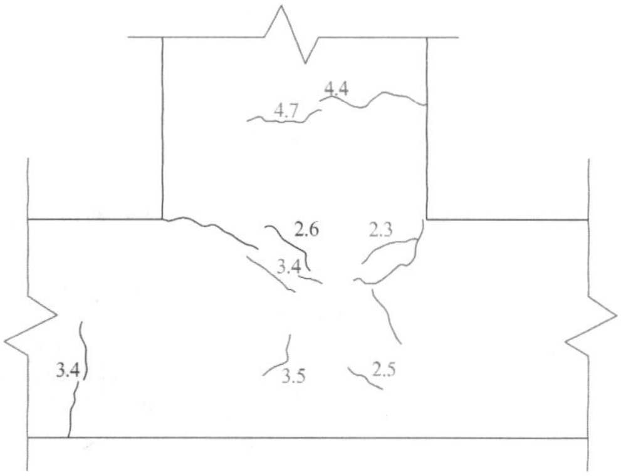

Figure 6. Loads Cracked Pattern 6 mm.

In monotonic loading, cracks occur in the area around the beam-column joint, which occurs continuously widening. The biggest cracks occur in the beam support area that has fused and continues to expand as it does loading. For a clearer variety of crack patterns that occur in monotonic loading is in figure 7. below.

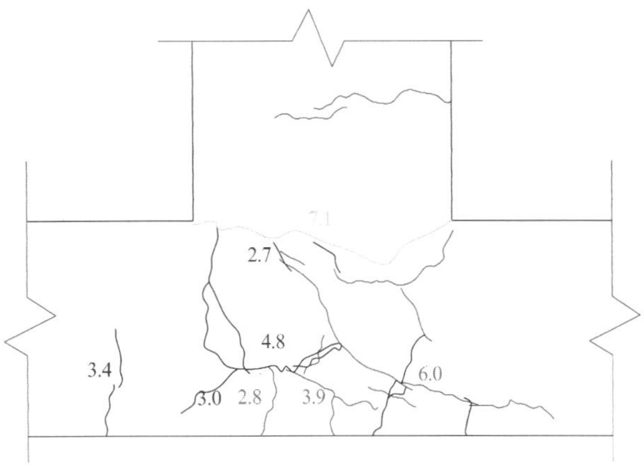

Figure 7. Monotonic loads cracked pattern.

4.2 Column Beam Secants Stiffness
The decrease in secant stiffness in column beam specimens is calculated based on section 2.9.1. The graph of the relationship of decreasing stiffness with lateral displacement can be seen in figure 8.

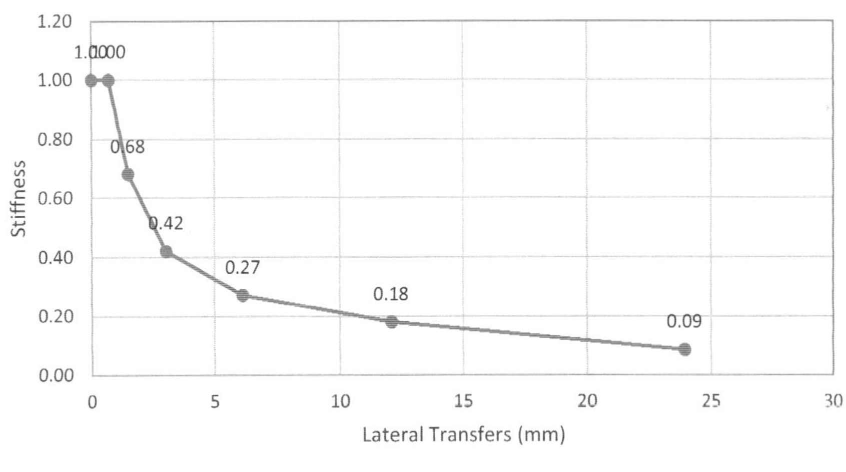

Figure 8. Graph of relative decreases relative securities stiffness.

Based on figure 8, the test specimen experienced a decrease in stiffness starting from the lateral displacement of 0.73 mm to the displacement of 1.515 mm at 0.68 mm. In the next stiffness the specimen continued to decrease stiffness in the lateral displacement of 6.14 mm by 0.27 to the final stiffness at the displacement of 23.93 mm at 0.09. The conclusion obtained from the graph above is that the SNI91 test object continues to decrease stiffness before the last loading cycle is carried out.

4.3 Relationship Load - Strain Reinforcement

Measurements of the main reinforcement steel and stirrup reinforcement on column specimens were carried out on 2 (two) reinforcing steel, columns and beams. Testing of beam-column joint test items with the SK SNI T-15-1991-03 regulation, greater reinforcement strain occurs than regulations that still use PBI, due to the addition of stirrup reinforcement in the column joint joint area, so that the reinforcement is more restrained resulting in a greater load.

5. Conclusion

Based on the results of research, data processing and discussion that has been done, a number of conclusions can be taken and suggestions include:

i. The addition of stirrup reinforcement in the beam-column joint area can increase the value of cyclic load capacity.
ii. Cyclic load capacity obtained when the maximum load is 6.97 tf (reject), 5.09 tf (pull).
iii. Displacement obtained when the maximum load is 50.98 mm.
iv. Decreased stiffness of column beam secant is 0.09 with displacement of 23.93 mm.
v. The biggest energy dissipation cycle occurs at a load of 24 mm. The biggest energy dissipation value is 13.25. The presence of stirrups on the joint can increase energy dissipation in the joint by giving stiffness to the joint.

References

[1] Schodek, D.L (1998) *Struktur*, Terjemahan Bambang Suryoatmono, Refika Aditama, Bandung.
[2] Dahl. K.K.B. (1992) : Uniaxial Stress Strain Curves for Normal and High-Strength Concrete, Departmenet of Structural Engineering, Lyngby, Technical University of Denmark.
[3] Bae, S., dan Bayrak, O., (2003) : Early Cover Spalling in High Strength Concrete Columns, *Journal of Structural Engineering*, V.129, No.3, March, pp.314-323.
[4] Ansari,F., dan Li, Q. (1998): High Strength Concrete Subjected to Triaxial Comression, *ACI Materials Journal* V.95, No. 6, November-December, pp747-755.
[5] Richard, P. & Cheyrezy, M. (1994) RPC with High Ductility and 200-800 MPa Compression Strength, *Proceeding of V. Mohan Malhotra Symposium*, pp. 507-518, Farmington Hills.
[6] R.Sudarmoko (1996), Diagram Perancangan Kolom Beton Bertulang, Jurusan teknik Sipil, fakultas Teknik, UGM, Yogyakarta.
[7] Chapalkar, Subhash, Khetre, Meshran (2015), Modal Analysis of Cantilever Beam Structure Using Finite Element Analysis & Experimental Analysis, in *American Journal of Engineering Research*).
[8] Sheikh, S.A., dan Khoury, S.S. (1993) : Confined Concrete Columns with Stubs, *ACI Structural Journal*, V.90, No.4, July-August 1993, pp.414-431.
[9] Imran, I., Moestopo, M., dan Suharwanto (1999) : Strength and Deformation of Confined High Strength Concrete, *First International Conference on Advance in Structural Concrete Engineering and Mechanics*, Seoul, August 1999, pp.373-378.
[10] Paultre, P., Légeron, F., (2008),"Confinement Reinforcement Design for Reinforced Concrete Columns", ASCE, *Journal of Structural Engineering*, Vol. 134, No. 5, May 1, pp. 738-749.
[11] Dipohusodo (1999), *Struktur Beton Bertulang*, PT. Gramedia Pustaka Utama, Jakarta.

Evaluation of Natural Ventilation Characteristics on the Sultanate of Ternate Mosque

Mustamin Rahim[1*] **and Firdawaty Marasabessy**[1]

[1] Department of Architecture, Faculty of Engineering, Universitas Khairun, Ternate, North Maluku, Indonesia

*Corresponding author: mustamin_rahim@yahoo.co.id

Abstract. Thermal comfort is very important in mosque buildings, especially during prayer times in order to increase the comfort and solemn in prayer. The usage of air conditioning for the mosque in Ternate City is not recommended because of the energy resources is limited (rotating blackout) and also to reduce its operational cost every month. Therefore, the application of natural ventilation is essential to improve thermal comfort in the mosque. This study aims to identify the characteristics of the natural ventilation system of Sultan Ternate Mosque by analyzing the orientation of inlet and outlet, ratio and dimension of openings, type and position of ventilation through the literature studies and field observations. The results show that the natural ventilation system of Sultan Ternate Mosque uses clerestory ventilation and cross ventilation systems, therefore the indoor thermal conditions can reach a comfortable level. Ventilation types are semicircular and vertical jalousie on the walls, and roof ventilation of 4-levels. The total area of ventilation holes are 110.24 m^2, with the ratio is about 1.04 (inlet): 1.00 (outlet).

Keywords: Sultan Ternate Mosque, Natural Ventilation, Cross Ventilation, Thermal Comfort

I. Introduction

Ventilation is a system in which the internal air is continuously replaced from an occupied space by the relatively fresh outside air through vents, windows, doors, and so forth. Three types of ventilation system are available such as forced ventilation, natural ventilation, and hybrid ventilation to improve the indoor air quality. Forced ventilation is served by powering a mechanical system such as fans, blowers, and so forth to push the external air into the space of interest. It is needless to note that this ventilation system consumes electric power in operation. On the other hand, the natural ventilation system neither consumes electric power nor needs any mechanical system [1].

In buildings, electric power is mostly consumed by heating, ventilation, and air-conditioning (HVAC) systems. According to Awbi that ventilation consumes 30–60% of the total energy consumption in modern and retrofit buildings. The fossil fuel and nuclear power based electricity generations are discouraged by modern science and the fact due to global warming, environmental pollution, and the safety matter [1-2]. Combustion of fossil fuels to generate electricity produces atmospheric CO_2 emissions. In 2005, total CO_2

emissions were 26.6 billion tons. More than 41% of it was produced from fossil fuels consumed in the world. It will increase to 46% by 2030. CO_2 gas in large amount released to the atmosphere as the primary cause of global warming [3-5]. Therefore, the application of natural ventilation is very important in reducing the global warming effects, and the application of natural ventilation in the building is necessary to evaluate in order to find out the effective natural ventilation system. This study aims to identify the characteristics of natural ventilation in the Sultanate of Ternate Mosque (STM).

2. Research Methodology

This research is qualitative research that emphasizes on the identification of the Natural ventilation system, through: (1) Field study: direct of observation and measurement in the field to obtain primary data; dimension, type, and orientation of ventilation. (2) Literature review: to complement secondary data information about the performance of clerestory ventilation and cross ventilation system. (3) Interviews: to obtain information relating to primary and secondary data to supplement data obtained through field studies and literature studies.

3. Natural Ventilation System

Natural ventilation is an energy efficient alternative for reducing energy use in buildings, achieving thermal comfort, and maintaining a healthy indoor environment [6-8]. Typically, the energy cost of a naturally ventilated building is 40% less than that of an air-conditioned building [9]. Natural ventilation has become a new trend in building design in the architectural community [10, 11]. Natural ventilation is an important and efficient passive technique to reduce building cooling energy need and improve indoor air quality [12].

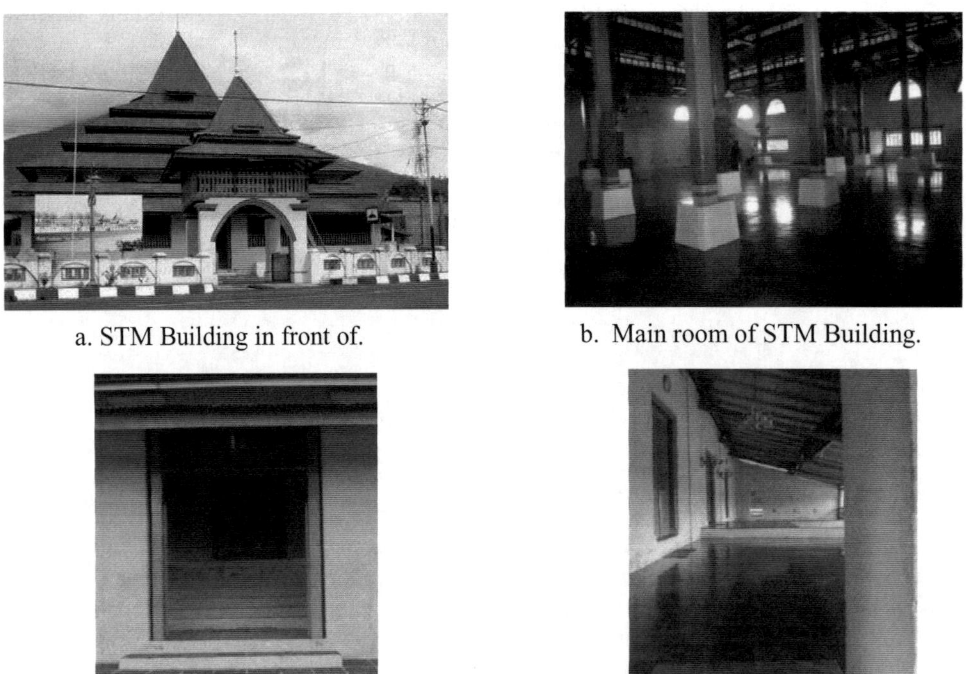

a. STM Building in front of.

b. Main room of STM Building.

c. Main Entrance

d. Terrace

Figure 1. STM Building Model.

There are a number of different natural ventilation air flow paths in buildings, the three main ones being: cross ventilation, single-sided ventilation, and passive stack ventilation. In cross ventilation system, the inlet and outlet are generally located in a different side, and for Single-side ventilation, the inlet and outlet are generally at the same side. Stack ventilation (outward air flow) is generally assumed to be driven by temperature differences between the hot air in the occupied space and the cooler external air. For STM Building uses the cross ventilation system with combination wall ventilation and roof ventilation (see figure 1).

Sultan Ternate Mosque is the oldest mosque in Ternate, North Maluku. It was rebuilt around 1807-1821 periods because of burned. The mosque is also known as the *Sigi Lamo* with the main room area (prayer room) of about 486 m^2. At the beginning construction of this mosque, the roof material used sago palm leaves, and it was replaced with zinc material on 1983 [13]. Model of STM building is unique with 7 levels roof and without ceiling (see figure 1). This model can improve the performance of natural ventilation.

4. Openings Type of STM Building

A naturally ventilated structure often includes an articulated plan and large window and door openings, while an artificially conditioned building is sometimes best served by a compact plan with sealed windows. The specific approach and design of natural ventilation systems will vary based on building type and local climate. However, the amount of ventilation depends critically on the careful design of internal spaces, and the size and placement of openings in the building [14].

a. Entrance (P1).

b. Side door (P2).

c. Roof ventilation (V1).

e. Wall Jalousie (J1).

d. Wall Ventilation (V2).

Figure 2. Openings type of STM Building.

Figure 2 portrays that the opening types of STM Building are consist of 5 models: double doors (P1), double doors-jalousie (P2), window/wall ventilation (V2), and jalousie in wall and roof (J1 & P1). According to Moore [15] that Jalousie (J1 & V1) can circulate air of about 15%, and 90% in V2 model. For double swing door of P1 and P2 models can circulate air of about 90% at the open condition, and P2 model (closed condition) can circulate air of about 15 % through jalousie on the door. It indicates that the performance of natural ventilation in the STM Building is optimal by a combination of clerestory ventilation and cross ventilation systems especially during prayer time (open doors condition).

5. Openings Orientation of STM Building

Cross ventilation relies on wind for cooling. When the wind blows on a building, a pressure difference is created between the windward (wind-facing) and leeward walls. Openings placed on opposite sides of a building will allow the cooler external air to enter the building while warmer internal air is sucked out from the leeward side openings. The degree of passive cooling is determined by the size and placement of the building and ventilation openings, as well as the regularity of wind [16]. Figure 3 portrays that the opening position and orientation of STM building, where the inlet is located in the east wall at the position of the wind in the east wall of the building so it will affect the increased flow of air inside the room past the area of prayer to the inlet on the west wall of the building. These conditions can influence the indoor thermal comfort at the prayer room.

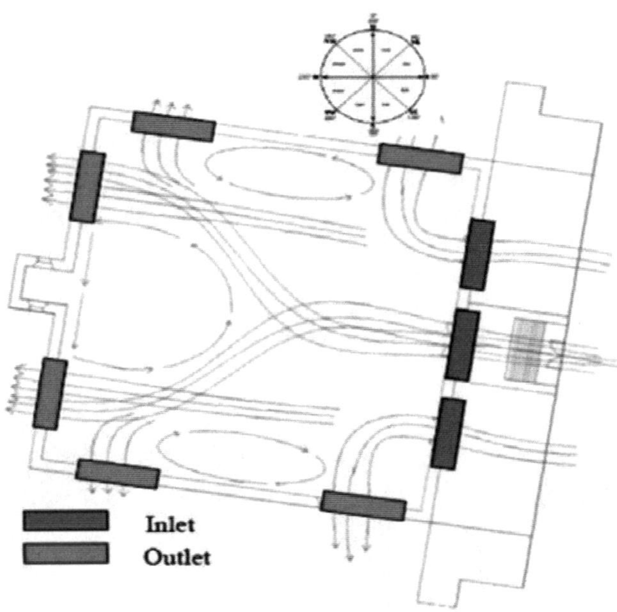

Figure 3. Openings position and orientation in the STM Building.

The ideal position of an inlet is about the height of human activity, whereas outlet position should be higher than the inlet position so that the hot air can be removed easily without mixed with fresh air coming through the inlet. The height of human activity in the room is around of 60-80 cm for prayer activity and around of 100-160 cm for standing activity.

Figure 4 shows the cross ventilation performance in STM Building, where the air flows from the door in east side (inlet) to window and jalousie in west side (outlet) flows through the height of human activity during prayer (see figure 4a). In addition, the roof ventilation that is located on four sides of the building can facilitate circulation of air inside the room through a system of cross ventilation. Hot air is at top of the room easily flow out through roof ventilation (see figure 4b). Therefore, the indoor thermal comfort can be archived especially during daylight time. According to Aynsley [17] that indoor thermal comfort can be achieved from natural ventilation more often during daylight hours of 103 out of 124 occasions; when wind speeds are higher and relative humidity is lower than during night time, and 79 out of 124 occasions in night time; when relative humidity rises and calms in wind are more frequent.

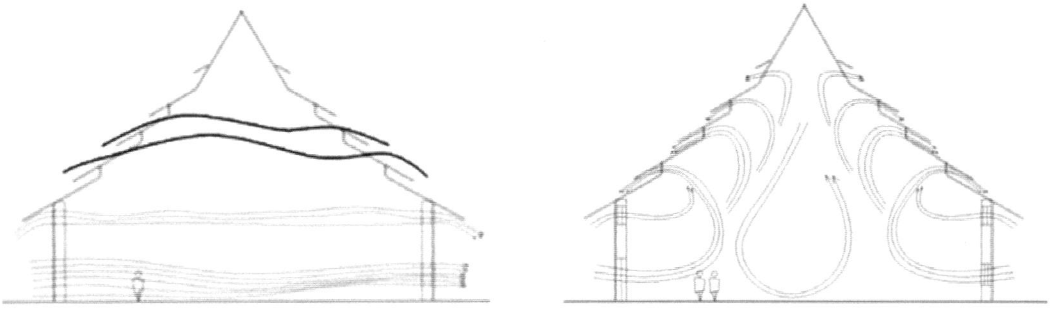

a. Horizontal circulation b. Vertical circulation

Figure 4. Cross ventilation system in STM Building.

Opening types in STM Building are door, window, wall ventilation (jalousie) and roof ventilation. They can improve the wind circulation in indoor so that the hot air in the room can flow to the outdoor optimally, and then thermal comfort in indoor can be achieved. According to Karava [18] that the internal airflow pattern has a significant impact on induced airflow rate and internal pressure distribution; this effect is more pronounced for configurations with large wall porosities (higher than 10%).

Area size of main room of STM Building is around 484 m^2 so that the minimum of openings is required of about 96,8 m^2 (openings standard is 20% of the total room size [19-20]). Figure 5 and table 1 show that the openings dimension of STM Building is around 87,52 m^2 in the wall (entrance and window/jalousie types) and 22,72 m^2 in the roof of building (roof ventilation type), with the total size of opening is about 110,2 m^2 or 22,78 % of the total room size. It indicates that the openings of the STM Building are more than the openings standard. Therefore, it can optimize cross ventilation in the building. Moreover, the size of the inlet is around 56,13 m^2 and outlet of about 54,13 m^2. It indicates that the inlet size higher than outlet size. This condition can increase air flow in the room so that the indoor thermal comfort can be achieved. Natural ventilation is useful for improving thermal comfort in hot climates [21-24]. According to Aldawoud [25], the cross-ventilation may offer significant potential to improve the indoor thermal environment and comfort conditions to acceptable levels, indoor temperature reductions of 4–8°C are possible, air flow rate and the indoor temperature are directly proportional with the size of inlets and outlets. Higher flow rates occur with large inlets and outlets that have a proper orientation as regards prevailing wind.

Table 1. Ventilation size of STM Building.

Type Code	Ventilation Type	Ventilation total	Ventilation size (m^2)
P1	Main Entrance	1	5,00
P2	Side entrance	2	8,00
J1	Window/Jalousie	6	26,52
V1	Roof ventilation	16	48,00
V2	Wall ventilation	16	22,72
		Total	110,24

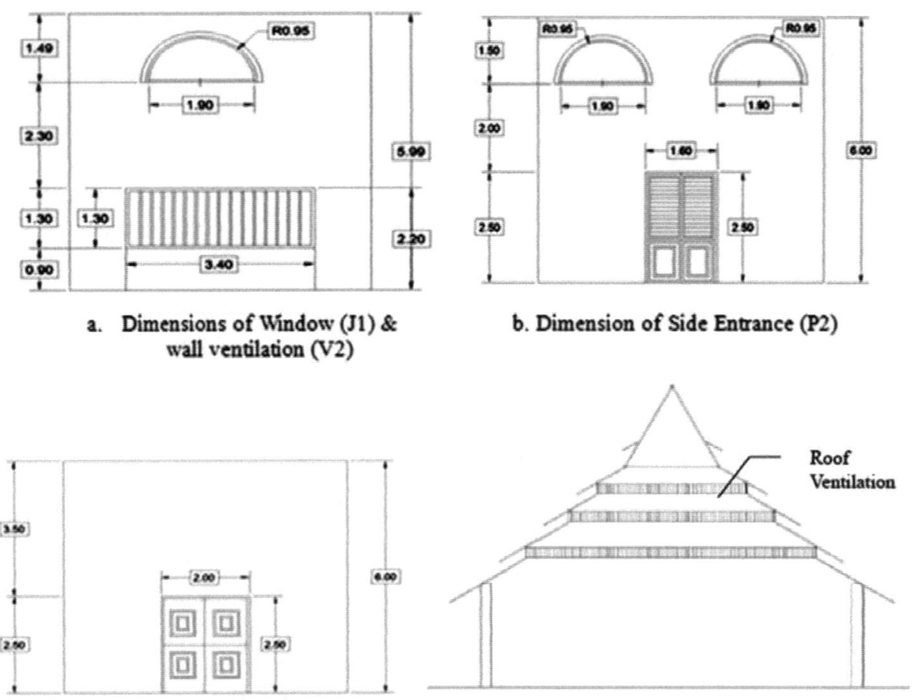

a. Dimensions of Window (J1) & wall ventilation (V2)

b. Dimension of Side Entrance (P2)

c. Dimension of Main Entrance (P1)

d. Roof Ventilation (V1)

Figure 5. Openings dimension of STM Building.

6. Conclusions

Based on the evaluation, it can be concluded that:

i. Natural ventilation system in STM Building uses clerestory ventilation and cross ventilation systems, so that, the indoor thermal conditions can reach a comfortable level.

ii. Ventilation types are semicircular and vertical jalousie on the walls, and roof ventilation of 4-levels.

iii. Total area of openings is 110.24 m^2 or around 22,78 % of the total room size, with the ventilation ratio is about 1.04 (inlet): 1.00 (outlet). It indicates that the natural ventilation in the STM Building is optimal to improve indoor thermal comfort, especially during prayer time.

IOP Conf. Series: Materials Science and Engineering **506** (2019) 012035 doi:10.1088/1757-899X/506/1/012035

References

[1] Bangalee, M.Z.I. at al. *Effects of Lateral Window Position and Wind Direction on Wind-Driven Natural Cross Ventilation of a Building: A Computational Approach*. Journal of Computational Engineering 2014, 1-15.

[2] Awbi, H.B. *Ventilation of Buildings*. Spon Press, London, UK, 2nd edition, 2003.

[3] Mondal, A.H.Md., Denich, M. *Assessment of Renewable Energy Resources Potential for Electricity Generation in Bangladesh*. Journal of Renewable and Sustainable Energy Reviews 2010,14,2401-2413.

[4] *World Energy Outlook 2007*. International Energy Agency (IEA), 2007.

[5] Fauzi, H., Rahim, M. *Penerapan Sel Surya pada Bangunan*. Rona 2007,4(2),85-90.

[6] Busch, J.F., *A Tale of Two Populations: Thermal Comfort in Air-Conditioned and Naturally Ventilated offices in Thailand*. Journal of Energy and Buildings 18 (3-4) (1992) 235-249.

[7] Zhao, R., Xia, Y. *Effective Non-Isothermal and Intermittent Air Movement on Human Thermal Responses*. Proceedings of Room Vent 1998, 2, 351-357.

[8] Finnegan, J.J., at al. *The Sick Building Syndrome: Prevalence Studies*, British Medical Journal 1994, 289, 1573-1575.

[9] *Energy Consumption Guide 19*. Energy Efficiency Office/HMSO, London, 1993.

[10] A. Krishan, *Climate Responsive Architecture: A Design Handbook for Energy Efficient Buildings*. Tata McGraw-Hill Pub. Co., New York, 2001.

[11] Willmert, T. *The Return of Natural Ventilation*. Architectural Record 2011, 189 (7), 137-148.

[12] Zhai, Z.J. *Review of Natural Ventilation Models*. Energy Procedia, 2015, 78, 2700-2705.

[13] Dero, Ridwan. *Masjid Agung Kesultanan Ternate* (*Sigi Lamo*), Majalah Parada Edisi 14; 2002.

[14] *Natural Ventilation*. https://www.wbdg.org/resources/natural-ventilation

[15] Moore, *Fuller, Environmental Control Systems*, McGraw-Hill Inc., 1993: 30, 179-187, 192-193.

[16] *Cross Ventilation*. http://challengeforsustainability.org/resiliency-toolkit/cross-ventilation/

[17] Aynsley, R. *Estimating Summer Wind Driven Natural Ventilation Potential for Indoor Thermal Comfort*. Journal of Wind Engineering and Industrial Aerodynamics 1999, 83, 515–525.

[18] Karava, at al. *Airflow Assessment in Cross-Ventilated Buildings with Operable Façade Elements*. Journal of Building and Environment 2011, 46 (1), 266–279.

[19] Satwiko, Prasasto. 2004. *Fisika Bangunan 2*. PT. Andi, Yogyakarta.

[20] Mangunwijaya, Y.B. 1988. *Pengantar Fisika Bangunan*. PT Jambatan, Jakarta.

[21] Omar S. Asfour, O.S., 2015. *Natural ventilation in buildings: An overview; In book: Natural Ventilation: Strategies, Health Implications and Impacts on the Environment*. Edition: 1, NOVA Publication, New York. 1-25.

[22] Tan, Z., Deng, X. *Assessment of Natural Ventilation Potential for Residential Buildings Across Different Climate Zones in Australia*. Journal of Atmosphere 2017, 8 (9) 1-17.

[23] Chen, Y., Tong, Z., Malkawi. *A. Investigating Natural Ventilation Potentials Across the Globe: Regional and Climatic Variations.* Journal of Building and Environment, 2017, 122, 386–396.

[24]. Haase, M., Amato, A. *An investigation of the Potential for Natural Ventilation and Building Orientation to Achieve Thermal Comfort in Warm and Humid Climates*. Journal of Solar Energy 2009, 83, 389–399.

[25] Aldawoud, A. *Windows Design for Maximum Cross-Ventilation in Buildings*. Journal of Advances in Building Energy Research 2017, 11, 67-86.

The Effects of Arbuscular Mycorrhiza Fungi (AMF) and Organic Fertilizer on Chemical Features of Soil Planted by Sunflower (*Helianthus Annuus* L.) on Critical Land of Iron Ore Mining

Suwarniati[1*], U Hanum[1] and Muslim[2]

[1] Department of Biology, Islamic Faculty, University Muhammadiyah Aceh, Banda Aceh, Indonesia
[2] Department of Informatics, Mathematics and Natural Science Faculty, University Syiah Kuala. Darussalam, Banda Aceh, Indonesia.

*Corresponding author: anik.unmuha@gmail.com

Abstract. The rate of land degradation is increasingly widespread and alarming, where the land becomes critical land. The transfer of forest into the iron ore mining area in Lhoong Subdistrict causes drying of river water. For that, it is needed a solution for the conservation of the land with the right and useful plants. This study aims to determine the effect of Arbuscular Mycorrhiza Fungi (AMF) and organic fertilizer on soil chemical properties planted with sunflower (*Helianthus annuus* L.) on critical land of iron ore mine in Lhoong, Aceh Besar. The soil sample was taken from Lhoong area, Aceh Besar District with a depth of 0-20 cm. The study was arranged according to a randomized block design of 2x3 factorial patterns with 4 replications. The first factor, AMF with 2 types, those are without Mycorrhiza and Mixed Mycorrhiza as much as 10 g each polybag. The second factor is organic fertilizer with 3 types those are Without Fertilizer, Manure, and Guano respectively as many as 30 tons each hectare area equivalent to 150 g/polybag. The results of the observations were tested by the variance analysis at $\alpha = 5\%$. The results of soil chemistry research showed that the provision of AMF had no significant effect on pH of H_2O, C-organic, N-total, P-available and CEC of soil. Organic fertilizer has a very significant effect on pH of H_2O, C-organic and N-total soil.

Keywords: Arbuscular Mycorrhiza Fungi (AMF), Organic fertilizers, Iron ore mining

1. Introduction

Aceh Government, particularly Aceh Besar District in 2006, granted mining concession rights to a mining company. The company was exploiting 500 hectares of land, while the exploration rights was 6000 hectares. The 500-hectare area is in the two settlements of Mukim Cot Jeumpa and Mukim Blang Me, located in four villages area; Jantang, Baroh Blang Me, Baroh Geunteut, and Gampông Teungoh Geunteut. [1]. Those villages will receive direct impact from the exploitation process up to seven other villages around. It wwould increase the amount of critical land and environmental degradation [2].

The Ministry of Forestry stated that approximately 1.2 million hectares of Indonesia's forests was damaged every year [3]. Critical land defined as land that is subjected to physical, chemical and biological damage due to their inadequate use and capability, which endanger the hydrological, orological, agricultural, socioeconomic and environmental functions [4]. Such soil has a low fertility due to low pH, nutrient deficiency especially P and N, low cation exchange capacity and poor organic material.

Critical land use can be renewed by planting sunflowers. These plants can be found in disturbed habitats, newly opened environments and various pastures [5]. It was able to grow in the presence of granulated slags and accumulated metals preferentially in above-ground tissues [6]. Sunflower can grow along the year in all places around the world and one source of biodiversity that acts as phytoremediation and biofuel producer. Phytoremediation use plants and microbes which is related more used as green technology to clean up the contaminated soil of mine [7], especially in an iron-mining sites [8, 9]. The ability of metal uptake differs depending on the type of AMF in remediating the soil [10].

Provision of AMF in host plants (sunflower plants) is able to reduce toxicity of contaminated mining land. Inoculation of AMF (R. *irregularis*) may be for phytoextraction Cd whereas F. mosseae can be useful for Cd and Zn fitostabilization [11]. In addition, , P-used of total organic fertilizers in the long term can increase the productivity of land and prevent land degradation and contribute substantially to the improvement of physical, chemical and biological soil [12]. Provision of organic fertilizer into ex-mining land can increase the pH value, organic C-total, CN available, P available, K, Ca, Mg, Na, K, and CEC [13].

Based on the study of the potential of AMF, organic fertilizer and sunflower benefits are quite promising. The researchers tried to use mycorrhizal and organic fertilizer to see how the influence of the AMF can and growing organic fertilizer plant sunflowers to changes in soil chemical properties. Futhermore, hopefully this study may be one of the solutions to improve the properties of the chemical on critical land of the former iron ore mine, spesically in Lhoong - Aceh Besar District.

2. Research Methods

The research was conducted in Meunasah Papeun Village, Krueng Barona Jaya Sub-district, Aceh Besar District. Initial soil chemical analysis of animal manure and Guano were conducted at the Laboratory Soil and Plants, Faculty of Agriculture, University Syiah Kuala. While soil chemical analysis was conducted at BPTP Aceh. The research activities started from January to December 2017.

The materials used were the critical soil taken from the former mining ore in Lhoong, Aceh Besar. Inoculum AMF used were Mixed Glomus (*Glomus*, *Gigaspora* and *Acaulospora*) with 100 spore density of each 10 g, obtained from the Soil Biology Laboratory of Soil Sciences, University Syiah Kuala. Local sunflower varieties obtained from Gampong Lhong Raya, Banda Raya District, Banda Aceh. Organic fertilizers consist of farm manures obtained from Area Usaha Shop, Diponegoro Street, Banda Aceh and guano from Gampong Lamdom, Sub-district Lampuuk, Aceh Besar. The basic fertilizer consists of KCl, TSP and Urea. Tools used are knives, weights and polybags (40 x 50 cm), hoes and buckets.

There are two factors studied. The first factor is the use of the AMF as much as 10 g each polybag which consists of 2 types AMF that is Without Mycorrhiza (Mo) and Mixed Mycorrhiza (M1). The second factor is an organic fertilizer with a dose of 30 tons each hectare area or equivalent to 150 g each polybag, consisting of 3 types of soil, that is Soil only (Po), soil with Manure (P1), and soil with Guano (P2). From those two factors, there were 6 treatment combinations and each treatment combination was repeated 4 times to 24 units of experiment. Randomized Block Design (RBD) 2x3 factorials pattern with 4 replications were applied. The treatment combination can be seen on the table 1 as follows.

Table 1. Combination arrangement of AMFs and Organic Fertilizers Treatment.

No	Treatment	Type of Mycorrhiza	Type of organic fertilizer
1	M0P0	Without Mycorrhiza	Without fertilizer
2	M0P1	Without Mycorrhiza	Manure
3	M0P2	Without Mycorrhiza	Guano
4	M1P0	Mixed Mycorrhiza	Without fertilizer
5	M1P1	Mixed Mycorrhiza	Manure
6	M1P2	Mixed Mycorrhiza	Guano

The mathematical math of Randomized Block Design The factorial group is as equation 1 follows.

$$Y_{ijk} = \mu + \beta_i + M_j + P_k + (MP)_{ijk} + \varepsilon_{ijk} \quad (1)$$

Where Y_{ijk} is the observations of the influence of mycorrhiza *to-j* and type of organic fertilizer *to-k* in a *to-i.,* μ is the average value, β_i is the effect of *i*- block (*i* = 1,2, and 3), M_j is the Effects of mycorrhizal species k e *j* (*j* = 1, and 2), P_k is the influence of the type of organic fertilizer *k* (*k* = 1,2, and 3), $(MP)_{ijk}$ is the effect of interaction of mycorrhiza *to-j* and type of organic fertilizer *to-k* d l m block *to-i. and* έ error $_{ijk}$ is the effect of treatment of mycorrhiza *to-j* and type of organic fertilizer *to-k* in the *i*- block .

Statgraphics for Windows software was used to analyze diversity of each observation with variance analysis (F test) at α = 5%. If there is a significant different effect on the parameters, the depth analysis will be done by using Least Significant Difference (LSD) 5% approach.

The research starts with initial soil analysis, analysis of organic fertilizers and preparation of soil for treatment. The soil properties analyzed were pH of H_2O, C-organic (Walkley & Black method), N-total (Kjeldhal method), P-available (Bray II method), K-dd, Ca-dd, Mg-dd, Na- dd (extraction method of 1 N NH4OAc pH7), CEC, alkaline saturation (NH4OAc pH7 extraction method), and EC (Electricity Conductivity). While the analysis of organic fertilizer includes C-organic content (Walkley & Black method), N-total (Kjeldhal method), total P-and total K (25% HCl destruction method). Determination of P and K in the extract solution was carried out each with a spectrophotometer and flame photometer. The soil material was taken at a depth of 0-20 cm, then cleaned and after it is dried and mashed and sifted with a hole diameter of 2 mm, then the soil is filled into a polybag of 10 kg each and watered to taste.

Giving organic fertilizer is done according to the treatment a week before planting. The method of administration is by mixing 150 grams of organic fertilizer with soil in each polybag. Then the uniform size and normal growth of seedlings are prepared. Furthermore, planting with in a hole made as deep as 5 cm on the surface of the ground. After that, 10 g of polybag-1 mycorrhiza was given (except the treatment without mycorrhiza) into the hole. The application of basic fertilizer was also carried out on each soil polybag, namely KCl as much as 50 kg ha-1 or 0.25 g polybag-1, TSP 100 kg ha-1 or 0.5 g polybag-1 and Urea 50 kg ha-1 or 0, 25 g polybag

Maintenance needs to be done with watering twice a day morning and evening except rain. Weeding is done every two weeks depending on weed growth. Cleared weeds are immersed back into the soil. At the end of the experiment several soil properties were analyzed, those are pH of H_2O, C-organic (Walkley & Black method), N-total (Kjeldhal method), P-available (Olsen P method) and CEC (NH4OAc pH7 extraction method).

3. Results And Discussion

3.1 pH of H_2O

The result of variance analysis (F test) showed that AMF did not have significant effect on pH of H_2O while organic fertilizer showed very significant effect on pH of H_2O. The combination of these two factors does not show interactivity. The mean value of pH of H_2O soil based on treatment of AMF and organic fertilizer can be seen in table 2.

Table 2. Average pH of H_2O land due to treatment of AMF and organic fertilizer.

Treatment	PH (H_2O)			Average
	Without Fertilizer	Manure	Guano	
Without Mycorrhiza	7.24	7.85	7.13	7.41 A
Mixed Mycorrhiza	7.30	7.89	7.22	7.47 A
Average	7.27 a	7.87 b	7.17 a	

BNT of fertilizer = 0.275. Description: P level ≤ 5%. The lower case is read horizontally and the uppercase is read vertically. "A" means statistical group between Mycorrhiza, meanwhile "a" and "b" are statistical group between fertilizers.

Table 2 shows that there is no significant different between Mixed Mycorrhiza and Without Mycorrhiza on pH of H_2O of soil. Nevertheless when compared to pH of H_2O before the treatment, there is an increase before and after the treatment, from 6.11 to 7, 41, means changing from little acid to be neutral. In the treatment, organic fertilizer highly significant increases pH of H_2O [13], where the manure has the highest influence on increasing pH of H_2O. The manure can increase pH, C-organic and availability of nitrogen, fosfor, kalium and other micro substances [14], followed by Without Fertilizer and Guano.

When compared to pH of H_2O soil before the treatment, there was an increase in pH of H_20 after the experiment, from more acid to be neutral in the treatment of Without Fertilizer and Guano and became slightly alkaline in the treatment of Manure. Organic substances in the soil have a rich *buffer capacity* in order to keep pH of soil relative stabile [15]. This pH increase is influenced by organic fertilizer decomposition assisted by AMF, resulting in the release of alkaline cations as a contributor of OH-ions into the soil. The more OH-ions increase, the more pH ofH_2O increase.

C-organic Furthermore in C-organic , the result of variance analysis (F test) showed that giving of AMF did not significant changing C-organic content in soil, but giving organic fertilizer showed it has significant effect in changing C-organic content in soil, even though no interaction effect. The average value of C-organic percentage in soil due to the treatment of AMF and organic fertilizer can be seen in table 3.

Table 3. The average content of -C-organic soil due to the treatment of AMF and organic fertilizer.

Treatment	C-organic (%)			Average
	Without Fertilizer	Manure	Guano	
Without Mycorrizha	0.21	0.43	0.79	0.47 A
Mixed Mycorrizha	0.30	0.43	0.82	0.51 A
Average	0.25 a	0.43 b	0.80 c	

BNT fertilizer = 0, 156.
Description: The numbers followed by the same letter in the same column and row means no significant statistical different at the P level ≤ 5%. Lower letters are read horizontally and uppercase letters are read vertically. Different letter means has different effect of the treatment between groups.
"A" means statistical group between Mycorrhiza, meanwhile "a" and "b" are statistical group between fertilizers

Table 3 shows that the treatment of Mixed Mycorrhizae has no significantly different from the Without Mycorrhiza on the soil C-organic content. However, when compared to C-organic soil before and after the experiment, there was an increase of C-organic soil after the experiment from 0.24% to 0.47% with Without Mycorrhiza and 0.51% with Mixed Mycorrhiza. Moreover, organic fertilizer has very significant effect in increasing C-organic content of the soil [13]. It shows that Guano is the highest effect on C-organic before manure, and lowest effect is Without Fertilizer. The effect indirectly on the addition of C-organic soil due to the increase of soil nutrients such as N (0.5-0.8%), P (0.07-0.12%), K (1.2-1.7%), S (0.05-0.10%) and Si (4 -7%) [16].

The role of AMF one of which is to improve the soil structure where its hyphae binds one particle of soil to another particle. In addition to the extension of external hyphae, the secretion of the polysaccharide compounds, organic acids and mucus produced by it, will be able to bind the soil micro-aggregate / grain into a secondary grain / macro aggregate. This organic agent is very important in stabilizing micro aggregate through the strength of adhesives and binding by organic acids and hyphae. They will form a solid macro aggregate[17]. So that the provision of organic fertilizer and AMF can increase the number and activities of soil microbes as well as growth of plant roots.

C-organic content indicates that the content of organic matter in the soil where the C-organic content is approximately 58% of the soil organic matter[18,19]. Some elements of C can also be lost to the atmosphere, therefore the addition of organic matter is absolutely necessary every year to maintain and increase the C-organic -content in the soil. This is proven in the treatment of Mycorrhiza and without fertilizers C-organic content decrease from 0.24% to 0.21%.Iit is true that the decrease of C-organic is due to the decomposition process of organic material that is converted into in organic compounds so that the C- organic decreases [20].

3.2 N-total

The result of variance analysis (F test) showed that AMF did not significantly affect the N-total content of soil while organic fertilizer showed very significant effect on N-total of soil. The combination of these two factors does not show interaction. The average value of percentage of N-total soil content due to treatment of AMF and organic fertilizer can be seen in table 4.

Table 4. Average N-total of soil due to the application of AMF and organic fertilizer.

Treatment	N-Total (%)			Average
	Without Fertilizer	Manure	Guano	
Without Mycorrhiza	0.02	0.06	0.17	0.08 A
Mikoriza Mixed	0.04	0.04	0.17	0.08 A
Average	0.03 a	0.05 a	0.17 b	

BNT fertilizer = 0.042. Chart n The numbers followed by the same letters in the same columns and rows do not differ significantly at the level of $P \leq 5\%$. The lower case is read horizontally and the uppercase is read vertically. "A" means statistical group between Mycorrhiza, meanwhile "a" and "b" are statistical group between fertilizers

Table 4 shows that Mixed Mycorrhiza was not significantly different from the Without Mycorrhiza to the N-total soil content, but when compared with N-total soil before the experiment of 0.13%, there was a decrease in total N-total of soil after experiment to 0.8 %. Probably, this occur because of a low C / N ratio of organic fertilizers, which has not yet met the minimum N nutrient content for organic fertilizers and also because of the use by plants for growth. In the treatment of different organic fertilizer is very significant effect on the content of N-total soil. Organic matter contains much nitrogen nutrients and the rate of nitrogen liberation process through mineralization process from the remaining organic material needed by microorganisms [21].

The results show that Guano has the best effect on N-total improvement. This is because of the N content of Guano higher than that of manure. When compared to N-total before Guano experiments also were able to increase the total N-total of the soil to 0.17%. The increase in N-total content of the soil is still low allegedly due to the use of N by plants for growth and microorganisms including mycorrhiza for its development and activity in soil, due to the rapid decomposition process and mineralization of organic fertilizers in soils in tropical climates. Other reasons can also occur because of the loss of N through washing. Nitrates are the result of mineralization processes that are mostly absorbed by most cultivated plants, but these nitrates are easily washed through water and evaporate into the inner atmosphere form of gas [17].

3.3 P-available

The result of variance analysis (F test) showed that the giving of AMF and organic fertilizer had no significant effect on P-available in soil and combination of the two factors did not show interaction. The average value of P-available soil due to treatment of AMF and organic fertilizer in table 5.

Table 5. Average P-available soil due to treatment of AMF and organic fertilizer.

Mycorrhiza	P-available (ppm)			Average
	Without Fertilizer	Manure	Guano	
Without Mycorrhiza	0.67	0.69	0.78	0.71 A
Mixed Mycorrhizae	0.59	0.68	0.60	0.62 A
Average	0.63 a	0.68 a	0.69 a	

Description: The numbers followed by the same letter in the same column and row are not significantly different at the P level ≤ 5%. The lower case is read horizontally, uppercase vertically. "A" means statistical group between Mycorrhiza, meanwhile "a" and "b" are statistical group between fertilizers

Table 5 shows that the treatment of Mixed Mycorrhiza was not significantly different from the mycorrhiza on the P-available soil, as well as the organic fertilizer application did not show any significant difference. Although statistically not showing any significant difference but from the mean value of table 5, it is seen that Guano administration is better in the improvement of P-available.

This is possible because the highest nutrient content in Guano is the phosphorus associated with calcium. In the form of Ca-P, so that in the presence of high phosphorus content guano commonly called phosphorite [22]. Additionally, the pH content of the soil given Guano pH is neutral. Phosphorus availability is strongly influenced by soil acidity, generally available at very low to neutral soil acidity[23]. The decomposition of organic matter yields humic acid and fulvate so that the bound P can be released and become available in the soil [24].

3.4 Cation Exchange Capacity (CEC)

Result of analysis of variance (F test) showed that giving of AMF and organic fertilizer had no significant effect to CEC in the soil and combination of both factors did not show interaction. The average value of CEC content of soil due to treatment of AMF and organic fertilizer in table 6.

Table 6. Mean of CEC for the treatment of AMF and organic fertilizer.

Mycorrhiza	CEC (cmol / kg)			Average
	Without Fertilizer	Manure	Guano	
Without Mycorrhiza	3.00	3.00	2.50	2.83 A
Mixed Mycorrhiza	4.25	3.00	3.00	3.42 A
Average	3.63 a	3.00 a	2.75 a	

Description: The numbers followed by the same letter in the same column and row are not significantly different at the P level ≤ 5%. The lower case is read horizontally and the uppercase is read vertically. "A" means statistical group between Mycorrhiza, meanwhile "a" and "b" are statistical group between fertilizers

Based on the initial soil analysis results , it can be seen that the land taken from the location of the former Lhoong iron ore mine, Aceh Besar including the class of dusty clay soil. This soil has a slightly acidic soil pH and low N, K, Na, Ca and CEC nutrients and C, P, Mg and H are very low. Likewise Electrical Power Supply and Base Saturation are also very low. This indicates that the soil is a critical soil where the soil fertility level is very low as well.

Based on the results of research and analysis of the soil after experiments showed that p emberian AMF and organic fertilizer well evident statistically or not, but on the value of the average showed an increase in the pH of H_2O soil increased to Neutral except for the provision of fertilizer K torch being bit alkalis, then against Corganic, N-total and P-available although small . This indicates a chemical soil improvement but takes a long time to land that has been degraded due to iron ore mining in Lhoong Aceh Besar.

4. Conclusions

Provision of AMF shows no statistical different effect on some soil chemical properties. But, organic fertilizer has significant effect on some chemical properties of soil. It has very significant effect on pH of H_2O, C-organic and N-total. Guano Fertilizer show the best results on the improvement of some soil chemical properties that are planted by the sunflower on the critical land of the former iron ore mine in Lhoong Aceh Besar. In the future it need an integrative research, such as phytoremedian, to find a proper method to rehab the area to recover from a damaged land to be a fertile land.

Acknowledgments

This research was funded by Ristekdikti and published supported by University Muhammadiyah Aceh (Unmuha). Thanks to Rector of Unmuha whom motivate us to present the result to the conference. We appreciate and thanks to all colleagues as well as all those who have contributed from research conducted until this paper was published.

References

[1] Online Source, 2014. Stop Konsesi Tambang untuk Menyelamatkan Bumi Aceh. http://www.waa-aceh.org/2014/03/stop-konsesi-tambang-untuk.html, akses Feb 2016.

[2] Online Source, 2011. Kisruh Keberadaan PT. Lhoong Setia Mining. http://www.jkma-aceh.org/kisruh-keberadaan-pt-lhoong-setia-mining/ diakses 20 Februari 2016.

[3] Zamzami, 2015. Menyelamatkan Hutan Kita dengan Moratorium. http://m.greenpeace.org/seasia/id/high/blog/menyelamatkan-hutan-kita-dengan-moratorium/blog/52454/ Diakses 22 April 2016.

[4] Suwarniati. 2014. Pengaruh AMF Dan Pupuk Organik Terhadap Sifat Kimia Tanah Dan Pertumbuhan Bunga Matahari (*Helianthus annuus* L.) Pada Lahan Kritis. *Jurnal Biotek*. UIN Ar-Raniry. Banda Aceh.

[5] Van Auken, O.W. and R. Freidrich. 2006. Growth and Micorrhizal Infection of Two Annual Sunflowers with Added Nutrients, Fungiside or Salts. Texas Journal Sciences. 58(3):195-218. San Antonio.Texas.

[6] Agnello A.C, Potysz A, Fourdrin C, Huguenot D, Chauhan P.S, 2018. Impact of pyrometallurgical slags on sunflower growth, metal acmulation and rhizosphere microbial communities. Journal of Chemosphere. Volume 208, October 2018, Pages 626-639.

[7] Yu Zhang, Junli Hu, Jianfeng Bai, Junhua Wang, Rui Yin, Jingwei Wang, Xiangui Lin, 2018. Arbuscular mycorrhizal fungi alleviate the heavy metal toxicity on sunflower (*Helianthus annuus* L.) plants cultivated on a heavily contaminated field soil at a WEEE-recycling site. Journal of Science of The Total Environment. Volumes 628–629, 1 July 2018, Pages 282-290.

[8] Caroline KrugVieira, Matheus Nicoletti Marascalchi, Arthur Vinicius Rodrigues, Rafael Dutrade Armas, Sidney LuizStürmer, 2018. Morphological and molecular diversity of arbuscular mycorrhizal fungi in revegetated iron-mining site has the same magnitude of adjacent pristine ecosystems. Journal of Environmental Sciences. Volume 67, May 2018, Pages 330-343.

[9] Sandra Rosa, Matias Marcela, Cláudia Pagano, Frederico Carvalho Muzzi, Christiane A.Oliveira, Sá Nadja Horta, Maria Rita Scotti. 2009. Effect of rhizobia, mycorrhizal fungi and phosphate-

solubilizing microorganisms in the rhizosphere of native plants used to recover an iron ore area in Brazil. European Journal of Soil Biology, Volume 45, Issue 3, May–June 2009, Pages 259-266.

[10] Faisal Danu Tuheteru, Asrianti Arif, Eka Widiastuti, & Ninis Rahmawati. 2017. Serapan Logam Berat oleh Fungi Mikoriza Arbuskula Lokal pada Nauclea orientalis L. dan Potensial untuk Fitoremediasi Tanah Serpentine. Jurnal Ilmu Kehutanan II, pages 76-84.

[11] Hassan S.E., Hijri M., St-Arnaut M. 2013. Effect of Arbuscullar Mycorrhizal Fungi on Trace Metal Uptake by Sunflower Plants Grown on Cadmium Contaminated Soil. Journal of Biotechnology. Volume 30, Number 6. September 2013. Institut de Recherche en Biologie Vegetale, Universite de Montreal. Canada.

[12] Kloepper, J.W. 1993. Plant growth-promoting rhizobacteria as biological control agents. p. 255-274. In F.Blaine Metting, Jr. (Ed.). Soil Microbiology Ecology, Applications in Agricultural and Environmental Management. Marcel Dekker, Inc., New York

[13] Suria Darma, Wawan Kustiawan, Daddy Ruhiyat, Sumaryono. 2017. Organic Fertilizers Improves Trembesi Samanea Saman Seedling Growth A Case Study Of The Implementation Of Post-Mining Land Reclamation And Revegetation Within The Forest Cultivation Zone. International Journal of Scientific & Technology Research, Vol 06, Iss 10, Pp 393-399.

[14] Flaig, W. 1984. *Soil Organic Matter as a Source of Nutrients*. Organic Matter and Rice. Los Banos Laguna, Philippines: International Rice Research Institute. p. 73-92.

[15] Utami S.N.H. dan Handayani, S. 2003. Sifat kimia entisol pada sistem pertanian organik. Ilmu Pertanian 10 (2), 63-69.

[16] Hanum, Hamidah *et al*, 2016. Prosiding Seminar Nasional Lahan Suboptimal 2016, Palembang.

[17] Subiksa, IGM. 2002. Pemanfatan Mikoriza Untuk Penanggulangan Lahan Kritis. Makalah Falsafah Sains Program Pasca Sarjana. IPB Bogor.

[18] Syukur, A. 2005. Penyerapan Boron oleh Tanaman Jagung di Tanah Pantai Pasir Bugel dalam Kaitannya dengan Tingkat Frekuensi Penyiraman dan Pemberian Bahan Organik. *Jurnal Ilmu Tanah dan Lingkungan* 5(2):20-26.

[19] Syukur, A., Harsono, E.S. 2008. Pengaruh Pemberian Pupuk Kandang dan NPK Terhadap Sifat Kimia dan Fisika Tanah Pasir Pantai Samas Bantul. *Jurnal Ilmu Tanah dan Lingkungan* 8(2):138-145.

[20] Sukarwati. S. 2011. Jerapan P Pada Tanah Andisol yang Berkembang dari Tuff VulkanGunung Api Di Jawa Tengah Dengan pemberian Asam Humat Dan Asam Silikat. Media Limbang Sulteng.

[21] Utami, S.M.H. 2004. Sifat Kimia Andisol pada Pertanian Organik dan Anorganik.Jurnal Ilmu Tanah. Skripsi. Institut pertanian Bogor.

[22] Hakim, N., Nyakpa, M.Y., Lubis, A.M., Nugroho, S.G., Saul M.R., Diha, M.A., Hong, G.G.B., H.H.Bailey. 1986. Dasar-dasar Ilmu Tanah. Universitas Lampung.

[23] Foth, D.H. 1995. Dasar-Dasar Ilmu Tanah. Cetakan Ketiga. Diterjemahkan oleh Endang D.P., Dwi Retno L., dan Rahayuning T. Editor Sri Andani B. Hudoyo. Fakultas Peternakan Universitas Diponegoro. UGM Press. Yogyakarta. Hal 336-365.

[24] Hastuti (2003). Pengaruh Berbagai Jenis Bahan Amelioran Terhadap Pertumbuhandan Hasil Tanaman Sorgum di Ja wa Timur. http://repository.ipb.ac. id/bitstream/handle/20IV_A93atr. pdf. Diakses 9 Desember 2017.

Visual Analysis of Satellite Landsat Images Multitemporal and GPS as a Geographic Information System for Mapping of Nugmet Plantations in Tapaktuan

Safridatul 'Audah[1*], Nazliyati[1], Bakruddin[1], Edi Saputra[2], Saiful Wathan[2], Muharratul Mina Rizky[3,4]
[1]Department Industrial Engineering, Politeknik of Aceh Selatan, Tapaktuan,Indonesia
[2]Department Mechanical Engineering , Politeknik of Aceh Selatan, Tapaktuan, Indonesia
[3]Department of Informatics Engineering , Politeknik of Aceh Selatan, Tapaktuan, Indonesia
[4]Department of Electrical Enginering, Universitas Syiah Kuala, Banda Aceh,Indonesia

*Corresponding author: Safridatul@yahoo.co.id

Abstract. Geographical information systems (GIS) are needed in planning and controlling land use as a basis for anticipating policy decisions in planning and structuring Tapaktuan territory. This study seeks to integrate GPS technology and remote sensing technology products such as Landsat-8 OLI satellite imagery, Landsat-7 ETM + using multitemporal methods using the supervised (Maximum Likehood) classification method. Based on the analysis of the processing of satellite images of nutmeg plantation area for several years, there has been an increasing spread of land use for nutmeg plantations.

Keywords: GIS, Remote Sensing Products, GPS, Classification Supervised, Nutmeg Plants

1. Introduction

Nutmeg is known as a spice plant that has economical and multipurpose value because every part of the plant can be used in various industries. Seeds, mace and nutmeg oil are export commodities and are used in the food and beverage industry [1]. Tapaktuan is the west coast region of Aceh, which is known as a place for producing spice plants such as nutmeg plants, this nutmeg plant thrives and is one of the sources of income for local people because it has economic and multipurpose value in every part of the plant that can be utilized in various industries, such as the food and beverage industry is used as a source of income for the people in South Aceh, especially in the Tapaktuan District area [2].
Remote sensing is the science that provides information about land areas and waters obtained from images at a distance [3]. Remote sensing provides multispectral and multitemporal data [4]. Remote sensing is an important and effective means of monitoring land cover with its ability to provide information about the spatial diversity on the surface of the earth quickly, widely, precisely, and easily [5]. Remote sensing data is an important factor in the success of land cover classification [6-7]. Data satellite imagery has been used to monitor limited land cover types according to spectral classification.

In addition, it has been used to estimate the biophysical characteristics of the soil surface through a linear relationship with spectral reflectance or vegetation index [4-8]. Plantation monitoring can be done by remote sensing and GPS method, which is based on the interpretation of land cover of remote sensing images that can provide fast and accurate spatial, spectral data [9]. This study focused on analyzing Landsat-8 OLI (Operational Land Imager) image data, Landsat-7 ETM + (Enhanced Thermal Mapper plus) and multitemporal GPS with recording time 2003, 2013 and 2017 as Geographic Information Systems (GIS) for mapping and knowing widely Plantation area at Tapaktuan area.

2. Background Study

South Aceh Regency is one of the Regencies in Aceh Province which is located in the West - South coast area with the Regency Capital is Tapaktuan. The land area of South Aceh Regency is 4,185.56 km^2 or 418,556 Ha, and geographically located at 02^0 23 '24 "- 03^0 44' 24" LU and 96^0 57 '36 "- 97^0 56' 24" BT. The topographic conditions of South Aceh Regency vary greatly, consisting of lowland, undulating, hilly, to very steep slopes. The research location was focused in several villages in Tapaktuan sub-district (figure.1).

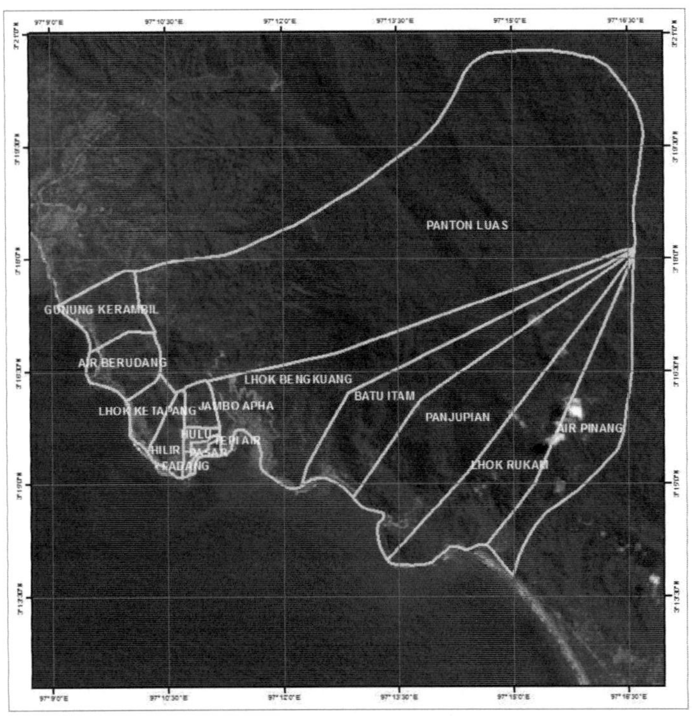

Figure 1. Map of Tapaktuan

3. Methodology

The method used is supervised classification (Maximum Likelihood) with channel combinations found in Landsat 7 Enhanced Thermal Mapper plus (ETM) and Landsat-8 Operational Land Imager (OLI). The imagery in a multi-temporal manner with different recording times and downloads in 2003, 2013 and 2017, which are sourced from the United States Geological Survey (USGS) website. The stages of processing Landsat image data can be seen in figure 2.

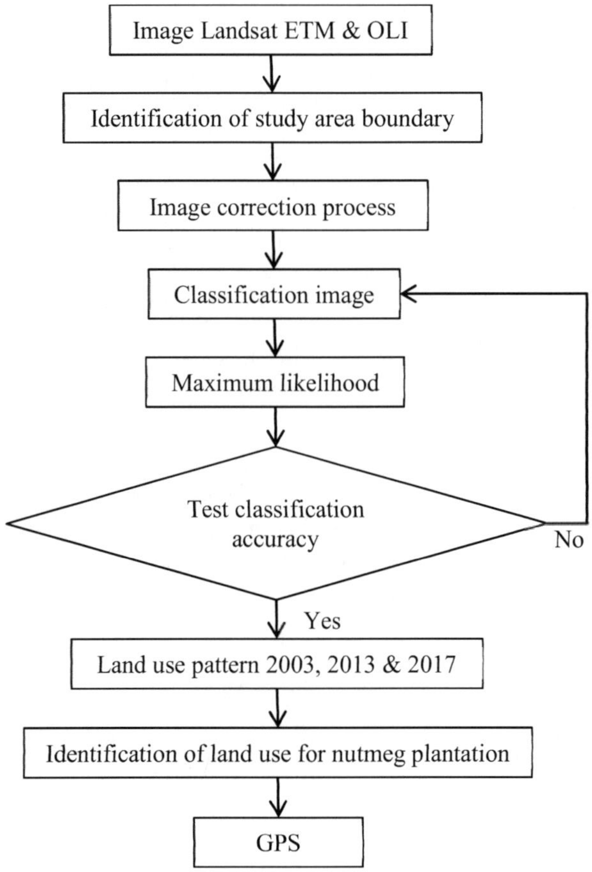

Figure 2. Flow chat of research

Based on the literature, many methods have been used to detect land changes, the method produces a variety of results [10]. Some researchers have used a fairly complicated approach in increasing the accuracy of image classification for land use [11-13]. Several stages of the image correction need to be done such for improving the image quality so that there is no error arises during analyzing the digital images. Furthermore, the image classification process use the supervised method (Maximum Likelihood) which is a method that is often used in image processing to determine land use. Image classification provides flexibility for computers to classify images themselves based on digital values. Digital classification in some images is the process where pixels with the same spectral characteristics are classified and arranged in several colors. In this study image classification is divided into six classes of land use classifications including nutmeg gardens, forests, settlements, vacant land, water and mixed forest plantation. After six classification classes, focus is given towards the land that is used for nutmeg plantations by identifying the locations where there is a spread of nutmeg plantations. The identification is done using GPS with groundcheck extracts from nutmeg plantation objects.

4. Result and Discussion

4.1. Results of Classification of Landsat Image Processing

Based on the interpretation of Landsat-7 ETM images recorded in 2003, Landsat-8 OLI images that are recorded in 2013 and 2017, the six classes of land use are identified using the best band combination. The classes consist of protected forest, settlements, open land, mixed forest plantations, nutmeg garden, water. The appearance of the type of land cover for each of the class stated above in the image is displayed with different class color. For example, the water body is represented in blue which indicates that there are water objects such as swamps, rivers and others. Vegetation is represented in bright green to dark colors. The degree of green brightness usually represents the vegetation density. High density forests will appear dark green when compared to low density forests or mixed forests. Built land and open land are represented in red while nutmeg plantations are represented in bright green. The classification results can be seen in figure 3a, 3b and 3c for each image. The figure shows the interpretation of the image in each part, such as the northern part of the protected forest, which is very broad. Settlements and water bodies are in the south. The southwest part of the built land which is the development of the city of Tapaktuan as well as mixed forests and nutmeg plantations. Whereas in the eastern part of Tapaktuan sub-district, there are forest land uses nutmeg plantations. Therefore, based on the multitemporal image analysis of the land use in the Tapaktuan area, it is discovered that the protected forest areas, the land used for nutmeg plantations and residential areas are increasing in a certain period of time.

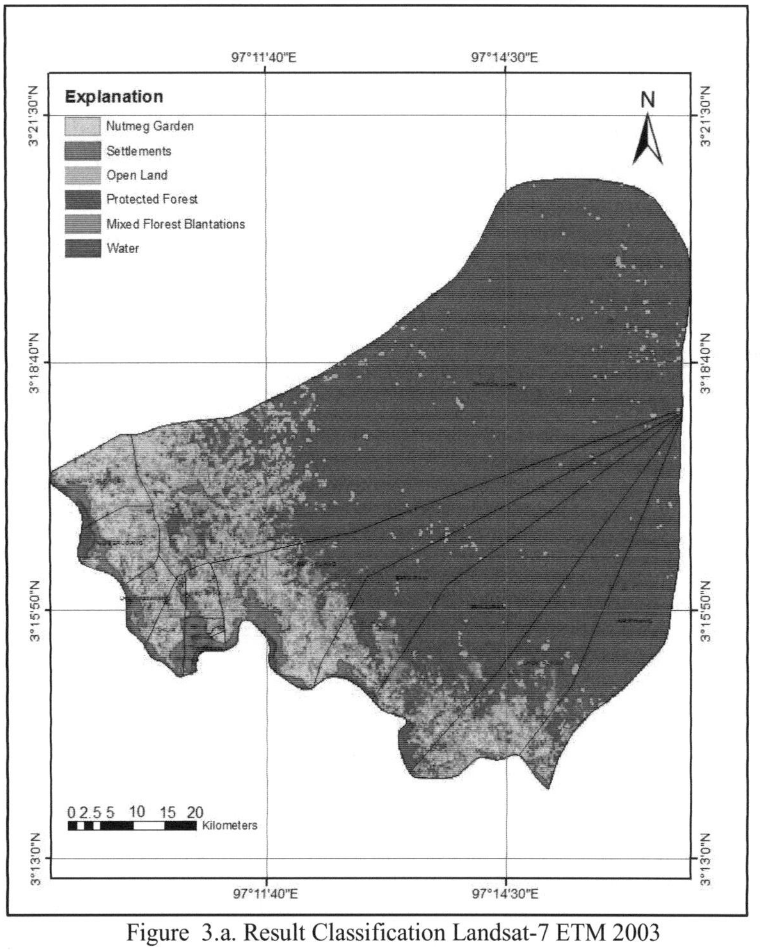

Figure 3.a. Result Classification Landsat-7 ETM 2003

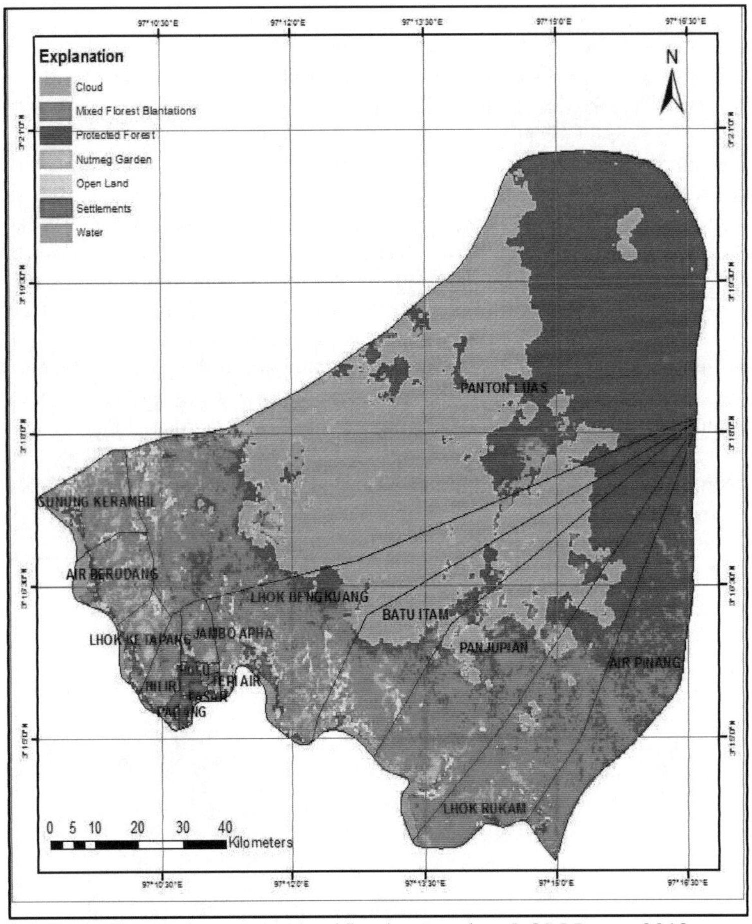

Figure 3.b. Result Classification Landsat-8 OLI Image 2013

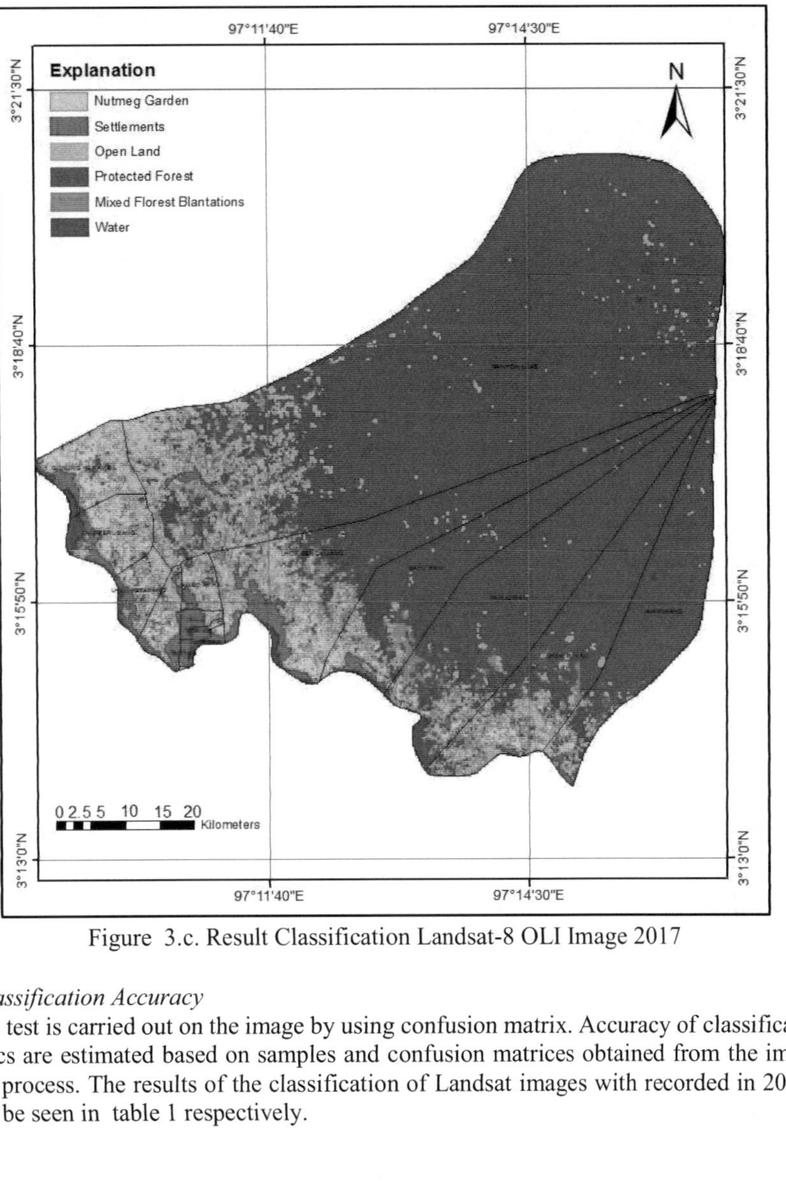

Figure 3.c. Result Classification Landsat-8 OLI Image 2017

4.2 Image Classification Accuracy

Classification test is carried out on the image by using confusion matrix. Accuracy of classification and kappa statistics are estimated based on samples and confusion matrices obtained from the image data classification process. The results of the classification of Landsat images with recorded in 2003, 2013 and 2017 can be seen in table 1 respectively.

IOP Conf. Series: Materials Science and Engineering **506** (2019) 012037 doi:10.1088/1757-899X/506/1/012037

Table 1. Value of calculation results of confusion matrix Landsat-7 ETM + in 2003

Classification	1	2	3	4	5	6	Total	User accuracy (%)	Commision (%)
Settlements	524	0	8	0	38	0	570	91.93	8.07
Protected forest	0	1021	29	0	1	0	1051	97.15	2.85
Mixed forest plantation	7	11	488	1	2	0	509	95.87	4.13
Nutmeg garden	2	0	1	311	15	0	329	94.53	5.47
Open land	93	5	10	5	240	6	359	66.85	33.15
Water	0	0	0	0	4	102	106	96.23	3.77
Total	626	1037	536	317	300	108	2924	-	-
Producer accuracy (%)	83.71	98.46	91.04	98.11	80.00	94.44	-	-	-
Omission (%)	16.29	1.54	8.96	1.89	20.00	5.56	-	-	-
Overall accuracy					95.194				
Kappa Coefficient					0.9365				

Table 2. Value of calculation results of confusion matrix Landsat-8 OLI in 2013

Classification	1	2	3	4	5	6	7	Total	User accuracy(%) (%)	Commision
Settlements	262	0	0	4	18	2	0	286	88.51	11.49
Protected forest	0	2082	0	4	2	0	30	2118	97.47	2.53
Mixed forest plantation	2	0	189	9	7	1	0	208	87.10	12.90
Nutmeg garden	2	2	29	251	0	1	15	300	81.64	18.36
Open land	0	11	7	0	152	1	7	178	80.92	10.08
Water	1	0	0	0	1	122	87	211	80.71	19.29
Cloud	1	13	0	0	2	0	2053	2069	99.21	0.79
Total	268	2108	225	268	180	127	2192	3301	-	-
Producer accuracy (%)	90.66	96.97	84.00	92.05	82.40	96.06	91.38	-	-	-
Omission (%)	9.34	3.03	16.00	7.95	17.60	3.94	8.62	-	-	-
Overall accuracy						96.46				
Kappa Coefficient						0.9465				

Table 3. Value of calculation results of confusion matrix Landsat-8 OLI in 2017

Classification	Settlements	Protected forest	Mixed forest plantations	Nutmeg garden	Open land	Water	Total	User accuracy (%)	Commision (%)
Settlements	70	0	0	0	0	0	70	100.0	0.00
Protected forest	0	285	0	0	0	0	285	100.00	0.00
Mixed forest plantation	0	0	22	2	0	0	24	91.67	8.33
Nutmeg garden	0	0	0	34	0	0	36	94.44	5.56
Open land	11	0	0	0	39	2	52	75.00	25.00
Water	0	0	0	0	0	233	233	100.00	0.00
Total	81	285	24	36	39	235	700	-	-
Producer accuracy (%)	86.42	100.00	91.67	94.44	100.00	99.15	-	-	-
Omission (%)	13.58	0.00	8.33	5.56	0.00	0.85	-	-	-
Overall accuracy					97.57				
Kappa Coefficient					0.9655				

Based on the classification test on Landsat-7 ETM imagery in 2003 as shown in table 1, the overall classification accuracy value was 95.19% and the kappa accuracy was 93.65% with matrix accuracy in the identification nutmeg plantations of accurasy producer 98.11% in the presence of error detection of nutmeg plantations as much as 6 points from 317 points of nutmeg gardens, these points were detected as mixed plantations and open land, while accurasy users were 94.53% where there were 17 points detected. The classification test was also conducted on Landsat-8 OLI images with the recording time of 2013 (table.2). The overall classification accuracy value was 96.46% and the kappa accuracy was 94.65%, the value of matrix accuracy in the identification accurasy producer nutmeg plantations was 92.05% with the detection error of nutmeg plantations as much as 17 points from 268 nutmeg garden points, these points were detected as mixed plantations, settlements and protected forests, while accurasy users were 81.64% where there were 49 points detected.

The matrix test of Landsat-8 OLI imagery in 2017 was also carried out (table.3), identification of nutmeg plantations with accurasy producer was 94.44% with the detection error of nutmeg plantations as much as 2 points from 36 points of nutmeg gardens, both of which were detected as mixed plantations, while the accurasy user value is 94.44 where there are errors of 2 points detected as mixed forests. Based on the provisions of the land cover classification system, image classification is considered correct if the confusion matrix calculation ≥ 80% [1], the accuracy value of the three multitemporal landsat image data provides high accuracy because it meets the requirements set by the USGS stating the accuracy of interpretation is more than 80% so the classification results of Landsat images in the Tapaktuan region can be used to determine the identification of the distribution of nutmeg plantations.

4.3 Analysis of Nutmeg Plantations Based of Land Use Imagery on Multitemporal

Land use is all forms of human activity on land to fulfill their daily needs. Land use in the research area obtained from the results of multitemporal image data processing, land use forms namely nutmeg plantations, protected forests, mixed forests, settlements, open land, bodies of water, and clouds. For more details about land use in the research area can be seen in the table 4.

IOP Conf. Series: Materials Science and Engineering **506** (2019) 012037 doi:10.1088/1757-899X/506/1/012037

Table 4. *Area of land use in Tapaktuan District*

Class Land Cover	Area In 2003	Area In 2013	Area In 2017
Water	13.626	10.243	7.281
Protected Forest	518.875	286.092	673.209
Settlements	27.774	56.335	72.783
Open Land	83.367	43.031	43.353
Mixed Forest Plantation	308.819	206.246	186.687
Nutmeg Garden	79.596	15.174	48.744
Cluod	-	414.936	-
Total	1.032.057	1.032.057	1.032.057

Based on table.1, the nutmeg plantation land is decreasing, namely in 2003 with a land area of 79.596 ha to 15.174 ha in 2013 and experienced an increase in 2017 of 48.744 ha. This happened because of the conversion of plantation land into residential land and other land uses. In addition to the use of plantation land in the Tapaktuan area, it is mostly used as protected forest covering an area of 673.209 ha, this can be seen from the results of processing image data recorded in 2017, but in 2013 the area of protected forest plants decreased to 286.092. it is predicted that in the event of burning of protected forests marked from a lump of smoke seen in the image. In addition to the protected forest of the Tapaktuan area, it is used for mixed forest which has decreased by 206.246 ha to 186.687 ha in 2017, Based on this description, it was found that the forest area in the Tapaktuan area had a high area compared to the other classes and based on a field survey that the community in Tapaktuan almost planted nutmeg on a large area.

4.4 Ground Check Data

Make a map of the location of the distribution point of nutmeg plantations carried out using GPS with taking points (groundchek). The overall measurement of position points in the field has represented a description of the situation from the research location. Each coded Pala (P01, P02, ..., P0n), identity (with numbers or letters), or the combination has information in the form of position information with a Geography projection system, and a datum based on WGS 1984. Because not all points / codes are easily recognizable landmarks at the location, only the code that represents a landmark point is displayed. Some of the location points of the Pala plantations in 2018 measured in the field can be seen in table 5.

IOP Conf. Series: Materials Science and Engineering **506** (2019) 012037 doi:10.1088/1757-899X/506/1/012037

Table 5. Points of location of Pala plantations coordinates in 2018.

No	Symbol	Coordinate		No	Symbol	Coordinate	
		X	Y			X	Y
1	P01	296295	361526	25	P25	305062	358003
2	P02	296308	361541	26	P26	305176	357733
3	P03	296312	361546	27	P27	3030732	358374
4	P04	296331	361554	28	P28	303495	358719
5	P05	300252	359386	29	P29	303167	358369
6	P06	305481	357437	30	P30	301886	359061
7	P07	303743	358373	31	P31	301545	359537
8	P08	302947	358642	32	P32	301464	359587
9	P09	303160	358372	33	P33	300526	359420
10	P10	301022	359963	34	P34	300268	359373
11	P11	300226	359418	35	P35	297085	360265
12	P12	299775	360496	36	P36	296525	361278
13	P13	298557	361333	37	P37	305110	358138
14	P14	297675	359631	38	P38	296360	361424
15	P15	296356	361421	39	P39	296359	361531
16	P16	296193	361798	40	P40	296042	361854
17	P17	295043	363667	41	P41	295869	362067
18	P18	296985	360311	42	P42	304745	357741
19	P19	297554	361325	43	P43	303230	357619
20	P20	297732	360965	44	P44	304444	358051
21	P21	297598	361022	45	P45	304425	358045
22	P22	300531	359420	46	P46	304422	358046
23	P23	303378	358831	47	P47	303958	358251
24	P24	303679	358406	48	P48	303735	358381

The results of the Groundcheck distribution point map made with ArcGIS can be seen in figure 4 (marked in red point). Groundcheck is done to get information in the field in the form of information about observations in the field and coordinates taken. These information can be used to check the truth of the visual classification results, the results of which can be used as an example area (training area) in the classification of digital images. Taking these coordinate points is 48 points using a tool in the form of a Global Positioning System (GPS).

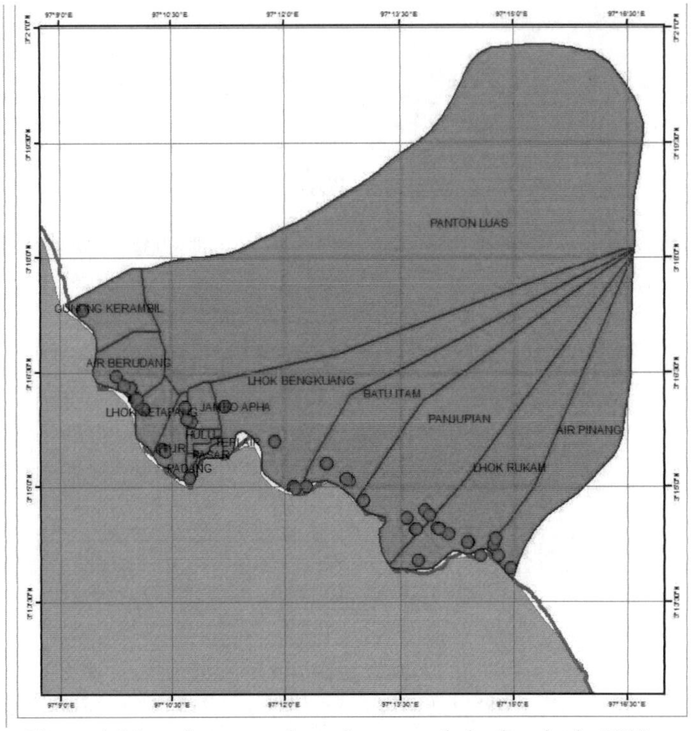

Figure 4. Map of nutmeg plantation ground check point in 2018

Figure 5. Nutmeg plantation in Tapaktuan sub-district

5. Conclusion

Based on the visual interpretation of Landsat-7 ETM images in 2003 and Landsat-8 OLI imagery in 2013 and 2017 using the best combination of bands for nutmeg plantation land cover classification, observing the truth of objects of land cover visually on the field was identified as 6 land cover classes. The classes consist of Protected Forest, Settlements, Open Land, Mixed Forest Plantation, Nutmeg Garden, Water. From the results of the classification that has been done, it can be concluded that some of the Tapaktuan region consists of protected forest areas, while the area of nutmeg plantation land use is decreasing, namely in 2003 amounting to 79.596 ha to 15.174 ha in 2013 but experiencing an increase in 2017 of 48.744 ha. This happened because the conversion of plantation land into residential land and other land uses while the results of the classification accuracy test showed a high degree of accuracy and fulfilled the requirements set by USGS (> 85%). With the results obtained, it shows that image classification maps can be used for mapping nutmeg plantations.

References

[1] McCullough E Pingali P Stamoulis K 2008 The Transformation of Agri-Food System *Routledge*

[2] Audah S Nazliyati 2018 Utilization of Satellite Landsat-8 Operational Land Imager (OLI) for Land Cover Classification Nutmeg Plantation In Tapaktuan Sub-District *Jurnal Inovasi Teknologi dan Rekayasa* vol 3 pp 23-28

[3] Campbell B James Introduction To Remote Sensing *The Guilford Press New York* 1994

[4] Nuarsa I Wayan Kanno S Sugimori Y and Nishio F 2005 Spectral Characterization of Rice Field Using Multi-Temporal Landsat ETM+ Data *International Journal of Remote Sensing and EarthSciences* vol 2 pp 65-71

[5] Thenkabail P S Biradar C M Noojipady P Dheeravath V Li Y J Velpuri M Gumma M Gangalakunta O R P Turral H Cai X L Vithanage J Schull M A Dutta R 2009 Global Irrigated Area Map (GIAM), Derived from Remote Sensing, for The End of The Last Millennium *International Journal of Remote Sensing* vol 30 pp 3679-3733

[6] Gumma M K Thenkabail P S Hideto F Nelson A Dheeravath V Busia D Rala A 2011 Mapping Irrigated Areas of Ghana using Fusion of 30 m and 250 m Resolution Remote Sensing Data *Remote Sensing* vol 3 pp 816-835

[7] Jia K Xiangqin W Xingfa G Yunjun Y Xianhong X Bin L 2014 Land Cover Classification using Landsat 8 Operational Land Imager Data in Beijing, China. *Geocarto International* vol 29 pp 941-951

[8] Steininger M K 1996 Tropical Secondary Forest Regrowth in TheAmazon: Age, Area and Change Estimation With Thematic Mapper Data *International Journal of Remote Sensing* vol 17 pp 9–27

[9] Hardiyanti F S Grandharum K T L dan Rambo 2006 Pemantauan Kerusakan Lingkungan Wilayah Meratus Kalimantan Selatan *Citra Landsat-TM dengan Kajian Geografis* pp 50-63

[10] Berberoglu S Akun A 2009 Assesing Different Remote Sensing Techniques to Detect Land Use/Cover Changes in The Eastrn Mediterranean *International Journal of Applied Earth Observation and Geoinformation* vol 11 pp 46-53

[11] Berberoglu S Lloyd C D Atkinson P M and Curran P J 2000 The integration of spectral and textural information using neural networks for land cover mapping in the Mediterranean *Computers and Geosciences* vol 26 pp 385-396

[12] Mendoza J E and Etter R 2002 Multitemporal Analysis (1940–1996) of Land Cover Changes in The Southwestern Bogota highplain (Colombia) *Landscape and Urban Planning*, vol 59(3) pp 147-158

[13] Stuckens J Coppin P R and Bauer M E 2000 Integrating contextual information with per pixel classification for improved land cover classification *Remote Sensing of Environment* vol 71 pp 282-296

IOP Conf. Series: Materials Science and Engineering **506** (2019) 012038 doi:10.1088/1757-899X/506/1/012038

The Study of Potential Use of PV Pump For The Rain-Fed Rice Land In Simeulue Island by Using Homer Simulation

Muhammad Yasar[1,2*], Mustaqimah[2], and Yuswar Yunus[2]
[1]Politeknik Aceh Selatan
[2]Agricultural Engeneering Department, Faculty of Agriculture University of Syiah Kuala Banda Aceh, Indonesia

[*]Corresponding author: yasar@unsyiah.ac.id

Abstract. Simeuleu is one of the districts in Aceh Province which is located in the outer territory of the Unitary State of the Republic of Indonesia. In general, communities in districts with island clusters consisting of 147 large and small islands rely on agriculture and fishermen. One of the agricultural sub-sectors that is the main concern of the government is food, especially rice as the staple food of the community. Limited land resources supported by appropriate and adequate infrastructure have resulted in this business experiencing serious challenges and constraints. Of the available rice field area, 43.11% is rainfed lowland. During this time to irrigate some of the land, the potential of rainfall owned is considered able to meet the water needs of rice crops. However, the emergence of global climate change symptoms has led to cultivation efforts that rely on rainwater alone can not be expected completely. The irrigation technology of PV pumps has excellent prospects for the future of agriculture especially the food sector. This technology is very cheap for long term and also simple so easy to make and modified by anyone. PV pump irrigation does not require the operator to operate it, it is only necessary once check and control. The main benefit of solar irrigation technology development is to maintain environmental sustainability because it does not produce air pollution so that it can suppress the increase of global warming.

1. Intoduction

Solar energy is most abundant source of energy in the world. Solar power is not only the answer to the current energy crisis but also an environmentally friendly form of energy. The photovoltaic generation is an efficient approach to using solar energy. Solar panels (the arrangement of photovoltaic cells) are now widely used to run the street lights, to turn on the water heater, and others. The domestic cost of solar panels continues to decline, which encourages increased use in various sectors. One application of this technology is used in irrigation systems for agriculture [1]. Solar irrigation system can be an appropriate alternative for farmers in a state of energy crisis in Indonesia which is environmentally friendly green energy.

The solar irrigation system is a system that can effectively contribute to the conservation of vacant and unproductive lands. It is usually used PV to produce direct energy from solar radiation to generate a pump irrigation system. Utilization of vacant land can increase productivity and income of local farmers. The application of solar irrigation is closely related to the condition of water resources and local area [2]. Thus, knowledge of water sources and geographic conditions of the site is necessary for a study. PV pumping is now being applied for different uses in many regions. For instance, a PV pumping project in South Morocco was begun in 1997, and by 2005, the project had already reached 18 villages, affecting 15,000 people with total involved photovoltaic power is 46 kWp, and the total volume of pumped water since the installation of the systems approaches 0.7 $x10^6$ m^3. PV-powered pumping is also evaluted in Bantul, Indonesia, where needed power 800 Wp to discharge 3.8 m^3/hr [3]. PV pumping technology also used in Saudi Arabia in supplying sufficient quality water [4]. Nowadays, PV pumping irrigation is being applied for the conversation of pastures in many parts of the world. At the same time, solar energy is being used in grassland pumping for over 25 years in Namibia [5]. The results show that PV pumping can operate longer and more effectively compared with traditional irrigation measures. The system also directly improves quality of life and promotes socio-economic development in that area [6].

The application of this technology that has application in Mexico and German also demonstrates that solar energy pumping produces more benefit and less pollution. PV systems for the pumping of groundwater are also used in Upper Egypt, proving that the cost of the water unit pumped by PV systems is significantly lesser than that pumped by diesel systems. [7] has discussed the potential for solar water-pumping systems in India and got the conclusion that solar photovoltaic systems are suitable in areas where cheaper sources of energy are not available. The research and application of solar irrigation system in China has not caused widespread concern. The application of solar powered irrigation has significant meanings in China, especially in Northwestern China. For the distribution of water resources is uneven and the precipitation is scare in that region. At 2016, it was evaluated performance of groundwater irrigation system by using solar water pump with 51 solar panels covering area of 120 m^2. Evaluation results of the technology application shows that solar energy can generate 7,873.5 watts. And maximum discharge 14.17 liter/second [8].

2. Literature Review

The solar power pump irrigation system is a system that can effectively contribute to the conservation of vacant and unproductive lands. Utilization of vacant land can increase productivity and income of local farmers. According [2] the application of solar irigation is closely related to the condition of water resources and local area. Thus, knowledge of water sources and geographic conditions of the site is necessary for a study. [2] also mention some indicators to be considered in solar irrigation systems. Among others are rainfall, land slope, temperature, solar radiation:

i. Appropriate rainfall for solar power irrigation.
ii. The need for irrigation water is very dependent on rainfall. Rainfall must be higher than the water requirement.
iii. Appropriate slope for solar irrigation systems
iv. Tilt is closely related to the feasibility of solar irrigation. Tilt should not be more than 2-5% for groove irrigation and it should not more than 0.2% for closed irrigation.
v. Temperature suitable for irrigation of solar power
vi. Temperature changes result in evaporation and transpiration changes and greatly affect crop water requirements. Higher temperatures will increase crop water requirements and increase pressure on photovoltaic pumps.
vii. Solar radiation suitable for solar power irrigation. Solar radiation is one of the important indicators for solar irrigation systems because it is an energy source for solar powered pumps. Large solar radiation is a guarantee for the application of solar irrigation systems. The sun exposure rate should be above 30%.

3. Methodology

This research was conducted in two stages of activity. The first phase is a feasibility survey of the development area includes the potential of solar radiation and ground water which is a very crucial consideration in this research. The second stage is a simulation of the utilization of solar energy as a source of energy driving the water pump. The simulation is done with case study of pump usage to pump water into the maximum 50 meter pump head storage tank. The simulation is created by using HOMER software. In the optimization process, HOMER simulates each configuration system in search space and is displayed in a table, sorted by total net present charge (NPC). Figure 1 illustrates the solar-water pump system and its components. The solar radiation data for the simulation on the program is taken from the results of direct measurements to the field and validated with data from NASA satellites.

3.1. Research procedure

1. Survey of the feasibility of the development area by reviewing the availability of rice fields in the research sites supported by: wetland area data and irrigation field data.
2. Potential of solar energy:
   a. Measurements of solar radiation at the study site for approximately 1 month using a solarimeter radiation instrument
   b. In one day, there are three times measurement, morning (8.00 s.d 9.00 pm), noon (20.00-13.00 pm), and afternoon (16.00-17.00 pm)
   c. Measures the average duration of solar irradiance in a day.
3. Potential groundwater; to identify the potential of ground water, geoelectric apparatus is needed to obtain the resistivity value of rocks and minerals in the soil at the study site.
4. Land suitability; measuring the land topography
5. Simulation of HOMER

   Beginning with the determination of pump loads, determination and selection of types and prices of solar panels, batteries and converters, then input solar radiation data. Input data as follows: interest rate 4.25%, 500 watt inverter = Rp. 600.000 = $ 45, Trojan battery T-105 = Rp. 2.650.000 = $ 197, solar panel 100wp = Rp. 1,200,000 = $ 89.

3.2. Configuration of PV pump Irrigation System

The solar powered pump irrigation systems designed in this study (figure 1) include solar cell modules, pumps, batteries, storage (tanks) and converters. The basic principle of this PV pump's irrigation system works is: solar cells absorb sunlight, then generate electrical energy in the form of DC current stored to battery, go into converter to be converted to AC current, so move pump to suck water in tank. Water in the tank is used to meet the needs of the suboptimal land.

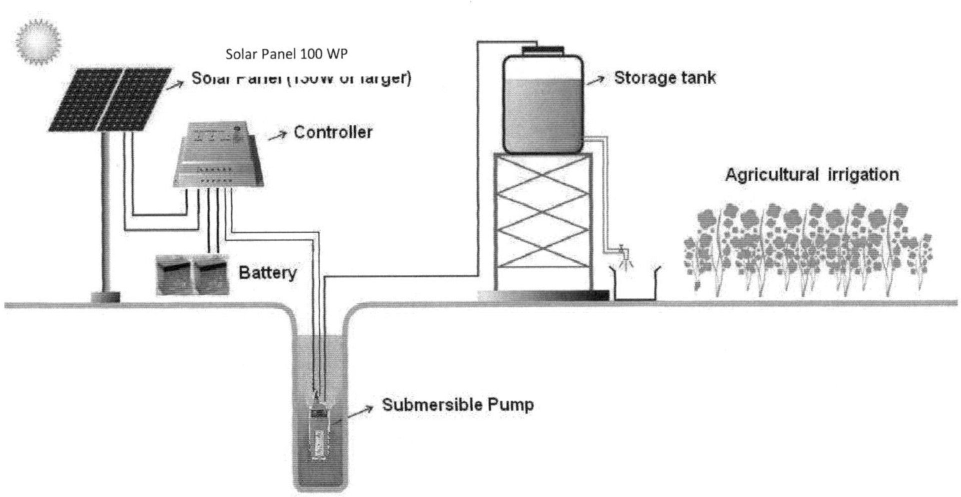

Figure 1. Scheme of solar power pump irrigation system (USDA and NSCR,2010).

3.3. Structural and functional design

3.3.1 Solar panels. Solar panels or modules serve to absorb sunlight and generate direct electric current (DC). Solar panel used is Solar Panel 100 Wp Shinyoku Polycrystalline with specification:

i. Maximum Power (Pmax): 100 W.
ii. Type Cell : Polycrystalline
iii. Voltage at Pmax (Vmp) : 17.5V

iv. Current at Pmax (Imp) : 5.71A
v. Short circuit current (Isc): 6.4 A
vi. Open circuit voltage (Voc): 21V
vii. Maximum system voltage: 1000V
viii. Number of cells : 36 cells
ix. Dimensions + - (mm) : 1085 x 675 x 25
x. Weight (kg) : 7.55

3.3.2 *Pumps and inverters*. The PV pump system consists of a PV generator, the electrical nominal Pel, a pumping motor unit that incorporates a motor and a pump (MP) and an inverter (INV) as conditioning power for generating PV.

3.3.3 Storage. Batteries are usually not recommended for solar-powered pump irrigation systems because they can reduce the overall efficiency of the system and increase maintenance costs and instead of storing electricity in batteries, it is generally simpler and more economical to install 3 to 10 days of water storage. This study designed two models, namely those that use batteries and do not use batteries considering the daily solar radiation research location that is not always there.

3.3.4 *Water tank, battery and solar charger controller.* Batteries were used in this study with consideration of daily solar radiation research sites that did not always exist. Battery used Battery Trojan T-105. Solar Charge Controller is a component in the PLTS system functioning as a regulator of current (Current Regulator) both to the incoming currents of the PV panel as well as outflow / use outflows. Works to keep the battery from overcharging (over charge), It regulates the voltage and current from the solar panel to the battery. Most Solar PV 12 Volt produces a voltage out (V-Out) of about 16 to 20 volts DC, so if there is no regulation, the battery will be damaged from excessive charging which is generally 12Volt battery requires charging (Charge) around 13-14, 8 volt (rely on battery type) to be fully charged.

4. Results and discussion

4.1 Feasibility study of the development area

According to data from Simeulue Agriculture Agency, the area of rice field in Simeulue Regency is 4,764,53 Ha with 56.8% is irrigated rice field and 3,610.67 Ha (43,11%) other is rainfed land. The large area of rainfed lowland rice field becomes a challenge and obstacles for the realization of food security programs. Rainfed lands are the kind of farm that completely depend on the need and availability of water to rainfall. Although it has a fairly high rainfall, but the emergence of global climate change symptoms also affect the distribution and intensity of rainwater in the Simeulue islands. Rainfall becomes unpredictable so that the rice cultivation program simultaneously becomes ineffective. Rainfed rice fields become dormant because they are not planted by farmers due to the lack of adequate irrigation water sources.

The location selected for this study was the Situbok village of the Tepah Tengah sub District of Simeulue island. This village has a rice field with an area of 32 ha which is entirely a rain-fed rice field. Utilization of groundwater so far only to meet the needs of households. Utilization of shallow groundwater from wells by the community, lifted by a pump powered by electricity and generator, will require high operational costs (electricity or fuel) so that sustainability of the pump business in the utilization of ground water can not be maintained. One of the energy that can be used free of charge and available at any time is the source of solar energy.

4.2 Potential of solar power

Solar radiation is an important indicator for building a solar energy irrigation system because it is a source of energy driving the pump on the irrigation system. So the adequacy of solar radiation will ensure the application of solar irrigation systems. [2] determined that the sun exposure rate should be more than 30% to eligible to develop a solar powered irrigation system. Based on the measurement data as shown in table 1, the sunlight level in Situbok Village is an average of 43.75% which means that this area is feasible for the development of solar irrigation system.

Tabel 1. Solar Radiation of Situbok Village.

Day	Observation time(pm) (mv)			Solar radiation
	08.00 - 09.00	12.00 - 3.00	16.00 - 17.00	(w/m^2)
1	1	4,1	5,8	189,49
2	1,8	11,3	3,9	299,36
3	6,5	10,8	1,2	398,09
4	2,7	10,2	3,3	300,96
5	4,3	8,9	3,8	339,17
6	5,2	3,2	4,9	294,59
7	3,8	8,5	5,7	347,13
8	3,5	7,5	3,2	281,85
9	5,6	8,9	4,2	386,94
10	2,6	4,2	3,2	200,64
11	3,2	11,5	5,2	367,83
12	6,4	13,3	4,1	480,89
13	3,1	7,2	2,2	248,41
14	7,5	11,3	2,4	457,01
15	5,4	11,8	6,2	458,60
16	4,3	7,2	2,3	288,22
17	5,8	7,6	6,2	404,46
18	7,8	13,4	/5,6	550,96
19	6,7	8,7	4,7	426,75
20	3,8	11,7	7,5	426,75
21	6,5	13,8	6,7	533,44
			Total	7681,53
			average	365,79

4.3 Potential groundwater

To optimize the role of rainfed land, one potential water source is through the use of ground water. Teroritically, based on its utilization, there are two types of ground water that is shallow groundwater and deep ground water. This grouping is closely related to the use of groundwater and its infrastructure needs. For areas with potential shallow groundwater sources, the utilization will be easier because the required infrastructure is simpler, so that it can be developed by local farmers independently or if support is still at a relatively limited level. Shallow groundwater sources are commonly found in soil layers that are not very deep, making it possible to be lifted to the surface using a pump. To know the depth of shallow groundwater in Situbok Village, survey and observation of existing wells in the village both open well and well covered. The results showed that the shallow groundwater depth of Situbok village ranged from 5 to 10 meters.

Depth data collection process "groundwater" in this study using resistivity Geolistrik Wenner-Schlumberger configuration are conducted using Resistivitimeter ARES. The data were collected on 1 west-east trajectory with a path length of 460 m and a space of 20 m. Coordinate of measurement path: starting point of path S0 (0 m) N 02°25'14.2 "/ E 96°18'12.2" and end point S46 (460 m) y N 02°25'43.9 "/ E 96°18'27.1". Figure 2 shows Cross-section of resistivity below Situbok track surface of Situbok village. Based on the description of cross-sectional resistivity on the track shows that the trajectory Situbok (STB) in general obtained a low resistivity value that is 2 - 100 Ωm which is alluvium. The lower resistivity value (conductive) is generally a layer containing ground water. In the cross section of the STB trajectory, the highly conductive layer of resistivity <16 μm can be seen in the range of spacing range 240 - 360 m with a depth range of ± 45 - 94 m.

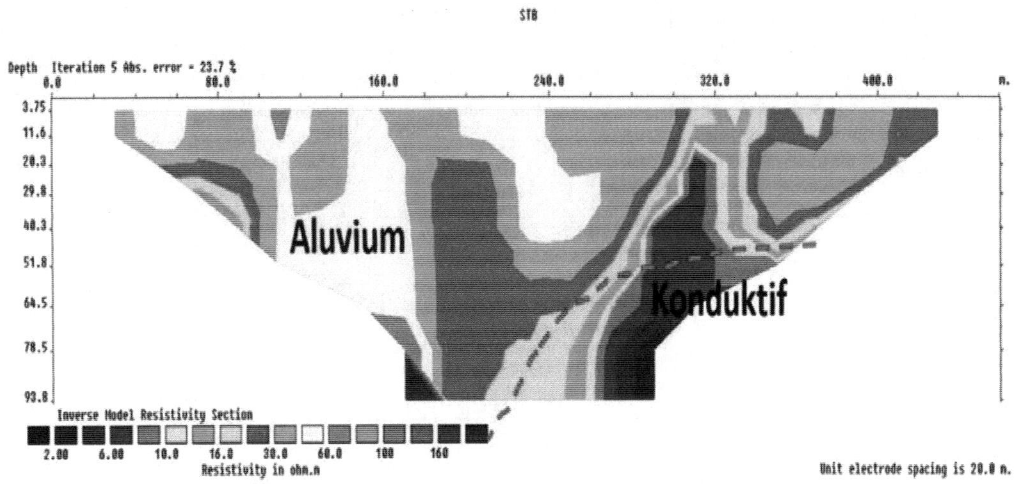

Figure 2. Cross-section of resistivity below Situbok track surface.

4.4 Land suitability
Land topography is also one of the indicators to consider in designing a solar power pump irrigation system. Optimum topography greatly determines the efficiency level of irrigation water use. If the topography is too sloping, it is estimated that the irrigation water flow range will be limited, and vice versa if it is too steep, it is estimated that there will be runoff. [2] stated the topography of land for surface irrigation 2 - 5%. Situbok land topography is 2%. It is mean that Situbok Village is suitable for developing PV pump irrigation system.

4.5 HOMER Simulation
4.5.1 Electrical load. Data Electrical loads are measured for daily hourly electrical load requirements of the pump. The main load is 2 kWh / d with a peak of 242 Wp. The hourly load consumed by the pump is presented in figure 3. The pump set requires electrical load for 9 hours (08: 00-17: 00).

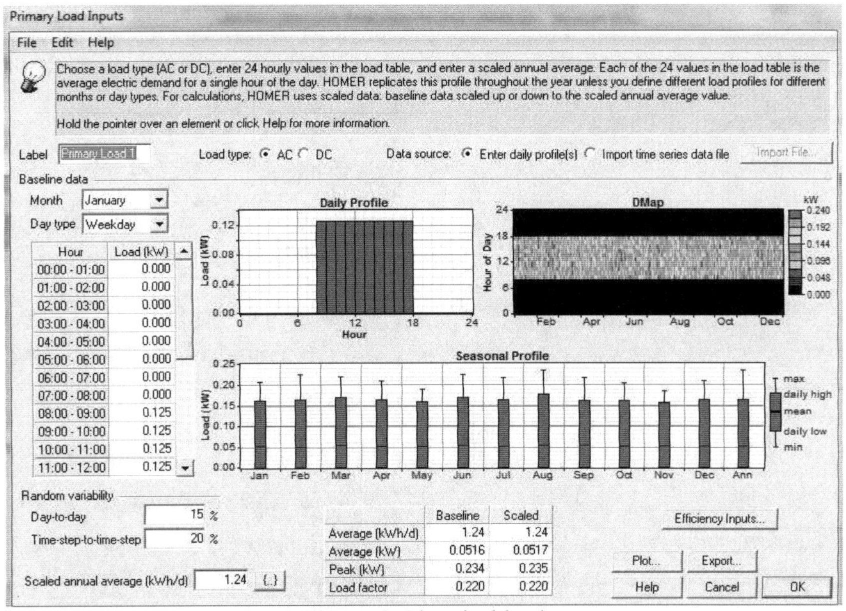

Hour	Load (kW)
00:00 - 01:00	0.000
01:00 - 02:00	0.000
02:00 - 03:00	0.000
03:00 - 04:00	0.000
04:00 - 05:00	0.000
05:00 - 06:00	0.000
06:00 - 07:00	0.000
07:00 - 08:00	0.000
08:00 - 09:00	0.125
09:00 - 10:00	0.125
10:00 - 11:00	0.125
11:00 - 12:00	0.125

	Baseline	Scaled
Average (kWh/d)	1.24	1.24
Average (kW)	0.0516	0.0517
Peak (kW)	0.234	0.235
Load factor	0.220	0.220

Figure 3. Electrical load.

4.6 Energy sources.

The energy source comes from photovoltaic solar panels. The total solar PV panel capacity is 0.1 kW. The initial cost of the 1kW panel is \$ 89, the estimated panel replacement is \$ 80. The panel's lifetime rate is 30 years. The monthly clarity index and daily radiation are shown in figure 6. Situbok location is located between latitude 2 and 5 in the north, and 96 and 7 east longitude. HOMER can help user to idendify the solar radiation from NASA satellite. Figure 4 shows the data of daily solar radiation in Stubok village.

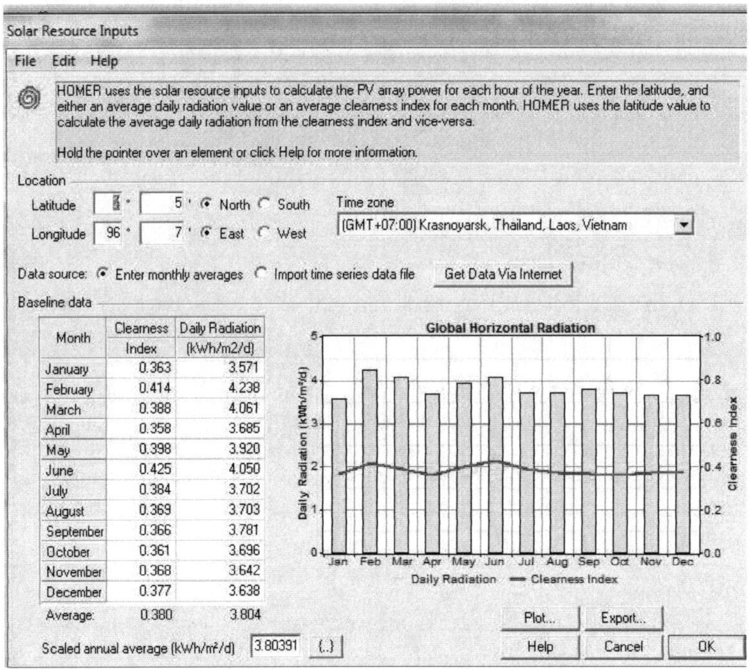

Figure 4. Daily radiation (source: NASA satellite).

4.7 System components

The battery acts as a storage of solar energy. The trojan 105 battery is considered a bank battery. Estimated cost of one battery \$ 197 with \$ 197 replacement. The battery requirement may contain a number of batteries of 1.2, 4.6 and 8 units.

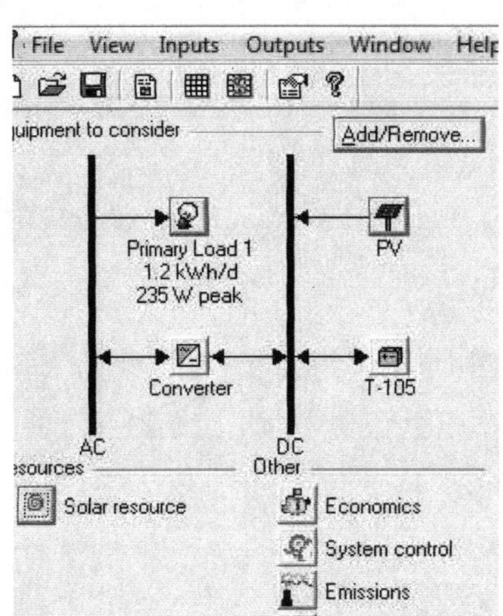

Figure 5. System components of solar power pump.

Converter is also required in this system to convert ac and dc energy. For pumping systems, installation and replacement costs are taken as $ 45 and $ 45. Various converter sizes (2, 4, 6 and 8 kW) are considered in this system. The life of the converter unit is estimated to be 15 years with 90% efficiency.

Simulation of power pump irrigation systems uses different number of components, related to energy cost and overall system performance. HOMER is used to optimize the size of components in the system, taking into account the technical characteristics of the operating system and minimizing the number of NPC systems. The optimization results for this analysis are shown in Figure 6. This system illustrates that the most optimum results obtained for this system are 2 units of solar panels, 1 battery unit and 2 converters with TNPC $ 2,380 COE 0.346 $ / kWh and operating costs $ 15 / year.

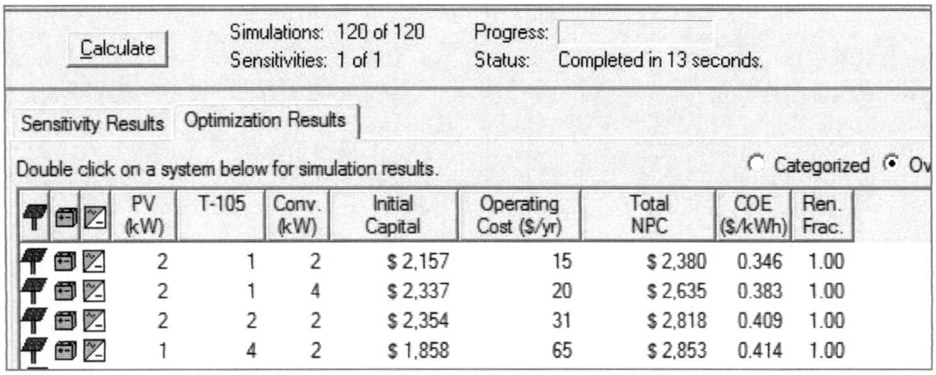

PV (kW)	T-105	Conv. (kW)	Initial Capital	Operating Cost ($/yr)	Total NPC	COE ($/kWh)	Ren. Frac.
2	1	2	$ 2,157	15	$ 2,380	0.346	1.00
2	1	4	$ 2,337	20	$ 2,635	0.383	1.00
2	2	2	$ 2,354	31	$ 2,818	0.409	1.00
1	4	2	$ 1,858	65	$ 2,853	0.414	1.00

Figure 6. Optimum simulation results.

5. Conclusions

i. Land area of Simeulue Regency is 4,764,53 Ha with 56.8% is irrigated rice field and 3,610.67 Ha (43,11%) other is rainfed lowland (suboptimal land).
ii. The location selected for this research is Situbok Village, Tepah Tengah subdistrict, of Simeulue Regency. This village has a rice field with an area of 32 ha which is entirely a (suboptimal land).
iii. Solar radiation in Situbok village is an average of 43.75% which means that this area is feasible for the development of solar irrigation system.
iv. Survey results and observations indicate that the shallow groundwater depth of Situbok Village ranges from 5 to 10 meters.
v. Based on the measurement of deep groundwater depth in this research using Geolistrik Resistivity configuration method of Wenner-Schlumberger which is done using Resistivitimeter ARES, highly conductive layer that is resistivity value <16 Ωm with potential water lies in depth of ± 45 - 94 m.
vi. The optimization results obtained for the PV pump irrigation system in this study are 2 units of solar panel, 1 battery unit and 2 converters with TNPC $ 2,380 COE 0.346 $ / kWh and operating cost 15 $ / year.

References

[1] Kashiv, A., Avi Bilala, Naseem Shirazi, Amol Dwivedi, and Rajesh Joshi. 2016 Solar DRIP Irrigation System. *International Journal of Emerging Technology and Advanced Engineering*, 6, Issue 4.
[2] Yu Yingdong, Jiahong Liu, Hao Wang, and Miao Liu 2011 Assess the potential of solar irrigation systems for sustaining pasture lands in arid regions – A case study in Northwestern China. *Jurnal Applied Energy*, 88: 3176–3182.
[3] Kumolosari E, Setiawan A A, and Suryopratomo K 2013 Comparative Evaluation Clean water supply system of electricity and solar power in remote areas (case study in Mangunun village, Dlingo district, Bantul). Jurnal Teknofisika, 2 (3): 61-68.

[4] Saleh A, Mohammed S S, Shah R 1995 PV-powered water pumping and desalination plant for remote areas in Saudi Arabia. Appl Energy, 52:283–9.n
[5] Anonimous 2008 A cost and reliability comparison between solar and diesel powered pumps global 2008.
[6] Elham M and Hossen E N 2003 Renewable energy and sustainable developments in Egypt: photovoltaic water pumping in remote areas. Appl Energy 2003;74:141–7.
[7] Bhave A 1994 Potential for solar water-pumping systems in India. Appl Energy, 48:197–200.
[8] Joubert M D, Ridwan D, Pratiwi R M 2017 Performance of groundwater irrigation system on drip irrigation using solar water pump. Jurnal Irigasi, 11: 125-132

The Impact of Buffer Queue Interface Size to The 802.11 Ad Hoc Network Performances

Rudi Arif Candra[1], Dirja Nur Ilham[1], S Suherman[2*], A Sani[2], D S Tarigan[2]

[1]Politeknik Aceh Selatan, Aceh Selatan, Indonesia
[2]Electrical Engineering Department, Universitas Sumatera Utara, Indonesia

*Corresponding author: suherman@usu.ac.id

Abstract. Radio technology for wireless local area network (WLAN) has been used widely as it is the most popular for internet access point both in houses and buildings. Technology development rapidly changes to improve connection speed from modulation techniques to application layer, such as buffering. Some buffering methods have been employed for application adaptation. Buffer management is able improving network performances. In order to measure how buffering influence the performance, this article examines buffer queue interface impact to 802.11 performance by using network simulator. The evaluation results show that the increment on buffer size from 10 packet to100 packet boosts packet delay and jitter about 121.96% and 17% subsequently. However, packet loss is reduced up to 59%.

1. Introduction

Technology development on wireless radio is aimed at increasing connection speed, by enhancing modulation speed, multiplexing capacity, multiple access performance, network coding to application layer techniques. Modulation techniques such as AM (Amplitude Modulation), FM (Frequency Modulation), and PM (Phase Modulation) exert limited speed[1]. Speed improvement is gained by using digital modulation from ASK (Amplitude Shift Keying), FSK (Frequency Shift Keying) and PSK (Phase Shift Keying), where current development is on M-ary PSK variations[2].

Besides modulation techniques, multiplexing techniques such as FDD (Frequency Division Duplex) and TDD (Time Division Duplex) combinations have significant transmission capacity improvement. Current OFDM (Orthogonal Frequency Division Multiplexing) technology enhances cellular and packet radio speed significantly [3]. OFDM has been used in WiMAX (802.16)[4], WiFI (802.11), LTE and other mobile technologies. Meanwhile, multiple access that manages point to multipoint connections reaches RTS/CTS technology [4].

Some existing techniques employ buffering techniques in medium access, transport and application layers. Wang et al [5]manages medium access layer buffer to tolerate network delay, while transport layer copies [6] increases TCP reliability. Meanwhile queue management in buffer [7]significantly improves 802.11.

The question is what is the basic relation between buffer capacity and network performances. Therefore, this article focuses on examining buffer capacity and network performances relation by evaluating packet delay, jitter and packet loss pattern when buffer queue interface on medium access layer changes. The evaluation is conducted on 802.11 ad hoc networks as multiple access technique on it increases collision probability that results sensitive network performances.

2. Research method

Figure 1 shows steps in evaluating the interface queue buffer. An ad hoc network with the number of users increases from 2 to 10 nodes is set up, link and background traffic models are adjusted, buffer variations from 10 to 100 packets are also implemented. The transmission scenario is set two ways, and transmission control protocol (TCP) is employed.

.

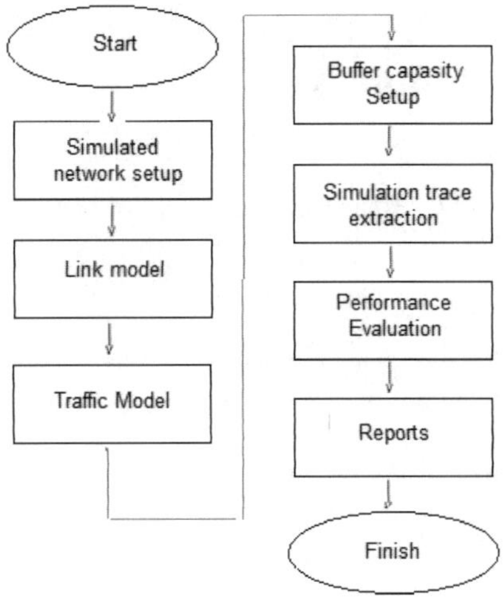

Figure1. Research outline.

Traffic is chosen based on the Evalvid framework with certain bit rate sent by nodes. The packet transmission and reception are recorded. Traffic is uniform in all nodes with video bit rate 539 bps. The transmitter is set to cover 1000 m with transmitting power 0,281838 W by using two-ray ground propagation model.

3. Experiment Results

3.1. Delay

The average delay to buffer capacity and delay to number of nodes are plotted in figure 2. Delay increases logarithmically to buffer increment as its value is 52 ms at 10 packet buffer, moves slowly to 115.3 msat 100 packet buffer. It increases about 121.96% or 10 folds buffer increment. The pattern is regressively approximated by using y=0.0268ln(x/10)+0.0618 where x is buffer capacity and y is delay.

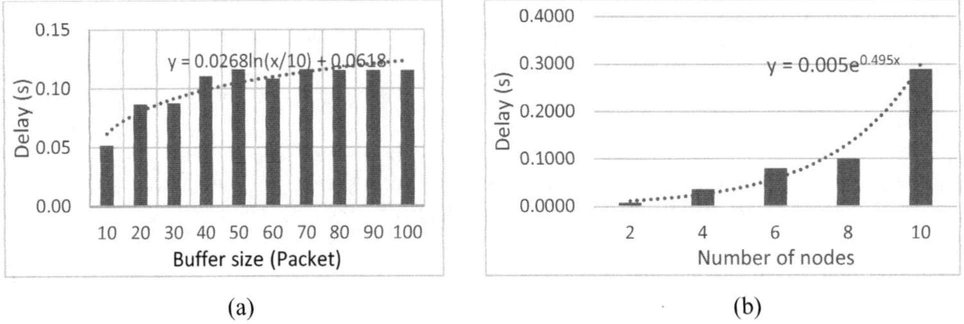

Figure 2. (a)Delay to buffer capacity, (b)Delay to number of nodes.

Increasing number of nodes produces more collision probability that causes packet retransmission. Retransmission induces higher delay as plotted in figure 2(b). Average delay increases significantly as number of nodes added. The approximated regression is $y = 0.005e^{0.495x}$.

3.2. *Jitter*

Jitter increases slowly as buffer capacity increases following trend $y = 0.0189\ln(x/10) + 0.2046$, where jitter increases 17.16% when buffer changes from 10 packets (41.7 ms) to 100 packets (48.9 ms). It is shown in figure 3(a). Meanwhile, jitter increases significantly when number of nodes added, following the regression pattern $y = 0.0013e^{0.435x}$. Jitter increases 4868.45% from 38 ms at number of nodes2 to 187 ms at node 10 (figure 3(b)).

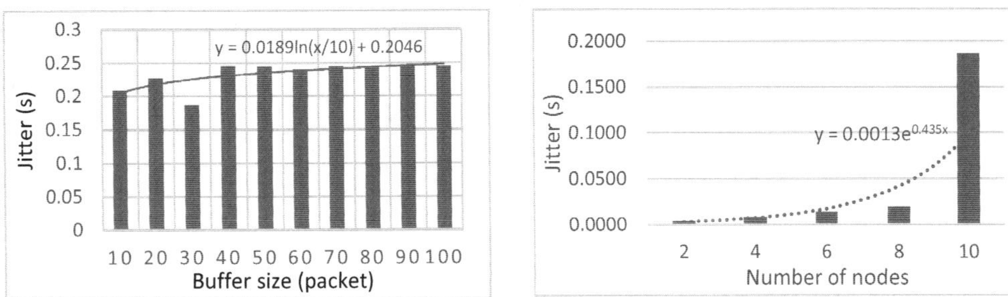

(a) (b)

Figure 3. (a)Jitter to buffer capacity, (b)Delay to number of nodes.

3.3. *Packet loss*

Packet loss increases from about 2.72% for buffer 10 packets to about 1.11% for buffer 100 packet. Losses decrease about 59%. This means, even delay and jitter increase when buffer added, lossess can be minimized. Losses increases when new node is introduced. The characteristics of losess are shown in figure 4.

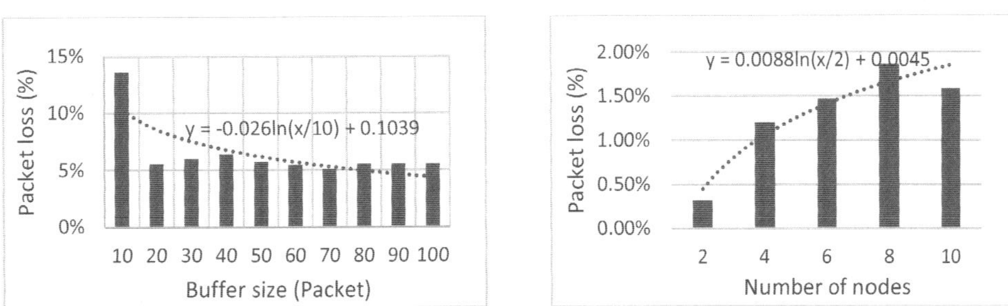

(a) (b)

Figure 4. (a)Packet loss to buffer capacity, (b) Delay to number of nodes.

4. Conclusions

To conclude, generally, the increase on buffer size results increments on packet delay and jitter. However, packet loss can be minimized. Delay and jitter increase about 121.96% and17% subsequently,but packet loss decreases about 59% when buffer size changed from 10 packetsto 100 packets.The increase on number of nodes causes network performances poorer.

This research shows that buffer size keeps packet loss low but expands delay. There should be further step to reduce latency, for instance, by applying proper scheduling. This could be of the future work.

Acknowledgement
This research has been supported by the Penelitian Produk Terapan, funded by DIKTI through the grant No.38/UN5.2.3.1/PPM/KP-DRPM/2018.

References

[1] Dan I, Rey S, Merkle T, Karner T, and Kallfass I 2017 Impact of modulation type and baud rate on a 300GHz fixed wireless link. In *2017 IEEE Radio and Wireless Symposium (RWS)* pp 86–89

[2] Chen J, Lee B, and Wu W 2018 Performance evaluation of BER for an Massive-MIMO with M-ary PSK scheme over Three-Dimension correlated channel *Comput. Electr. Eng.* Vol. 65, Pages 196-206

[3] Başar E 2015 OFDM with index modulation using coordinate interleaving. IEEE Wireless Communications Letters, 4(4) 381-384

[4] Xu K, Gerla M, and Bae S 2002 How effective is the IEEE 802.11 RTS/CTS handshake in ad hoc networks. In Global Telecommunications Conference GLOBECOM'02. IEEE (Vol 1 pp 72-76) IEEE

[5] Wang E, Yang Y, and Wu J 2015 A knapsack-based buffer management strategy for delay-tolerant networks. *Journal of Parallel and Distributed Computing*, 86 1-15

[6] Suherman, Mubarakah N, Wiranata O, and Kasim S T 2018 Energy-signal quality trade-offs in a WiMAX mobile station with a booster amplifier in *IOP Conference Series: Materials Science and Engineering* vol 309 no 1

[7] Huang J, Wang J, and Ye J 2014 A buffer management algorithm for improving up/down transmission congestion protocol fairness in IEEE 802.11 wireless local area networks. *International Journal of Communication Systems* 27(10) 2228-2240

The Analysis of Tsunami Risk Based on The Building Vulnerability and Community Preparedness in Kuta Alam Sub-District Banda Aceh

W Soviana[1*] and A Fatimah[1]
[1] Faculty of Civil Engineering, University Muhammadiyah Aceh, 23245 Banda Aceh, Indonesia

[*]Corresponding author: widya_soviana@yahoo.com

Abstract. Kuta Alam Sub District was one of the worst impacted by Tsunami in 2004. More than 50% of the area was covered by the wave more than 1 meter. The research is using qualitative and quantitative approach through surveying and scoring. Qualitative methods are carried out to assess the level of disaster risk faced by the community. While quantitative methods are used to assess the level hazard, vulnerability and community preparedness. Data analyzing is conducted spatially through geographic information system. The result show that the level of vulnerable building classified can be into very high, high, fair, and low. From the total buildings in Kuta Alam Sub District, which have a very high vulnerability of 1.79%, high of 55.17%, fair of 35.21% and low of 7.83%. The preparedness of the community in avarage is low of 32.17%. However, high vulnerability and low preparedness will increase the potential risk of tsunami disaster.

Keywords: Tsunami, Geographic information system, Qualitative method

1. Introduction

A region is potentially disastrous impacted if the community and environment are quiet vulnerable toward various disaster whether it's natural disaster or not [1]. As the region that is surrounded by the coastline, Banda Aceh city is vulnerable to be affected by disaster strike from the ocean such as wave of tidewater, tornado, and tsunami. Based on the history of paleo tsunami, in 1450 a tsunami disaster occurred in Aceh. Tsunami destroyed buildings during the sultanate of Aceh Darussalam [2]. The National Disaster Management Agency has designated Aceh province as one of the regions that has a high level of risk to the tsunami disaster [3].

Kuta Alam was one of sub district impacted by tsunami in December 2004. It is located directly with the coastline that has completely flat area with only more than less 0.5 meters from the sea level. Consequently, there were huge number of people killed in tsunami tragedy, beside devastatedly destroyed the infrastructure and the buildings. According to recorded data, losses of life in Kuta Alam due to the tsunami were 8.227 people or almost 15% from the total number of the people [4]. The buildings that were in about 2 kilo meters from the coastline generally devastated and colapsed.

After the 2004 Aceh tsunami disaster rehabilitation and reconstruction can be seen housing built with a simple residential building model. The house is very risky for the tsunami disaster. While the tsunami height is above 3 meters, then it is certain that the house will suffer major damage. Thus, community preparedness is vital for dealing with disasters that can occur at any time.

2. Building tsunami vulnerability analysis

This study aims to know the degree of tsunami risk in Kuta Alam area by conducting the study toward the vulnerability of building infrastructure and the community preparedness. The result would recognise the level risk of Tsunami in Kuta Alam sub-district. The level of risk would be low if the building vulnerabilty is low and community capacity awareness is high. However, The risk level would be high when vulnerability of the building is high and community capacity awareness is low. The studies were conducted in Kecamatan Kuta Alam, Banda Aceh. The area is located in the northern Banda Aceh that is bounded with Malacca Strait from north, Lhueng Bata Sub-district from south, Syiah Kuala and Titi Panjang river from east, and from west with Kuta raja sub-district and Krueng Aceh river.

The target of study is the preparedness of the people and infrastructures to deal with tsunami in Kuta Alam area. Data collection for susceptibility of the building conducted by direct survey toward all the constructions while that for the community preparedness through direct interview and questionaire. The results are important for identifying the risk level of tsunami catasthrope. There are three parameters for analysing the grade of susceptibility toward the threat of tsumani as displayed in the following table 1.

Table 1. Parameter of building susceptibility toward tsunami strike [5].

Parameter	Score
Building Condition	$F_{wb} = 3$
Inundation Zone	$F_{wi} = 2$
Sea Defence	$F_{ws} = 1$

3. Community preparedness analysis

The preparedness is a series of organised activity conducted to cope with disaster properly [6]. According to Head of Disaster Relief Agency in 2012, communities are the main actor to relieve the disaster. In this case, people are expected independently able not only to adapt and overcome hazardness but also could relieve immediately from the disaster impact consequences. The calcification of the level of community preparedness against tsunami disasters is shown in the table 2.

Table 2. Community level tsunami disaster preparedness and readiness [7].

Community Preparedness	Level	Score
Low	Bad	< 35
Medium	intermediate	35 – 50
High	good	> 50

Preparedness is a planning process for disaster management. Scenarios and objectives are set and agreed upon to prevent and reduce the impact of disasters [8]. The purpose of community preparedness is to reduce danger, reduce the vulnerability of the community, reduce the impact of disasters and can establish cooperation between communities [9].

4. Experimental procedure

There are three parameters for determining the level of building vulnerability those are (i) physics condition of the building, (ii) the height of inundation zone, (iii) Natural or artificial protection (sea defence) such as coastwise woods and sea wall [10-14]. Based on the parameter above, the approachment that can be conducted to calculate the level of building vulnerability toward the tsunami as followed below.

$$BTV\,\% = \frac{(Fwb\ x\ Fcb)+(Fwi\ x\ Fci)+(Fws\ x\ Fcs)}{\sum_{k=1}^{3}(Fcmaks\ x\ Fw)k}\ \text{x}\ 100 \tag{1}$$

In which Fwb is the building weigh factor, Fcb is the building condition factor, Fwi is the inundation weigh factor, Fci is the inundation factor, Fws is the sea defence weigh factor, Fcs is the sea defence factor, k is the constanta and Fcmaks is the indicator maximum.

In order to know the community level preparedness and readiness in each area, the questions of questionaire are classified into five parameters group. They are (i) legislation and contiguity, (ii) organization, (iii) funding, (iv) capacity development, (v) disaster relief and recovery. Each question would be responded “Yes” or “No”. Respondent would be scored “1” for “yes” and scored “0” for responding “No”.

For community preparedness level, the process of interviewing data collected conducted with : (i) percentage of question indicator, which is every question has equal scoring system so that the scores between 0-100 %, (ii) scoring for each question from each indicator with score “1” for responding “yes” and score “0” for answering “No”. The levels of community preparedness and readiness to deal with tsunami are classified into three levels, low, medium, and high [15]. Defines risk marking as one of methodology to determine the scope of risk through analysing conduct toward pontential disaster and evaluate existing vulnerable conditions that causes a threat or disadvantage for the population, properties, lives, and local enviroinment [16]. Another approachment risk can be identified by observing threat, vulnerability and capacity.

$$Risk = \frac{Hazard \ x \ Vulnerability}{Capacity} \tag{2}$$

5. Results and discussion

Kuta Alam subdistrict is one of the areas prone to tsunami disasters because the area is side by side with the sea. The level of vulnerability of the buildings varies from low to high. The level of community preparedness is relatively low. The level of disaster risk is potentially low for the Beurawe and Kota Baru areas. Potentially moderate to high level of disaster risk for the Lampulo, Lamdingin, Lambaro Skep, Peunayong, Keuramat, Mulia, Bandar Baru and Laksana areas. The experience of the tsunami disaster makes the level of community preparedness better

Analysis result toward the physics of buildings, the height of tsunami swamp in 2004 that appeared in figure 1a [15], and condition of sea defence acquired level of building tsunami vulnerability (*BTV*) in Kuta Alam sub-district shown in table 3. Spreading each building tsunami vulnerability seen in building tsunami vulnerability map (figure 1b). Community preparedness and awareness in each area in Kuta Alam sub-district clearly shown by the map of community capacity in figure 1c.

Table 3. Building Tsunami Vulnerability in Kuta Alam District.

Building Tsunami Vulnerability Level	Amount (unit)	Ratio (%)
Very High	197	1,79
High	6.088	55,17
Medium	3.885	35,21
Low	864	7,83
Summary	11.034	100

Based on analysis result, building vulnerability and community tsunami awareness above, the level of disaster risk in Kuta Alam relatively high, whereas nine villages still potentially risk of disaster at medium to high except for Gampong Beurawe and Kota baru that were not impacted by the water of last tsunami on 26 December 2004. Tsunami risk map for Kuta Alam based on vulnerable building analysis and community tsunami awareness level presented in figure 1d.

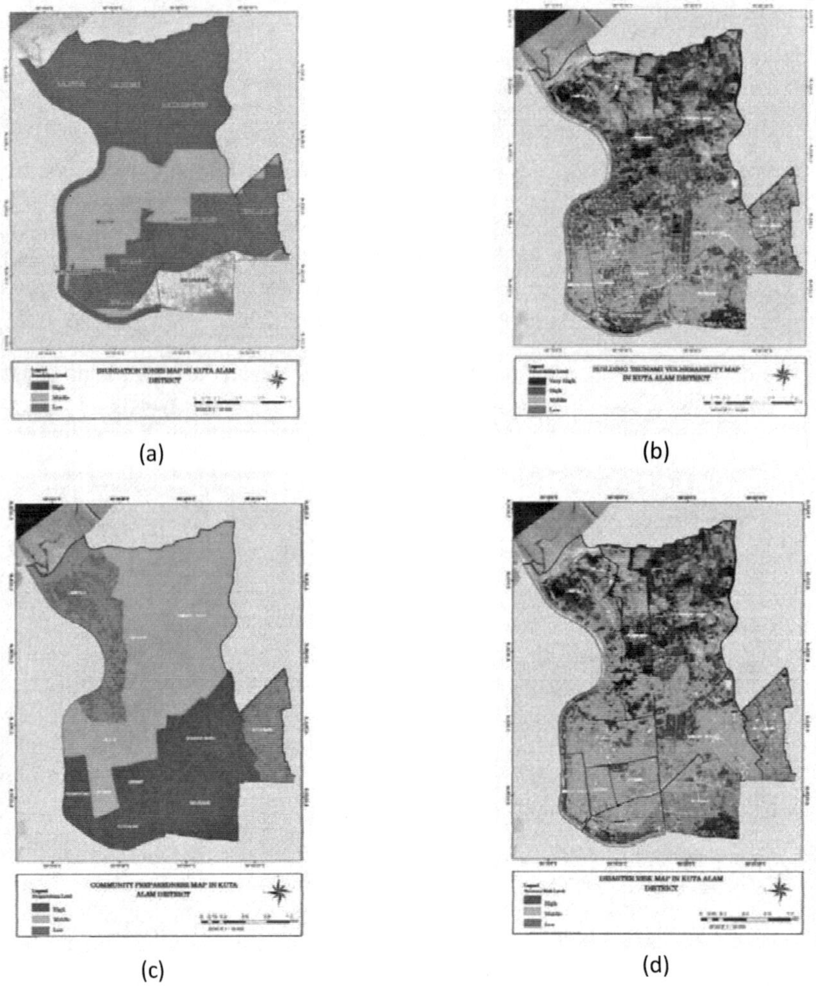

Figure 1. Kuta Alam sub district map; (a) inundation zone, (b) building tsunami vulnerability, (c) community capacity, (d) tsunami risk.

6. Conclusion

It is concluded that the level of tsunami risk in Kuta Alam low to high. The level of building tsunami vulnerability (BTV) in Kuta Lama is various, the percetages for very high with 1,79%, high 55,17%, medium 35,21 % and low 7,83 %. The community preparedness and awareness of the tsunami relatively low, that is average 32,17 %. The experience of survivors determines the tsunami disaster preparedeness and awareness. The worse impact of the disaster experience in 2004 consequently to the community higher awareness level, and vice versa.

Acknowledgments

The authors would like to be obliged to Muhammadiyah Aceh University for providing laboratory facilities and financial assistance under project.

References

[1] Mardiatno, D., Marfai, M.A., Rahmawati, K. 2012. Penilaian Multirisiko Banjir dan Rob di Kecamatan Pekalongan Timur. Magister Perencanaan dan Pengelolaan Pesisir dan Daerah Aliran Sungai (MPPDAS), Universitas Gadjah Mada. Yogyakarta

[2] Natawidjaja, D. H., 2015. Siklus Mega Tsunami di Wilayah Aceh-Andaman dalam Konteks Sejarah. Riset Geologi Pertambangan 2015. Jrisetgeotam.com

[3] BNPB. 2012a. Menuju Indonesia Tangguh Menghadapi Tsunami. Master Plan Pengurangan Risiko Bencana Tsunami. Jakarta.

[4] GTZ-SLGSR. 2007. *Rencana Detail Tata Ruang Wilayah Kecamatan Kuta Alam Kota Banda Aceh Tahun 2007-2016*. Banda Aceh: Pemerintah Kota Banda Aceh

[5] Omira, R., Baptista, M.A., Miranda, J.M., Toto, E., Catita, C., Catala, J. 2010. Tsunami vulnerability assessment of Casablanca-Morocco using numerical modelling and GIS tools dalam Jurnal Nat Hazards. Vol.54, pp 75-95 : SpringerLink.

[6] Undang - Undang Republik Indonesia Nomor 24 Tahun 2007 Tentang Penanggulangan Bencana : Pemerintah Republik Indonesia.

[7] BNPB. 2011. Panduan Perencanaan Kontinjensi Menghadapi Bencana. Jakarta

[8] GTZ-International Services. 2010. *Perencanaan Kontinjensi Tinjauan tentang beberapa Pedoman Perencanaan dan Rencana Kontinjensi*. Jakarta.

[9] IDEP. 2007. *Buku Panduan Umum Penanggulangan Bencana Berbasis Masyarakat*. Yayasan IDEP.

[10] Omira, R., Baptista, M.A., Miranda, J.M., Toto, E., Catita, C., Catala, J. 2010. Tsunami vulnerability assessment of Casablanca-Morocco using numerical modelling and GIS tools dalam Jurnal Nat Hazards. Vol.54, pp 75-95 : SpringerLink.

[11] Ismail, H., Wahab A. K. A,. dan Alias N. E. 2012. *Determination of Mangrove Forest Performance in Reducing Tsunami Run-Up Using Physical Models* dalam Jurnal Nat Hazards. Vol. 63, pp 939-963 : SpringerLink.

[12] Amri, S. 2006. *Teknologi Audit Forensik, Repair dan Retrofit untuk Rumah dan Bangunan Gedung*. Buku 1. Jakarta: Yayasan John Hi-Tech Idetama.

[13] Iemura, H., Takahashi, Y., Pradono, M. H., Sukamdo, P., Kurniawan, R. 2006. *Earthquake and Tsunami Quessionaires in Banda Aceh and Surrounding Areas*. Disaster Prevention and Management Vol. 15 No. 1, 2006 pp. 21-30. Emerald Group Publishing Limited.

[14] Tanaka, N., Sasaki, Y., Mowjood, M.I.M., Jinadasa, K.B.S.N., Homchuen, S. 2007. Coastal *Vegetation Structures and Their Functions In Tsunami Protection: Experience of The Recent Indian Ocean Tsunami* dalam artikel Landscape and Ecological Engineering and Springer.

[15] BNPB. 2012b. Peraturan Kepala Badan Nasional Penanggulangan Bencana No. 01 Tahun 2012 tentang Pedoman Umum Desa/Kelurahan Tangguh Bencana. Jakarta.

[16] Pine, J.C. 2008. *Natural Hazards Analysis, Reducing the Impact of Disasters*. CRC Press, Taylor & Francis Group

[17] Soviana, W. 2015. Jurnal Penanggulangan Bencana. Vol. 6, No.1 , pp 54-63 : Pusat Data Informasi dan Humas BNPB. Jakarta

Effect of Type and Maturity on Water Content of Three Varieties of Hot Chilli (*Capsicum Frustencent* L, Catas, Segana End Domba Variety)

Khairiah[1*], Imas Siti Setiasih[2], Een Sukarminah[2] Kusumiyati[2]
[1]Department of Food Technology, Faculty of Food Technology, Muhammadiyah Bandung University, Jl. Palasari, no 9A, Bandung 40615, Indonesia
[2]Department of Food Industrial Technology, Faculty of Agro-industrial Technology, Universitas Padjadjaran, Jl. Raya Bandung-Sumedang Km 21. Jatinangor, Sumedang, 40600, Indonesia

[*]Corresponding author : Khairiahria86@yahoo.com

Abtract. Variety and ripeness are two factors that will influence water content and vitamin C of hot chilli (*Capsicum frustencens* L). This reseach was carried out to observe the influence of varieties and rippeness of hot chilli toward water content and vitamin C. The results showed that varieties and ripeness give a significant different influence towards water content and vitamin C of hot chilli. The amount of water content decreases and the amount of vitamin C increase when hot chilli becomes red. From the there varietes, the green ripeness level of catas, segana and domba showed that the hignest water content of each variety is not so different. Meanwhile in orange variety ripeness level, catas and domba have much amount of water content. Each has 67.88% and 72.99% respectively. Domba has the highest amount of vitamin c when it is at green ripeness at 52.864mg/100g and Catas has 47.859mg/100g in stage. Segana possess the highest amount of vitamin C at orange ripeness level at 71.990mg/100gr. The highest vitamin C is found on Catas variety at red ripeness level at 68.558mg/100gr.

Keyword : *hot chilli, water content, vitamin C, rippeness*

1. Introduction

Hot chilli is horlicultural commodity that is consumed by many people inindonesia,especially by spiced food lover. Some varieties of hot chilli are consumed in fresh condition by Indonesian. Hot chilli is used as cooking spice to add flavor. Hot chilli has been used by many food and farmacy industries. As any other horticulture comodities, the length of its storage is short. After the chili has been harvested, metabolism activities are still occurring and later the process will affect the decrease quality of chili.

The freshness of agricultural product is determined by the amount of water content. Metabolism like respiration and tranpiration will cause the reduction of water content that will cause the lower of chili quality. It can be shown by withered, wizened, and hard chili.

Variety and the phase of harvest will influence the water content of commodity. This information will be valuable in order to how to treat and store it after being harvestted. The research on how to store hot chilli has been done to keep it fresh until it reach the end user hand. Water content of hot chilli varies from one to another. It depends on kinds and ripeness levels. Information of ripeness level is important to know the proper time to harvest so that water content and the freshness are getting optimal. The longer water content lasts in the chili, the longer freshness and quality can be maintained too.

Chili has different amount of water content and vitamin C on each different level of their maturity. Most Indonesain consume catas and sagana in green ripeness level, while domba is mostly consumed red ripeness level. This study has observed on three variety of hot chilli. The are catas, segana, domba. This research intent to find out water content, vitamin C on three of red hot chili varieties, catas, segana, domba, based on theirs level of ripeness.

2. Material and method

2.1 Water content

Hot chilli was supplied from farmer's field of Lembang, West Java, Indonesia. It was picked on various kinds of ripeness or maturity. The samples were catas variety (small hot chilli), segana (green hot chilli) and domba (white hot chilli). Green ripeness level is harvested after getting 50 days of its plantation, orange ripeness is in 60 days and red ripeness is in 70 days. Measurement procedur of water content sample can done in several steps. First, empty clean bowl is dried at the temperature of 105^0 C for on hour. Then, it is cooled in desicator for 30 minute (w_1). The bowl is weighed afterward. Next, 2 grams of sample are needed. The sample is put into the bowl (w_2). The bowl contained sample of 2 grams is put into furnace at the theperature of 105^0 C untill the weight is constant. The bowl containing sample is than cooled desicator. Finally, the bowl is weighed again (W_3) to get its water content calculation.

2.2 Vitamin C content

The vitamin c is measured by using iodometri method. 50 ml of sample solution of chilli is taken by using pipette and it is put into erlenmeyer. The solution of 10% H2SO4 is added and 1% amilum solution is added into Erlenmeyer. They are given a titration process with 1_2 standard solution.

3. Result and Discussion

3.1 Water content

Withered of hot chilli is caused by transpiration activities that later result a wrinklet and lower fresh of the chili. Water content affects the freshness of hot chilli. Water content in hot chilli is different from each varienty and its ripeness. Green rippenes, orange rippeness, and red rippeness have different water content. In order to keep good quality of hot chili from the harvest time to it is on the customers' hand, it is important to maintain the water content.

The result of research of three varienties of hot chilli catas (small hot chilli), sagana (green hot chilli), and domba (white hot chilli) shows that water content of each different variety and level of ripeness gives different result. The result reveals that from the three varienties of hot chilli, domba, sagana and catas contains the highest water content when they are on green ripeness phase. Each has 79.82%, 78.30%, 79.63% of water content respectively. Orange ripeness stage of catas and domba were significantly different from segana. Domba has the highest degree of water at 72.99%, catas has 67.90%. However, Segana has the lowest value at 61.62%. The result also shows that at red ripeness level, Segana was significantly different from the two others. Segana has the highest of 73.32% water content. While catas and domba has lower rate of 57.35% dan 61.69% each. Data of the three varieties were shown on the following figure.

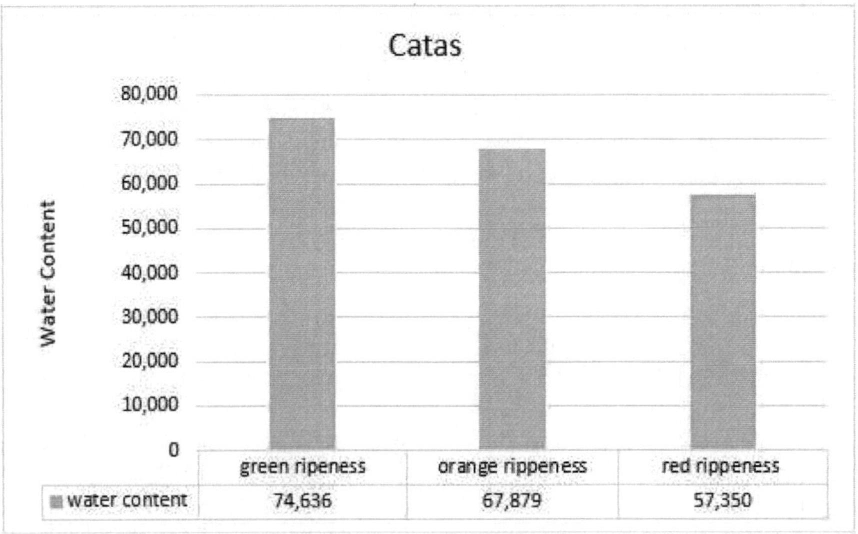

Figure 1. Water Content of Sample in Catas

Figure 1 tell us the different percentage of water content of Catas on each level of ripeness. Green ripeness percentage differs significantly from those of orange and red do. Green has 74.63%, orange has 67.88%,and red has 57.35%. We can conclude that the largest amount of water content was in the green ripeness phase. On the other hand, the lowest one is red ripeness phase.

Both green and red ripeness phase of Segana has the highest water content and they differ significantly compared to orange ripeness phase. Green has 78.31%, red has 73.32% and orange has 61.61%. Data of this variety were shown in Figure 2.

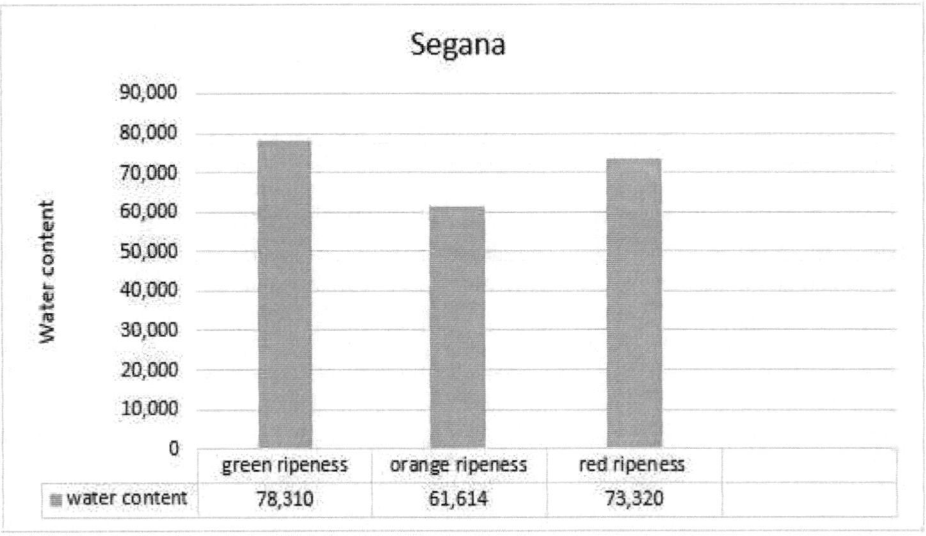

Figure 2. Water Content of Sample in Segana

The green ripeness level of domba's water content is different from orange and red one. The highest water content wass found in green ripeness phase which is 79.82%. The second higher rate was in orange ripeness phase. While the least percentage of water content in the Domba belongs to red ripeness phase. Data of the variety were shown on Figure 3.

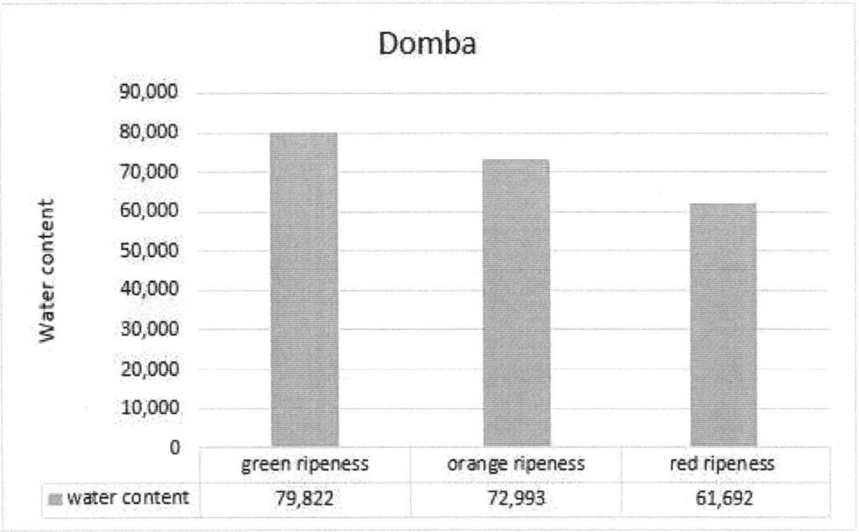

Figure 3. Water content of sample in Domba.

The previous elaboration on ripeness level affecting the water content of green, orange and red phase is related to the research that has been done by Adhytya et al. in their work that revealed the water content decreased as the fruit is ripe. The maturity showed the significant difference on the water content. The water content in hot chilli is about 49.55% - 61.46% [1]. The reduction of water content is not so different between orange fruit and dark red fruit. However, the reduction and the increase of water content in hot chilli is caused by the humidity of environment on harvest time [4]. Water content of hot chili seed that is harvested on various ripeness phase have a significant drop from green phase to its maturity phase showed by dark red color [5]. Water content in hot chilli is about 49.55 % - 61.46%.

3.2 Vitamin C content

The content of vitamin C of three varieties of hot chilli of various maturity levels shows significant difference. Therefore 5 % of duncan test is continued. The result showed that the content of vitamin C is different on each maturity level. The more ripe the higher vitamin C maturity level is. How ever Segana variety has the highest of vitamin C when orange maturity and decreases its vitamin C when it is at red maturity. At green level Catas, Segana, and Domba have no significant difference of vitamin C content. While at orange level, the content of vitamin C of the three varieties of hot chilli shows different content. The highest content was found in Segana at orange maturity level 71.99mg/100g. However, Catas and Domba reveal no significant difference at red maturity level. The highest content of vitamin C was found in Catas with 68.56mg/100g while Catas and Segana shows no significant different. The content were about 61.22mg/100g in Catas. The content of vitamin C at red maturity was different with green and orange one. Orange and green levels indicate the significant difference. The highest vitamin C content of Catas was found at red maturity level with 68.56mg/100g, orange with 53.06mg/100g, and green with 47.86mg/100g. The data were shown in Figure 4.

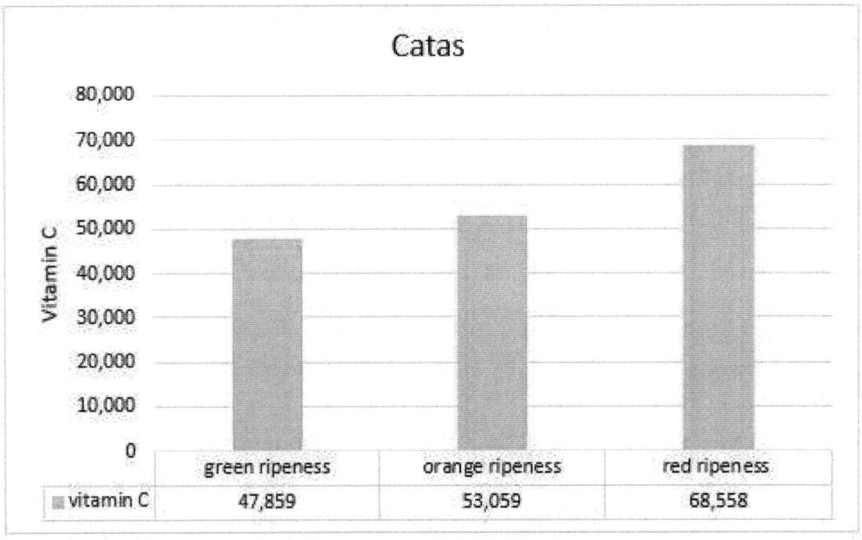

Figure 4. Vitamin C content of Catas

Meanwhile, orange ripeness of Segana show significant difference from green ripeness and red rippenes. The highest vitamin C content was found in orange ripeness which is about 71.99mg/100g whereas the lowest vitamin C was found in green ripeness which is about 63.82mg/100g. The detail about the vitamin C content of each group level of maturity were revealed in the following Figure 5 below.

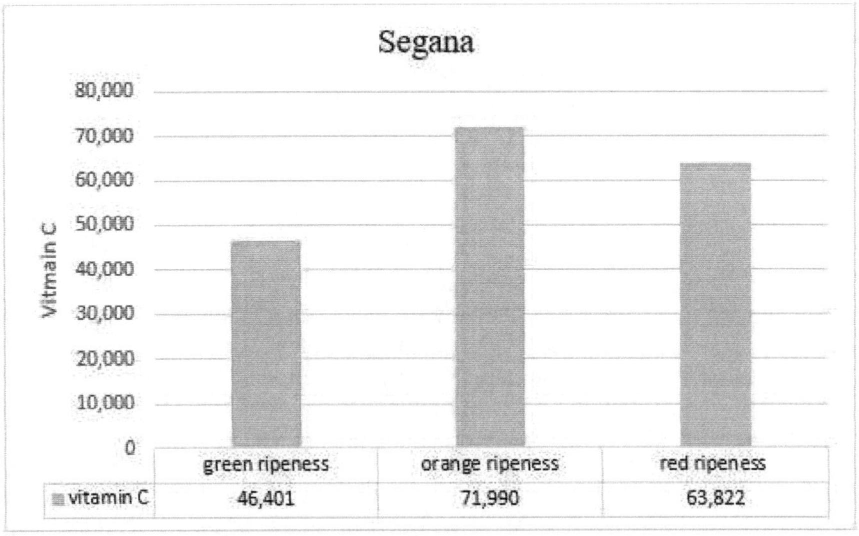

Figure 5. Vitamin C content of Segana

Vitamin C content in the domba at red and green maturity level indicates the significant difference. However, the difference between the red and orange maturity level were not significant. The highest content was found in red maturity level with 61.22mg/100g, orange with 57.70mg/100g, and green with 52.86mg/100g. The data were shown in Figure 6.

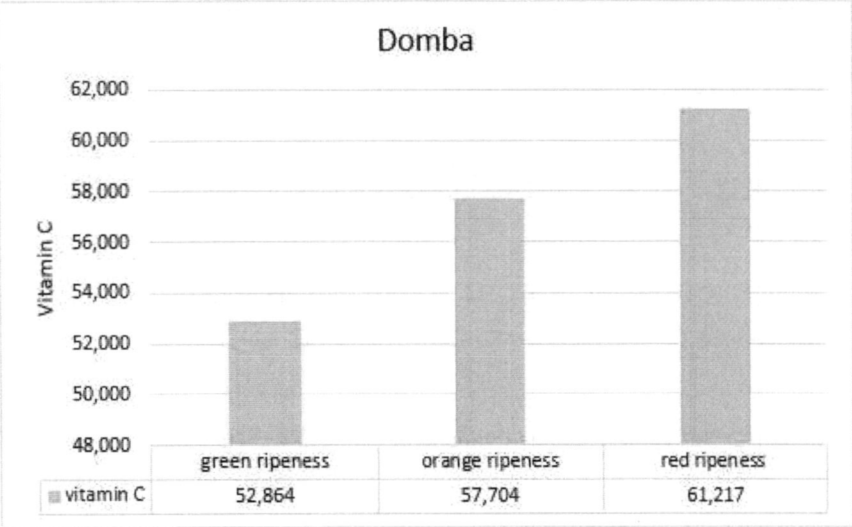

Figure 6. Vitamin C content of Domba

The result obtained based on analyzing vitamin C content in the three varieties of hot chilli is nearly the same as the study that has been conducted before. Temperature and storage process gives influence on vitamin C. The content of vitamin C in fresh green hot chilli before the chili is kept is 59.9mg/100ml [2]. In addition, the other study reveals that the content of vitamin C in fresh hot chilli before the chili was kept is 68 mg [3].

4.0 Conclusion

Varieties and maturity level of hot chilli give influence toward the water content and their vitamin C. The water content in hot chilli will decrease when the fruit is ripe. Moreover, the vitamin C content in the fruit will increase when it is ripe.

Acknowledgement

The authors would like to express gratitude to Indonesia endowment fund for education for the research dissertation grant.

References

[1] Cahya A D Respatijarti Soetopo L 2014. Effect of seed maturity on growth and produktion of hot pepper (*Capsicum frutescent L.*) on comexio variety *Journal of plant production* Volume 2 Number 4 pp 339-346.

[2] Rachmawati R Ria M D Luh N S 2009. Effect of temperature end length of storage on vitamin C concentration of chilli (Capsicum frustencens L) *Biologi journal XIII* (2) pp 36-40

[3] Wulandari S Bey Y dan Desyani K T 2012 Pengaruh jenis bahan pengemas dan lama penyimpanan terhadap kadar vitamin C dan susut berat cabai rawit (Capsicum frustencens L) *Biogenesis journal* vol. 8 Nombor 2.

[4] Villela F A 1998. Water Relations in Seed Biology *Sci.Agric* vol 55 pp 98-101.

[5] Valdes V M and D Gray 1998 The influence of stage of fruit maturation on seed quality in tomato (Lycopersicum esculentum (L.) Karsten) *Seed Sci & Tech* 26 pp 309-318

IOP Conf. Series: Materials Science and Engineering **506** (2019) 012042 doi:10.1088/1757-899X/506/1/012042

Optimizing solar panel output by using light sensors, driving motors and fuzzy controller

Dirja Nur Ilham[1], Fifi Fadhila[2], Erna B.Nababa[2], Suherman[2*] and Rudi Arif Candra[1]

[1]Politeknik Aceh Selatan, Indonesia
[2]Universitas Sumatera Utara, Medan 20155, Indonesia

[*]Corresponding author: suherman@usu.ac.id

Abstract. Indonesia has limited electricity supplies and often experience blackout. On the other hand, this country is rich with sunlight. Sunlight is always available for the entire year due to the tropical weather and earth position is near to Equator. However, solar cell usage is still limited to be applied. This paper presents solar panel controller design to maximize the power output by using light sensors, two driving motors and fuzzy logic controller. The prototype is modelled and simulated based on measured data. Simulations have been made between the designed systems versus static system. The recorded data shows that by controlling solar panel following the position of the sun, the power output increases about 15.69 % over the static system.

1. Introduction

Solar cell is the semiconductor component converting solar energy to electric current. This cell has been used widely to generate the alternative energy source both for small and large scales. Solar cell or photovoltaic cell is basically a semiconductor diode that devised to produce 0.5 to 1.0 volt per cell with current flow depending on the light intensity. The solar power from the sun reaching the earth surface may reach 4.8 kWh/m^2/day [1]. However, as the solar cell efficiency is limited, it can only generate up to 25% according to the new solar cell achievement [2]. Therefore, in order to maximize energy conversion, the cell surface should receive as maximum as possible of sunlight exposure.

The power magnitude depends on the intensity of received sunlight. Weather at this point and shading influence the system the most. Fog and cloud easily degrade power level as light intensity decreases. Besides, designing cell surface that is able to trap as optimum as possible the incoming rays, the other rational solution is to expose cell surface as optimum as possible. Many works have been performed to ensure maximum sunlight is on track. Some techniques are proposed by using intelligent systems [3-6]. The dual axis controller is the most used method on how motor moves the panel [7]. This paper studies solar cell implementation in Medan, Indonesia by recording the potential energy resulted from the sun in this area. Sunlight intensity is surveyed, system is modelled, and the automatic tracking is simulated to measure how optimum the system can be. None of the novel system is proposed; however, as sunlight position is mostly in southern or northern hemisphere, the works are researching the impact of sunlight tracking to energy generations in this area. In order to do so, two motor controls are employed. The first one is to follow solar movement from east to west, other motor to adjust north and south solar position. The controller employs the simple fuzzy neural network. The works combine real measurement and simulated evaluation. The following sections outline the research methodology and the outcomes.

IOP Conf. Series: Materials Science and Engineering **506** (2019) 012042 doi:10.1088/1757-899X/506/1/012042

2. Methodology

In order to assess the performance of solar panel after the improvement had been optimized by sensors, actuators and controller, system is modelled as in figure 1. To obtain the maximum position, four LDR sensors are located in east, west, north-south directions. Sensors outputs are fed to fuzzy controller to decide how two motors are actuated. The two motors are employed to move solar panel either to east-west or north-south directions.

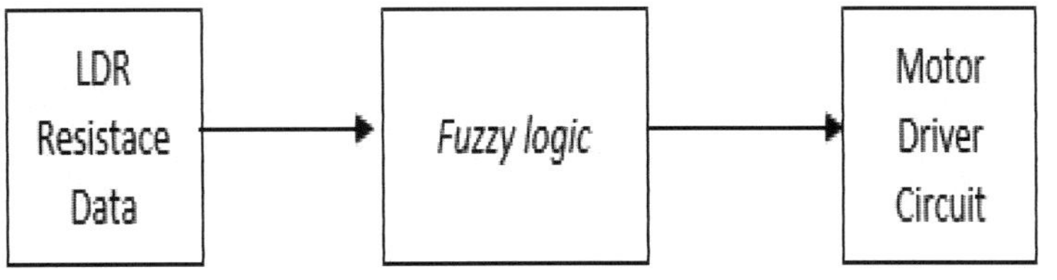

a). Sensors, controller and actuators

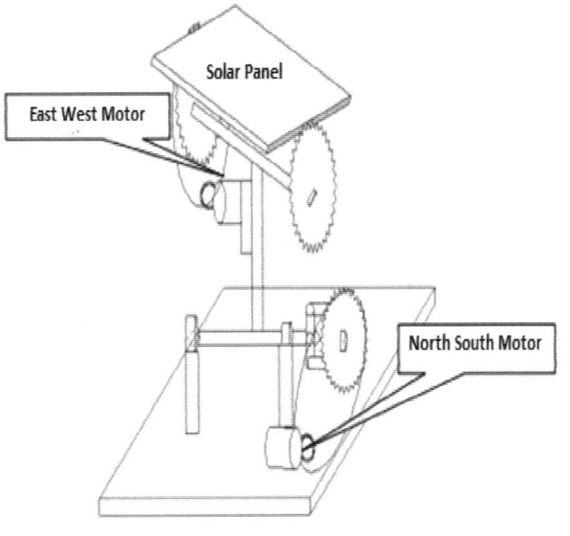

b). Mechanical model

Figure 1. System model

Sensor responses are important to determine the maximum solar radiation intensities. The higher the radiation, the lower the LDR resistance. Figure 2 shows the LDR responses to sunlight for both east-west and north-south sensors. This data is fed to simulations to determine the right position of solar panel. However, the information should go to fuzzy controller first before moving the actuators. This can be shown in figure 2 below;

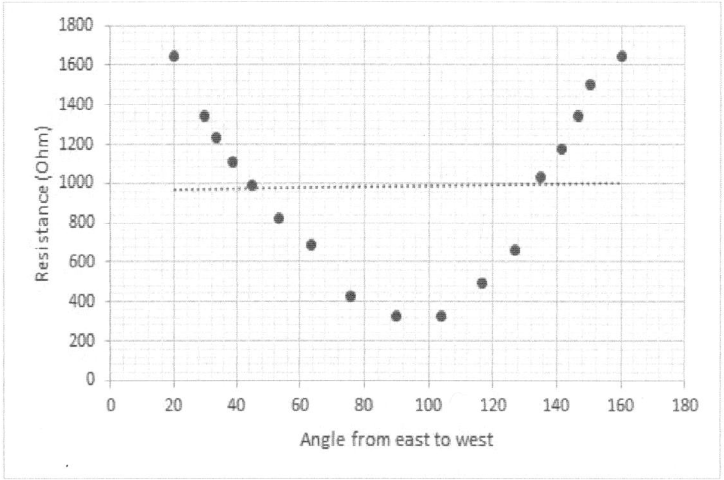

a). East-west LDR sensor

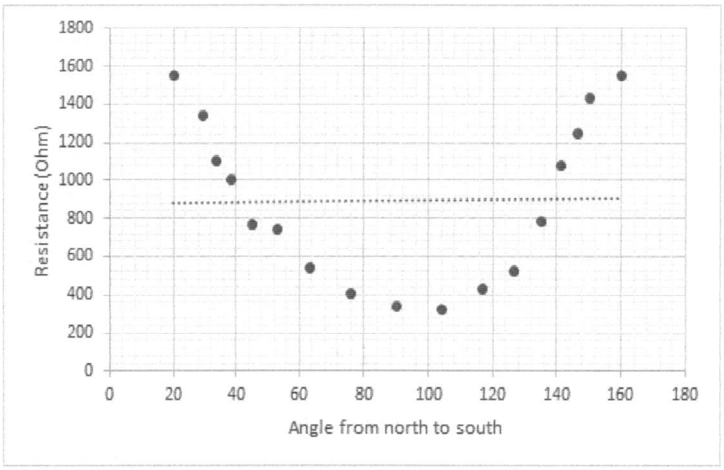

b). North-South LDR sensor

Figure 2. Sensor responses

Fuzzy rulers are set to obtain whether sunlight intensity is bright or dark. This requires sets of rules such as bright is when resistance is 0-320 Ohm, slightly bright if it is 300-800 Ohm, dimmed if 500-1400 Ohm. For dark if 1000-1600 Ohm. Meanwhile, the motor driver is set as minimum rotation or maximum rotation, the rules are set for instance a minimum if the position from 100-650, close if position is from 600-1200, far if position from 1000-1500 and maximum if position larger than 1400. The fuzzy output is obtained to actuate both motors. For instance, if east sensor in bright, the west sensor is dimmed, north sensor is bright and south sensor is dimmed, motor is then set to close.

The whole processes are implemented by using programming script depending the microcontroller type employed in system. Figure 3 shows the software flow to implement the system. Every time panel moves, the new position is searched and the new output is compared to the maximum result. In order to avoid oscillated system, fuzzy rules are adjusted accordingly. This paper employs MATLAB Software to evaluate and simulate the system. It can visualise in figure 3.

IOP Conf. Series: Materials Science and Engineering **506** (2019) 012042 doi:10.1088/1757-899X/506/1/012042

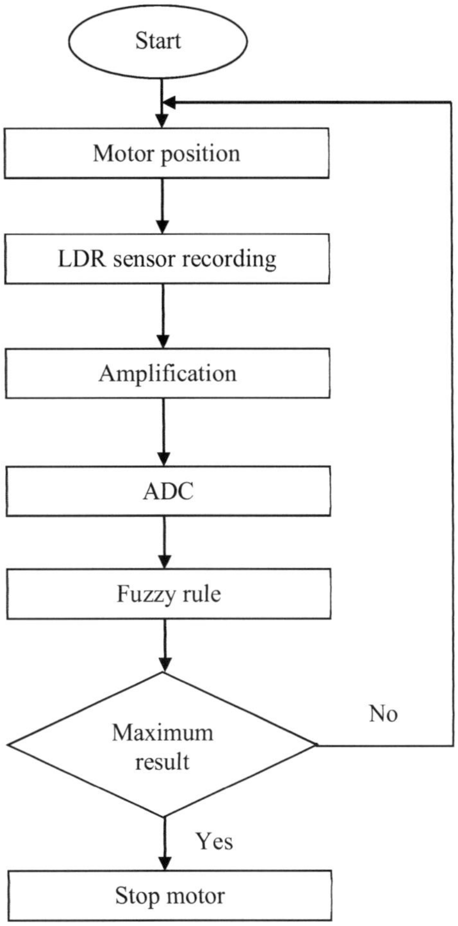

Figure 3. Controller software flow

3. Result and Discussion

The dynamic positioning of solar panel is proven to increase the open circuit output voltages for all hours. Figure 4 shows the results. The maximum voltages are obtained from about 0900 am to 1500 am. The dynamic solar panel is consistently increasing the output voltages. Although the dynamic system results some unstable output, this will not be an issue as stabilizer is easily implemented.

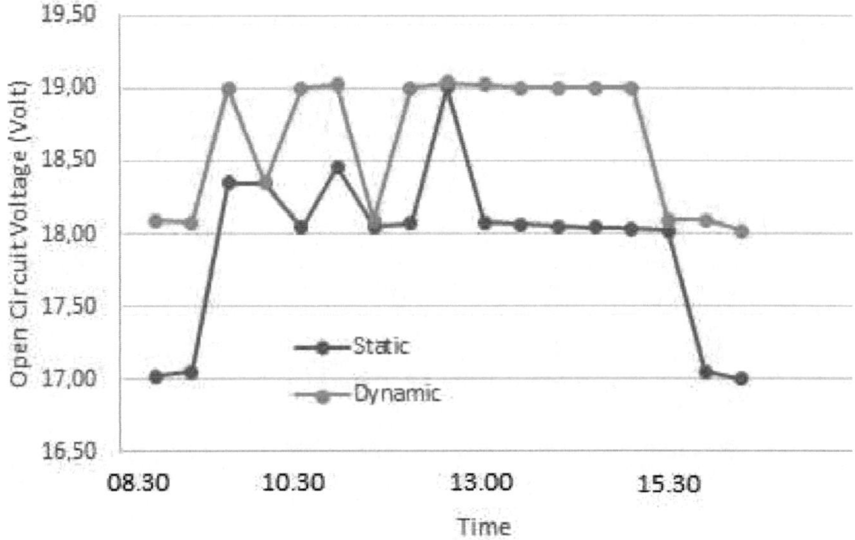

Figure 4. Open voltage output

The increasing voltage is oscillated a little bit when solar panel connected to the load. However, the level is maintained to the level of the static system generated. Figure 5a and 5b show the voltage and current characteristics when system is loaded.

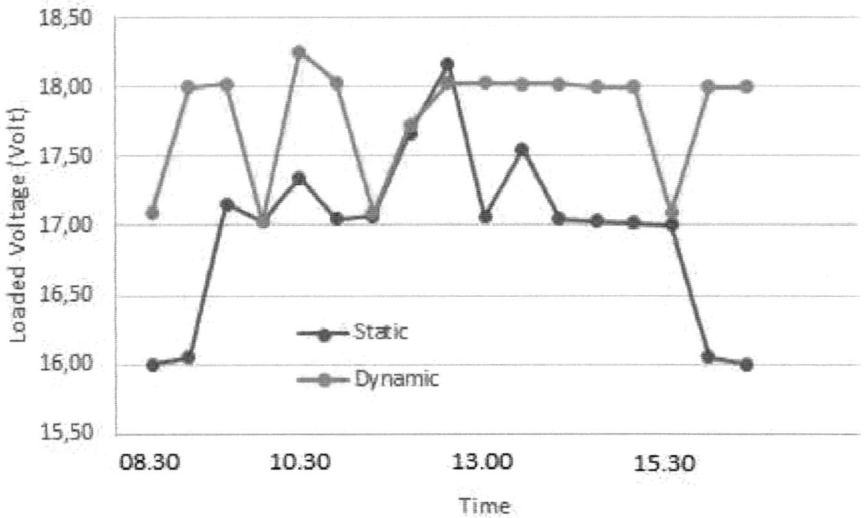

5a). Loaded voltage

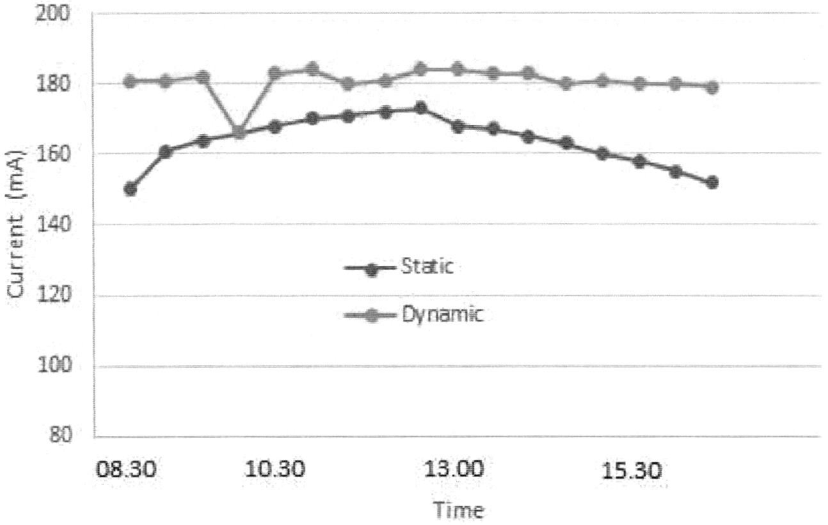

5b). Loaded currents

Figure 5. Loaded output comparisons

In average, loaded voltage is boosted about 4.9 % and current flow jumps to more than 10.0 % higher. This results increase in the average output wattage about 0.5 watt. The output power is shown in figure 6.

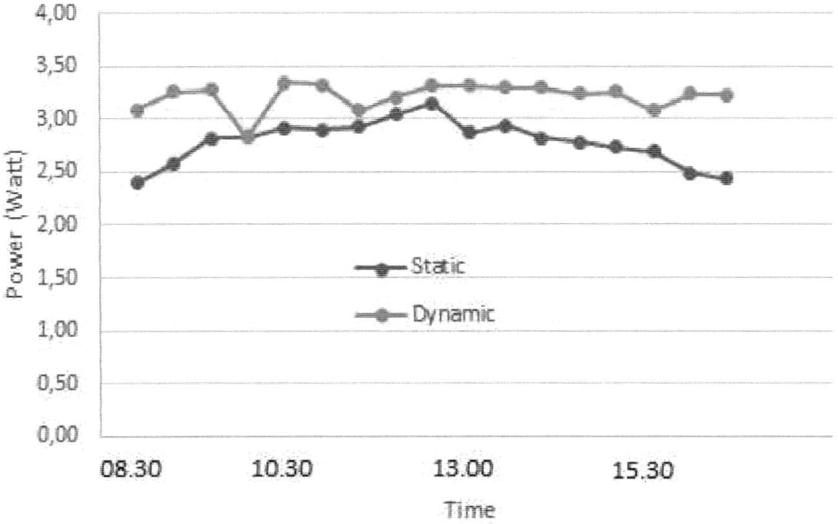

Figure 6. Output power comparisons

IOP Conf. Series: Materials Science and Engineering **506** (2019) 012042 doi:10.1088/1757-899X/506/1/012042

4. Conclusion

This paper has examined the output optimisation of solar panel by using two set of sensors: East-west and north-south where their outputs are fed to fuzzy controller to determine how east-west and north-south motors are actuated. System is evaluated through data recording and system simulation. The results show that the dynamic solar panel outperforms the static system, increasing open load voltage by 3.99 % from 17.93 Volt in the static system to 18.64 Volt in dynamic system. The loaded voltage and current of dynamic system are also better producing power improvement of 15.69 % in average.

References

[1] Indonesia Energy Outlook, 2013

[2] Masuko Keiichiro, Masato Shigematsu, Taiki Hashiguchi, Daisuke Fujishima, Motohide Kai, Naoki Yoshimura, Tsutomu Yamaguchi et al. 2014 *Achievement of more than 25% conversion efficiency with crystalline silicon heterojunction solar cell.*" IEEE Journal of Photovoltaics 4,**no. 6** 1433-1435

[3] Sani A., Warman E.and Depari, A. P. S. 2018, September. *Measuring the commercial solar panel performance.* In IOP Conference Series: Materials Science and Engineering. **Vol. 420, No. 1**, p. 012051

[4] Gupta A., Chauhan Y. K. and Pachauri, R. K. 2016. *A comparative investigation of maximum power point tracking methods for solar PV system.* Solar energy **136** 236-253

[5] Ali M., Usman, F. and Yousaf A. 2017, March. *Design and Simulation of Power Electronic Controller for Grid Connected PV Array with maximum power point tracking (MPPT).* In Renewable Energy Congress (IREC) 2017 8th International IEEE. pp. 1-4

[6] Mohapatra A., Nayak B. And Misra B. 2014, March. *Model validation and maximum power point tracking of photovoltaic module.* In Power and Energy Systems Conference: Towards Sustainable Energy IEEE. pp. 1-4

[7] Benzekri A. *FPGA-Based intelligent dual-axis sola tracking control system.*" PhD diss 2015

Investigation on the effects of suspension stiffness using experimental modal analysis and finite element model updating

A R Bahari[1], M A Yunus[1*], M N Abdul Rani[1], A Nalisa[2], and M A S Aziz Shah[1]

[1]Structural Dynamics Analysis and Validation (SDAV), Faculty of Mechanical Engineering, Universiti Teknologi MARA, 40450 Shah Alam, Selangor, Malaysia
[2]Faculty of Mechanical Engineering, Universiti Teknologi MARA, Cawangan Johor, Kampus Pasir Gudang, Bandar Seri Alam, 81750 Masai, Johor, Malaysia

*Corresponding author: mayunus@salam.uitm.edu.my

Abstract. The model validation process known as the finite element model updating method aims to improve the accuracy of the predicted results by reducing the discrepancies between the initial finite element model and the experimental data. The objective of this paper is to investigate the effects of the suspension spring stiffness to the experimental data. Experimental modal analysis with free-free boundary conditions is conducted on the slender beam to measure the modal parameters. The finite element analysis is performed to obtain the natural frequencies and the mode shapes. The results from both methods are compared to investigate the discrepancy of the finite element model of the suspension spring. Firstly, the initial model of the slender beam is updated to reduce the discrepancy between the experiment data and the initial finite element model of the slender beam. Finally, model updating method is performed to the suspension stiffness of the spring and the result shown effectiveness finite element updating method in reducing the discrepancies between the initial model of the stiffness of the suspension spring and the measured result from 33.23 % to 3.05 %.

Keywords: suspension stiffness, boundary condition, experimental modal analysis, finite element model updating

1. Introduction

Structural engineering analysis requires accurate finite element model to produce high confidence result to be used for the subsequent analysis. Model updating is a systematic adjustement method than can be used to improve the finite element model as close as possible to a real structure or system. It is based on modifying the properties assign in the model so that the result from experiment data match the numerical results [1]. The accuracy is depend on the type of finite element model used to represent the structure and the properties assigned to the elements in the model [2].

In the finite element model updating, the measured data of a structure is used as a reference parameters. The modal parameters such as natural frequencies, damping ratios and mode shapes can be obtained by finite element analysis and experimental modal analysis [3]. Analytic simulations through finite element analysis in the free-free boundary conditions are more preferable than the boundary conditions with fixed conditions [4-8]. However, in the experimental modal testing set-up, the structure must be supported in some manner to approximate the free-free boundary conditions. It is execute by using such as soft spring to support the structure [9]. Understandably, the dynamic characteristics of the test structures are significantly sensitive to the suspension system [10-14,]. The effects of suspension stiffness towards the natural frequencies may lead to significant errors in the measured modal frequencies of the test structures [15].

In this present paper, model updating is performed to produce accurate model of the finite element model of the slender beam. Suspension system for free-free boundary conditions in the experimental modal testing is applied by using the soft springs and the effects of the suspension springs are investigated. Comparison is made between the initial finite element with and without suspension stiffness consideration.

2. Experimental modal analysis

The test set-up as shown in Figure 1 is designed to investigate the effects of suspension stiffness to the modal parameters, namely natural frequencies and mode shapes. The slender beam used in this study is made from stainless steel. Table 1 shows the detail dimensions of the slender beam. In the experimental work the slender beam is suspended at both ends using soft springs to represent free-free boundary conditions and by considering the bending modes of the slender beam.

In the the experimental work the impact hammer Dytran model 5800 was used to excite the slender beam and the excitation direction was performed in the Z-direction. This vertical direction is referred to the direction of the deflection of the soft springs. This experimental modal analysis was performed based on the roving accelerometers method. It is functions to avoid mass loading issue to the test slender beam. The location of the excitation is fixed to one point only. Meanwhile the acceleration of the slender beam as an output response was measured using two accelerometers Dytran model 3032A. The average measurement for each node considered in this experimental modal testing is 10 readings.

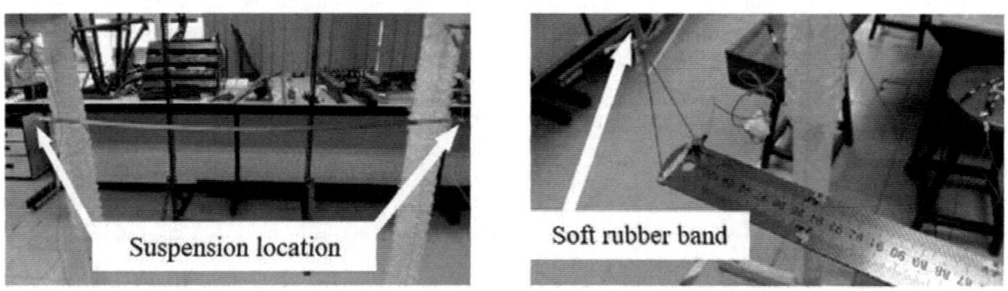

Figure 1. Experimental modal analysis set-up for the slender beam

Table 1. Geometry properties of the slender beam

Material	Stainless steel
Length (mm)	1024
Width (mm)	32
Thickness (mm)	1.5

3. Finite element modelling and model updating

The finite element model of the slender beam has been created using HyperMesh simulation software. Figure 2 shows the finite element model of the slender beam that are consisted of 3813 elements and 8350 nodes. The determination of the modal parameters, which are the natural frequencies and the mode shapes was performed using the normal modes analysis that is available in HyperMesh. The natural frequencies and mode shapes of the beam were obtained by solving the equation of motion for an undamped free vibration [16-17] as in Eq. (1).

$$\mathbf{M}\ddot{x}(t) + \mathbf{K}x(t) = 0 \tag{1}$$

where M and K are the mass and stiffness matrices. These matrices are the square $n \times n$ system. $\ddot{x}$ and x are the $n \times 1$ acceleration and displacement vectors, respectively. The results from normal mode analysis are the modal parameters which are the natural frequencies and mode shapes of the structure using with ω and $\emptyset$ denoting the undamped eigenfrequency and eigenvector [18].

$$(\mathbf{K} - \omega^2\mathbf{M})\emptyset = 0 \tag{2}$$

The purpose of model updating is to obtain a better finite element model by modifying the model parameters which are mass, stiffness and damping matrices based on the experimental data in term of the modal characteristics (natural frequencies and mode shapes) [19]. The optimisation algorithm, which is a sensitivity-based iterative procedure, allows the objective function (J) to be minimised by adjusting the eigenvalues of the initial finite element model until the objective function is converged. The objective function based on eigenvalues is defined as follows [20],

$$J = \sum_{i=1}^{n} W_i \left(\frac{\lambda_i^{fe}}{\lambda_i^{exp}} - 1\right)^2 \tag{3}$$

where λ_i^{exp} is the ith experimental eigenvalue and λ_i^{fe} is the ith predicted eigenvalue from the finite element model and n is the number of eigenvalues involved in the updating procedure. The similarity between the experimental mode shape $\emptyset_m$ and the numerical mode shape $\emptyset_a$ was obtained by using Modal Assurance Criterion [21] as in Eq (4).

$$\text{MAC} = (\emptyset_m \emptyset_a) = \frac{\left|\emptyset_m{}^T \emptyset_a\right|^2}{\left(\emptyset_a{}^T \emptyset_a\right)\left(\emptyset_m{}^T \emptyset_m\right)} \tag{4}$$

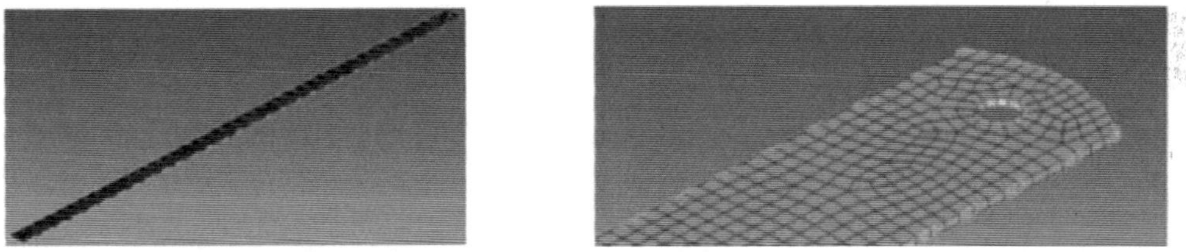

(a) (b)

Figure 2. Finite element model (a) isometric view of the CAD model (b) meshed model

4. Results and discussion

4.1 Experimental and simulation results

Experimental modal analysis was conducted to obtain the natural frequencies and the mode shapes of the slender beam. Finite element analysis by using normal mode has been performed to calculate the natural frequencies and the mode shapes by employing the equation of motion for an undamped free vibration as shown in Eq. (2). Table 2 presents the results of the first five natural frequencies and MAC values from the experimental modal analysis and the initial finite element analysis. As can be seen, there are considerable differences between the two results. By comparison, the total error (column III) is 33.23

percent and the average MAC value is above 0.88 (column IV). It is also found that the largest error is in the first mode, which is 16.89 percent.

Table 2. Comparison of the results between the measured and the initial FE.

Mode	I Experiment Frequency (Hz)	II Initial FE Frequency (Hz)	III Relative Error (%) [I-II/I]	IV Initial FE MAC
1	8.82	7.33	16.89	0.95
2	21.35	20.22	5.29	0.96
3	41.25	39.64	3.90	0.91
4	68.02	65.55	3.63	0.93
5	101.51	97.95	3.51	0.89
Total error			33.23	

4.2 Model updating the material properties

In order to improve the initial finite element model of the slender beam, the initial model need to be updated as close as possible to the experimental counterpart. This can be performed by adjusting systematically the uncertainties of the finite element modelling parameters, which potentially influences the modal properties of the slender beam [22]. The material properties that influenced the modal properties are Young's modulus, density and Poisson's ratio, namely. For this experiment, the density parameter was calculated manually from the measured mass and volume of the slender beam. Therefore, the Young's modulus and Poisson's ratio were selected as an updating parameters. These two potential parameters were used as the updating process (Young's modulus and Poisson's ratio) which are based on the most sensitive parameters listed in the sensitivity analysis via NASTRAN optimisation code (SOL 200). Element of the sensitivity matrix was calculated until the updating parameter is converged to the final value.

Table 3. Comparison of the results between the measured and the updated FE model (Young's modulus and Poisson's ratio).

Mode	I Experiment Frequency (Hz)	II Initial FE Frequency (Hz)	III Relative Error (%) [I-II/I]	IV Initial FE MAC	V Updated FE (Hz)	VI Error (%) [I-IV/I]	VII Updated FE MAC
1	8.82	7.33	16.89	0.95	7.62	13.61	0.96
2	21.35	20.22	5.29	0.96	20.99	1.69	0.97
3	41.25	39.64	3.90	0.91	41.15	0.24	0.93
4	68.02	65.55	3.63	0.93	68.04	0.03	0.93
5	101.51	97.95	3.51	0.89	101.67	0.16	0.90
Total error			33.23			15.72	

The discrepancies between the updated finite element model of the slender beam and the measured data is shown in Table 3. It is found that the total error was reduced from 33.23 to 15.72 percent. In this first updating, the initial value of the Young's modulus has increased from 193 GPa to 208 GPa while the initial value of the Poisson's ratio has decreased from 0.31 to 0.30. It is note that by updating these two material properties alone is not capable to reduce the discrepancies especially the error in the first mode in (Column VI). The natural frequency in the first mode still contributes to the highest error in the updated results. The values of the updated parameters are listed in Table 4.

Table 4. Parameters before and after updated FE model

Parameter	Initial Value	Updated Value	Unit
Young's modulus	193	208	(GPa)
Poisson's ratio	0.31	0.30	-

4.3 Model updating the suspension stiffness

In this paper, free-free boundary conditions is applied for the modal parameters measurement. For the experimental modal testing, the soft springs were used to support the slender beam to approximate the free-free boundary conditions. Impulse excitation was performed by using an impact hammer while the acceleration responses of the slender beam were recorded by using accelerometers with roving accelerometers method. It is generally known that the different type of boundary conditions affect the natural frequencies and the mode shapes of the structure or system. Therefore, in the second model updating stage, the suspension stiffness of the soft springs was considered in the sensitivity analysis.

Figure 3 shows a stiffness parameter that is available in the simulation software known as PELAS was included in the finite element model of the slender beam. The assigned location of PELAS in the model is refers to the same location of the soft springs used to support the slender beam in the experimental modal testing as described in Figure 1. PELAS is the elastic property of CELAS elements to represent the suspension stiffness of the soft springs.

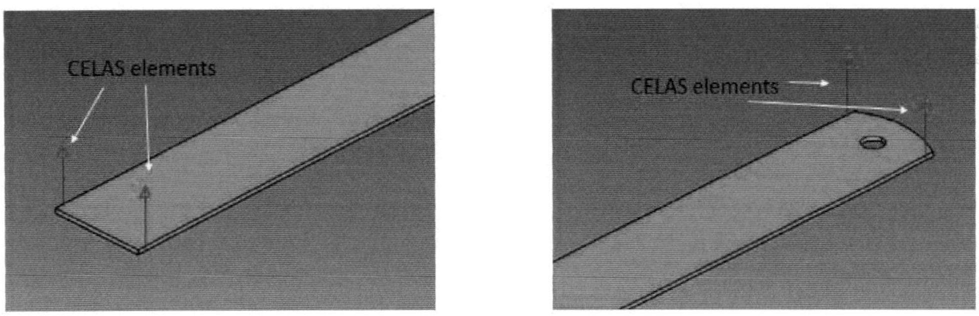

Figure 3. Finite element modelling of the slender beam with the suspension springs

Normal mode analysis has been performed to obtain the modal parameters in free-free boundary conditions. The initial value of PELAS stiffness (suspension soft spring) assigned to the model is 0.01 N/mm. By the sensitivity analysis during the model updating process, the final value for PELAS stiffness was increased to 0.015 N/mm. Table 5 lists the natural frequencies and MAC values after updating the PELAS stiffness. A significant reduction was found at the first mode of the slender beam from 8.82 Hz to 8.61 Hz. The initial error for mode 1 which is 16.89 percent was reduced to 2.38 percent. By statistical calculation, the total error between the updated and the measured natural frequencies (column VI) was reduced from 33.23 percent to 3.05 percent. From this result, it indicates that the significant improvement has been achieved as a result of including the suspension stiffness of the soft springs in the finite element model. The mode shapes and the corresponding natural frequencies for the updated model and experimental modal analysis are displayed in Table 7.

IOP Conf. Series: Materials Science and Engineering **506** (2019) 012043 doi:10.1088/1757-899X/506/1/012043

Table 5. Comparison of the results between the measured and the updated FE model (Young's modulus, Poisson's ratio and spring stiffness).

Mode	I Experiment Frequency (Hz)	II Initial FE Frequency (Hz)	III Relative Error (%) [I-II/I]	IV Initial FE MAC	V Updated FE (Hz)	VI Error (%) [I-IV/I]	VII Updated FE MAC
1	8.82	7.33	16.89	0.95	8.61	2.38	0.98
2	21.35	20.22	5.29	0.96	21.36	0.05	0.97
3	41.25	39.64	3.90	0.91	41.33	0.19	0.93
4	68.02	65.55	3.63	0.93	68.15	0.19	0.93
5	101.51	97.95	3.51	0.89	101.75	0.24	0.90
Total error			33.23			3.05	

Table 6. Parameters before and after updated FE model

Parameter	Initial Value	Updated Value	Unit
Spring stiffness	0.01	0.015	N/mm

Table 7. Comparison of the mode shapes between the experimental and the updated FE model.

Mode	Experimental Modal Analysis	Updated FE Model	MAC value
1	 8.82 Hz	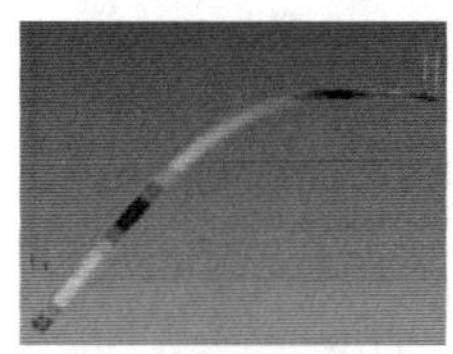 8.61 Hz	0.98
2	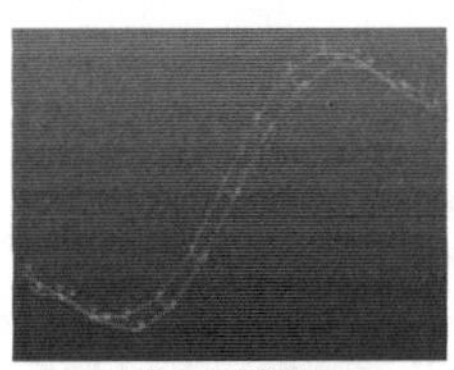 21.35 Hz	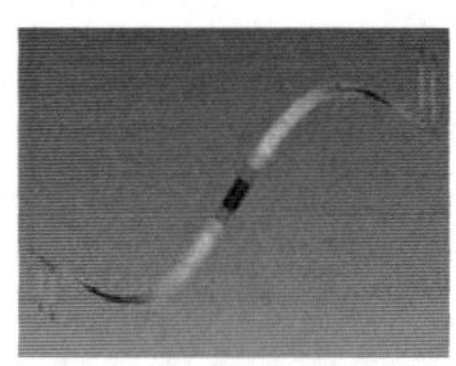 21.36 Hz	0.97
3	 41.25 Hz	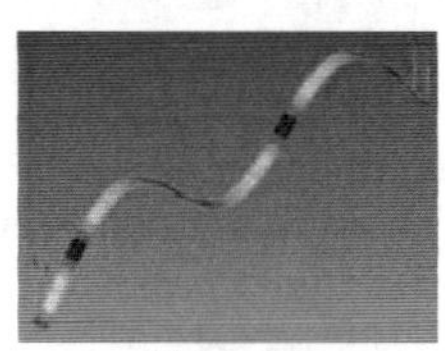 41.33 Hz	0.93

4	 68.02 Hz	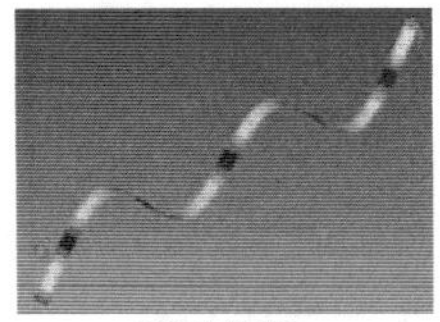 68.15 Hz	0.93
5	 101.51 Hz	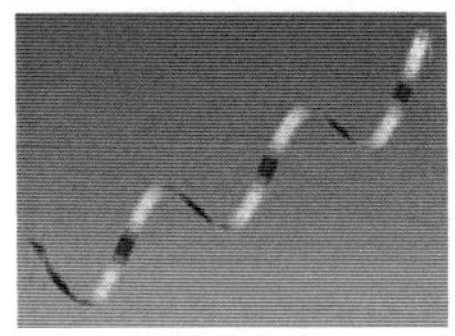 101.75 Hz	0.90

5. Conclusions

The present work investigates the effects of the suspension spring stiffness in experimental work of the slender beam in the free-free boundary conditions. The modal parameters of the slender beam, which are the natural frequencies and the mode shapes are measured by numerical simulation and experimental modal analysis. Next, the finite element model updating is performed by modifying the material properties, which are Young's modulus and Poisson's ratio in order to match the results between the finite element analysis and the experimental data. However, the obtained results show a large discrepancy. Therefore the second model updating processes is then performed , it is found that a better correlation is obtained by including the effect of the suspension springs in the updating parameters. The CELAS element is used as the updating parameter to represent the suspension springs as a support for free-free boundary conditions in the experimental modal testing. A significant improvement in the finite element model of the slender beam by the reduction of the total error from 33.23 percent to 3.05 percent. It is also revealed that the suspension spring stiffness influenced the low frequency of the slender beam.

Acknowledgment

The research has been funded by Research Management Institute (RMI) of Universiti Teknologi MARA (UiTM) for the LESTARI Research Grant with the file number 600-IRMI/MyRA 5/3/LESTARI (088/2017). Besides, the authors would like to ex-press thankfulness to the group members of the Structural Dynamics Analysis and Validation (SDAV) UiTM for their technical assistance and laboratory facilities.

References

[1] Zhao, K., Cheng, L., Li, S., Liao, A.: A new updating method for the damped mass-spring systems. Applied Mathematical Modelling 62, 119–133 (2018).
[2] Zárate, B.A., Caicedo, J.M.: Finite element model updating: Multiple alternatives. Engineering Structures 30, 3724–3730 (2008).
[3] Lee, So-Young, Huynh, Thanh-Canh, Kim, Jeong-Tae.: A practical scheme of vibration monitoring and modal analysis for caisson breakwater. Coastal Engineering 137, 103-119 (2018).
[4] Carne, T. G., Griffith, D. T., Casias, M. E.: Support conditions for experimental modal analysis. Sound and Vibration 41, 10-16 (2007).
[5] Sulaiman, M. S. A., Yunus, M. A., Abdul Rani, M. N., Mohd Saman, A., Mohamed, Z., Aziz Shah, M. A. S.: Characterisation of the dynamic characteristics of a complex jointed structure. MATEC Web of Conferences: ICMAA 108 12001 (2017).
[6] Yunus, M. A., Ouyang, H., Abdul Rani, M. N., Mat Isa, A. A.: Modal test and model updating for a welded structure made from thin steel sheets. 20th International Congress on Sound and Vibration, 2, pp. 1161-1167 (2013).

[7] Yunus, M. A., Ouyang, H., Abdul Rani, M. N., Mat Isa, A. A.: Finite element modelling and updating of bolted joints in a thin sheet metal structure. 20th International Congress on Sound and Vibration, 2, pp. 1146-1152 (2013).
[8] Hassin, N. M., Yunus, M. A., Abdul Rani, M. N., Sulaiman, M. S. A., Wan Iskandar Mirza, W. I. I., Mat Isa, A. A.: Experimental and finite element analysis of riveted joints structure of a simplified model of aluminium crash box. Journal of Engineering and Applied Sciences 11(7), 1451-1455 (2016).
[9] McConnell, K. G.: The elusive free-free boundary condition. Proceedings of the XXXV International Modal Analysis Conference (2016).
[10] Awad, Z. K., Aravinthan, T., Zhuge, Y.: Investigation of the free vibration behaviour of an innovative GFRP sandwich floor panel. Construction and Building Materials 37, 209-219 (2012).
[11] Buchanan, G. R., Rich, B. S.: Effect of boundary conditions on free vibration of thick isotropic spherical shells. Journal of Vibration and Control 8, 389-403 (2002).
[12] Mitra, A., Sahoo, P., Saha, K.: Free vibration analysis of initially deflected stiffened plates for various boundary conditions. Journal of Vibration and Control 17(14), 2131-2157 (2011).
[13] Abdul Rani, M. N., Ouyang, H., Yunus, M. A., Aminudin, B. A.: Model updating for a thin steel sheet welded structure. 20th International Congress on Sound and Vibration, 1138-1145 (2013).
[14] Aydogdu, M.: Free vibration analysis of angle-ply laminated beams with general boundary conditions. Journal of Reinforced Plastics and Composites 25(15), 1571-1583 (2006).
[15] Wolf, Jr. J. A.: The influence of mounting stiffness on frequencies measured in a vibration test. SAE Paper 840480, Society of Automotive Engineers, Inc. (1984).
[16] Hu, Y., Chen, M.Z.Q., Smith, M.C.: Natural frequency assignment for mass-chain systems with inerters. Mechanical Systems and Signal Processing 108, 126–139 (2018).
[17] Nanda, B., Maity, D., Maiti, D.K.: Modal parameter based inverse approach for structural joint damage assessment using unified particle swarm optimization. Applied Mathematics and Computation 242, 407–422 (2014).
[18] Ouyang, H., Zhang, J.: Passive modifications for partial assignment of natural frequencies of mass–spring systems. Mechanical Systems and Signal Processing 50-51, 214–226 (2015).
[19] Biswal, S., Ramaswamy, A.: Finite element model updating of concrete structures based on imprecise probability. Mechanical Systems and Signal Processing 94, 165–179 (2017).
[20] Yan, S., Li, B., Li, F., Li, B.: Finite element model updating of liquid rocket engine nozzle based on modal test results obtained from 3-D SLDV technique. Aerospace Science and Technology 69, 412–418 (2017).
[21] Eiras, J.N., Payan, C., Rakotonarivo, S., Garnier, V.: Experimental modal analysis and finite element model updating for structural health monitoring of reinforced concrete radioactive waste packages. Construction and Building Materials 180, 531–543 (2018).
[22] Mottershead, J. E., Mares, C., Friswell, M. I., James, S.: Selection and updating of parameters for an aluminium space-frame model. Mechanical Systems and Signal Processing 14, 923-944 (2000).

Various layer techniques to improve video transmission on 20 GHz radio link in a rainy environment

Hardisal[1], Rudi Arif Candra[1], Dirja Nur Ilham [1], Naemah Mubarakah[2], Suherman[2]
[1]Politeknik Aceh Selatan, Indonesia
[2]Universitas Sumatera Utara, Medan 20155, Indonesia

Corresponding author: suherman@usu.ac.id

Abstract. The transmission rate via radio increases to transmitting frequency increment. The higher frequency the higher transmission rate. Video traffics require radio with large bandwidth. This requirement may be fulfilled by the use of millimetre radio transmission on 20 GHz. However, radio signal on millimetre band decreases significantly in presence of rain mainly caused by link attenuation. This paper evaluates the performance of video transmission on 20 GHz in presence of rain on tropical environment. As various layers techniques are available, some techniques such as APA, Java Servlet Alias (JSPA), Digital Subtraction Angiography (DSA) and mac-application layer scheduling are implemented to help radio dealing with rain attenuation. The evaluation results through designed simulations proves that the video quality degrades significantly to rain. The combined technique is able to increase video quality, namely; physical technique up to 0.395 dB, MAC scheduler up to 6.5dB and upper layer technique achieves 1.17 dB.

Keyword : *APA, Java Servlet Alias (JSPA), Digital Subtraction Angiography (DSA), mac-application layer scheduling*

1. Introduction

Millimetre band on frequency range of 20-40 GHz offers high data rate transmission [1]. Potential high data rate on non-licence frequency spectrum may be useful for terrestrial video transmission link either for public services or commercial video transmission. Video transmission for remote monitoring [2], for instance, observation of eruption of Mount Sinabung as well as traffic surveillance require unshared frequency spectrum so that high definition video can be transported.

The biggest challenge in using radio working on frequency from 20 GHz is the rain that significantly degrades the radio signal. Rain causes fading where the received signal is under receiver threshold. The received video may not be able to be decoded at this fading presence.

The matter becomes emergency as Indonesia has tropical weather where rain fall is very intense. Rain attenuation achieves 80dB for 5.7 km link on frequency 29 GHz [3]. Some techniques have been proposed [1, 3, 4, 7] to deal with rain attenuation, including the use of orthogonal multiplexing that narrowing the orthogonal carriers, by using dynamic channel allocation, power control and upper layer solutions. This paper will examine the rain impact to the video transmission on 20GHz frequency through the measurement of peak signal to noise ratio (PSNR).

2. Methodology

The intensity of the rain were measured by using Hellman measurement unit [3] which is in millimeter per hour (mm/h) and conducted in several regions that were located in Medan,Indonesia. The rain attenuation was modelled by using Equation 1, where ITU-R Rec.P.530-10 was employed for the validation [3].

$$A = \int_{o}^{L} aR(x)^{b}\, dx \tag{1}$$

The radio was assumed to use orthogonal frequency division multiplexing (OFDM) as it is currently the most effective technique. The slotted model of the radio frame was shown in Figure 1 [7]. Uplink and downlink frames were assumed at the same length, which bandwidth request of piggybacking mechanism was assumed to be allocated in a specific channel.

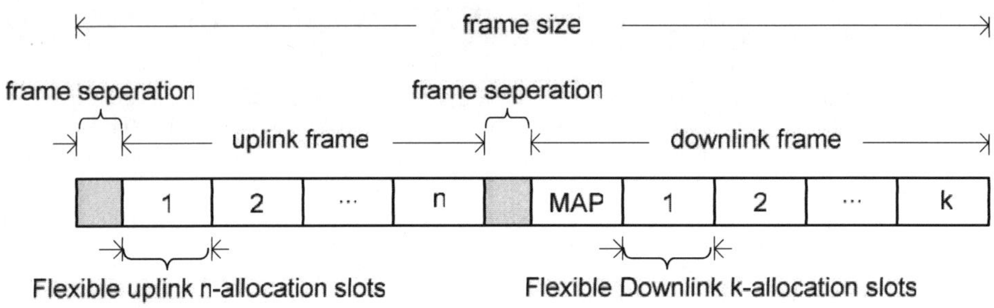

Figure 1. OFDM model

The evaluated video traffics were generated by using Akiyo cif video traces [8] with various starting point in every simulation iteration. The peak signal to noise ratio (PSNR) was used to measure the video quality [10]. The PSNR calculation was modelled by using Equation 2 approximation, $\varphi_{i,m}$ is the received bytes of the mth I-frame, $T_{i,m}$ is the transmitted bytes and Max_I is the maximum PSNR.

$$PSNR = \frac{\sum_{m}\left(\frac{\varphi_{i,m}}{T_{i,m}} Max\_I\right) + \sum_{n}\left(\frac{\varphi_{p,n}}{T_{p,n}} Max\_P\right)}{m+n} \tag{2}$$

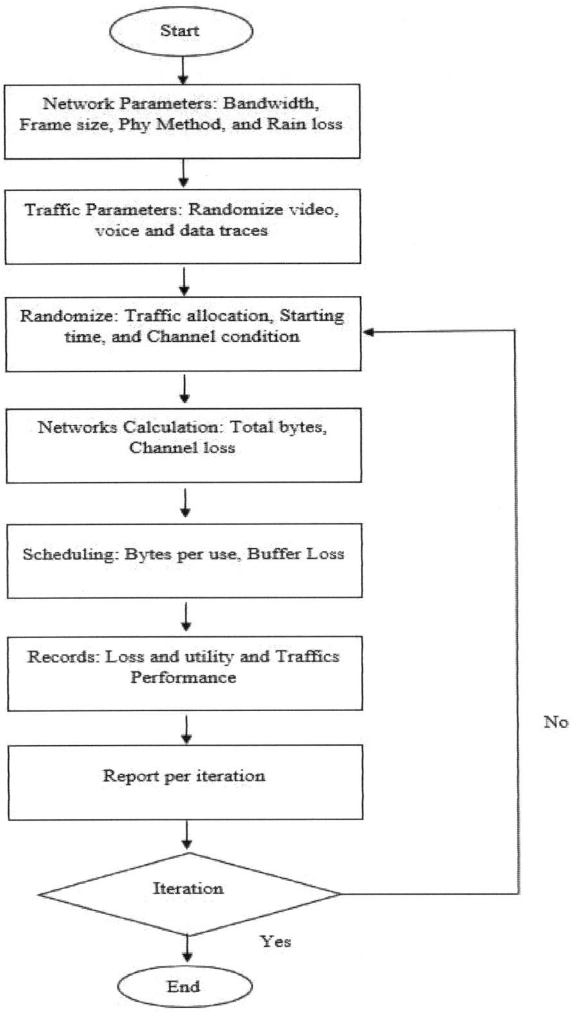

Figure 2. The simulation design flow diagram

Figure 2 shows the simulation flow. Radio transmission time was divided into some transmission slots. The rain loss and the possible available bandwidth are approximated by hundreds iterations. In each iteration every node has random starting video. Nodes transmit the bytes according to the allocated bandwidth. Channel introduces link loss, base station introduces buffer loss. Both losses recorded, the total of successfully received bytes within the same allocation frame were calculated. Network utility and total loss were then determined. Number of mobile users and the total bandwidth were randomly chosen based on the fact that the generated traffics are low bit-rates. Mobile users were set from 5 to 35 and bandwidth 1 MHz which results maximum 10 Mbps in clear sky.

3. Results and discussion

Simulations run for a-thousand iterations, record the successful bytes and calculate the PSNR. It was found that the rain degrades the radio signal, hence, will resulting the video quality decreased by 0.73 dB in average. The APA, DSA and JSPA techniques [3] are then applied and the improvements of the video quality about 0.19dB, 0.4 dB, and 0.39 dB were observed subsequently. The upper layer scheduling techniques such as round robin (RR), first in first out (FIFO), minimum signal to noise ratio (mSNR), PF and BABS scheduling [5] are then used to prioritize frame transmission when the link is better. Figure 3 below shows the overall PSNR improvements. The DSA combination to these scheduling resulting various achievements where the FIFO technique increase the quality about 0.08 dB, mSNR technique improved about 8.3 dB, and both of PF and BABS techniques enhance about 2.75 dB and 8.32dB.

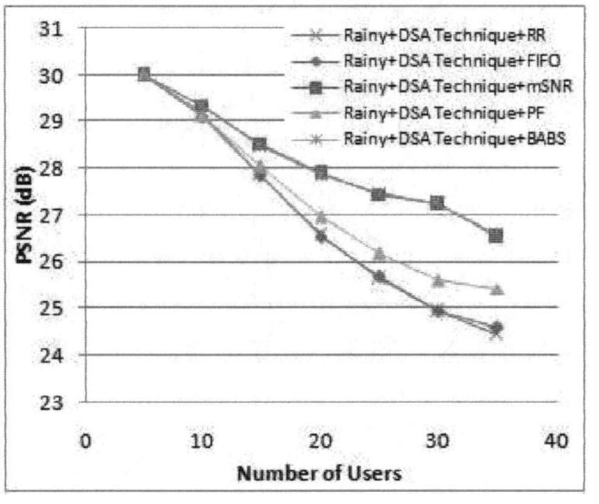

Figure 3. Various techniques impacts

4. Conclusion

This paper has assessed video quality degradation to rain presence in 20 GHz video transmission. The PHY and MAC techniques such as APA, JSA and DSA increase quality at most 0.4 dB.by prioritizing frames (application layer) using mac scheduler which means mac-application layer technique, increasing video quality significantly. The mSNR and BABS frame scheduling's produce up to 8.3dB improvement.

References

[1] J Y Zang B K Letaief 2004 Cross-layer Adaptive Resource Management for Wireless Packet Network with OFDM Signaling *IEEE Transactions on Wireless Communications* vol.5 No.11

[2] R Golembewski M Rossberg G Schaefer 2011 Towards Smart Infrastructures for Modern Surveillance Networks.*Technical report, TU-Ilmenau.*

[3] N Mubarakah R W Arsianti and M Mulyadi 2012 Performance of Subcarrier and Power Allocation OFDM on Millimeter Wave. *TELKOMNIKA (Telecommunication Computing Electronics and Control)* pp 167-174.

[4] C Mohanram S Hashyam 2007 Joint Subcarrier and Power Allocation in Channel-Aware Queue-Aware Scheduling for Multiuser OFDM. *IEEE Transaction on Wireless Communications.*vol.6 No.9.

[5] N Qiang 2010 Performance analysis of contention based bandwidth request mechanisms in WiMAX networks. *IEEE Systems Journal,* 4.4 477-486.

[6] S Suwadi G Hendrantoro and W Wirawan 2011 An area segmentation strategy for adaptive transmission to achieve near-uniform high-quality coverage in 30 GHz fixed wireless cellular systems in tropical regions. *WSEAS Transactions on Communications* vol 10.8 pp 243-253.

[7] S Suherman and M Al–Akaidi. 2013 Increasing uplink broadband video streaming protocol performance in WiMAX network *International Journal of Internet Protocol Technology* vol 7.3 pp 176-185.

[8] CH Ke C K Shieh W S Hwang A Ziviani 2008 An Evaluation Framework for More Realistic Simulations of MPEG Video Transmission *Journal of Information Science and Engineering*

Mechanical Properties Characterization of Marble and Resin Composite Materials

Husaini[1]*, Ihsan[2], M. Nizar Machmud[1], Sandi Yudha BZ[3] and M P Anhar[2]

[1] Mechanical Engineering Department, Engineering Faculty, Universitas Syiah Kuala, Banda Aceh, Indonesia
[2] Mechanical Engineering Department, Polytechnics of South Aceh, Tapaktuan, Indonesia
[3] Mechatronics Engineering Department, Polytechnics of Aceh, Banda Aceh, Indonesia

*Corresponding author: husainiftm@unsyiah.ac.id

Abstract–The process of forming furniture and house construction, the marble used produces a lot of waste in the form of marble powder. One solution offered to overcome this problem is to make a composite of marble's powder and resin. The purpose of this study was to determine the tensile strength and hardness of resin composites and marble's powder and determine the optimal value of the composition between marble's powder and resin. These composites were then tested to obtain mechanical properties such as yield strength, elastic modulus and hardness. Fracture surface observation by using SEM on the specimens was also conducted. The results showed that the optimal composition for composites is the composition between 60/40 and 70/30.

Keywords: Composite materials, Marble, Resin, SEM

1. Introduction

Marble is a rough crystalline rock derived from limestone or dolomite. Pure white marble and mainly compose of calcite minerals [1]. In general, marble is used for interior or exterior design of a house or building. In addition, marble can also be used as raw material for crafts or sculpture.

Aceh province is a producer of marble stone, especially in the district South Aceh (Tapaktuan). At this time, marble in this area is widely used as flooring, household appliances, and so on. The process of producing raw materials into products causes a lot of waste. Gradually, this waste will accumulate and should be reduced by the waste into useful products [2]. One solution is to make the marble powder waste into a composite material. This composite can later be used into form and needed by the community.

Several studies related to marble powder have been carried out. Sugiarto Danu et al [3] conducted a study by utilizing marble waste into composites after adding polyester polymer as a matrix and abaca fiber as reinforcing material with the drying process using γ Co 60 radiation and peroxide catalyst. Mahayatra et al [4] conducted a study to determine the effect of variations in marble particle size on the mechanical properties of composites. Tjaronge M. Andi Mizwar et al [5] conducted a study on the utilization of marble mud, Styrofoam waste and coal fly ash for the manufacturing of hollow concrete bricks. Rania A. Hamza et al [6] conducted a study comparing the mechanical strength of marble and granite powders. Husaini et al [7] have conducted research on the mechanical properties of composite Marble/Resin only in the composition of 50/50.

Purposes of this study is to determine the mechanical properties of marble and resin composites in four types of compositions. The method used to find mechanical properties in this research are by applying tensile and hardness testing. The results of this test is the comparison of each composition. At the end, the results is to obtain the best compositions of resin composite and marble's powder. It was found that the composition with optimal tensile strength and hardness is occurred in between the compositions from 60/40 to 70/30.

2. Basic Concept

Marble is a rough crystal stone derived from limestone or dolomite. Pure white marble and mainly composed of calcite minerals. Marble is a rock from the process of metamorphosis or transition from limestone. The effect of temperature and pressure produced by endogenous forces causes recrystallization of these rocks to form foliation or not foliation.

The resin produced from polymerization addition process on heating with an amino catalyst. These resins have wide uses in the chemistry, mechanical and civil industries as adhesives, coating paints, cord printing and molded objects. The most produced epoxy resin is from bisphenol A and epiklohidrin. Composites are materials that are composed of a mixture of two or more materials with different chemical and physical properties, and produce a new material that has different properties with its constituent materials.

3. Methodology

This research uses several supporting tools. For the manufacture of specimens, a mold made of acrylics is used as shown in figure 1.

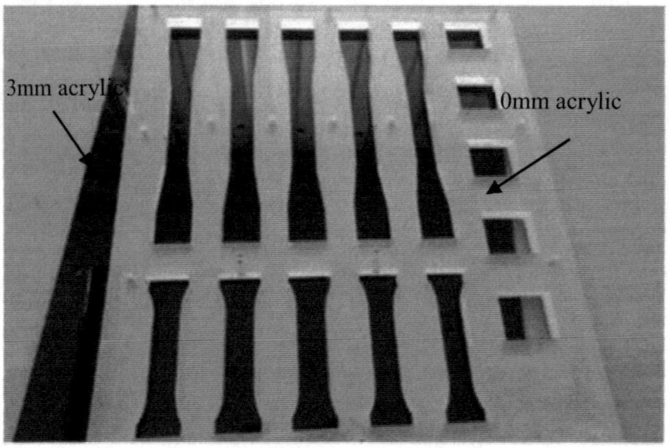

Figure 1. Acrylic molds used in the experiment.

Marble powder is obtained by sieving powder using a 150 µm (0.15 mm) mesh sieve. The equipment used to make composites can be seen in Figure 2. Making composites is done in ways: Pour resin into a measuring cup. Then pour it into a container. Next enter the marble powder that has been sifted and measured using a measuring cup little by little into a container that already contains resin using a spoon. Next add enough catalyst (ratio 1:10 with resin) and stir it slowly so that the marble powder can be mixed perfectly using the stirrer. The volume composition between marble powder and resin is planned as shown in table. 1

Table 1. Composition of marble powder and resin.

No	Resin (%V)	Marble Powder	Identification
1	50	50	50/50
2	60	40	60/40
3	70	30	70/30
4	80	20	80/20

Apply wax to the mold specimen using a brush, then pour the composite mixture that has been mixed evenly throughout the specimen mold. The composite mixture on the mold is then dried at room temperature for 3 days. After dried, the composite is removed from the mold and then the surface is smoothed using sandpaper. After that, clean the mold using 96% alcohol. Figure 2 shows the equipment and materials used in the experiment.

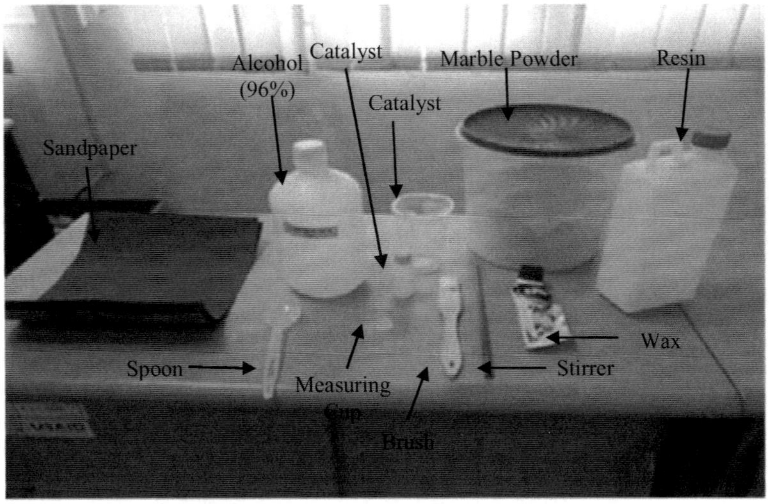

Figure 2. Equipment and materials used.

Tensile testing were carried out for 3 pieces of specimens for each composition. Examples of specimens to be tested can be seen in Figure 3.

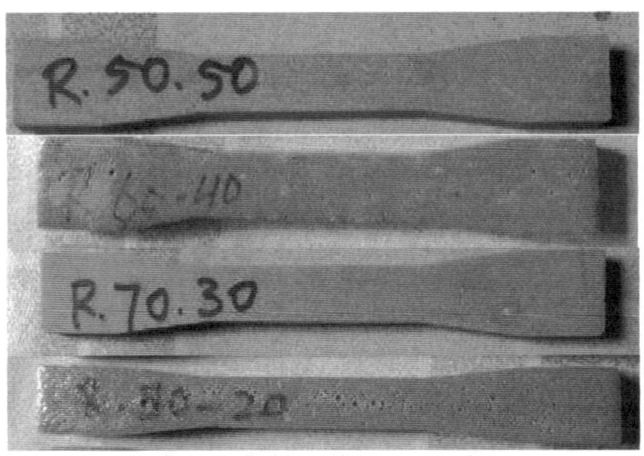

Figure 3. Tensile test specimen.

This tensile test uses the ASTM D 638-03 tensile testing standard [8]. In addition to the tensile test, specimens were also subjected to a hardness test as shown in Figure 4. Testing of this hardness uses the ASTM standard D 785-03 [9]. After all tests have been completed, then the fracture surface of the tensile test was observed using a SEM (Scanning Electron Microscope).

Figure 4. Hardness test specimens.

4. Results and Discussion

4.1 Tensile strength

The tensile test results, can be seen in figure 5, where the greatest yield stress occurs in the composition of 70/30. The overall results of tensile testing can be seen in table 2.

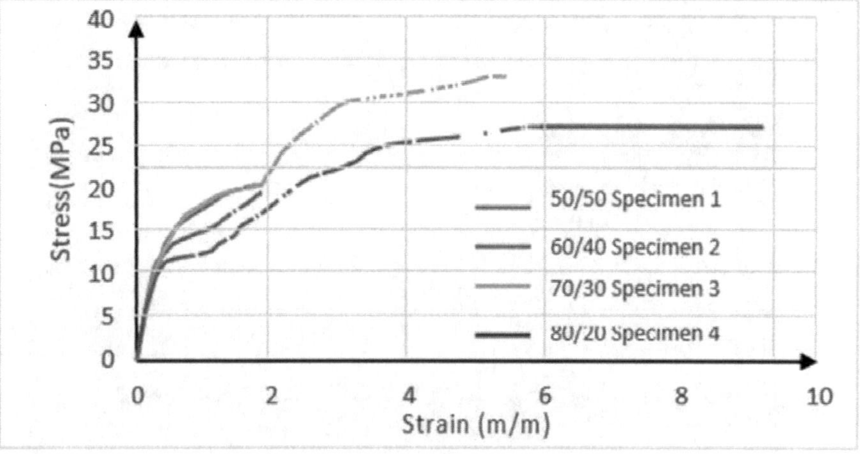

Figure 5. Tensile test result.

Table 2. Yield stress σy and modulus young E obtained from tensile test each composition.

No	Composition	σ_y	E (MPa)
1	50/50	663	39
2	60/40	1051	4375
3	70/30	1077	3474
4	80/20	879	3381

4.2 Hardness test

The hardness test was carried out on specimens by taking 3 sampling points. Hardness test using the Rockwell method with an R scale (HRR). The test results are shown in Figure 6.

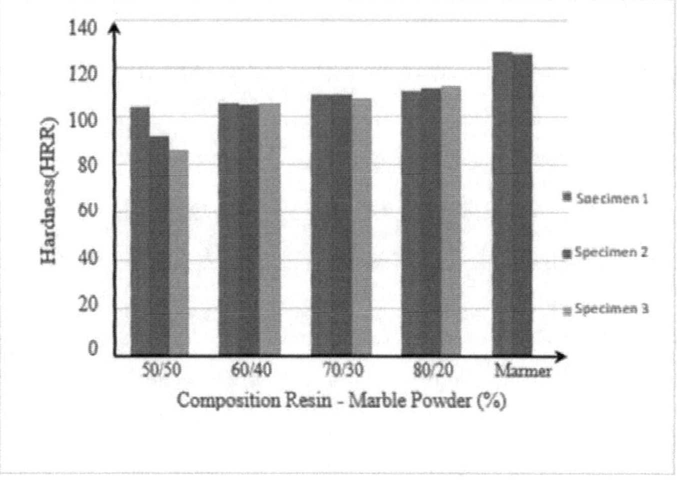

Figure 6. Hardness value for each composition.

The results of the hardness test can be seen in table 3 below. The results of this test are compared with the hardness of marble stone averaging 126.75 HRR.

IOP Conf. Series: Materials Science and Engineering **506** (2019) 012045 doi:10.1088/1757-899X/506/1/012045

Table 3. Hardness value for each composition.

No	Composition	Hardness (HRR)
1	50/50	93.99
2	60/40	105.2
3	70/30	108.24
4	80/20	111.71

4.3 Observation using SEM

Observations using SEM were carried out on broken surfaces of specimens from tensile test. Figure 7 shows the results of the fracture of the specimen after experiencing a load of drag on the composition 50/50. In the previous study, the propagation of crack propagation in ABS resin and PC/ABS blends was also observed with SEM [10-12].

Figure 7. Fracture result of the specimens after tensile testing on a composite with composition of 70/30

This 70/30 composition has a lower hardness compared to 80/20 composition but has a higher hardness than 50/50 or 60/40 composition. SEM observation (figure 8) shows that the marble powder is spread evenly in all parts of the resin and the surface is rougher than the other compositions.

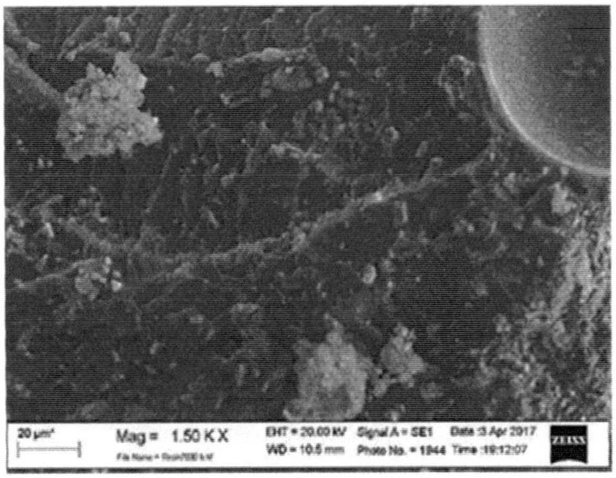

Figure 8. SEM image on the fracture surface on the specimens with the composition 70/30.

5. Conclusion

The results of testing and SEM observations, provide several conclusions: the largest value of yield stress, σ_y = 10.77 MPa, obtained by 70/30 composition. The largest Young's modulus value, E=43.75 MPa, obtained by 60/40 composition and the greatest value of hardness (111.71 HRR), obtained by 80/20 composition. The composition with optimal tensile strength and hardness is between the compositions from 60/40 to 70/30.

Acknowledgement
The authors gratefully acknowledge support from Computational Mechanics Laboratory, Mechanical Engineering Department, Universitas Syiah Kuala, Mechanical Engineering Workshop, South Aceh Polytechnic, and also the University of North Sumatra which has helped carry out this research.

References
[1] http://www.dapurkitchenset.com/2015/02/pengertian-dan-jenis-batu-marmer-di.html online, accessed on 14-3-2016.
[2] polesmarmerjakarta.co.id/manfaat-dan-kegunaan-batu-marmer online access on24-2-2016
[3] Danu S, 2012, Pemanfaatan Limbah Marmer Untuk Pembuatan Komposit Campuran Marmer-Poliester-Serat Abaca, *Prosiding Simposium Nasional Polimer IV* (*in Indonesian*), pp. 67-73.
[4] Mahayatra, I G, 2013, Pengaruh Variasi Ukuran Partikel Marmer Statuari Terhadap Sifat Mekanik Komposit Partikel Marmer Statuari, *Jurnal Fema (in Indonesian)*, vol. 1, No. 4, pp. 13-17.
[5] Mizwar A, 2015, Pemanfaatan Lumpur Marmer, Limbah Styrofoam dan Abu Layang Batubara Untuk Pembuatan Bata Beton Berlubang, *Jurnal Intekna (in Indonesian),* vol 12(1), p. 1-8.
[6] Hamza, R. A. (2011),Utilization of Marble and Granite Waste in Concrete Bricks, *International Conference on Environment and BioScience, IPCBEE* vol.21,pp. 115-119.
[7] Husaini, M. Nizar Machmud, Ihsan, Sandi Yudha B.Z. (2017), Studi Tentang Sifat Mekanik Komposit Marmer/ResinPada Komposisi 50/50, *Proseding Seminar Nasional Teknologi Rekayasa (SNTR) IV (in Indonesian)*, Poltas Aceh Selatan, pp 95-97.
[8] ASTM D 638-03, 2004, Standard Test Method for Tensile Properties of Plastics.
[9] ASTM D 785-03, 2004, Standard Test Method for Rockwell hardness of Plastics and Electrical Insulating Materials.
[10] Husaini, Kishimoto K, 2001, Mixed mode fracture behavior of PC/ABS blends, *Proceedings of SPIE - The International Society for Optical Engineering*, Vol. 4317, No. 1, pp.111-116.
[11] Husaini, Notomi M., Kishimoto K, Shibuya T, 1997, Crack initiation behavior of ABS resin under mode I and mixed mode loading, *Materials Science Research International*, Vol. 46, No.9, pp. 158-165.
[12] Husaini,Kishimoto, K.,Notomi, M. (2004), Finite element analysis of rubber particles size distribution on fracture toughness of rubber-modified polymer, *Key Engineering, Trans Tech Publications*, Ltd.,Vol. 261, pp. 711-716.

IOP Conf. Series: Materials Science and Engineering **506** (2019) 012046 doi:10.1088/1757-899X/506/1/012046

Implementation of GSM based Flood Data Communication in the Flood Disaster Location Information System

Dedi Satria[1*], Syaifuddin Yana[2], Rizal Munadi[3], Saumi Syahreza[4] and Munawir[1]

[1]Department of Computer Engineering, Universitas Serambi Mekkah, Aceh, Indonesia
[2]Department of Industrial Engineering, Universitas Serambi Mekkah, Aceh, Indonesia
[3]Department of Electrical Engineeringg, Universitas Syiah Kuala, Aceh, Indonesia
[4]Department of Physics , Universitas Syiah Kuala, Aceh, Indonesia

*Corresponding author: dedisatria@serambimekkah.ac.id

Abstract. The frequency of floods that often occur in Indonesia has made floods one of the national disasters designated as national priorities. The flood disaster has caused many casualties and material losses. For the government to anticipate more flood victims, a flood information system is needed that can provide real-time flood information in the form of location information and the level of flood level. Based on these problems, this research article aims to build a flood disaster information system in real-time location-based information using data communication based on Global System for Mobile Communications (GSM). The system design is built in two systems, namely the client system or flood detector consisting of ultrasonic sensor components as input modules, Arduino Uno microcontrollers as processing modules and SIM900 modules as output components. Next is a server system consisting of Wavecom GSM Modem as input component and PHP Engine software, mySQL database, Apache Web Server as data processing. In processing flood information, the client system sends location and water level data using SIM900 in the form of SMS to the server system. The server system processes the data so that it produces information on the location and height of the flood that can be distributed through the website. This study produced a location-based flood disaster information system prototype with data communication between flood detectors and flood information system servers using Global System for Mobile Communications (GSM) media.

Keywords: Information System, Flood Disaster, GSM, Web, Early Warning System.

1. Introduction

Flood Disaster has become a national disaster in Indonesia. Floods have given many victims and property. Floods that occur in general are caused by two factors, namely factors that are caused by illegal logging, resulting in land being unable to hold water in the forest. Water that cannot be detained on the ground will flow as a whole to the river so that the water level rises and flows to the population area. The second factor is due to waste being dumped in the drainage drains of the population resulting in clogging of the waterways and becoming a flood. The cause of these two factors is due to human error. To anticipate the two factors that cause flooding due to human actions a long-term program is needed in the form of making socialization of the importance of public awareness of the environment not to cut trees illegally and littering. Whereas to anticipate flooding in short-term programs is to build an information system that can provide flood hazard warnings and flood monitoring systems.

Based on the short-term program needed for flood disaster management, this paper explains the research aimed at creating a flood location information system that implements a GSM communication system. The information system development program was developed because of government programs that pay attention to government and private agencies to be able to develop research focused on the development of disaster mitigation systems so as to minimize future disaster victims [1].

2. Related Works

Disaster information systems have now been developed from various research centers in various countries. Therefore, a number of studies related to disaster mitigation have been summarized. A study has been conducted to anticipate a house fire disaster that was built using internet communication. In this study, fire information is sent to homeowners through real-time website information using the concept of clients and servers using the internet network [2][3]. Whereas in other studies a similar system has been developed which focuses on web-based flood information using the concept of Internet of Things (IoT) [4][5]. In addition to the web-based disaster information system several studies have developed information systems for victims of mobile-based disasters that are focused on data collection of disaster victims for the needs of regional disaster management agencies [6]. In previous studies the use of internet networks in sending information was the main communication system in the process of distributing disaster information.

In addition to internet communication there is GSM communication that utilizes SMS services in sending information. Information systems using SMS services are also called SMS Gateways. the role of SMS gateway has been applied to the construction project's information progress system [7]. In the information system the interface between the database and modem uses Gammu as its intermediary. Besides that the application of SMS gateways can also be integrated with an RFID-based embedded system for student attendance detection applications [8]. And some of them are used in security monitoring systems [9]. From the application built using the SMS gateway it can send information of an object remotely to the user via GSM communication [8]. For the application of disaster information systems, it has used GSM communication systems such as fire and flood information systems. has provided benefits in the flood disaster information system process using SMS services [10,11,12].

3. Method

The research method of making a flood location information system uses two stages, namely the system analysis stage and the system design stage. At the stage of the system analysis explains the work process of the system as a whole starting data collection to the information that will be displayed to the user. While the second stage is a system design that aims to provide a description of the required components and a description of the integration between each component in the form of a block diagram.

3.1. System Analysis

Analysis of the location information system begins with a sensor that detects the height of the flood through a flood detector system. The data received by the flood detector is processed to produce a data format for the purposes of sending data in the form of SMS to the information system server via a GSM modem. Altitude data received by the flood location information system is stored in the database server. The results of the flood data processing are sent to the flood monitoring officer as seen in figure 1.

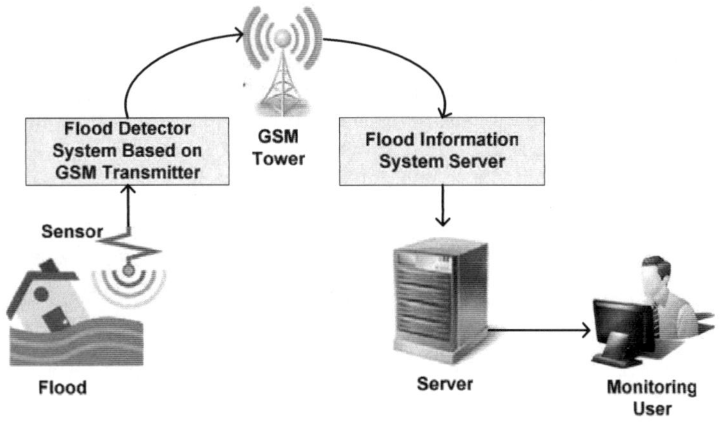

Figure 1. Description of the flood location information system.

3.2. System Design

For the design of the flood detector system and flood location information system, it is made using block diagrams as seen in figure 2. In figure 2, it can be seen that the system is divided into two, namely flood detector systems and flood information systems. In the system detector there is an ultrasonic sensor as an input block that sends data to the Arduino Uno Microcontroller as a processing block. Data that has been processed in the processing block is sent to the GSM modem to send data using the SMS Gateway service to the flood information system. In the flood location information system the data will be received by the GSM mode as the input block. Data from the modem is forwarded to the Gammu application and MySQL database. Flood data on MySQL databases is processed using PHP Engine and displays in the form of information via a web server. In the process on Gammu, MySQL, PHP and Web Server are activities on the exposure block. While the output block is a web browser as the interface between the information system and the user.

In this study the components used by the flood detector system are the GSM SIM900 Module, Arduino Microcontroller, HC-SR04 Ultrasonic Sensor. Whereas the flood location information system uses the GSM modem Wavecom as data receiver, Gammu application, MySQL, PHP and Apache Web Server.

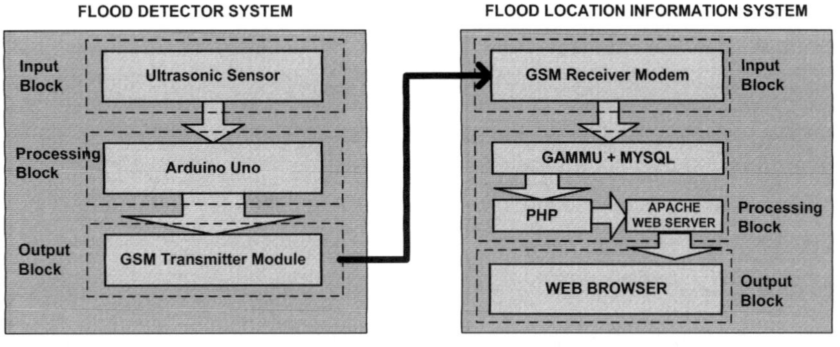

Figure 2. Block diagram of flood location information system.

IOP Conf. Series: Materials Science and Engineering **506** (2019) 012046 doi:10.1088/1757-899X/506/1/012046

4. Result

Based on system analysis and system design for flood location information systems that have been created, this study produced a prototype flood location information system consisting of two prototypes namely flood detector system and flood location information system. In the flood detector system shown in figure 3 that the system was built using an ultrasonic sensor, Arduino Uno Microcontroller and GSM SIM900 modem. Whereas in the flood location information system there is a GSM modem Wavecom as the data receiver that is connected to a computer that is used as a server.

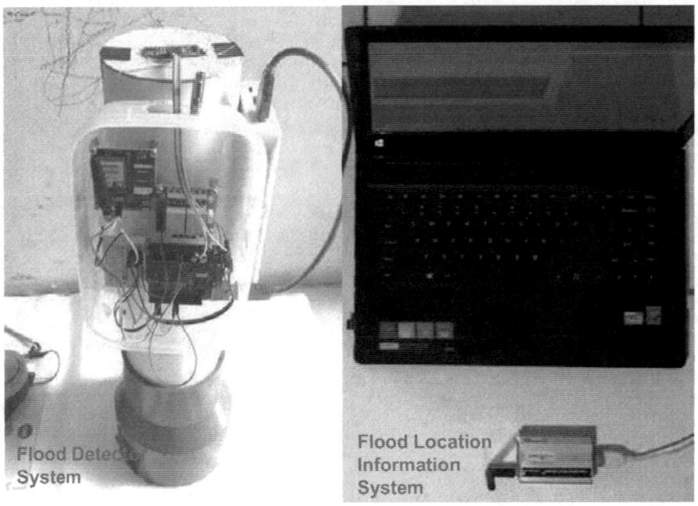

Figure 3. Results of prototype flood detector system and flood alarm.

In testing the flood detector system, an Arduino IDE is used through a serial monitor to see the effect of simulation of increasing flood water. This test is done by giving different heights starting with the height of 0 cm, 2 cm and 7 cm as seen in figure 4. At an altitude of 7 cm, the flood detector system will send altitude and location data using the SMS gateway to the flood information system. In this test the algorithm used is if the height> = 5 cm then send an SMS. Furthermore, if the height> = 5cm and the next height data with the previous height data, there will be no sending of SMS.

Testing with a height of 11 cm will send data via SMS at one time and at a fixed height at 11 cm there will be no SMS sending as seen in figure 4. Location detector systems have permanently stored location variables when the flood detector system program is uploaded. And in testing and simulation using two detectors with different location variables.

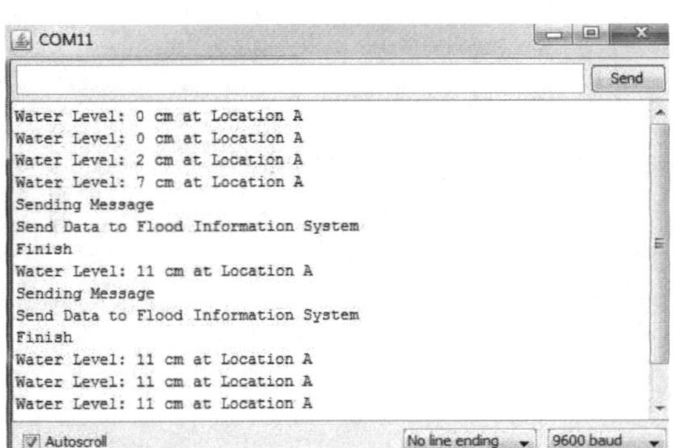

Figure 4. Flood detector system prototype test results.

In the flood location information system prototype generated web-based interface with information in the form of date and time when the flood height change data is stored in the database. Next the system interface displays the location of the flood and the height of the flood in centimeters. In this study testing uses two flood detectors with different locations as shown in figure 5. Information displayed on the flood information system will change if the flood level changes and is followed by the date and time.

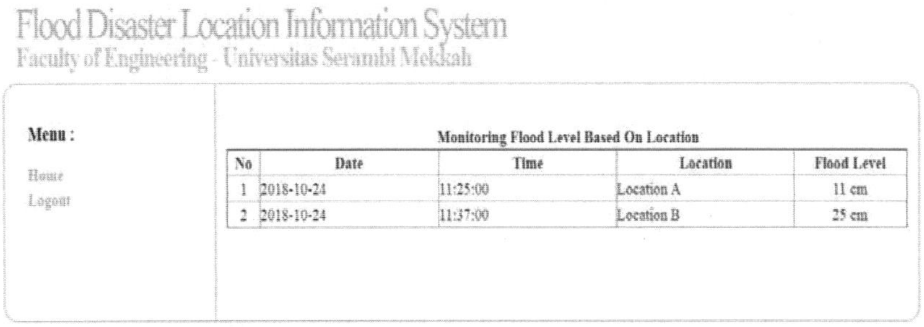

Figure 5. Results of flood information system testing.

5. Conclusion

Based on the results, the flood location information system that uses GSM communication has been successfully tested by sending flood height data from the flood detector system to the flood location information system through GSM using SMS services so that users can see changes in flood height based on location through the flood location information system web-based remotely. From the testing of the two prototypes produced this can provide benefits to government agencies that have areas of flood disaster mitigation in anticipating flood victims in flood-prone areas. the system that has been built is limited by the system being able to send data as long as the SMS quota is available, the flood level is no higher than the construction of a flood detection system and at the time of experiment resources use a power bank or battery. As for future research, it is necessary to develop a higher flood detector using internet of things (IoTs) technology which is accompanied by resources using solar cells.

Acknowledgments

This research was funded by the Ministry of Research and Higher Education (Ristekdikti) through a collaborative university research grant (PKPT). And We thank the joint Team of Researchers from Universitas Serambi Mekkah and a partner team from Universitas Syiah Kuala who have provided technical research consultancy and laboratory facilities.

References

[1] R. J. M. Mercado, "Design of wireless sensor networks using embedded Programmable System-on-Chip (PSoC) as applied to community-based flood early warning systems (CBFEWS)," in *2016 International Conference on Advances in Electrical, Electronic and Systems Engineering (ICAEES)*, 2016, pp. 214–223.

[2] S. S. Dewi, D. Satria, E. Yusibani, and D. Sugiyanto, "Design of Web Based Fire Warning System Using Ethernet Wiznet W5500," in *Malikussaleh International Conference on Multidisciplinary Studies (MICoMS 2017)*, 2018, pp. 437–442.

[3] S. S. Dewi, D. Satria, E. Yusibani, and D. Sugiyanto, "Prototipe Sistem Informasi Monitoring Kebakaran Bangunan Berbasis Google Maps dan Modul GSM," *J. JTIK (Jurnal Teknol. Inf. dan Komunikasi)*, vol. 1, no. 1, pp. 33–38, 2017.

[4] D. Satria, S. Yana, R. Munadi, and S. Syahreza, "Sistem Peringatan Dini Banjir Secara Real Time Berbasis Web Menggunakan Arduino dan Ethernet," *J. JTIK (Jurnal Teknol. Inf. dan Komunikasi)*, vol. 1, no. 1, pp. 1–6, 2017.

IOP Conf. Series: Materials Science and Engineering **506** (2019) 012046 doi:10.1088/1757-899X/506/1/012046

[5] D. Satria, S. Yana, R. Munadi, and S. Syahreza, "Design of Information Monitoring System Flood Based Internet of Things (IoT)," in *Malikussaleh International Conference on Multidisciplinary Studies (MICoMS 2017)*, 2018, pp. 629–639.

[6] Bahagia, D. Satria, and H. Ahmadian, "Perancangan SIstem Informasi Manajemen Data Korban Bencana Berbasis Mobile Android," *J. Manaj. dan Akunt.*, vol. 3, no. 2, pp. 22–30, 2017.

[7] H. Ahmadian, D. Satria, and S. Kurniawan, "Progress of Construction Project Information System Based on SMS Gateway," *J. Nas. Komputasi dan Teknol. Inf.*, vol. 1, no. 2, pp. 114–122, 2000.

[8] D. Satria, T. Hidayat, M. A. Hidayat, and Zakaria, "Application of SMS Gateway on Attendance Detection Systems using RFID," *J. Nas. Komputasi dan Teknol. Inf.*, vol. 1, no. 2, pp. 106–113, 2018.

[9] D. Satria and H. Ahmadian, "Designing Home Security Monitoring System Based Internet of Things (IoTs) Model," *J. Serambi Eng.*, vol. III, no. 1, pp. 255–261, 2018.

[10] S. Azid, B. Sharma, K. Raghuwaiya, A. Chand, S. Prasad, and A Jacquier, "SMS Based Flood Monitoring and Early Warning Systems," *ARPN J. Eng. Appl. Sci.*, vol. 10, no. Vol. 10, No.15, pp. 6387–6391, 2015.

[11] E. Kuantama, P. Mardjoko, and M. A. Saraswati, "Design and Construction of Early flood warning system through SMS based on SIM300C GSM modem," *Proc. 2013 3rd Int. Conf. Instrumentation, Commun. Inf. Technol., Biomed. Eng. Sci. Technol. Improv. Heal. Safety, Environ., ICICI-BME 2013*, pp. 115–119, 2013.

[12] O. Piller, J. Deuerlein, D. Gilbert, and J.-M. Weber, "Installing Fixed Sensors for Double Calibration and Early-warning Detection Purposes," *Procedia Eng.*, vol. 119, pp. 564–572, 2015.

IOP Conf. Series: Materials Science and Engineering **506** (2019) 012047 doi:10.1088/1757-899X/506/1/012047

Risk-based Contract Management on the Design and Build Construction to Minimize Disputes in Infrastructure Projects

Kemala Hayati[1*], Yusuf Latief[2], Achmad Jaka S[3]
[1]Civil Engineering Department, Faculty of Engineering, University Muhammadiyah Aceh, 23245 Banda Aceh, Indonesia
[2]Civil Engineering Department, Faculty of Engineering, Universitas Indonesia, 16424 Depok, Indonesia
[3]University Djuanda, Bogor, Indonesia

*Corresponding author email: k.hayati@yahoo.com

Abstract. A poor contract management is considered the main cause of disputes in construction contracts. While dispute in a project oftenly unavoidable, it is however can be minimized by applying a proper contract management which should be in line with the project's nature. A common understanding to the roles and intricacies of the contract will lead the involved parties in the project to be open to negotiate, hence produces a better construction contract in a balanced manner. This study aims to determine the management phases in the construction contracts in which incorporate the approach of input, process, output, risk and impact. All activities in every phases of the contract management will be identified for it influence to dispute. The findings will suggest a strategic management strategy that is manageble in minimizing dispute in construction industry.

1. Introduction

Infrastructure project is aimed to facilitate and cover community needs, government as its owner or sometimes private parties that is granted a concession from the government, thus requires specific attention in administer its phases especially in terms of social and regulatory aspects. Nevertheless the target of the massive-planned infrastructure development have been defined, some calls for innovation on current methods applied for effectivity in achieving its framed goals. One of the issues is to review the application of the methods in the contract agreement. Infrastructure projects often use a design and build contract that is an arrangement of contract in which the design and construction part are made as an unseparated contract of the project [1-[2]. The concept of the contract management is a process in construction industry in which the phases of the construction works are deemed for its appropriateness and detailed activities at each stages of construction are defined, that is the phase of input, process, and output of the construction project incorporated. Whereas poor contract management has been suggested as one of causes of dispute on the construction projects, previous studies also considered it as the main trigger of dispute in the construction industry over the years [3]. Nevertheless, to put some efforts thus to understand the nature of contract, the involved parties whether the project owners, contractors, and consultants may negotiate for better and balanced contract agreement, hence could minimize the dispute among them during the project implementation. The activities at each phases of the contract management should be identified in such ways, so that reflects possible strategies to minimize dispute on the design and build infrastructure projects [4].

Contract management defined as an activity to manage a contract agreement in a proper manner so that the document of agreement can be used as guidance of control of the project work. Design and build contract is the contractual arrangements in which the design and construction phase are presented on one document of contract, thus not separate the design phase of a project and its construction phase into different contracts. Contract management includes [5-6]:

1. Preparation for contracting;
2. Developing contract for project guidelines;
3. Contract as a means of project control;
4. Administering and recording as it is drafted, as guidelines, and as a controlling tool of the project.

Dispute determined as a non-conformities circumstances between parties that are going into or in an agreement for partnership and thus to cooperate. In the construction industry, it may occur between two or more parties; it generally takes a very long time to be resolved, and it has a high complexity. Eventhough the parties had mutually agreed to each other on a partnership agreement. In this regard, dispute between parties usually occurs at the project implementation phase of the construction contract [5]. There are three causes of dispute on the construction contract, which are (i) caused by behavior, (ii) caused by contractual term, and (iii) caused by technical.

2. Methodology

In order to justify study problems on the introduction chapter, an initial review on literature conducted that is to formulate study problems and thus to framed the research question. Furthermore, in-dept studies were run in order to find proper methods in achieving the research objectives that have been set previously. A set of questionnaire was also distributed to respective respondents at the initial phase of the research, and their responses have been collected. The questionnaire applied was closed questionnaire using Likert scale, and conducted in four stages as the following in Figure 1:

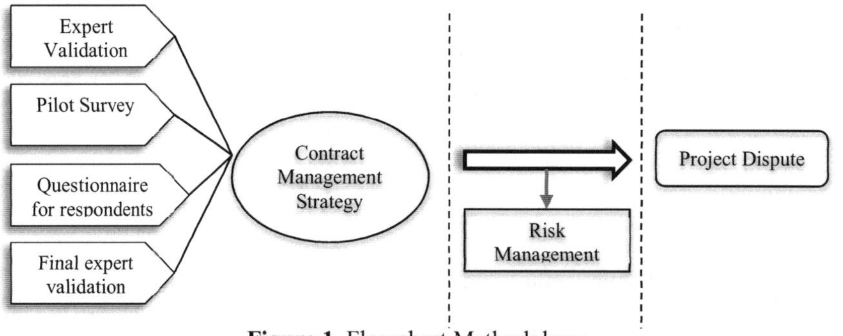

Figure 1. Flowchart Methodology

The variables in this research consist of sub-activity in design and build contract management in infrastructure projects considered affects the dispute and risk factors of the sub-activity. The suggested results were analized by performing Relative Importance Index (RII) in order to determine the ranking of the most influential sub-activities on the dispute. While, as for the results of risk factor variables, risk analysis was applied to justify the highest risk of the factors which influence to dispute on design and build contract management in infrastructure projects. For analysis purpose, validity and reliability tests were applied for the validity of the questionnaires responses. While, correlation and regression tests used to see the relation and effect of the independent variables to the dependent variables.

3. Result and Discussion

Based on the findings of the literature study, there were 118 variables of sub-activity with its risk factors on the design and build contract management on infrastructure projects that affects the dispute. Following the first stage of the Initial Expert Validation performed to five relevant experts, 103 variables were approved as the sub-activity with its risk factors. The 103 variables were applied for the Pilot

IOP Conf. Series: Materials Science and Engineering **506** (2019) 012047 doi:10.1088/1757-899X/506/1/012047

Survey and Respondents Questionnaire. Furthermore, those variables were tested by running the Validity and Reliability tests followed by the RII analysis to determine the ranking of the most influential sub-activity to the dispute. After Final Expert Validation as its last stages, there are nine top ranking thus suggested as the most influential sub-activity on dispute. Table below indicates those nine most influential sub-activity on the dispute based on the Final Experts Validation stage.

Table 1. The Most Influential Sub-activity to the Dispute

Rank	Sub Activity
1	Signing the contract
2	Detail Engineering Design (DED) arrangement
3	Obtaining initial information
4	Internal meeting to follow-up the initial information
5	Developing pre-design proposal/conceptual design
6	Preparing quotation
7	Issuing the first hand-over report
8	Ensuring the amendment order are officially in writing
9	Ensuring the return of warranties

Based on the results obtained from the above qualitative risk analysis, there are four sub-activities which then identified as high risk factors, which will be further analized for its causes, possible preventive actions, impact and its corrective action. These will be analized following the Final Expert Validation stage. From the identified preventive and corrective actions, there were actions that considered as the new sub-activities thus will be incorporated to the design and build construction contract management strategy in order to minimize dispute on the risk-based infrastructure projects. In addition, from 103 identified risk factors from all sub-activity on the design and build contract management that affect the dispute which analized by risk analysis, there were initially 11 high risk factors determined on the design and build contract management sub-activities on the infrastructure projects. These 11 sub-activity then also validated by the experts through the data collection of stage four; Final Experts Validation, that resulted ten high risk factors from the sub-activity on the design and build contract management that affect the dispute, as listed on the table below:

Tabel 2 . Rangking of High Risk Factors to the Dispute based on the Final Expert Validation

Rank	Risk Factors
1	Inaccuracy in Detail Engineering Design (DED) arrangement
2	Misintepretation or failure reviewing draft contract/agreement
3	Misunderstanding or misperception in signing the contract
4	Inaccuracy in proposing quotation
5	Inaccuracy, failure and incomplete data in reviewing the contractual aspects of claims or anti-claims
6	Inaccuracy, insufficient preparation of prerequirements, and incomplete data or evidence for claims or anti-claims documents, including its chronology duly signed by service users, engineer, and contractor
7	Uncooperative attitude from service users, or contractors negligence to obtain an official amendment order in writing
8	Failure to obtain initial information
9	Improper development of the proposed technical proposal
10	Disagreement in negotiation of the draft of contract

The results of design and build construction contract management strategies to minimize dispute on risk-based infrastructure projects were then validated by experts through the final interview. The following new sub-activities as the outputs are added to the contract management.

Table 3. Recapitulation Data as the Final Interview Result

No.	Sub-activity	Output
1	The acquisition of the land is entirely the service user's responsible	Prove of acquisition of the projects' sites, including its utility are fully under service user's responsibility.
2	Preparing work insurance	Professional Indemnity Insurance (PII) and Contractor's All Risk insurance (CAR).
3	Conducting a full review to tender documents comprising all related functions comprehensively	A thorough and comprehensive tender document report reviewed by related team with its relevant functions.
4	Provide detailed and thorough information to top level management regarding opportunity and risk of the contract when it signed	Detailed and complete information to the top management about the opportunity and risk of the contract and its follow-up action plan.
5	Referring to company database	To having reference for similar work which uses the design and build contract, as well as other types of work.

Based on the research findings, it was suggested four processes of construction design and build contract management that will minimize dispute on infrastructure project from contractor point of view: contract planning, contract formation, contract administration, and contract monitoring. It can be concluded that there is no significant difference in the phases of construction contract management, either on conventional construction project or on design and build. The differences were only on activities undertaken in each processes and kind of activities need to be considered in carrying out those activities.

3.1 Activities on the Process of Construction Contract Management on Design and Build that Affects Dispute on Infrastructure Projects

The results of the RII analysis which further finalized by the experts pointed that there are nine sub-activities o the design and build construction contract management which influence the dispute. Below are those nine sub-activities and discussions about the results of its validation:

a. Signing the contract

According to the experts, sub-activities of signing contracts can affect the dispute when the signed contract is not well understood by the contractor or involved parties. There also circumstances when contractors are forced to sign an unbalanced contract due to various considerations.

b. Detail Engineering Design (DED) arrangement

In the nature of design and build contract, contractors are requested to prepare the DED before the construction starts. According to the experts, this activity in its practice is very prone to cause dispute between parties. Problems that usually arise in this sub-activity are: lack of competence in DED where contractors have no experience on the design and build project; incomplete and inaccurate data of the project sites either provided by the service users at the time of tender or by the contractor after the commencement of work; contractor failed in interpreting employer's requirement; late of design approval from service user; different level of understanding about the design between involved parties. Incomplete and inaccurate data about the project sites is the most common problem on infrastructure project due to the nature of the infrastructure project that requires more adequate and sufficient field surveys. Hence, this sub-activity becomes very critical thus influential on the design and build contract since it can lead to a dispute between contracting parties, and therefore has great potential to cause an enormous impact for the contractor, such as cost overrun, time overrun, rejected by the users, even design failure.

c. Obtaining initial information

According to the experts, the sub-activity of obtaining early information affects the dispute as at this stage the contractor has to consider the project feasibility, at the same time has to decide whether or not to continue the participation in the tender. In the event of failed in obtaining sufficient information, it can be ascertained that contractors will experience losses due to aspects that are not taken into account in their offer.

d. Internal meeting to follow-up the initial information

Conducting internal meetings in order to decide the follow-up action of the initial information is also suggested to have an effect on the dispute. This is due to at this stage the decision to go or no go have to be made. This stage must also involve all related functions within the organization with all its competence and responsible person to authorize and decide.

e. Developing pre-design proposal/conceptual design

Sub-activity of developing pre-design proposal/conceptual design is also an activity that affects the dispute. At this point, however is then determine the next phase, namely preparation of DED, on which will bring direct impact of DED quality, price and time. Failure at this stage will automatically cause errors on DED that will, however cause dispute with the service users and lead to contractor losses and even the design failures.

f. Preparing quotation

On the design and build contract, the most ideal price offer is made in the form of cost center. Whereas the bidder does not include the volume in their offer to service user than the costs for each work items or lumpsum for each work items. This is also, however indirectly affirmed in FIDIC in which grouping the design and build contracts by lumpsum payment. Nevertheless, the experts indicated this sub-activity as one of the most influential on the dispute since in general the design and build contract uses the lumpsum payment method. The price stated in the bid is suggested very important because it will be considered as the contract price without any re-measurement of volume proposed. Thus, it is strongly advised in the process of calculating price offer, the working methods of the service user's and contractor's rights, scope of work, project duration and the project funding must be clearly defined.

g. Issuing the first hand-over report

Still according to the experts, the sub-activities of issuing the first hand-over report (BAST I) has an effect on the dispute. In the matter of fact, however, at this stage the dispute between parties is often occurred. Problems that usually arise at this stage includes when the BAST I has been issued, but there are on-processes of the unfinished claims. This circumstance however will cause the contractor's claim be void due to the BAST I document has issued, while the service user may use the constructed facilities eventhough there still dispute with the contractor. Hence, before the BAST I is prepared for issued, the contractor must first make sure that all related matters resolved.

h. Ensure the amendment order is in official writing

To avoid any problems that end up as dispute with the service user, the contractor should also make sure that all work orders from the service user, including any of its amendment, are properly provided in official writing. Should there are any oral commands given by the service user, the contractor needs to confirm in writing to the service user. Thus, the amendment of order will be officially recorded.

i. Ensuring the return of warranties

Ensuring the return of warranties is an activity that the contractor needs to undertake in running the construction contract management.

3.2 Contract Management Strategy on Design and Build Construction to Minimize Dispute on Infrastructure Projects

The design and build construction contract management strategy to minimize dispute on infrastructure projects begins by identifying risk factors from all sub-activities to the follow-up of causes, preventive actions, impacts and corrective actions on the high risk factors of the most influential sub-activities against dispute. From the identified preventive and corrective actions, it is determined which actions can be incorporated as the new sub-activities of a design and build construction on risk-based infrastructure projects. The new sub-activity in the construction contract design and build strategy to minimize dispute on the project presented below.

a. The acquisition of the land is entirely the service user's responsible

It is undoubtly that the total acquisition of the land as the project site is fully under the responsibility of the service users. This is nevertheless one of the most important on the construction contract management to minimize dispute, particularly on infrastructure projects. This sub-activity is carried out in the early stages of the tender and should be recorded in minutes of explanatory notification/Berita Acara Pemberian Penjelasan (BAPP), and duly signed by involved parties.

b. Preparing work insurance

Preparing work insurance is also an activity that needs to be considered and implemented as for the uncertainty and risks in construction projects, in particular on design and build projects. In this regard, eventhough the contractor's all risk insurance (CAR) commonly used on conventional contracts, however it is also necessary to insure the design work to mitigate the occurrence of design errors, i.e. with the Professional Indemnity Insurance (PII).

c. Conducting a full review to tender documents comprising all related functions comprehensively

In response to the risk of errors in reviewing tender documents, it is necessary to ensure that the review is conducted by all relevant teams in a comprehensive manner due to the broud aspects that need to be considered in preparing the offer. In this case, the management team needs to make policy and appoint appropriate person in charge, thus the review of this tender document can be conducted thoroughly, i.e. to coordinate among functions so the risk of missing work and in quantifying the contractor's offer can be avoided.

d. Provide detailed and thorough information to top level management regarding opportunity and risk of the contract when it signed

According to the expert, in the event of a policy or circumstance in which the contractor have to sign an unbalanced contract, team member is obligated to review the drafted contract and the tender team shall provide detailed and complete information to the top level management about the opportunity and risk of the contract; if it is signed.

e. Referring to company database

In minimizing the occurrence of failure in the design and construction work, it is necessary for the contractor to have reference by seeking the company database for similar work previously implemented uses the design and build contract, as well as other types of work. This is urged to avoid similar problems that may occur on the related projects. The Project Manager and the Management Team are expected to coordinate with the project team and other departments to find out what issues are faced in handling the design and build contract.

4. Conclusion

From all of design and build construction contract management process, there are thirteen variables of activity that affects to dispute. The RII analysis result which has been validated by the experts suggested nine sub-activity variables as the most influential on the design and build contract management. In addition, the study result indicated that there are four high risk factors from the most influential sub-activities to the dispute:

- Detail Engineering Design (DED) arrangement
- Signing the contract
- Preparing quotation
- Obtaining initial information on the project.

From this point, the follow-up on the causes, preventive action, impact and corrective action then defined. The identified preventive and corrective actions were then incorporated as the new sub-activity to the design and build construction contract management. While there are five new sub-activities for the design and build construction contract management strategy based on follow-up results of risk analysis, the result can then be regarded as a risk-based contract management strategy on design and build construction to minimize disputes on infrastructure projects.

Acknowledgments
The author would like to acknowledge the support by University Muhammadiyah Aceh and Universitas Indonesia.

References

[1] Jaffar, N., A. H., et al. (2011). *Factors of Conflict in Construction Industry: A Literature Review.* The 2nd International Building Control Conference 2011, Universiti Teknologi Mara Perak and Institution of Surveyor Malaysia (ISM), Malaysia.
[2] Ling, Florence Yean Yng (2006). *Design–build: How to Increase its Usage, and its Impact on Architects and Contractors in Singapore.* Department of Building, National University of Singapore, Singapore.
[3] Chan, Edwin H.W. dan Ann T.W. Yu (2005). *Contract Strategy for Design Management in the Design and Build System.* International Journal of Project Management, Department of Building and Real Estate, The Hong Kong Polytechnic University, Hong Kong.
[4] Lam, Gary T W, et al. (1999). The Effectiveness of Design and Build Conditions of Contract. 15th Annual ARCOM Conference, John Moores University, Liverpool.
[5] Wall, Colin J. (1993). *Dispute Prevention and Resolution for Design and Build Contracts in Hong Kong.* Commercial, Mediation & Arbitration Services Ltd., Hong Kong.
[6] Ilter, Deniz dan Attila Dikbas (2009). An investigation of the factors influencing dispute frequency in construction projects. RICS COBRA Research Conference, University of Cape Town.

IOP Conf. Series: Materials Science and Engineering **506** (2019) 012048 doi:10.1088/1757-899X/506/1/012048

An Experimental Investigation of a Diesel Engine Using Carbon Nanotubes Blended with Biodiesel

R A Alenezi[1, *], A M Norkhizan[2], H A Rani[3], M I Maulana[4] and M Yasar[5]
[1]Department of Chemical Engineering, College of Technological Studies, Public Authority for Applied Education and Training, PO. Box 42325, Shuwaikh 70654, Kuwait
[2]Advanced Fluid Focus Group (AFFG), Faculty of Mechanical Engineering, Universiti Malaysia Pahang, 26600 Pekan, Pahang, Malaysia
[3]Civil Engineering Department, Universitas Muhammadiyah Aceh, Banda Aceh, Indonesia
[4]Faculty of Mechanical Engineering, Universitas Syiah Kuala, Aceh, Indonesia
[5]Faculty of Agriculture, Universitas Syiah Kuala, Aceh, Indonesia

[*]Corresponding author: rsn249@gmail.com

Abstract. The implementation of nano-particles in various application tremendously rise due to its unique characteristics in enhancing the effectiveness of final products. The main purpose of this study is to investigate the effects of adding 50, 100 and 150 ppm carbon nanotubes in biodiesel blend (B10) on the performance, combustion characteristics and emissions of single cylinder diesel engine. The experiment was conducted on Yanmar Model TF120M diesel engine under constant engine speed of 1800 rpm and low, medium and full engine loads. The performances of the engine were analysed in terms of indicated power, fuel consumption, indicate specific fuel consumption and indicated mean effective pressure. While, in-cylinder pressure, in-cylinder temperature and combustion duration were analysed to investigate the combustion behaviour of fuel blends. From the study, addition of carbon nanotubes in biodiesel blends did improved the performances of engine, which reduce the indicated specific fuel consumption especially during low load condition. The peak in-cylinder pressure for CNTs fuel blends were higher than conventional diesel fuels (D100) in all load conditions. However, the in-cylinder pressure for B10C100 was slightly comparable with D100 during 14 Nm load condition.

1. Introduction

Biodiesel is considered as a clean renewable energy that come from vegetable oil and animal fats. Biodiesel had become the most promising alternative to diesel fuel due to its almost identical properties with conventional diesel fuel. Lot of research on biodiesel have been conducted for the purpose of exploring the potential of biodiesel to enhance the performance and reduce the exhaust emissions [1–3].The advantage of biodiesel to be used in diesel engine is it can be used without any special engine modification [4]. However, high proportion of biodiesel still cannot challenge the performance of conventional diesel fuel in certain engine conditions. Therefore, various techniques have been implemented to enhance the performance and combustion characteristics of the engines. Fuel additives is one of the promising strategies that acts as catalyst in improving fuel quality, combustion and reduce pollutant emissions. Numerous additives have been introduced as catalyst to enhance the combustion characteristics and reduce exhaust emissions such as alumina (Al_2O_3), cupric oxide (CuO), cerium oxide (CeO_2), carbon nanotube (CNT) and many others [5–9]. Carbon nanotube (CNT) is one of the mineral additive which used in various engineering fields such as in composite materials as well as in biofuel application. The application of CNT in automotive could act as a potential nano-additive for the fuel to enhance the burning rate of the fuel, improve the cetane number which act as anti-knock and promote clean burning [10–13]. The investigation on CNT is widely explored by previous researchers as an effort of improving engine performance, combustion and reducing undesirable gases [8–17]. Hosseini et al. [16] conducted a study on the effect of adding 30,60 and 90 ppm of CNT in B5 and B10 biodiesel blends on the performance and emissions of diesel engine. The authors had found that the addition of CNT did enhanced the engine performance and reduce the exhaust emissions. The large surface area of CNT did enhance the chemical reaction and reduce the ignition delay which improve the degree of fuel-air mixing and leads to better combustion are the main cause of lower CO emission. Studies conducted by El-Seesy et al., [17] indicated that the recommended amount of MWCNTs in B20 was 40 ppm which shown significant enhancement in engine performance and emissions. Ghafoori et al. [10], added 2.5, 5, 7.5, 10, 15, 20, 30 ppm of CNT in B20 biodiesel blends and tested on six cylinders, four strokes diesel engine under different engine speed condition. The authors had concluded that the addition of CNT in biodiesel blend did enhance the performance of diesel engine in terms of power, torque and brake specific fuel consumption (BSFC) up to 17%, 18% and 38.5%, respectively. Furthermore, the UHC and CO emission is decrease after the addition of CNT in biodiesel blends with percentage decrease of 22% and 18% lesser than pure biodiesel. Similar to Hosseini et al. [16], the authors also reported that the improvement of fuel properties and engine performance after addition of nano-additives are due to their enhanced surface area/volume ratio, quick evaporation and shorter ignition delay characteristics. According to the literature review, it is obvious that a limited scientific information on the effects of high MWCNTs content of 50, 100 and 150 ppm. The experiment was conducted on single cylinder CI engine under various engine loads and constant engine speed of 1800 rpm. On these runs of those working conditions, no engine modification is needed to use biodiesel-diesel with addition of MWCNTs.

2. Experimental apparatus and procedure

Pure diesel was obtained from local fuel station in Pahang, Malaysia. Biodiesel was obtained from Malaysia Palm Oil Berhad, located in Pahang. Carbon nanotubes (CNT) with 90-95% purity was delivered by local supplier. The blend fuels were prepared by mixing together diesel, biodiesel and CNT at different composition as shown in **Table 1** using ultrasonicator. **Figure 1** indicates the TEM images of CNTs added to the biodiesel-diesel blended fuels.

Table 1. Blend configurations and testing conditions.

No.	Blend	Diesel (Vol%)	Biodiesel (Vol%)	MWCNT (ppm)	Engine Speed (RPM)	Load (Nm)
1	D100	100	0	-	1800	7
2	B10C50	90	10	50		14
3	B10C100	90	10	100		21
4	B10C150	90	10	150		28

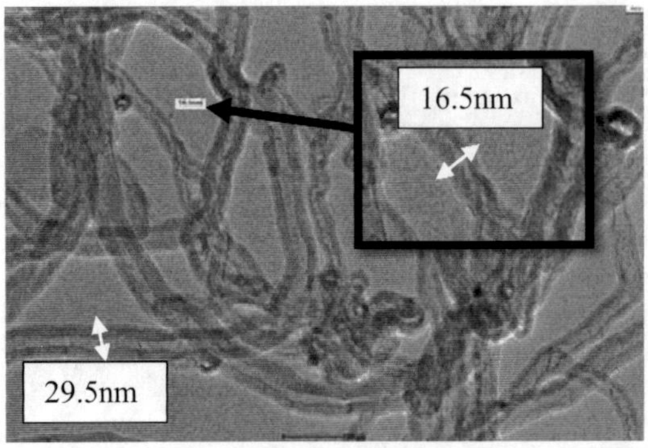

Figure 1. TEM images of CNTs in biodiesel-diesel blended fuels

2.1 Engine Setup
Experiment were conducted using single cylinder four stroke Yanmar Model TF120M Direct Injection Diesel Engine. **Table 2** shows the general specification of engine use in this test. The schematic layout of the engine setup is illustrated in **Figure 2**. The engine is coupled with eddy current dynamometer capable to deliver power of 15 kW (model Focus Applied Technologies BD-15KW). The in-cylinder pressure was measured using Optrand pressure sensor which linked with crank encoder. Dewesoft Data Acquisition was used to analyse the combustion characteristics of diesel engine. In this test, all thirteen fuel samples were tested under constant engine speed and throttle position. **Table 1** shows the test list and condition that involved in the test.

Table 2. Engine specifications.

Model	Yanmar Model TF120M
Type	Horizontal single-cylinder 4-stroke diesel engine
Bore (mm)	92
Stroke (mm)	96
Displace volume (cm^3)	638
Compression ratio	17.7:1
Continuous rating output	10.5 HP @ 2400 rpm
Maximum rating output	12.0 HP @ 2400 rpm
Fuel injection type	Direct Injection
Injection timing	17° BTDC
Max power	7.7 kW @ 2400 rpm
Max torque	161 Nm @ 4500 rpm
Cooling system	Water-radiation
Fuel tank capacity	11L

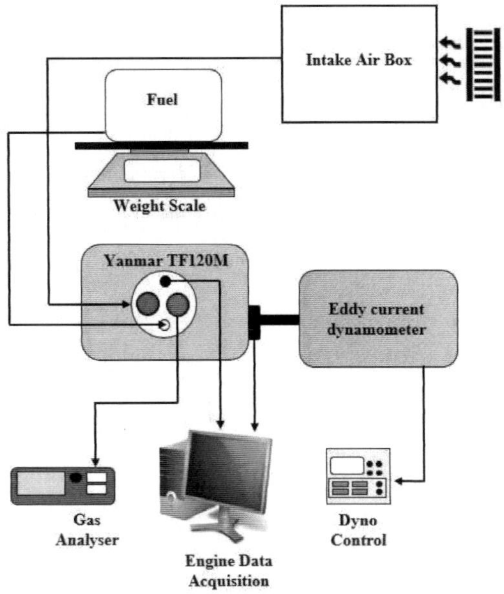

Figure 2. Schematic diagram of engine setup.

3. Result and Discussion

3.1 Performances of CNT-Biodiesel blends

The indicated power of diesel engine with CNT-biodiesel blends under constant engine speed of 1800 rpm and varies engine load of 7, 14, 21, 28 Nm is shown in **Figure 3**. As observed in **Figure 3**, at all engine loads, the indicated power for B10C50, B10C100 and B10C150 were lesser than baseline diesel fuel (D100). The lower engine power for biodiesel blends than baseline diesel is attributed to lower heating value of biodiesel than pure diesel [16]. Rizwanul Fattah et al. [18] studied the performance and emission characteristics of a diesel engine using palm oil based biodiesel blends. They reported that, the brake power for B20 and B100 were 1.02% lower than B0 due to its lower energy content, lower calorific value and higher viscosity. In addition, study done by Buyukkaya [19] indicate that the addition of 5% biodiesel in diesel fuel did not affect the engine power. The engine power was seen to be affected for B20, B70 and B100. They also explained this power reduction with lower heating value of the biodiesel.

Figure 4 shows the indicated specific fuel consumption (ISFC) for all fuel blends under different engine loads of 7, 14, 21 and 28 Nm. The ISFC for all blends were decreasing as the engine load increased that is closely related to the indicated power of the engine. From the figure, ISFC for all biodiesel-CNT blends were lower than baseline diesel. Under low load of 7 Nm, the addition of CNTs in biodiesel blends did affects the ISFC of the engine. However, the optimum content of CNTs for lowest ISFC was 100 ppm with value of 685 g/kWh that is 25% lower than D100. Furthermore, during high load of 28 Nm, less ISFC were recorded on B10C150 with 79% lower than D100. The presence of biodiesel and CNTs in the fuel blends helps in improving fuel economy which provide oxygen for combustion and accelerate the combustion. Studies done by Arul Mozhi Selvan et al. [6] indicates that at full load, the specific fuel consumption for diesel-biodisel-ethanol with addition of 50 and 100 ppm of CNT + CERIA were comparable and lower than 25 ppm CNT + CERIA content. Similar to present study, the ISFC for 50 and 100 ppm CNTs blends were slightly comparable.

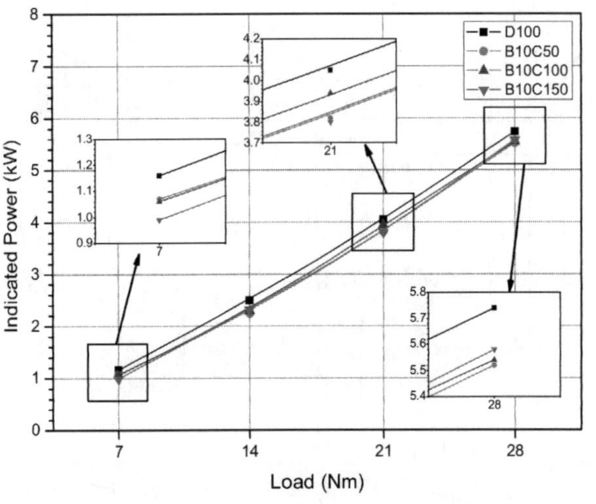

Figure 3. Indicated power under various engine loads

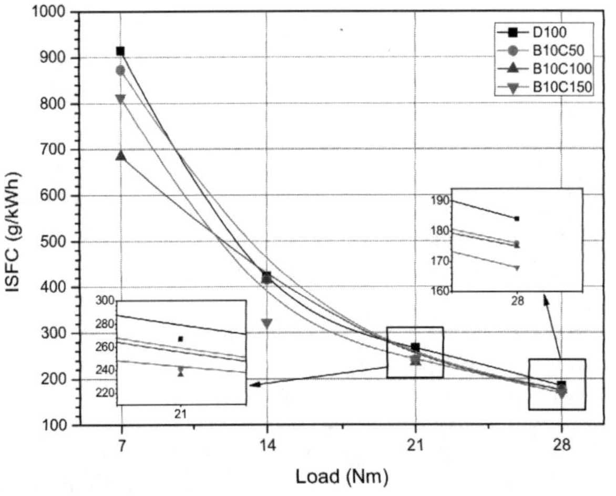

Figure 4. Indicated specific fuel consumption under various engine loads

3.2 Combustion characteristics of CNT-Biodiesel blends

This research work covers the combustion characteristics based on in-cylinder pressure and the crank angle data. **Figure 5-7** show the combustion characteristic of diesel engine in terms of in-cylinder pressure, peak in-cylinder pressure and combustion duration, respectively. **Figure 5a**, **5b**, **5c** and **5d** show the in-cylinder pressure curve at maximum pressure cycle from 300 consecutive cycles for D100, B10C50, B10C100 and B10C150 at different engine loads of 7, 14, 21, 28 Nm, respectively. **Figure 5a** shows the effect of different CNT content on the in-cylinder pressure of all fuel blends at low engine load of 7 Nm. As observed in **Figure 5a,** B10C50, B10C100 and B10C150 shows similar pressure rise behaviour that is slight increase in pressure after reaching maximum pressure in crank angle range of 3 °CA to 5 °CA. The D100 reaches maximum pressure earlier that others that occur between 2 °CA to 3

°CA. Therefore, the addition of 10% biodiesel and CNT did affect the behaviour of pressure rise of D100. **Figure 5b** shows the effect of CNT addition in B10 blends under 14 Nm engine load. The pressure rise behaviour for sample B10C50 and D100 was identical. It can be seen in **Figure 5b** that the peak in-cylinder pressure for D100 and B10C50 occurs at crank angle of 2.5 °CA to 3.5 °CA. While maximum pressure for B10C100 and B10C150 occur at crank angle of 3 °CA to 4 °CA. The peak pressure for B10C50, B10C100 and B10C150 was higher than D100 due to the shorter ignition delay resulting in higher peak in-cylinder pressure (**Figure 6**). The effect of different CNTs content on B10 blended fuels under 21 Nm engine load shows that pressure rise behaviour for D100, B10C100 and B10C150 was comparable. Sample D100, B10C100 and B10C150 reach peak in-cylinder pressure at crank angle range of 2 °CA to 3 °CA. While sample B10C50 reach peak in-cylinder pressure at 3 °CA to 4 °CA. Therefore, the ignition delay for sample B30C50 was longer than other samples. **Figure 6** indicates the peak in-cylinder pressure for all blended fuels. It can be seen from **Figure 6** that the addition of CNT in B10 did increase the in-cylinder pressure compared to non-CNT diesel fuel (D100). During low engine load (7 Nm), B10C150 blend shows the highest In-cylinder pressure at 3% higher than D100. However, under 14 Nm engine load, the in-cylinder pressure for B10C100 was slightly comparable with D100 and lower than other blends. The in-cylinder pressure for D100 under engine load of 21 Nm was 4.2% lesser than B10C50. Najafi [20], studied the effect of 40, 80 and 120 ppm of CNTs in biodiesel on the combustion characteristics, performance and emissions of diesel engine. The author had found that the addition of CNT in biodiesel did increase the peak in-cylinder pressure. In addition, increasing amount of CNTs content did increase the peak in-cylinder pressure of biodiesel when compared to pure diesel and pure biodiesel.

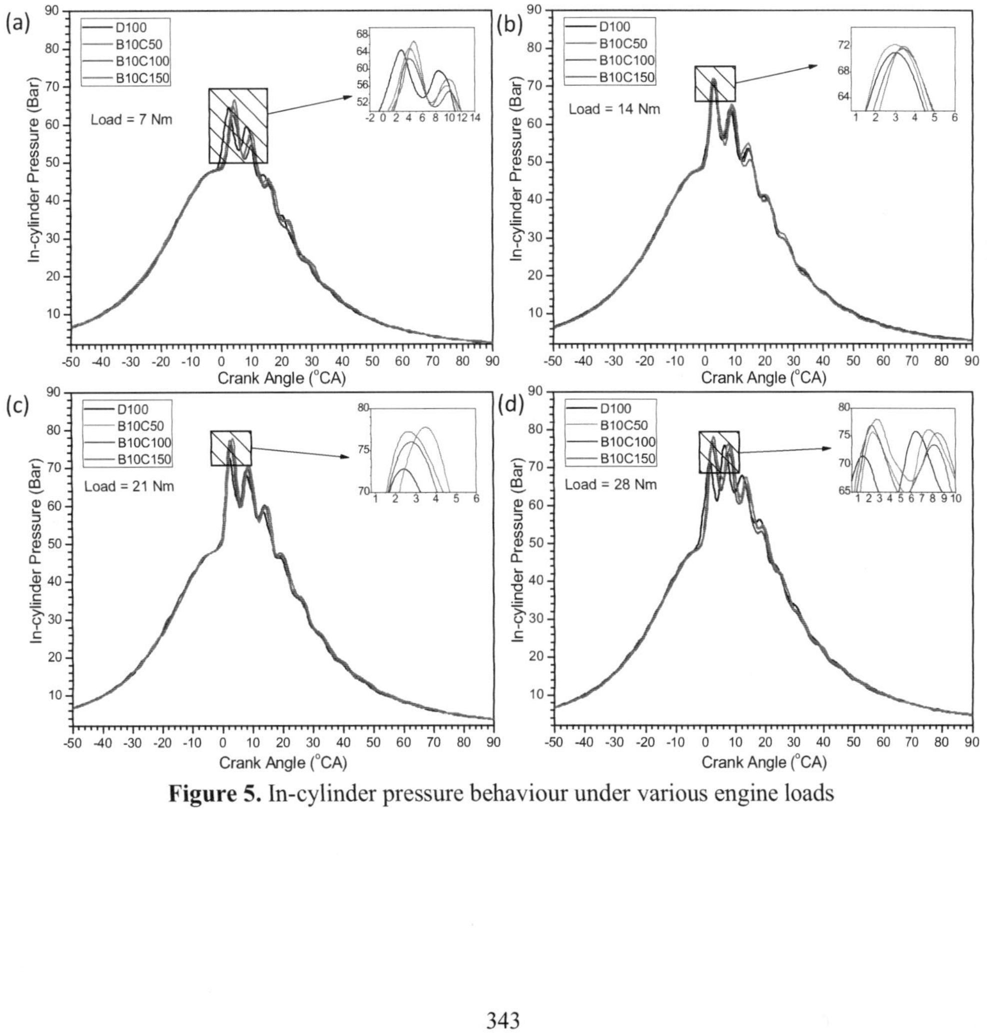

Figure 5. In-cylinder pressure behaviour under various engine loads

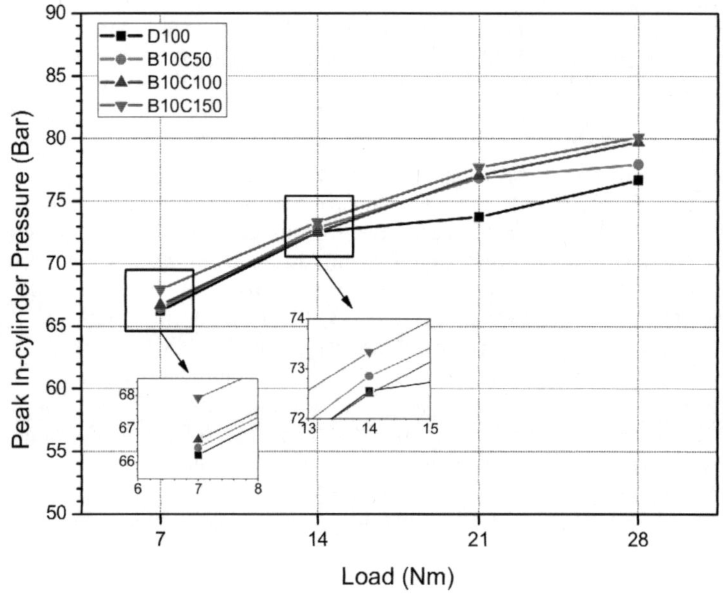

Figure 6. Peak In-cylinder pressure behaviour under various engine loads

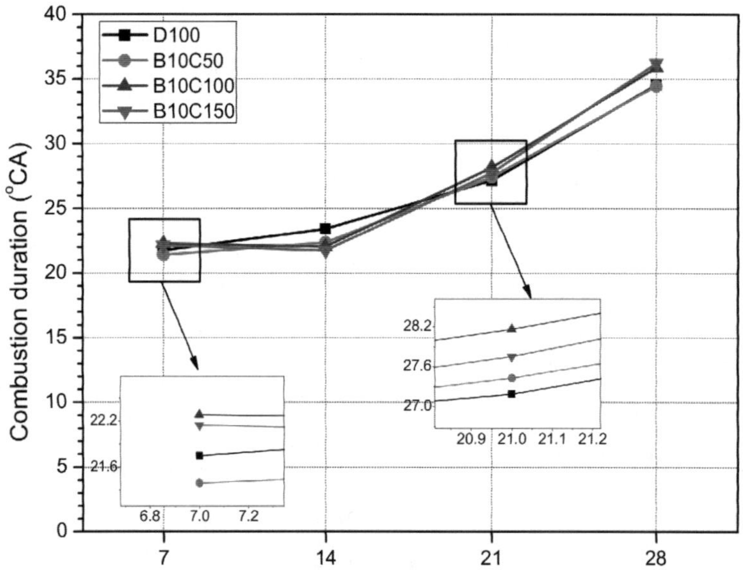

Figure 7. Combustion duration under various engine loads

4. Conclusion

The performances and combustion characteristics of single cylinder diesel engine using biodiesel-diesel blended with various CNTs content under constant engine speed and varies engine loads were investigated. Based on the experimental investigation, the following conclusions are drawn:

1) Increasing amount of CNTs in B10 fuel blends did improved the ISFC of engine. Especially during low engine load condition and the optimum content of CNTs for lowest ISFC was 100 ppm with value of 685 g/kWh that is 25% lower than D100. The presence of biodiesel and CNTs in the fuel blends helps in improving fuel economy which provide oxygen for combustion and accelerate the combustion.
2) During low engine load (7 Nm), B10C150 blend shows the highest In-cylinder pressure at 3% higher than D100. However, under 14 Nm engine load, the in-cylinder pressure for B10C100 was slightly comparable with D100 and lower than other blends. The in-cylinder pressure for D100 under engine load of 21 Nm was 4.2% lesser than B10C50.

From the result, it was clearly proved that the addition of CNTs in biodiesel-diesel blends did improved the engine performances and combustion characteristics. The engine run smoothly without any modification along the experiment.

Acknowledgments

The Authors would thank the Public Authority for Applied Education and Training, PAAE – Kuwait, for funding Sabbatical leave to University Malaysia Pahang, UMP. We thank our colleagues from Automotive Engineering Center (AEC) and Advanced Automotive Liquid Laboratory (ALL), who provided insights and expertise that greatly assisted the research.

References

[1] Chang Y C, Lee W J, Wu T S, Wu C Y and Chen S J 2014 Use of water containing acetone-butanol-ethanol for NOx-PM (nitrogen oxide-particulate matter) trade-off in the diesel engine fueled with biodiesel *Energy* **64** 678–87

[2] Yasin M H M, Paruka P, Mamat R, Yusop A F, Najafi G and Alias A 2015 Effect of Low Proportion Palm Biodiesel Blend on Performance, Combustion and Emission Characteristics of a Diesel Engine *Energy Procedia* **75** 92–8

[3] Mahmudul H M, Hagos F Y, Mamat R, Adam A A, Ishak W F W and Alenezi R 2017 Production, characterization and performance of biodiesel as an alternative fuel in diesel engines – A review *Renew. Sustain. Energy Rev.* **72** 497–509

[4] Tan P Q, Ruan S S, Hu Z Y, Lou D M and Li H 2014 Particle number emissions from a light-duty diesel engine with biodiesel fuels under transient-state operating conditions *Appl. Energy* **113** 22–31

[5] Annamalai M, Dhinesh B, Nanthagopal K, SivaramaKrishnan P, Isaac JoshuaRamesh Lalvani J, Parthasarathy M and Annamalai K 2016 An assessment on performance, combustion and emission behavior of a diesel engine powered by ceria nanoparticle blended emulsified biofuel *Energy Convers. Manag.* **123** 372–80

[6] Arul Mozhi Selvan V, Anand R B and Udayakumar M 2014 Effect of cerium oxide nanoparticles and carbon nanotubes as fuel-borne additives in diesterol blends on the performance, combustion and emission characteristics of a variable compression ratio engine *Fuel* **130** 160–7

[7] Sajith V, Sobhan C B and Peterson G P 2010 Experimental investigations on the effects of cerium oxide nanoparticle fuel additives on biodiesel *Adv. Mech. Eng.*

[8] Mirzajanzadeh M, Tabatabaei M, Ardjmand M, Rashidi A, Ghobadian B, Barkhi M and Pazouki M 2015 A novel soluble nano-catalysts in diesel-biodiesel fuel blends to improve diesel engines performance and reduce exhaust emissions *Fuel* **139** 374–82

[9] Banapurmath N R, Sankaran R, Tumbal A V., N. N T, Hunshyal A M and Ayachit N H 2014 Experimental investigation on direct injection diesel engine fuelled with graphene, silver and

multiwalled carbon nanotubes-biodiesel blended fuels *Int. J. Automot. Eng. Technol.* **3** 129
[10] Ghafoori M, Ghobadian B, Najafi G, Layeghi M, Rashidi A and Mamat R 2015 Effect of nano-particles on the performance and emission of a diesel engine using biodiesel-diesel blend *Int. J. Automot. Mech. Eng.* **12** 3097–108
[11] Sadhik Basha J and Anand R B 2014 Performance, emission and combustion characteristics of a diesel engine using Carbon Nanotubes blended Jatropha Methyl Ester Emulsions *Alexandria Eng. J.* **53** 259–73
[12] Thakur V K and Thakur M K 2015 *Chemical Functionalization of Carbon Nanomaterials: Chemistry and Applications* (CRC Press)
[13] Yetter R A, Risha G A and Son S F 2009 Metal particle combustion and nanotechnology *Proc. Combust. Inst.*
[14] Sadhik Basha J and Anand R B 2011 An experimental investigation in a diesel engine using carbon nanotubes blended water-diesel emulsion fuel *Proc. Inst. Mech. Eng. Part A J. Power Energy* **225** 279–88
[15] Tewari P, Doijode E, Banapurmath N R and Yaliwal V S 2013 Experimental Investigations on a Diesel Engine Fuelled With Multiwalled Carbon Nanotubes Blended Biodiesel Fuels *Int. J. Emerg. Technol. Adv. Eng.* **3** 72–6
[16] Hosseini S H, Taghizadeh-Alisaraei A, Ghobadian B and Abbaszadeh-Mayvan A 2017 Performance and emission characteristics of a CI engine fuelled with carbon nanotubes and diesel-biodiesel blends *Renew. Energy* **111** 201–13
[17] El-Seesy A I, Abdel-Rahman A K, Bady M and Ookawara S 2017 Performance, combustion, and emission characteristics of a diesel engine fueled by biodiesel-diesel mixtures with multi-walled carbon nanotubes additives *Energy Convers. Manag.* **135** 373–93
[18] Rizwanul Fattah I M, Masjuki H H, Kalam M A, Mofijur M and Abedin M J 2014 Effect of antioxidant on the performance and emission characteristics of a diesel engine fueled with palm biodiesel blends *Energy Convers. Manag.* **79** 265–72
[19] Buyukkaya E 2010 Effects of biodiesel on a di diesel engine performance, emission and combustion characteristics *Fuel* **89** 3099–105
[20] Najafi G 2017 Diesel engine combustion characteristics using nano-particles in biodiesel-diesel blends *Fuel* **212** 668–78

A Testing Packet Delay Variation and Packet Loss Problem on Local Area Network Based on ITU-T Standard

Arie Budiansyah[1*], M.Iqbal[2]
[1] Network Peer Group, Informatics Department, University of Syiah Kuala, Indonesia
[2] MikroTik Study Group, Informatics Department, University of Syiah Kuala, Indonesia

*Corresponding author: arie.b@unsyiah.ac.id

Abstract. A good computer network will always be the mainstay of its users. All traffic are aiming to come in/out to device with shortest delay. However, there will be times where our router, switch or access point (AP) start dropping packets because the traffic coming to fast in/out that device can process. The Quality of Service (QoS) computer network traffic is an effort to maintain or manage propagation traffic from source to destination. There are several telecommunication organizations such ETSI and ITU. They have been released recommendation standards. According to TIA/EIA-568-A, the maximum allowed length Cat 5e, Cat 6 cable is 100 m solid horizontal cable between the patch panel and wall jack. However, our research showed the length Cat5e above 130m for connecting router gateway to bridge free from Packet Delay Variation (PDV) and PL problem. The shortest length below 100m does not guarantee will free from PDV and PL problem.

Keyword: Computer network traffic, telecomunication, PDV, PL, ETSI standard, ITU standard

1. Introduction

A good computer network will always be the mainstay of its users because there are benefits as being able to share resources sharing data, information and devices. In terms of efficiency, it is able to save financing like purchasing a device because the devices in the network can be used together. Besides that, the devices in the network can be controlled remotely and are able to be used in field such as management and security information system. Nowadays many computer networks or Local Area Networks (LANs) has been connected to the internet. This situation will absolutely increase the value of resource sharing, efficiency and high reliability of computer networks.

To maintain their high reliability, it is necessary to manage network traffic such as configuration of firewall, routing, Network Address Translation (NAT), Domain Name System (DNS), bandwidth management and so on. All traffic are aiming to come in/out to device with shortest delay but there will be times where our router, switch or access point (AP) start dropping packets because the traffic coming to fast in/out that device can processed. On this situation our streaming are going to suffer both of the situation really need the QoS helps.

The first priority of managing traffics network is to shorten the delay come in/out of the devices. After that, the device was monitored whether it has the ability to process the traffic. High specification of device such as quad core Central Processing Unit (CPU), high memory, or might be Gigaport Ethernet are highly needed. Since it is not always working with perfect situation such low specification of device, finite amount of bandwidth, a lot user in network and other issues, the traffic need to be prioritized.

2. Standard QoS Traffic Telecommunication

The QoS computer network traffic is an effort to maintain or manage propagation traffic from source to destination in shortest delay. There are two telecommunication organization has release recommendation standards. First standard was released by European Telecommunication Standard International (ETSI) who issued a European standards for for global use [1]. The other standard was released by International Telecommunication Union (ITU-T). The study groups of ITU's Telecommunication Standardization Sector (ITU-T) assemble experts from around the world to develop international standards known as ITU-T that recommends an act to define elements in the global infrastructure of information and communication technologies (ICTs). Standards are critical to the interoperability of ICTs and so that whenever the exchange of voice, video or data messages are done, standards may enable global communications by ensuring that countries' ICT networks and devices are speaking the same language [2]. These two telecommunication organizations have released recommendation in telecommunication areas. Several of them have range of response time in millisecond for delay, latency, throughput and so on. By referring the recommendation, we hope interoperability of ICTs meet confidence to deliver a good network service.

As mentioned above, the QoS is the process of managing propagation network traffic between node to node. First of all, traffic testing needed to be done some as sample in order to figure out the real condition of network. In this research, ITU-T recommendation is followed since it was agreed about what ITU-T mentioned about Packet Delay Variation (PDV) definition in recommendation Y.1540 [3].

Since measurement of PDV with kind type of packet selection is not specified in RFC 3393[4] assumption could be made that the packet with impact the largest variation will be in delay in selected time period. The example of the packet such as packet Internet Control Message Protocol (ICMP) traffic, packet User Datagram Protocol (UDP) traffic, packet Transmission Control Protocol (TCP) traffic and so on. ITU-T mentioned that the delay is specified from the start of packet being transmitted at the source to the start of the packet being received at destination, or the packet arrival time at destination instead of the time the end of the packet is received. This situation figures out about the difference between PDV traffic with successive packet.

There are two objective traffic that are used in this research. The first one are PDV traffic and Packet Loss (PL) traffic. Definition of PDV as mentioned in above paragraph. The next objective is PL traffic which occurs when one or more packets traffic from source fail to reach their destination which may be caused by many reason. For example, it might be due to the error transmission on wireless network [5-6] and network congestion [7].

In this research, ICMP and UDP traffic were used to figure out PDV and PL. Both of traffic definitely has largest variation of delay in period of time by the arrangement the size of traffic packet from the smallest to the biggest The ICMP and UDP traffic easily generated using ping and iperf3 application. The main purpose of generated the traffic is to observe propagation of PDV and PL from node to node in network. The summarized of PDV based ITU-T written in table 1 and PL written in table 2.

Table 1. Recommendation For PDV ITU-T Y.114 [8].

Value (ms)	Status
0 < x < 20	Good
20 < x < 50	Accepted
x > 50	Worst

Table 2. Recommendation For PL ITU-T Y.541. [3].

Value (%)	Status
0 < x < 0,5	Accepted
0,5 < x < 1,5	Worst
1,5 < x	Unaccepted

3. iKOS.NET Hotspot Internet

The iKOS.NET Hotspot Internet is an internet hotspot network to obtain internet access for people using Wi-Fi technology. The iKos.NET network implements integration multi-hop wireless and wired network. The Multi-hop routing was chosen due to the focus on efficiency and low cost infrastructure but fast deployed access network in rural and suburban area [8]. Then, the wired node is connected to the radio POE AP 2.4GHz and radio POE 5GHz while some indoor AP has Power Supply Unit (PSU) 12/24V DC is placed inside the house. Figure 1 below shows the topology overview of iKOS.NET Hotspot.

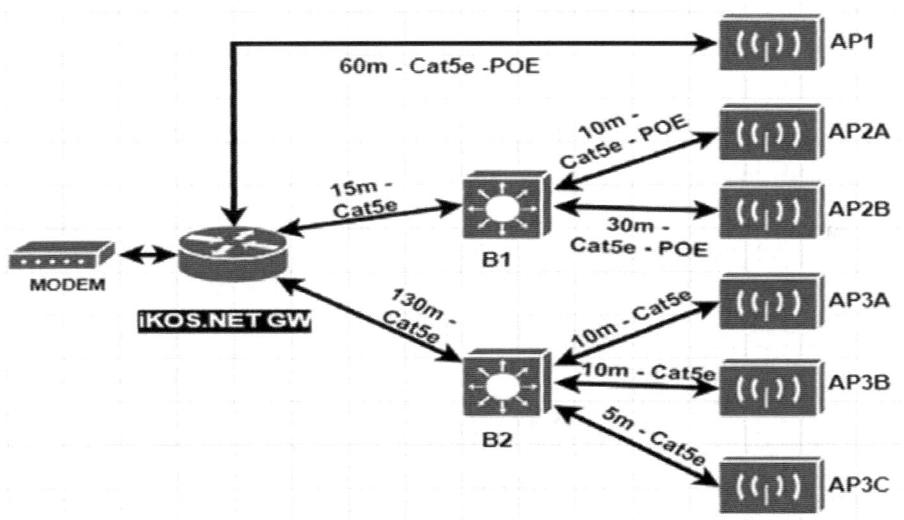

Figure 1. Topology of iKOS.NET Hotspot

4. The QoS Testing Process

The QoS testing is the proses of observing propagation of traffic from source to target destination. The observation is conducted inside iKOS.NET network or LAN because of the needs to check the PDV and PL problem whether they faced our network path or not. This is for having more reliable test. Generally, the QoS process is about calibration process after maintain daily routine.

In order to check PDV and PL problem, ICMP and UDP traffic are generated using ping and iperf3 application. The QoS process was divided into two process. First, ICMP and UDP traffic are generated using ping and iperf3 from user device to the gateway to observe ICMP and UDP PDV. Secondly, UDP traffic was generated from user device to user device to observing PL. Each QoS testing taken in 6 location of AP. Each testing took 5 times to get the average of response time (millisecond).The next process is to correlate the QoS data with ITU-T standard.

The main reason of taking ICMP and UDP traffic are because both of traffic are sensitive to delay. Also, they are vulnerable to being contaminated with daily routine jobs such as configuring firewall, routing, Domain Name System (DNS), and so on. Both of traffic usually propagated simultaneously from node to node. The example of propagation from Gateway to bridge, Gateway to user device, or user device to user device is shown in figure 2. According to ITU-T, the most reliable network usually have small number of PDV, latency and PL.

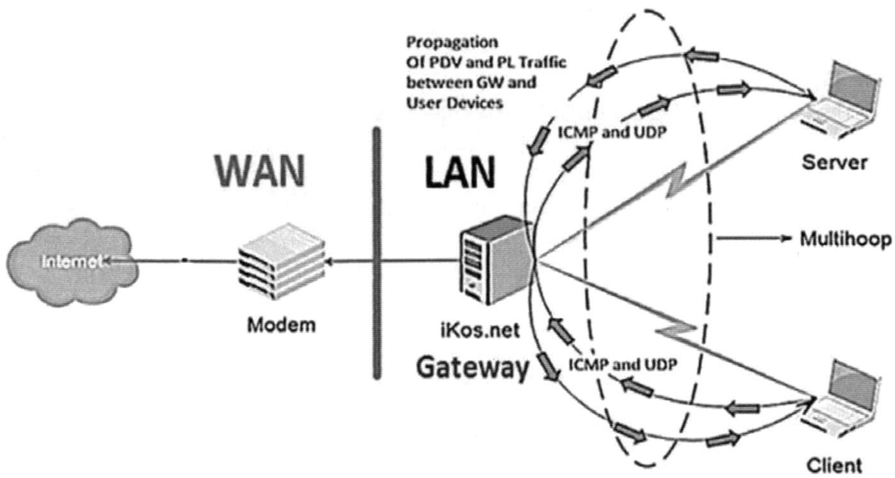

Figure 2. Propagation of ICMP and UDP Traffic Testing inside of iKOS.NET Hotspot.

Table 3 shows a CLI command that sends ICMP traffic from the user's device to the iKOS.NET gateway device. This CLI command is done through the terminal window of the operating system.

Table 3. CLI command that sends ICMP traffic from the user's device to the iKOS.NET gateway device

	Gateway iKOS.NET	User Devices
IP Address	172.16.*.1	172.16.*.*
CLI	ping 172.16.12.1 -n (n) -l (l)	
Example	Ping 172.16.12.1 –n 10 –l 64	
Note:	*-n count* *Number of echo requests to send.* *-l size* *Send buffer size.*	
Explanation	*ping to 172.16.12.1 for 10 times by sending 64 bytes*	

Table 4 shows a CLI command that sends UDP traffic from the user's device acts server to another user's device acts as client on the iKOS.NET network. This CLI command is done through the terminal window of the operating system using the iperf3 software.

IOP Conf. Series: Materials Science and Engineering **506** (2019) 012049 doi:10.1088/1757-899X/506/1/012049

Table 4. CLI command that sends UDP traffic from the user's device acts server to another user's device acts as client on the iKOS.NET network.

	User Devices	User Devices
IP Address	172.16.*.*	172.16.*.*
CLI	iperf3 -s	iperf3 -c 172.16.12.5-u
	-s *as server*	
Note:	*-c* *as client*	
	-u *UDP Type*	
Explanation	*Send UDP traffic to 172.16.12.5*	

The TIA/EIA-568-A [9-10], shows set of telecommunication standard from telecommunication Industry standard (TIA). The length of cable for telecommunication product and services when using 10/100/1000BASET, state the maximum allowed length for Cat 5e, Cat 6 cable to be 100 m (328 ft). This consists of 90 m (295 ft) of solid horizontal cable between the patch panel and wall jack. It is must be 100 m or less in length to certified.

5. The iKOS.NET Propagation Traffics

The summary of all propagation traffic of PDV and UDP QoS from six locations of iKOS.NET AP Hotspot are shown in table 5. Based on the table 5, code P is the abbreviation of Problem and NP is the abbreviation of No Problem. Each AP Hotspot related to their distance to the gateway local.

The radio AP2A, AP2B and AP1 having PDV problem while AP3A, AP3B does not having PDV problem except for radio AP3C that only have PDV UDP problem. The radio AP3B having PL UDP problem. The distance of AP2A, AP2B and AP1 to the gateway is the shortest compared to another iKOS.NET AP. The power supply of AP2A, AP2B and AP1 are using CAT5e POE 12 volt DC but another iKOS.NET AP is using 12 volt DC adaptor power supply.

Table 5. Summarized of PDV and UDP QoS from six location iKOS.NET AP.

	QoS Results			
Area	*PDV ICMP*	*PDV UDP*	*PL ICMP*	*PL UDP*
AP2A **25 m**	P	P	NP	NP
AP2B **45 m**	P	P	NP	NP
AP3A **147,8 m**	NP	NP	NP	NP
AP3B **140 m**	NP	NP	NP	P
AP3C **140 m**	NP	P	NP	NP
AP1 **60m**	P	P	NP	NP

IOP Conf. Series: Materials Science and Engineering **506** (2019) 012049 doi:10.1088/1757-899X/506/1/012049

Based on this findings, the perspective of users in each iKOS.NET Hotspot AP were asked. The users in location of AP2A, AP2B and AP1 felt sometimes the internet connection is disconnected and becomes slower. The users in location of AP3B felt sometimes internet is disconnected but the internet speed is going fast.

From this finding, it can be assumed the shortest length of CAT 5e cabling does not guarantee the connection will be free from PDV and PL problem in network, but those radios, which are having problem; they used POE mechanism to power up the radio. The problem might come from POE mechanism and might be not because of the tested radio AP1 is supplied by 12 volt DC adaptor and the problem still come up. This situation needs to be rechecked if further research is to be conducted. The results if PDV in ICMP and UDP protocol are shown in figure 3 and figure 4 respectively. Figure 5 and 6 shows the results of PL in ICMP and UDP while figure 7 shows the locations of the propagation problems.

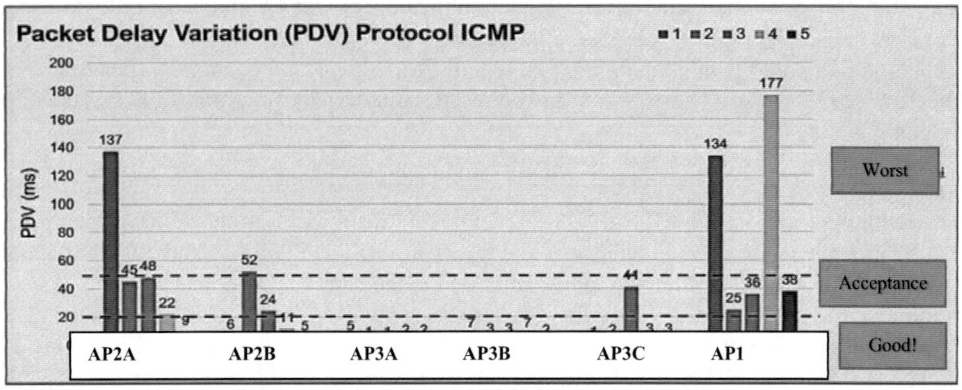

Figure 3. The result of PDV in protocol ICMP.

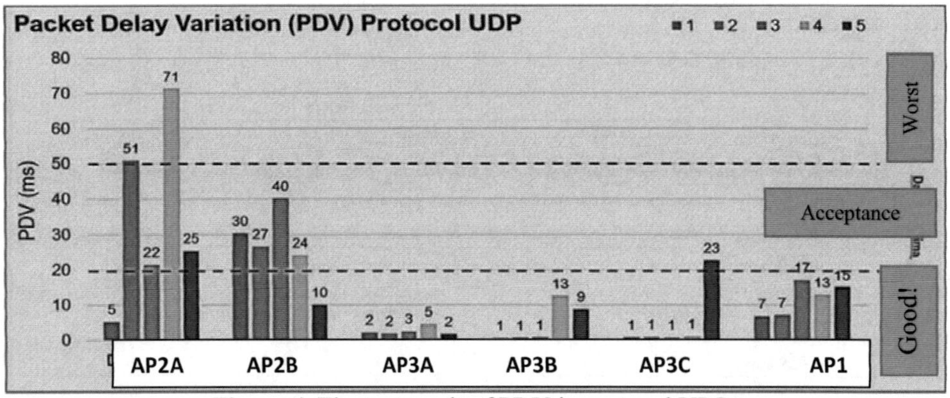

Figure 4. The test result of PDV in protocol UDP.

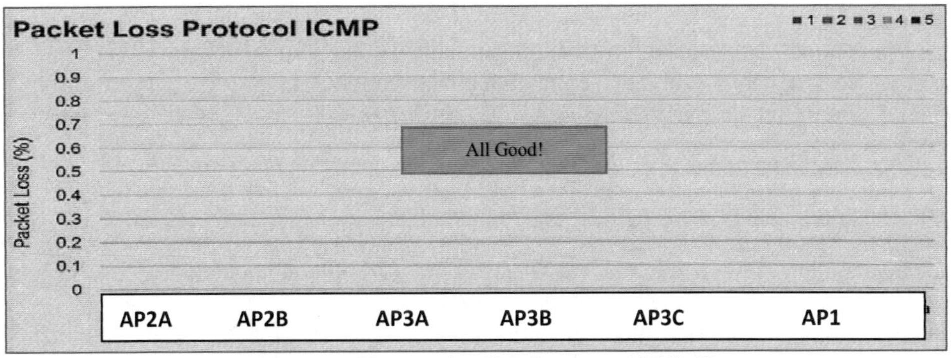

Figure 5. The result of PL in protocol ICMP

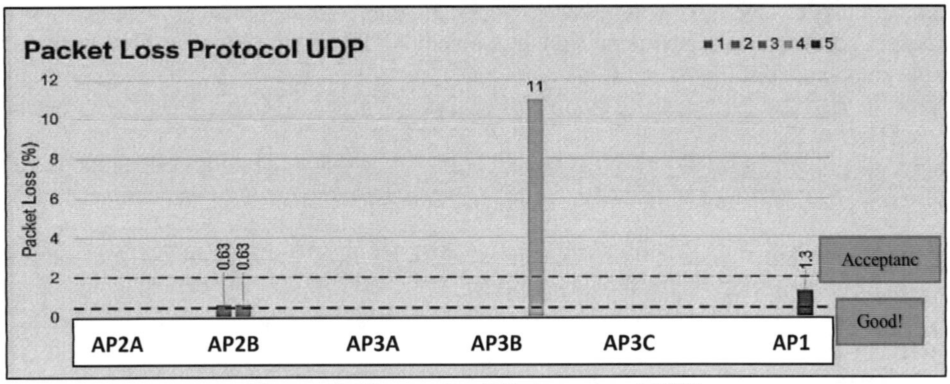

Figure 6. The result of PL in protocol UDP.

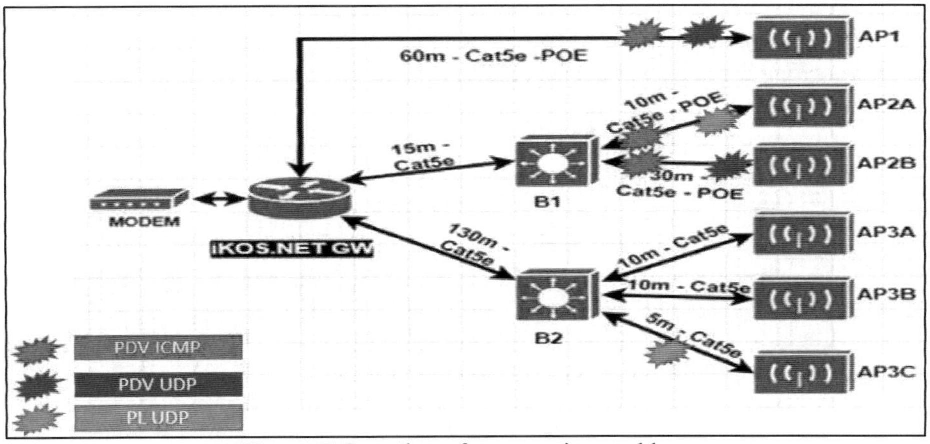

Figure 7. Location of propagation problem.

Acknowledgments
The authors would like thank the informatics department for supporting iKOS.NET Hotspot in Rukoh Gampong as network research. Truly, this research topic was initially from arguing some people about the length of UTP cabling for deliver internet services must below 100 meters otherwise each device wont connected.

References
[1] Dahmen-Lhuissier S 2019 ETSI - About us, what and who we are, what we do - standardization work activities, how we work, annual report, work programme (online) ETSI Available at *https://www.etsi.org/about* (Accessed 19 Nov 2018)
[2] Itu.int 2019 ITU Telecommunication Standardization Sector (Online) Available at *https://www.itu.int/en/ITU-T/Pages/default.aspx* (Accessed 19 Nov 2018)
[3] Itu.int 2011 ITU-T recommendation Y.1541: Series Y: Global information infrastructure, internet protocol aspects and next-generation networks: Internet protocol aspects - quality of service and network performance: Network performance objectives for IP-based service (Geneva) *International Telecommunication Union* pp 66
[4] Demichelis C and Chimento P 2002 IP Packet Delay Variation Metric for IP Performance Metrics (IPPM) *RFC Editor*
[5] Salyers D C Striegel A D and Poellabauer C 2008 Wireless reliability: Rethinking 802.11 packet loss *2008 International Symposium on a World of Wireless, Mobile and Multimedia Networks.* (IEEE)
[6] Ye T Kai X and Ansari N 2005 TCP in wireless environments: problems and solutions *IEEE Communications Magazine* vol 43(3) pp S27–S32
[7] Kurose J and Ross K 2013 Computer networking: A Top-down Approach (Sixth Edition) *Pearson*
[8] Bejerano Y 2004 Efficient Integration of Multihop Wireless and Wired Networks With QoS Constraints *IEEE/ACM Transactions on Networking* vol 12(6) pp 1064–1078
[9] ANSI/EIA/TIA-568 2001 Commercial Building Telecommunications Cabling Standard Part 1: General Requirements *SpringerReference*
[10] TIA Publishes New Cabling Standards Designed to Improve Efficiency for Designers, Installers and End Users (2009) *CPI-Chatsworth Product, Inc.*

The Bending Stress on GFRP Honeycomb Sandwich Panel Structure for a Chassis Lightweight Vehicle

E S Arbintarso[1,2*], H F Datama[2], R N W Aviyanto[2], B Prasetya[2], U S A B Purwokusumo[2], B Aditiawan[2], P P Sangkoyo[2], and R B S Adi[2]
[1]Jurusan Teknik Mesin, IST AKPRIND Yogyakarta, Jl. Kalisahak 28, Yogyakarta 55222, Indonesia
[2]Tim IST AKPRIND 1, Kalisahak28 Urban Concept Team, Laboratorium Pusat IST AKPRIND, Jl. I Dewa Nyoman Oka 32, Kotabaru, Yogyakarta 55232, Indonesia

[*]Corresponding author: ellyawan@akprind.ac.id

Abstract. This paper discusses the bending stresses of the hexagonal honeycomb sandwich panel structure made of glass fiber reinforced polymer (GFRP) composite. The 1D axial hexagonal honeycomb sandwich panel structure consists of a honeycomb core and thin wallsheets. The plastic steel epoxy resin, polyaminoamide-bisphenol-A resin, and thermosetting resin were used as adhesive in every layer of the sandwich panel structure. The initial design is based on specific requirements that the panel will use as a chassis of the lightweight vehicle to achieve better fuel efficiency (Urban Concept with maximum vehicle weight is 300 kg). The axial load was dominantly on the chassis. The specimen is tested using the universal testing machine by three points bending method. The analysis is carried out through the response of adhesive between honeycomb core and wallsheets. The results show the bending stress of the honeycomb sandwich panel structure with three adhesives is more than enough for a lightweight chassis vehicle.

1. Introduction

Honeycomb construction has been used extensively in the automotive, marine and aviation industries, as a structure with high flexibility and lightweight [1-3]. In general, honeycomb construction is a sheet used as a cover panel. The panel consists of a honeycomb core, and two surface layers call facesheets. The panel is only allowed to accept bending and shearing load, in the other hand this honeycomb construction becomes weak against the impact and buckling loads that usually occur in axially loaded construction [4].

The need for lightweight materials for rigid construction is increasing. Examples of rigid construction are chassis and bumper vehicles on energy-efficient future vehicles [5]. One key to fuel savings is to reduce the weight of the vehicle. This research aims to investigate the lightweight construction (honeycomb structure) which accept axial loads like a rigid structure. Fiber reinforced polymer composite materials have been shown to have high strength and low weight. The weakness of this material needs a complicated manufacturing process and have high flexibility. Therefore, they required a method to increase the stiffness. There is still a limited amount of data about the axial load capability of a honeycomb sandwich panel structure. The construction alternative was needed to support the demand for lightweight materials that can be used in construction with axial and bending loads.

The results of this study can be used as a reference in using the honeycomb structure in general and especially on lightweight chassis vehicle. This research is important to get answers on how the ability of the honeycomb structure can be loaded with the axial and bend loads safely.

1.1. Honeycomb structure

The honeycomb is a structure formed from the arrangement of hexagonal patterns [6] at first (now can be made of rectangular [7-8], polygonal and curve [3] patterns). The honeycomb structure is generally used as a core of panel construction. The honeycomb core material has been produced with a variety of shapes and types of materials (metals, plastics, ceramics, and composites). The honeycomb structure is commonly used on sheet panels, where the structure is very strong and low weight. One disadvantage of using this honeycomb structure is susceptible to bending loads because the Poisson's ratio is positive, usually negative [9]. The simple form of the honeycomb structure as shown in figure 1 below.

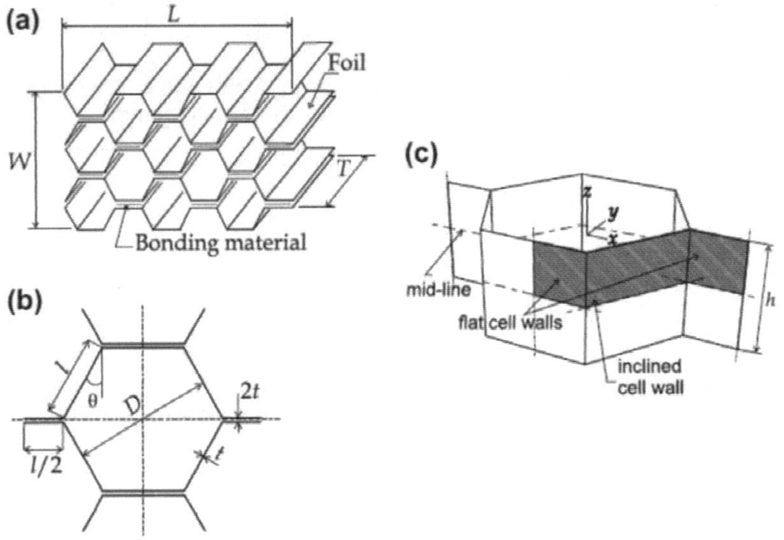

Figure 1. (a)Honeycomb core structure, (b)general view, (c)unit cell, (d)the unit cell parameters [10].

The process of making composites with glass fiber as a honeycomb core is almost the same as making composite sheets in general. The difference in the process of making this honeycomb is the hexagonal shape. Two of the hexagonal half-finished (corrugated) sheet is glued together with polymer adhesive and pressed into a honeycomb construction (hexagonal). The sequence of processes can be seen in figure 2 below:

(a) (b)

Figure 2. (a)Hexagonal half-finished sheet, (b)final honeycomb core

The honeycomb core provides out-of-plane stiffness to carry through-thickness compression and shear loads [9-10]. The overall in-plane stiffness of the structure also increases but ignored because less than the facesheets. The in-plane stiffness prevents the honeycomb core from bowing outwards as wrinkles form in the facings [10]. The core will behave like a spring in extension, without facesheets, and has no significant stiffness. The facesheets increases the effective in-plane stiffness [10]. The in-plane deformation of a cell is straightforward, without facesheets and the honeycomb cores cell walls will bend when they are loaded in-plane, giving linear elastic behaviour.

1.2. Bending stresses

Consider a simply supported beam with the total length L_f, span length L, thickness h and width b loaded in three-point bending, as shown in figure 3. The three-point bending used cylindrical loading head and supports.

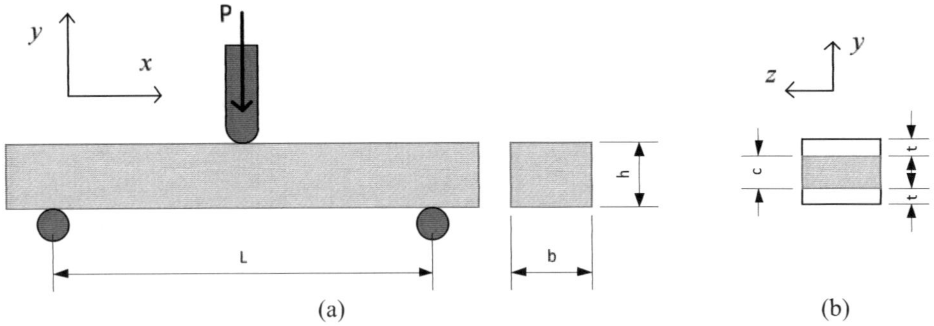

Figure 3. (a)A sketch of a pure beam loaded in three-point bending, (b)cross-section of the sandwich beam.

An expression for normal stress (σ_x) regarding the applied bending moment (M_b) is:

$$\sigma_x(y) = \frac{M_b(x).y}{I} \tag{1}$$

Where I is a moment of inertia about the z-axis of cross-section beam and y is vertical distance from the neutral axis. A composite beam has two or more elemental structural forms, that can be different or same materials, bonded or joined together. If the composite has two different property of the material, for instant E_f is Young's modulus for the facesheet (note in subscript f), and E_c is Young's modulus for the core (note in subscript c), then the normal stress of facesheet and core were written below:

$$\sigma_{xf} = -E_f.(y/\rho) \; and \; \sigma_{xc} = -E_c.(y/\rho) \tag{2}$$

$$M_b(x) = \frac{[E_f I_f + E_c I_c]}{\rho} \tag{3}$$

1.3. Chassis of lightweight vehicle

The lightweight vehicle is a concept in the automotive industry to build vehicles that are less heavy as a way to achieve better fuel efficiency and handling without lessen overall safety for driver and passengers. Automotive manufacturers are constructing lightweight vehicles for two reasons, i.e., meeting vehicle emissions standards and meeting fuel efficiency standards [11]. The chassis is the mainframe in a vehicle that supports all the components and occupants. The chassis should be rigidity in bending and torsion. Most of the chassis has a 24% of total weight [12] and reducing the weight of chassis means less mass on the moving axle parts that give a better engine response of the vehicle to the road. The new fiberglass reinforced polymer composite material will reduce the weight of the chassis more than 30% compared to the steel [12].

The honeycomb sandwich panels have been used for structural applications and performance automotive industries. The honeycomb is one of lightweight core material types, which have been used in various combinations with skins of fiber reinforced polymer composite as well as aluminium [13]. Honeycomb is a part of a composite – a core material, most of them lightweight that maintains the distance between two facesheets, which are stiff concerning in-plane extension and contraction [10]. Table 1 shows the properties of aluminium (5052), aramid fiber/phenolic resin (HRH), and fiberglass/phenolic (HFT) honeycomb core.

Table 1. The density, compressive and plate shear (W direction) strength of hexagonal honeycomb core (adopted from [14]).

Type	Cell size (inch)	Density (lb/ft^2)	Compressive Strength (lb/in^2) [MPa]	Plate shear Strength (lb/in^2) [MPa]
5052	1/8	3.1	285 [1.96]	130 [0.89]
	3/16	2.0	160 [1.10]	70 [0.48]
	3/8	3.0	285 [1.96]	125 [0.86]
HRH	1/8	3.0	300 [2.07]	100 [0.69]
	3/16	2.0	120 [0.83]	60 [0.41]
	3/8	3.0	290 [2.00]	95 [0.65]
HFT	1/8	3.2	310 [2.14]	95 [0.65]
	3/16	2.0	170 [1.17]	60 [0.41]
	3/8	4.0	500	195

2. Materials

2.1. Glass Fiber Reinforced Polymer (GFRP)

GFRP is a composite material that uses a glass fiber as a fiber and a polymer as a matrix. GFRP composite was widely used in the manufacture of composite material as aircraft components, marine applications, piping, automotive components, and industrial constructions. Table 2 shows the most important of glass fiber types from a tensile strength perspective.

Table 2. Common Glass Fiber Types Used in GFRP [13]

Glass type	Description
E	Alumino-borosilicate family of glasses, use in most GFRP
E-CR	Corrosion-resistant E-glass, better mechanical properties than E
R	High-strength glass with performance intermediate to E and S
S	S-glass was developed for high strength, modulus superior thermal and corrosion performance. A family of glasses composed primarily of the oxides of magnesium, aluminium, and silicon.

The glass fibers were formed or fiberized by pulling fibers through tiny orifices in a precious metal bushing then rapidly cooled, gathered, coated with sizing, and in the case of continuous fibers, fed to a winder that configure to product roving fibers [13]. Woven roving (WR) structure firstly has used on composite armor type. Rovings were kept independent of each other until reached the loom to minimize interaction between rovings [13].

WR 200 is a glass fiber woven roving fabric that made of E-glass direct roving by bidirectional interweaving and has an area weight of 200 g/m^2 [15]. It was suitable for hand lay-up, mold pressing, pultrusion, filament winding and continuous laminating processes to manufacturing automotive, industrial, and marine components [15]. The features of WR 200 are compatible with multiple resin systems as a matrix, excellent corrosion resistance, moldability and laminate property, and easy operation with less resin consumption. WR 200 has a tensile strength of 820 N on Warp (machine direction) and 700 N on Weft (cross-machine direction) each 25 mm length [15].

2.2. Adhesive Bonding

The strength of adhesively bonded honeycomb sandwich panel structure is the key to mechanical behaviour and performance in-service of this structure. This bond must exist between the surface of a facesheet and the edge of a thin flexible cell wall of the honeycomb core. The joint is known as a tee or butt joint that can be effective in transmitting the shear and normal loads for sandwich panel structure [16]. The adhesive can be classified by the way they set or by chemical type. Some of the classes are Toughened Acrylics, Epoxies, and Polyurethanes [17]. The Toughened acrylics can be used in minimal surface preparation and bond well to a wide range of materials. They are fast curing and offer high strength and toughness [17].

Epoxy adhesive is common as plastic steel adhesive consists of an epoxy resin and the hardener. This adhesive can be used to join most material and have good strength, low shrinkage and do not produce volatiles during curing [17]. Polyurethane adhesive provides strong impact-resistant joints and has better low-temperature strength than any other adhesive [17]. The adhesives are useful for bonding glass fiber reinforced polymer because of their fast curing.

In this study, the type of adhesives are selected to be plastic steel epoxy resin (Devcon), polyaminoamide-bisphenol-A resin, and thermosetting resin to study the effect of the type of adhesive on the bending strength of the studied specimens, and the corresponding specimens were named types A, B, and C specimens, respectively.

2.3. Honeycomb Sandwich Panel

Generally, honeycomb sandwich panel structures consist of three layers: the high-modulus high-density facesheets, the low-density core, and the core-to-facesheet bonding adhesives. The facesheets, in the sandwich panel structures, are spaced to provide most of the in-plane tensile, compressive loads and bending rigidity, i.e., the applied edgewise loads and flatwise bending moments (figure 4(a) and 4(b)).

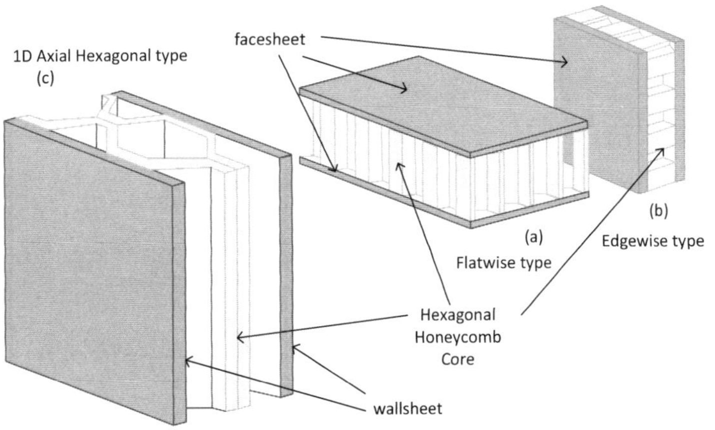

Figure 4. Honeycomb sandwich section (a)flatwise, (b)edgewise, (c)1D axial hexagonal types.

The core is designed to keep the facesheets (or wallsheets) a desired distance apart and transmits shear between them and provides most of the shear rigidity of the construction. The adhesive that bonding them together must be adequate to transfer the stresses from the facesheets (or wallsheets) to the core.

IOP Conf. Series: Materials Science and Engineering **506** (2019) 012050 doi:10.1088/1757-899X/506/1/012050

3. Experimental Setup

3.1. GFRP composite characterization

GFRP composite was made with WR 200 glass fiber and Eposchon Bisphenol A-epichlorohydrin epoxy resin mixed with Eposchon Polyaminoamide epoxy hardener matrix. Hand lay-up method was used during the composite molding with three-layer each composed. The hexagonal honeycomb core manufactured by slotting together woven glass fiber sheets reinforced polymer composite of thickness t_c = 3 mm and density (ρ) =1216 kg/m^3. The wallsheet comprised glass fiber WR200 (of diameter 5.5–6.5 µm) made of E-glass direct roving by bidirectional interweaving. The fibers were embedded within an Eposchon polyaminoamide-bisphenol-A epoxy resin. The two wallsheets manufactured from the same material of the core, three plies were stacked to obtain wallsheet thickness about 3 mm, and the density (ρ) is 751 kg/m^3. By bonding two thin wallsheets to a thicker core material, a layered 1D axial honeycomb sandwich panel structure is formed (figure 4(c)).

3.2. Three points bending test

The test specimens are made rectangular in cross-section. The width (*b*) of the specimen is taken as 45 mm, and the height (*h*) of the specimen is selected as 50, 90 and 180 mm, where the 180 mm height is the maximum height of the chassis of the lightweight vehicle. The length (L_f) of the specimen is taken as 240 mm where 60 mm is the total unsupported length, and span length (*L*) is 180 mm as seen in Figure 5. All the axial edgewise bending tests were carried out at a loading rate of 5 mm/min on a Hung Ta universal testing machine.

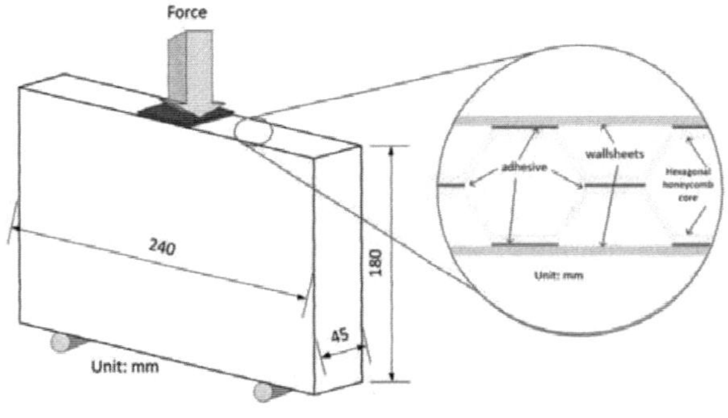

Figure 5. Schematic drawing of three-point bending of edgewise honeycomb sandwich panel.

The moment of inertia of hexagonal hollow is determined by:

$$I_h = \frac{A}{12}\left[\frac{d^2(1+2c\ ^2 30^o)}{4cos^2 30^o}\right] \quad (4)$$

$$A = \frac{3d^2 tan30^o}{2} \quad (5)$$

$$y = \frac{d}{2cos30^o} \quad (6)$$

$$Z = \frac{I}{y} = \frac{A}{6.9}\left[\frac{d(1+2co\ ^2 30^o)}{4co\ ^2 30^o}\right] \quad (7)$$

Where *d* is the distance from flat to the flat of hexagon shape (cell size), *A* is an area of hexagon and *Z* is section modulus. The distance cross-section of hexagon shape is 40 mm, then the result of I_h, *A* and

Z are 153,600, 1385,6 mm^2 and 6656, respectively. Equation (3) shows the bending moment theoretical when in practice, the applied bending moment is calculated with:

$$M_b(x) = \left(\frac{P}{2}\right)\left(\frac{L}{2}\right)\left(\frac{h}{2}\right) \tag{8}$$

$$\sigma_x(y) = \frac{M_b(x)}{Z} = \frac{P.L.h}{8\,Z} \tag{9}$$

4. Results and Discussion

1D axial hexagonal bending load-displacement curves of (A) plastic steel epoxy resin (Devcon), (B) polyaminoamide-bisphenol-A resin, and (C) thermosetting resin are shown in Figure 6. There is three specific height of specimens, i.e., 50 mm, 90 mm, and 180 mm. Figure 6 indicates the load time the displacement plots for different adhesive bonding at a specific height of 180 mm. The bending results demonstrated similar load profiles for all adhesive bondings. The ultimate load for every profile has different displacement; the longer displacement indicates the higher the modulus. All of the specimens have yield point at above 1000 N.

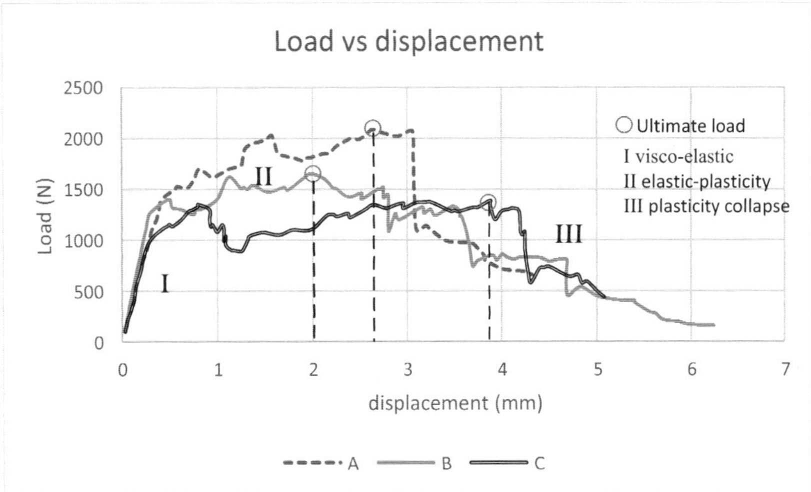

Figure 6. The load-displacement curve of honeycomb sandwich panels subjected to the static three-point bending test with different adhesive bonding.

The curves of 1D axial hexagonal bending load can be divided into three stages: first, visco-elastic stage, load increase almost linearly with displacement until the critical yield point; second, elastic-plasticity stage, is indicates by fluctuating load with displacement increase; third, plasticity collapse stage, the load decreases as displacement increases. Burlayenko and Sadowski [18] mentioned the optimum strains as large as 10% in bending where the honeycomb core behaves like a spring, and only small strains happened in the beam, although the geometry allows large distortion of the structure. When the strain reached 10%, the core begins to exhibit non-linear behaviour, and then the deflection curve increasing rapidly [18].

There are three failure modes appear on the bending of the honeycomb sandwich panel in this study, i.e., wall crushing (WC), wall debonding (WD), and core crushing (CC). The adhesion strength between the hexagonal honeycomb cores and wallsheet appeared to be the limiting factor in several cases; debonding (figure 7(a)) is the dominant failure mode.

Table 3. Failure loads and collapse modes.

Adhesive Type	height (h) (mm)	Displacement (max) (mm)	Fail Force (max) (N)	Fail Mode
A	50	3.45	273.0 ± 6.96	CC[1], WC
A	90	3.68	596.2 ± 16.87	CC[1], WD
A	180	4.32	2114.8 ± 29.63	CC[1], WD
B	50	4.44	243.4 ± 3.05	CC[1], WC
B	90	5.16	458.2 ± 9.55	WD
B	180	6.25	1682.6 ± 20.98	CC[1], WD
C	50	3.55	169.0 ± 4.47	CC[1], WC
C	90	4.24	378.2 ± 6.02	CC[1], WD
C	180	5.08	1391.6 ± 4.28	CC[1], WD

*WC = wall crushing, WD = wall debonding, CC = core crushing, [1]at the bottom

Table 3 shows the failure loads and collapse modes that occur on 1D axial honeycomb sandwich panels. Almost all the beams failed by the core crushing mode, especially at the bottom panel (figure 7(b)), and some wall crushing mode (figure 7(c)) can occur at the same time. The strength of the hexagonal honeycomb cores is much greater than the adhesive and the thin wallsheet since there is less failure in the honeycomb core materials.

(a) (b) ©

Figure 7. Failure modes on honeycomb sandwich panels (a)WD, (b)CC, (c) WC.

Debonding is the most dominant failure mode in the experiments for axial edgewise hexagonal honeycomb sandwich panels due to the higher shear strength of core materials compared to that of the adhesive. Figure 8 shows the applied bending stress on axial edgewise honeycomb sandwich panels with a different specific height of the specimens. The result shows that increase the height of the specimen will increase the bending stress, and the plastic steel epoxy adhesive (A) has higher strength than the others.

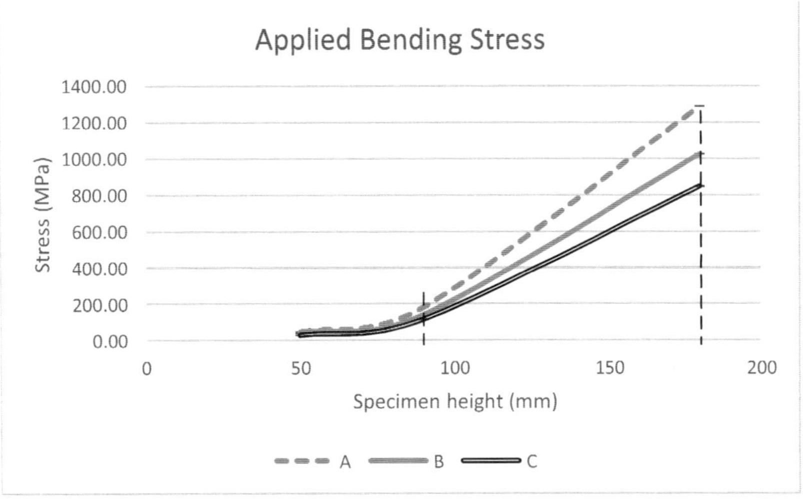

Figure 8. The applied bending stress of 1D axial hexagonal type.

Kalisahak28 is ethanol urban concept car (city car) that developed by IST AKPRIND 1 team. The main goals to design and develop the Kalisahak28 are a part of fuel efficiency and use renewable energy, also a part of to fulfill the regulation of fuel efficiency world competition lightweight vehicle. The chassis Kalisahak28 and the illustration of full-frame chassis are shown in Figure 9.

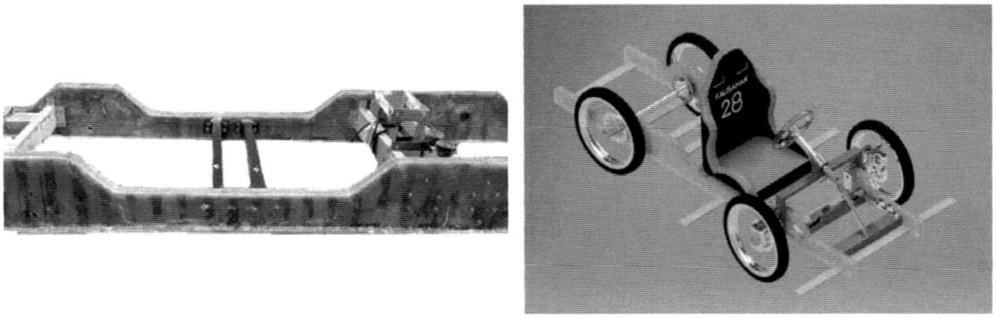

(a) (b)

Figure 9. (a)The honeycomb chassis of Kalisahak28 vehicle, (b)design of Kalisahak28 chassis.

Figure 10 shows the bending load case for Kalisahak28 structure. The loads that cause this bending load case scenario are body car weight (Wfb), driver weight (Wd), and engine weight (We). In this case, the bending and shear loads are calculated only on one frame, where loads of each part are assumed homogeneous and a half of their weight.

The experimental results indicate that all of the 90 mm and 180 mm height specimens can handle more than 378 N and 1391 N, respectively. The bending stress reached a minimum of 0.46 MPa and 1.71 MPa for 90 mm and 180 mm height specimens, respectively. It means all of the type adhesives can be used in this chassis. The best adhesive is A type (plastic steel epoxy).

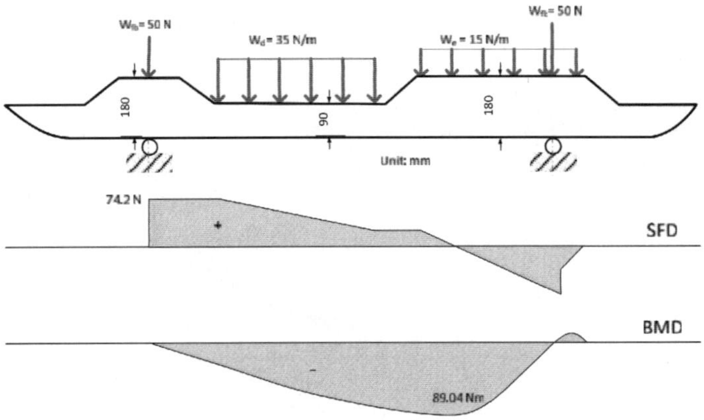

Figure 10. Bending moment diagram (BMD) and Shear force diagram (SFD) of Kalisahak28 Chassis.

This result is compliance with Manalo et al. [19] and Zhang et al. [20], wherein beams and similar applications, structural components are used in the edgewise orientation for higher strength and stiffness, the composite sandwich beams tested in the edgewise position failed at a higher load with less deflection.

5. Conclusions

The 1D axial hexagonal honeycomb GFRP composite sandwich panels with hexagonal honeycomb cores have been designed and manufactured using adhesion method. Three-point bending test was carried out to study the bending stress of the composite sandwich panels. Failures and theoretical static loads were analysed and calculated to predict the bending stress response all of the specimens with different adhesive type. The bending stress of 1D axial hexagonal honeycomb sandwich panel with three adhesives (plastic steel epoxy, polyaminoamide-bisphenol-A resin, and thermosetting resin) is more than enough for Kalisahak28 lightweight chassis vehicle.

Acknowledgment

The authors would like to thank profusely to Institut Sains & Teknologi AKPRIND that has given funding part of the research activities, and to all member of Kalisahak 28 Urban Concept Team for all of their support during the project.

References

[1] Cahyono S I, Widodo A, Anwar M, Diharjo K, Triyono T, Hapid A and Kaleg S 2016 Light-weight sandwich panel honeycomb core with hybrid carbon-glass fiber composite skin for electric vehicle application *AIP Conference Proceedings* **1717** 040025 doi.org/10.1063/1.4943468

[2] Crupi V, Epasto G and Guglielmino E 2013 Comparison of aluminium sandwiches for lightweight ship structures: Honeycomb vs. foam *Marine Structures* **30** pp 74-96 doi.org/10.1016/j.marstruc.2012.11.002

[3] Zhang Q, Yang X, Li P, Huang G, Feng S, Shen C, Han B, Zhang X, Jin F, Xu F, and Lu T J, 2015 Bioinspired engineering of honeycomb structure - Using nature to inspire human innovation *Prog. Mater. Sci.*, **74**, pp. 332–400

[4] Matta V, Kumar J S, Venkataraviteja D and Reddy G B K 2017 Flexural Behaviour of Aluminum Honeycomb Core Sandwich Structure *IOP Conf. Ser.: Mater. Sci. Eng.* **197** 012046 doi.org/10.1088/1757-899X/197/1/012046

[5] Fan J and Njuguna J 2016 An introduction to lightweight composite materials and their use in transport structures *(Lightweight Composite Structures in Transport Design, Manufacturing, Analysis and Performance)* ed J Njuguna (Woodhead Publishing, Duxford UK) Part one chapter 1 p 28

[6] Zhang T and Ding K 2013 A new proof of honeycomb conjecture by fractal geometry methods *Front. Mech. Eng.* **8** (4) pp 367–370 doi.org/10.1007/s11465-013-0273-7

[7] Jiménez L F and Triantafyllidis N 2013 Buckling of rectangular and hexagonal honeycomb under combined axial compression and transverse shear Dedicated to Prof. S. Kyriakides on the occasion of his 60th birthday *Int. J. Solids Struct.* **50** (24) pp 3934–46.

[8] Ansari Z, Tan C W, Rejab M R M, Bachtiar D, Siregar J, Zuhri M Y M and Marzuki N S D M 2017 Crushing behavior of composite square honeycomb structure: a finite element analysis *Journal of Mechanical Engineering and Sciences* **11** (2) pp 2637-49 doi.org/10.15282/jmes.11.2.2017.7.0241.

[9] Correa D M, Seepersad C C and Haberman M R 2015 Mechanical design of negative stiffness honeycomb materials *Integr. Mater. Manuf. Innov.* **4** (1) pp 1-11 doi.org/10.1186/s40192-015-0038-8

[10] Landesmann A, Seruti C A and Batista E de M 2015 Mechanical Properties of Glass Fiber Reinforced Polymers Members for Structural Applications *Mater. Res.* **18** (6) pp 1372–83 doi.org/10.1590/1516-1439.044615

[11] Rodgers M 2016 *How lightweight vehicles are transforming the automotive industry* Mentor Works Ltd https://www.mentorworks.ca/blog/market-trends/automotive-manufacturing-lightweight-vehicles/ access at 04/07/2018

[12] Deisser O, Friedrich H E and Kopp G 2014 Fibre reinforced plastic concepts for structural chassis parts *Transport Research Arena* pp 1-10 doi.org/10.1.1.843.6687

[13] Sathishkumar T P, Satheeshkumar S and Naveen J 2014 Glass fiber-reinforced polymer composites - A review *J. Reinf. Plast. Compos.* **33** (13) pp 1258–1275 doi.org/ 10.1177/0731684414530790

[14] Hexcel Corporation 2016 HebWeb® honeycomb attributes and properties *Manuals* http://www.hexcel.com/user_area/content_media/raw/HexwebHoneycombAttributesandProperties.pdf

[15] Sichuan Chang Yang Composites Company, "CYC fiberglass woven roving (ECY-WR)," 2018. [Online]. Available: http://www.cyccomposites.net/sale-8446015-cyc-fiberglass-woven-roving-ecy-wr.html. [Accessed: 30-Jun-2018].

[16] Keshavanarayana S R, Shahverdi H, Kothare A, Yang C, and Bingenheimer J, The effect of node bond adhesive fillet on uniaxial in-plane responses of hexagonal honeycomb core, *Composite Structures*, **175** pp. 111–22 doi.org/10.1016/j.compstruct.2017.05.010

[17] Composites UK, "Adhesive bonding of composites," 2018. [Online]. Available: https://compositesuk.co.uk/composite-materials/public-library/documents. [Accessed: 30-Jun-2018].

[18] Burlayenko V N and Sadowski T 2010 Effective elastic properties of foam-filled honeycomb cores of sandwich panels *Composite Structures* **92** pp 2890-2900 doi.org/10.1016/j.compstruct.2010.04.015

[19] Manalo A C, Aravinthan T, Karunasena W and Islam M M 2010 Flexural behaviour of structural fibre composite sandwich beams in flatwise and edgewise positions *Composite Structure* **92** (4) pp 984-995 doi.org/10.1016/j.compstruct.2009.09.046

[20] Zhang F, Liu W, Fang H, and Jia Z 2018 Flexural behavior of composite sandwich beams with different kinds of GFRP ribs in flatwise and edgewise positions *Composites Part B: Engineering* **156** pp. 229–39

Identifying Determinants of Child Malnutrition Using Spatial Regression Analysis

K M Zurnila[1*], M Saiful[1] and M Selvi[1]

[1]Statistics Department, Faculty of Mathematics and Natural Science, Syiah Kuala University, Banda Aceh, Indonesia

[*]Corresponding author:kesumaku@yahoo.com

Abstract. As a public health problem, the nutritional status of childrens under five years old is a sensitive indicator of a country's health status. Malnutrition affected by many factors, including spatial dependence factors. This factor indicates the value of observations from a region affected by the value of observations in other areas. It assumed that child health factor, maternal health, environmental health, and region influenced the prevalence of malnutrition among under-five children. The objective of the study was to determine factors that associated with malnutrition among under-five children in Sumatera Island and the spatial regression model for the case. The study used Basic Health Research (Riskesdas) and Public Health Development Index (IPKM) survey 2013 from 125 districts on Sumatera Island. . Spatial regression analysis and Spatial Autoregressive Model (SAR) applied in this study to obtain the significant determinants of child malnutrition. The results of this study revealed that the lack of chronic energy in women of childbearing age, the proportion of family planning users, households that having clean and healthy life behaviour (PHBS), and access to clean water were the significant factors of child malnutrition.

Keywords: malnutrition, Sumatera Island, spatial dependence, Spatial Autoregressive Model (SAR)

1. Introduction

Nutrition problem is a public health problem that can occur in all age groups. The most age group that should be concerned is 0 to 5 years old age group (infants and toddlers). The underweight condition of children under five causes them having a higher risk factor in the face of growth disorders than the adult age group. The disruption of growth during this period cannot be fulfilled in the future and will negatively affect the quality of future generations [1].

Nutrition status of children under five measured by measuring height and weight. These indicators based on body weight, height and ages mostly used to assess the population nutritional status. The nutritional based on the anthropometry status of children fewer than five could be divided into three indices, namely stunting (length or height not appropriate for age), wasting (unsuitable weight to height) and underweight (lack of weight against age). In this study, the nutritional status used was malnutrition or underweight based on body weight index by age [1].

By 2015, World Health Organization (WHO) found 250 from 667 million children in the world suffering from malnutrition. Public health problem is considered serious if malnutrition prevalence less between 20-29% and considered very high prevalence if ≥ 30%. Prevalence is the number of incidence of disease within one year compared with the population. The Nutrition Status monitoring survey 2016 [2] conducted at 514 districts in 34 provinces of Indonesia showed that malnutrition prevalence in

Indonesia was 17.8%. Based on WHO criteria, child malnutrition in Indonesia is a public health problem that fall into the medium category. While on the Sumatra Island, there are three provinces that have child malnutrition prevalence less than the national prevalence figures namely Aceh, North Sumatra and Riau Province [1, 3].

Generally, statistical analysis that describes the cause and effect relationship between one dependent variable with one or several independent variables is linear regression analysis. One type of data used in linear regression analysis is cross-sectional data. Cross-sectional data can also be referred as spatial data or area data, which is data that not only states the value of an observation, but also indicates the location where the data resides. There is possibility of spatial dependence in the data meaning that observations in a region can influence by observations in other areas. As Tobler states that everything is related to one another, but something adjacent is more influential [4]· Linear regression analysis ignores the existence of this spatial dependence so that the data analysis for spatial data is a spatial regression analysis. The fundamental component of the spatial model is the spatial weighted matrix reflecting the relationship between one region and another. This area aspect is important to be studied because inter-region certainly has different characteristics [4, 5].

To determine the factors that influence child malnutrition on Sumatra Island by considering aspects of the region, the authors conducted a study to identify child malnutrition using spatial regression analysis.

2. Spatial Regression Analysis

The first stage of the study was to determine spatial association among areas or districts. Spatial associations or spatial autocorrelation indicate an inter-regional relationship to observational values. There are several methods used to detect spatial autocorrelation, they are spatial weighted matrix, Global and Local Moran Index and Moran's spotted plot [6].

Spatial weighted matrix (**W**) is a matrix that shows the inter-regional relationships in the observation. The adjacent area called as neighbour. The general form of the **W** matrix is as equation (1) follows:

$$\mathbf{W} = \begin{pmatrix} w_{11} & \cdots & w_{1n} \\ \vdots & \ddots & \vdots \\ w_{n1} & \cdots & w_{nn} \end{pmatrix} \quad (1)$$

Where w_{ij} denotes the spatial weights for the region-i and region-j, with $i = 1, 2\ldots n$ and $j = 1, 2\ldots n$. Each element in row i is 0 for a region which is not a neighbour of region i, and 1 for a region which is a neighbour of region i. Based on that definition, it is not allowed region i to be a neighbour against itself. Therefore, the diagonal element in the matrix **W** is 0 [7].

There are several methods to express the relationship structure of the region using a spatial weighted matrix (**W**). One of the methods is the k-nearest neighbours method used for areas case. The number of nearby areas (neighbours) selected using the distance approach. Furthermore, the **W** matrix is standardized by row normalized weights method. If the weights in each row are normalized, then each row must have a value which if summed to 1. This can show in equation (2)

$$\sum_{j=1}^{n} w_{ij} = 1, \; i = 1, \ldots, n \quad (2)$$

Any normalization of the weighing line, w_{ij} can be interpreted as the fraction of all spatial influences on region i caused by region j. In the nearest neighbours K method, the normalization of spatial weights using weights of $\frac{1}{k}$ is performed [7].

One of the most important to do for spatial analysis is determining global and local moran index. The Moran Index is used to test spatial dependencies that indicate the circumstances in which the observation value of a single location depends on the value of neighbouring observations territories at adjacent locations. The Global Moran Index can be obtained through the following equations (3):

$$I = \left[\frac{n}{\sum_i^n \sum_j^n w_{ij}}\right]\left[\frac{\sum_i^n \sum_j^n w_{ij}(y_i-\bar{y})(y_j-\bar{y})}{\sum_i^n (y_i-\bar{y})^2}\right] \quad (3)$$

where I is Moran Index value, n is the number of observations, $\bar{y}$ is the average value of y_i from n location, y_i is observation value of location i, y_j is observation value of location j and w_{ij} is a spatial weighted matrix element.

Furthermore, the Local Moran Index is useful for detecting hotspots or coldspots in area data. The local Moran index with a spatial weighted matrix defined asequations (4) follows:

$$I_i = \frac{(y_i-\bar{y})}{\left(\sum_{i=1}^n \frac{(y_i-\bar{y})^2}{n-1}\right)^{1/2}} \sum_j w_{ij} \frac{(y_j-\bar{y})}{\left(\sum_{j=1}^n \frac{(y_j-\bar{y})^2}{n-1}\right)^{1/2}} \quad (4)$$

Visual exploration analysis capable of detecting spatial autocorrelation is Moran's scatter plot. The resulting output is not the original data but the data that has been standardized in Z-score. In visualization, the standardized observed value of a region on the x-axis and the observed value of the average observation of its neighbours on the y-axis. There are four quadrants in the Moran scatter plot, namely [5]:

i. Quadrant I. Indicates a positive autocorrelation (high-high).
ii. Quadrant II. Indicates a negative autocorrelation (low-high) / *coldspot.*
iii. Quadrant III. Indicates a positive autocorrelation (low-low).
iv. Quadrant IV. Indicates a negative autocorrelation (high-low) / *hotspot.*

The Moran's I test is a spatial dependency test on the OLS linear regression model's error. Moran's I statistics can write as equations (5):

$$I = \frac{\boldsymbol{\varepsilon}'\mathbf{W}\boldsymbol{\varepsilon}}{\boldsymbol{\varepsilon}'\boldsymbol{\varepsilon}} \quad (5)$$

where $\boldsymbol{\varepsilon}$ is a n x 1 error vector error of the regression model with OLS estimation method with N of observation and $\mathbf{W}$ is a standardized spatial weighted matrix of size n x n. If there is a spatial dependence, then it can be continued in the manufacture of spatial regression models [8].

A method used to model a data that contains location or spatial information is spatial regression model. Spatial Autoregressive Models (SAR) is a spatial regression model whose dependent variables are spatially correlated meaning that this model has dependence on one observation in a region with observations in its neighbouring region. The general SAR model [6] is as equations (6)

$$\mathbf{y} = \rho\mathbf{W}\mathbf{y} + \mathbf{X}\boldsymbol{\beta} + \boldsymbol{\varepsilon} \quad (6)$$
$$\boldsymbol{\varepsilon} \sim N(0,\sigma^2\mathbf{I})$$

Where y is the vector of dependent variable with size n x 1. X is the matrix of independent variables with size n x $(k+1)$; β is the vector coefficient of regression parameters with size $(k+1)$ x1; ρ is the spatial parameter coefficient lag dependent variable. Besides, W is the spatial weighted matrix by size n x n; ε is the error vector of size n x1. n is the number of observations or location and k is a number of independent variables (k=1, 2, ..., l). Lastly, I is the identity matrix by size n x n.

The spatial autoregressive lag parameter (ρ) indicates the degree of correlation of the spatial component of an area against the other region around it. Estimator of β and ρ parameters for SAR model using maximum likelihood method [8]. Hence, equations (7-9) is shown as follows;

$$\hat{\beta} = (\mathbf{X}'\mathbf{X})^{-1}\mathbf{X}'(\mathbf{I} - \rho\mathbf{W})y \quad (7)$$

$$\hat{\sigma}^2 = \frac{(Y-\rho\mathbf{W}y-\mathbf{X}\hat{\beta})'(Y-\rho\mathbf{W}-\mathbf{X}\hat{\beta})}{n} \quad (8)$$

$$\hat{\rho} = (\mathbf{y}'\mathbf{W}'\mathbf{W}\mathbf{y})^{-1}\mathbf{y}'\mathbf{W}'\mathbf{y} \quad (9)$$

One of the basic principles of Maximum Likelihood estimator is asymptotic normality, meaning that the larger the size of n the curve will be closer to the normal distribution. Statistical test significance of the parameters used is as equations (10)

$$Z_{counting} = \frac{\hat{\theta}}{s.b_{(\hat{\theta})}} \quad (10)$$

Where $s.b_{(\hat{\theta})}$ is asymptotic standar error. Hypothesis for partial parameter test is:
$H_0 : \hat{\theta} = 0$ (The regression coefficient is not feasible to use)
$H_1 : \hat{\theta} \neq 0$ (The regression coefficient is feasible to use)

Where $\hat{\theta}$ is spatial parameter regression (β, λ, and ρ). If $Z_{hitung} > Z_{(\alpha/2)}$ or *p-value* $< \alpha/2$, then the decision reject H_0, meaning regression coefficient feasible to be used on the model [9].

3. Research Methodology

The data used in this study is secondary data obtained from the results of Basic Health Study (Riskesdas) and Public Health Development Index (IPKM) in 2013. This study uses data from 8 provinces with 125 districts or cities on the island of Sumatra. Data analysis is done by using software *R* and *Arc-GIS*. The variables used in this study consist of dependent variable and independent variable in scale of measurement ratio (%), more can be seen in Table 1.

Table 1. Study variable

Main Factor	Dependable (Y)/ Independent (X) Variable	Operational definition
	Y	Prevalence of child malnutrition
Child Health	X_1	Prevalence of low birth weight
	X_2	Infant Percentage initiated early breastfeeding
	X_3	Percentage of children aged 0-23 months who were fed complementary food
	X_4	Percentage of children aged 0-23 months who are still breastfed
	X_5	Percentage of under five children who received vitamin A
	X_6	Percentage of toddlers weighed
	X_7	Full immunization Percentage
Maternal Health	X_8	Antenatal visit Percentage (K4)
	X_9	Prevalence of chronic energy deficiency in women of childbearing age
	X_{10}	Proportion of Family Planning users
Environmental Health	X_{11}	Percentage of households are living clean and healthy
	X_{12}	Percentage of access to clean water
	X_{13}	Percentage of sanitation access

4. Result And Discussion

The prevalence of child malnutrition in Sumatra Island is still quite alarming. The reasons there are still some districts or cities that have a prevalence rate above 30%. That is, public health problems on Sumatra Island are quite high. The spread of child malnutrition prevalence in eight provinces in Sumatera Island can be seen in Figure 1.

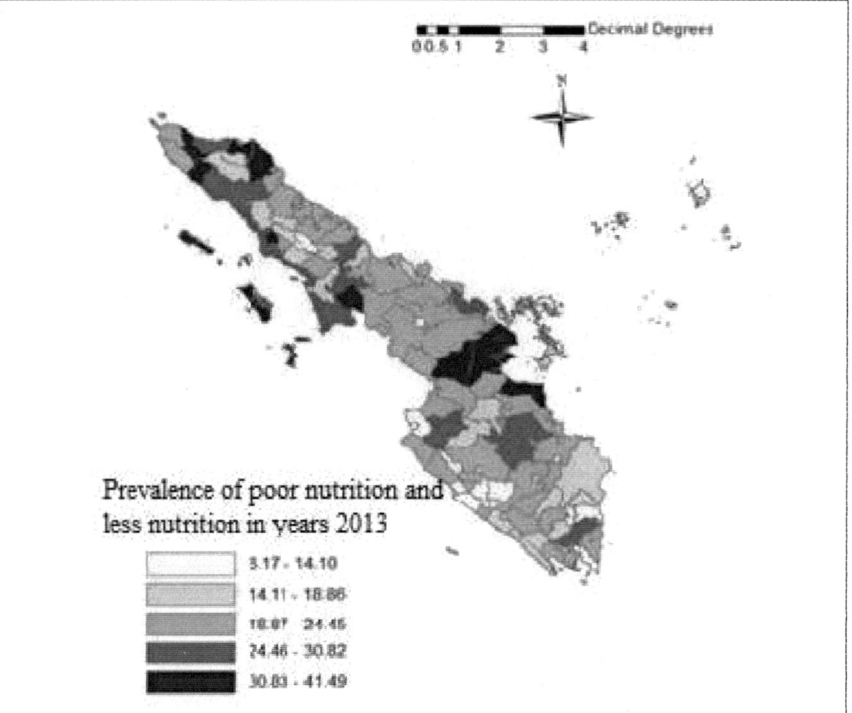

Figure 1. Distribution of child malnutrition's prevalence in Sumatera Island.

Figure 1 shows that the province of Aceh is the province with the highest number of child malnutrition's prevalence East Aceh District is at 35.51% while the district with lowest prevalence of child malnutrition were Banda Aceh City at 8.17% in Aceh Province.

Before a spatial weighted matrix is established, keep in mind the neighbours of each district on the Sumatra Island. The method used to determine nearest neighbour in this study is k-nearest neighbor method. Given the wide coverage of the Island of Sumatra, the nearest number of k selected is $k = 4.0$. This means that each district will have four nearest neighbours with a spatial weighted matrix that formed is not symmetrical with the size 125 x 125.

The Global Moran Index's value for child malnutrition prevalence in Sumatera Island was 0.297937 (P-value = 1.72×10^{-7}). These results indicate that there was a positive spatial autocorrelation or adjacent districts having similarities in the prevalence of child malnutrition. In another word, we can say that the neighbouring regions influenced the prevalence of child malnutrition in the district of Sumatra Island.

If globally it is known that there is a positive spatial autocorrelation on the data of malnutrition prevalence and underfive nutrition in infants on the island of Sumatra, it is necessary to analyze locally using Local Moran Index. The aim is to know which districts are significant to have spatial autocorrelation or which districts affected by the neighboring regions.

Moran Scatter Plot for child malnutrition prevalence in Sumatera Island can be seen in Figure 2. There are 13 districts with positive autocorrelation (quadrant I), 7 districts with negative autocorrelation (quadrant III), and no districts in quadrant II (hotspot) and quadrant IV (coldspot). The districts in quadrant I had positive autocorrelation, then Pidie Jaya District, Nias Districts, Mandailing Natal

District, South Nias District, Padang Lawas Utara District, Padang Lawas Regency, North Nias District, West Nias District, Padang Sidempuan City, Indragiri Hulu and Indragiri Hilir District had a high prevalence of malnutrition and were surrounded by districts (neighbouring regions) which also had high prevalence rates.

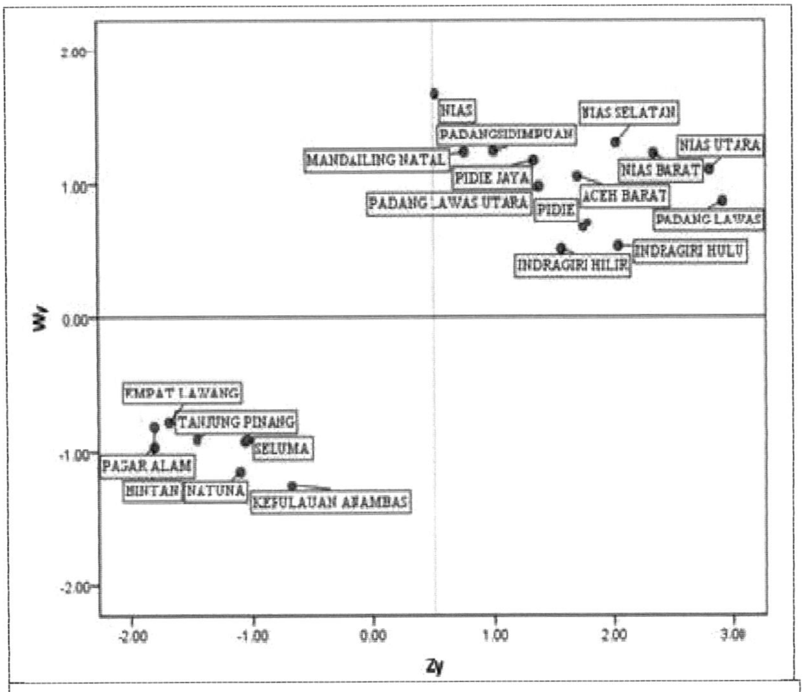

Figure 2. Moran Scatter Plot.

While the districts in quadrant III had negative autocorrelation, meaning that the district of Empat Lawang, Kota Pagar Alam, Seluma District, Bintan District, Natuna District, Anambas Islands District and Tanjung Pinang City had low malnutrition's prevalence in children under five and surrounded by districts (neighbouring regions) which also had low prevalence rates.

Before performing spatial regression analysis, spatial dependence test is performed. Before testing spatial dependence to determine the spatial regression model, a spatial dependence test was performed on the error of linear regression using Moran Index. The value of Moran Index (I) obtained is 0.103858 (P-value = 0.01768). By using $\alpha = 0.05$, it can be concluded that there is positive spatial autocorrelation in linear regression error so that need to be continued at making of spatial regression model.

One of the assumptions that must be met before creating a spatial regression model is that the range of errors in the linear regression model should be homogeneous. Testing homogeneity of various errors was done by Breusch-Pagan (BP) test statistic. BP test statistic value is 16.081 (P-value = 0.2448). Using $\alpha = 0.05$, it can be concluded that the homogeneous error is so that the assumption of homogeneity of various errors is met.

To determine which model is appropriated with the data, it is necessary to test Lagrange Multiplier. The result of Lagrange Multiplier test can be seen in Table 2.

Table 2. Lagrange Multiplier Test.

Statistic	Value	P-*value*
Lagrange Multiplier (SAR)	5.4921	0.0191*
Lagrange Multiplier (SEM)	3.0231	0.08208
Lagrange Multiplier (SARMA)	5.5067	0.06371

Table 2 showed the results of the Lagrange Multiplier test against three spatial regression models. The test results show that SAR model is significant at $\alpha = 0.05$ meaning that there is spatial dependence on lag or dependent variable. Therefore, the spatial regression model used in this study is Spatial Autoregressive Model (SAR). The result of spatial regression estimation tabulated in Table 3.

Table 3. The result of spatial regression parameter estimation of SAR model

Main Factor	Variable	β	Z-value	P-value
	Intersept (β_0)	47.157	5.0523	4.36×10^{-7}***
	ρ	0.24335	2.626	0.0086**
Child Health	X_1	-0.131	-1.4368	0.151
	X_2	-0.011	-0.3322	0.739
	X_3	0.016	0.5699	0.569
	X_4	0.025	1.026	0.305
	X_5	-0.067	-1.592	0.111
	X_6	0.039	1.9573	0.338
	X_7	0.036	1.4497	0.147
Maternal Health	X_8	-0.015	-0.3987	0.690
	X_9	0.139	2.5151	0.012*
	X_{10}	-0.370	-4.7213	2.34×10^{-6}***
Environmental Health	X_{11}	-0.136	-2.4117	0.016*
	X_{12}	-0.254	-2.9968	0.002**
	X_{13}	-0.044	-1.2589	0.208

*Note: level of significance *(0.05), ** (0.01), *** (0.000)*

Based on the results in Table 3, the SAR model for child malnutrition cases on the Sumatra Island written as equation (11):

$$\hat{y}_i = 47{,}157 + 0{,}24335 \sum_{j=1, j\neq 1}^{125} w_{ij} y_j + 0{,}139 X_9 - 0{,}370 X_{10} - 0{,}136 X_{11} - 0{,}254 X_{12} \quad (11)$$

The informations that can explained in the equation above is $\hat{y}_i$ as the prevalence of children under-five malnutrition in district *i*, where *i*=1, 2... 125. y_j as the prevalence of children under-five malnutrition in district *j* which is the neighbor of region *i*, where *j*=1, 2, ..., 125 and *j*≠*i*. In addition, X_9 as the prevalence of chronic energy deficiency in women of childbearing age; X_{10} as the proportion of family Planning (KB) users; X_{11} is the percentage of households are living clean and healthy (PHBS); X_{12} as percentage of access to clean water; w_{ij} is the spatial weighted matrix by size 125 x 125 and lastly ρ = 0.24355.

The result of estimation and testing of parameters of SAR model shows that there are four significant independent variables at $\alpha = 0.05$ that is prevalence of chronic energy deficiency in women of childbearing age (X_9), proportion of family planning (X_{10}), percentage of households are living clean and healthy (PHBS) (X_{11}), and percentage access to clean water (X_{12}).

Based on equation stated, the coefficient value β_0 is 47.157. This means that if the prevalence of chronic energy deficiency in women of child-bearing age (X_9), the proportion of family planning (X_{10}), percentage of households with PHBS (X_{11}), and coverage of access to clean water (X_{12}) is 0, and no spatial dependence in the model, then the prevalence of malnutrition and under nutrition in children will be 47.157%. This suggests that the prevalence of child malnutrition will be very high although there is no influence of the independent variables in the model.

Furthermore, the β_9 coefficient value of 0.139 has significance if the prevalence of chronic energy deficiency in women of reproductive age increases 1%, hence the prevalence of child malnutrition will increase by 0.139%, while other variables are considered constant. This result is in accordance with the statement of Ministry of Health of Republic Indonesia (2016) that women who are lacking energy or nutrition are at risk of delivering babies with low birth weight or malnutrition as well.

The value of β_{10} coefficient of -0.370 has a meaning if the proportion of family planning users increased by 1%, it will reduce the prevalence of child malnutrition by 0.370%. This indicates that the family planning program has succeeded in reducing the prevalence of child malnutrition in Sumatera Island. If the family planning program can be run with the maximum proportion, then the child nutritional problems can be prevented.

The coefficient of β_{11} is valued at -0.136, meaning that if the percentage of households is living clean and healthy PHBS increases by 1%, then the prevalence of child malnutrition will decrease by 0.136%. If the number of households with PHBS increases, the prevalence of child malnutrition can be reduced. It also shows that clean and healthy living behavior for family members is very important to prevent various health problems.

The coefficient value of β_{12} is -0.254, meaning if coverage of access to clean water increased by 1% it will decrease the prevalence of child malnutrition by 0.254%. These results prove that access to clean water is quite important given the large number of malnutrition prevalence rates in under-fives children on the Sumatra Island. Therefore, it is necessary to make every district have sufficient access to clean water so that malnutrition problem in children can be discharged.

Furthermore the coefficient ρ on the SAR model is significant at $\alpha = 0.05$ with a value of 0.24335. This means that if a district surrounded by four neighbouring regions, then the influence of these areas is 0.24335 multiplied by the average prevalence of child malnutrition in neighboring areas.

Through a pre-established SAR model, 22 spatial regression models established for each district or city with its neighbouring regions. For example, the SAR model for Pidie Jaya District in Aceh Province are as equation (12) and (13):

$$\hat{y}_{18} = 47{,}157 + 0{,}24335\,[0{,}25(y_9 + y_{10} + y_{16} + y_7)] + 0{,}139X_9 - 0{,}370X_{10} - 0{,}136X_{11} - 0{,}254X_{12} \quad (12)$$
$$\hat{y}_{18} = 47{,}157 + 0{,}06y_9 + 0{,}06y_{10} + 0{,}06y_{16} + 0{,}06y_7 + 0{,}139X_9 - 0{,}370X_{10} - 0{,}136X_{11} - 0{,}254X_{12} \quad (13)$$

SAR model for Pidie Jaya regency can be interpreted that if the prevalence of malnutrition and underfive nutrition in Pidie (Y_9) increased 1%, the prevalence of malnutrition and under-nutrition in under-five children in Pidie Jaya district will increase by 0.06%, with assume another variable is constant. Likewise, for the influence of Bireun (Y_{10}), Aceh Jaya (Y_{16}), and Aceh Barat (Y_7) District towards Pidie Jaya District can be interpreted in the same way. While the interpretation of the factors that influence the prevalence of child malnutrition the same as the previous explanation.

5. Conclusion

Based on the results of the study, the conclusions of this study are:

i. Factors affecting malnutrition and nutritional status among under-fives in Sumatera Island are prevalence of chronic energy deficiency in women of childbearing age, proportion of family planning users' percentage of clean and healthy households and percentage of access to clean water.
ii. There are 22 spatial regression models for each district or city with neighbouring regions using SAR model. For Pidie District (Y_9) for example, we conclude that if the prevalence of malnutrition among infants increase by 1%, the prevalence of malnutrition in children under five in Pidie Jaya District will increase by 0.06%, assuming other variables are constant. Likewise, for the influence of Bireun District (Y_{10}), Aceh Jaya (Y_{16}), and Aceh Barat (Y_7) on Pidie Jaya District, it can be interpreted in the same way.

References

[1]. Republic of Indonesia Ministry of Health. 2016. *Indonesian Health Profile 2015*. Republic of Indonesia Ministry of Health
[2]. Republic of Indonesia Ministry of Health. 2016. *Handbook on Nutrition Status Monitoring and Nutrition Performance Indicators in 2015*. Republic of Indonesia Ministry of Health.
[3]. WHO. 2010. *Nutrition Landscape Information System (NLIS):* Country Profile Indicators. World Health Organization, Geneva.
[4]. Tobler W. 1979. *Philosophy in Geography, Reidel, Dordrecht*. Cellular Geography, inS. Gale and G. Olsson (eds 379-386).
[5]. Anselin L. 1995. *Local Indicators of Spatial Association*. Study Paper 9331 Regional Study Institute West Virginia.
[6]. Ward M. D and Gleditsch, K. S. 2008. *Spatial Regression Model*. Sage Publication, Inc, United States.
[7]. Sarrias, Mauricio. 2016. *Lecture 1: Introduction to Spatial Econometrics*. Universidad Catolica del Norte, Chile.
[8]. Anselin L and Bera A.K. 1998. *Spatial Dependence in Linear Regression Models with an Introduction to Spatial Econometrics.* Dekker, New York.
[9]. Anselin L. 2003. *Spatial Econometrics: A Companion to Theoretical Econometrics.* Blackwell Publishing Ltd.

IOP Conf. Series: Materials Science and Engineering **506** (2019) 012052 doi:10.1088/1757-899X/506/1/012052

A Simple Candlenut Shell Carbonization Tool

R Darma[1] and A Amri[1*]
[1]Industrial Engineering Department, South Aceh Polytechnic, Aceh, Indonesia

[*]Corresponding author: asbahrul.arl@gmail.com

Abstract. In order to produce charcoals from the candlenut shell, a tool is need such as candlenut carbonization tool using a vertical multi chambers (CCT-VMC). The tool is indicated some weaknesses; the oxygen kept flowing into the chambers when cooling process are carried out, so oxidation occurred in the last phase of producing charcoal. This was due to some holes on the tools are not being able to close perfectly. The aims of the research are to produce a carbonization tool that are able to reduce the oxidation when cooling process being performed by reducing the openings on the tool and to determine the tool's capacity. The tool was designed by utilizing Autodesk Inventor 2018 software. The result shows that the tool has two vertical chambers that provide space for adequate air supply when performing a burning process. The main components are a half metal drum, a chamber barrier, and a cover. It is also found that the tool needs 9.5 hour to process 24 kg of candlenut which consist of 2 hours for burning process and 7.5 hours for cooling process.

Keyword *: candlenut seeds, biomass, carbonization tool, vertical multi-chambers*

1. Introduction

South Aceh, a district of Aceh Province, riches for natural resources, however Candlenut seeds, derived from Aleurites moluccana plants, are one of important commodities from the place. Candlenut (Aleurites moluccana) is one type of industrial plant from the family of Euphorbiaceae [1-3]. Candlenut seed and candlenut shell were shown in Figure 1. The seeds were taken by breaking the shells as the solid waste from the candlenut fruit [4]. Candlenut shell is one of biomass waste which is a renewable energy source [5-7]. However, today the biomass waste can be converted as charcoals and briquettes. The briquette has been used for individual and industrial purposes as alternative energy whereas the charcoal for industrial purposes such as cosmetics, purifying agent [6], and many others. Since the need of the such charcoal has increase as the increase in industrial need, charcoal become a promising business.

Figure 1. Candlenut and its shell

Charcoal is a carbon rich product that result from the thermal treatment of natural organic feedstock in an oxygen-limited environment [8]. The charcoal is produced by slow heating biomass waste called carbonization [6]. Carbonization is a crucial step because of its porous structural forming [9] which define the amount of porosity on the charcoal. However, to produce the best quality of charcoal, it is crucial to fulfil the standard values of several physical and chemical characteristics such density, porosity, water content, carbon content and so forth.

To produce the charcoal, a good tool is required to obtain best quality of it. Some tools have been produced, but they need other sources as the energy to burn the shells such palm shell pyrolysis drum reactor [10]. Moreover, some tools allow oxygen enter carbonization chambers while cooling process is taking place. It causes the decrease in the quality of the charcoal. From previous study regarding to the effectivity of the candlenut carbonization tool using a vertical multi chamber (CCT-VMC) on the quantity and quality of the charcoal [11]. It is indicated that there were some weaknesses of the tool; the oxygen kept flowing into the chambers when cooling process are carried out, so oxidation occurred in the last phase of producing charcoal. This was due to some holes on the tools are not being able to close perfectly, especially the holes where the carbonization chamber barriers are placed. Figure 2 is showing the CCT-VMC design.

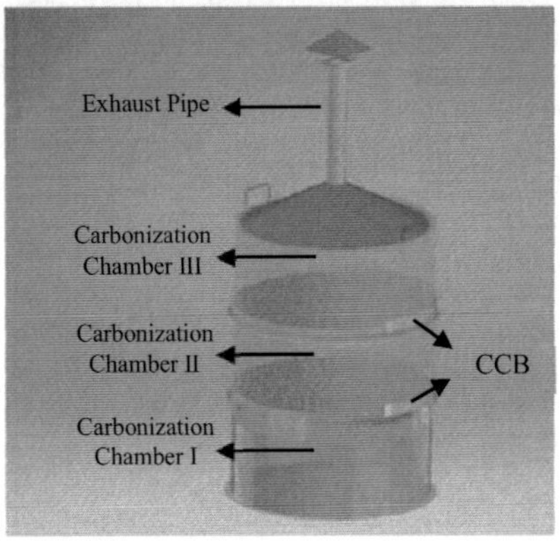

Figure 2. Candlenut shell Carbonization tool using a vertical multi chambers (CCT-VMC)

The carbonization chamber barriers (CCB) are being placed from the side of the drum. Those are pushed to slide into the chamber. The hole on the side wall of drum remains imperfectly closed, especially the one downside of the CBB which hardly closed using sticky mud. Thus, a new method and tool are needed to develop in order to attain a good quality of charcoal. In addition, the research is also aimed to define the Simple Candlenut Carbonization Tool's time consumed for a carbonization process by performing some experiments.

2. Method

2.1 Design

Considering minimizing the oxidation process when the cooling process taking place, a simple candlenut carbonization tool is designed containing holes as less as possible and providing an easy operational process of carbonization. If some of the holes cannot be dismissed, then some features may develop in order to stop the air kept flowing to the carbonization chambers. However, some features of the candlenut carbonization tool using a vertical multi chambers are still utilized. They are human operational power, fully uses the energy from the candlenut itself in carrying out carbonization process, low material and manufacturing process cost. Therefore, the tool is expected to be afforded by the user in terms of cost and produces a better quality of charcoal.

By utilizing Autodesk Inventor 2018, simple candlenut carbonization tool is designed using 2 vertical carbonization chambers, carbonization chamber I and II, which aims to provide space for adequate air supply when performing a burning process. It will attain a perfect candlenut burnt that generates enough heat to facilitate carbonization chamber II having carbonized. The carbonization chambers were shown in Figure 3.

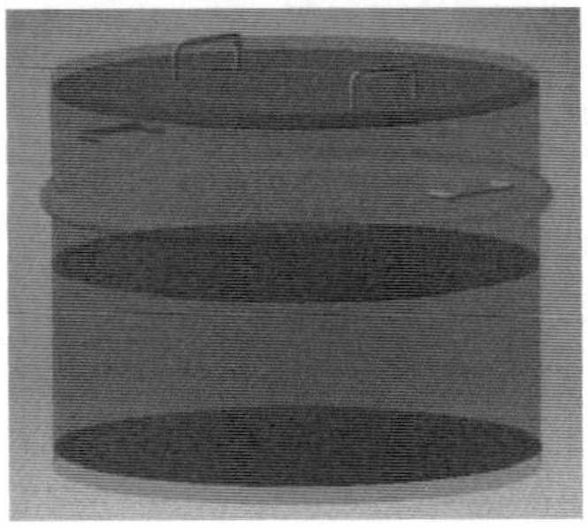

Figure 3. A simple candlenut carbonization tool

The tool mentioned above requires 3 main components, namely drum as the main frame to constrain the outside and inside environment and also as a base for initial burning, carbonization chamber barrier and a cover to shield the top part of the tool. Figure 4 below shows the detail of the 3 main components.

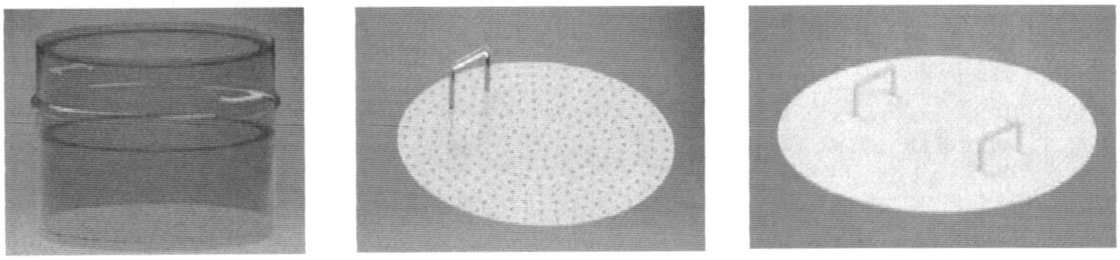

a. Drum b. Cover c. CCB

Figure 4. Three main components of a simple candlenut carbonization tool

2.2 Tools and Material
A Simple Candlenut Shell Carbonization Tool required some material to be manufactured and tools to be used in manufacturing process. The materials are half of 200 liters metal drum, L-Shape angle bar, 2 mm thick iron plate, 1 mm thick aluminum plate, and Ø 10 mm iron bar. The tool is designed by utilizing Autodesk Inventor 2018 software that allowed designer experiencing a real-looked of the product designed. Welding, grinding, and drilling machines and others appropriate tools such as hammers and scales are used in performing manufacturing process.

2.3 Tool Experiments
The experiments are performed to determine time consuming in a carbonization process using a Simple Candlenut Shell Carbonization Tool. In order to gain enough data, the experiments are done in three times. Every carbonization chambers filled with 12 kg of candlenut shell that will be carbonized. Tool's time consuming is defined by measures the temperature during the carbonization process which consist of burning and cooling process.

In burning process, temperature will be increased as time increased whereas in cooling process conversely. Burning process time is measured from the initial burnt in carbonization chamber I until the burning stage completed, which was indicated by the little amount of and bluish smoke [9]. Cooling process is starting when the burning process and completing when the temperature inside and outside carbonization chamber are similar. Temperature inside and outside carbonization chamber are measured using thermocouple. The measured temperatures will be used to determine time processing of the tool.

2.4 Flowchart of Manufacturing and Designing Tool

The following is a flow diagram in designing and manufacturing of a simple candlenut shell carbonization tool as shown in Figure 5.

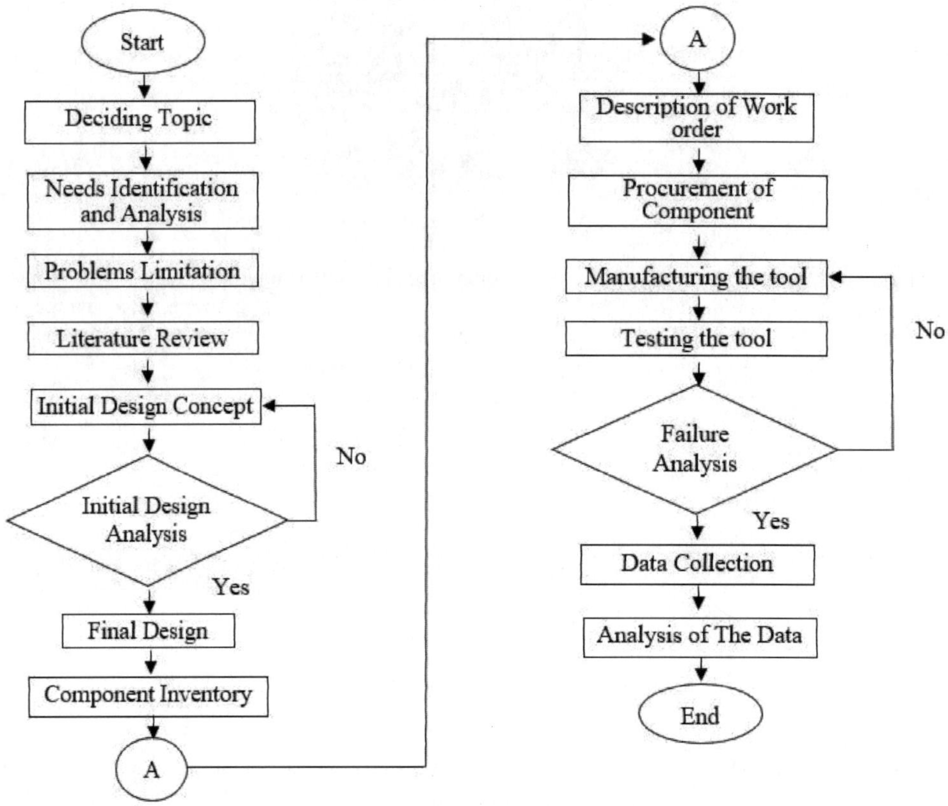

Figure 5. Flowchart of tool design and manufacture

3. Results and Discussion

3.1 A Simple Candlenut Shell Carbonization Tool

The final design of the simple carbonization Tool is shown in Figure 3 and 4. The tool has been designed with lesser holes than the previous tool, the tool with vertical multiple chambers. There are two ways to eliminate the high oxidation which was by taking out the slide mechanism and an attached reservoir beneath the tool.

Manufacturing process of the tool followed the design shown in Figure 3 and 4. Through all the process of manufacturing, all main parts of the tool and the complete one shown in Figure 6 respectively.

Figure 6. Manufacturing outcome which is A Simple Candlenut Shell Carbonization Tool.

3.2 Results of the Experiment

The temperature data in burning and cooling stage were attainted in all carbonization chambers (CC) every 30 minutes using thermocouple. The data are shown in Table 1 and Table 2.

Table 1. Temperature of candlenut shell in burning process

Time (Minutes)	Experiment I			Experiment II			Experiment III		
	Temperature (°C)			Temperature (°C)			Temperature (°C)		
	CC I	CC II	Avg.	CC I	CC II	Avg.	CC I	CC II	Avg.
0	32	32	**32**	32	32	**32**	32	32	**32**
30	263	145	**204**	301	169	**235**	253	123	**188**
60	411	304	**357.5**	356	243	**299.5**	399	323	**361**
90	435	354	**394.5**	429	274	**351.5**	435	352	**393.5**
120	323	220	**271.5**	312	209	**260.5**	323	221	**272**

Table 2. Temperature of candlenut shell in cooling process's

Time (Minutes)	Experiment I			Experiment II			Experiment III		
	Temperature (°C)			Temperature (°C)			Temperature (°C)		
	CC I	CC II	Avg.	CC I	CC II	Avg.	CC I	CC II	Avg.
0	90	84	**87**	92	90	**91**	98	99	**98.5**
0.5	65	56	**60.5**	77	71	**74**	83	87	**85**
1	55	47	**51**	62	60	**61**	66	82	**74**
1.5	49	45	**47**	58	53	**55.5**	61	75	**68**
2	46	43	**44.5**	54	50	**52**	53	63	**58**
2.5	44	43	**43.5**	50	46	**48**	53	63	**58**
3	43	42	**42.5**	47	45	**46**	51	58	**54.5**
3.5	40	40	**40**	43	41	**42**	47	53	**50**
4	39	40	**39.5**	40	39	**39.5**	44	48	**46**
4.5	39	40	**39.5**	39	39	**39**	39	46	**42.5**
5	38	39	**38.5**	38	39	**38.5**	38	39	**38.5**
5.5	38	37	**37.5**	38	37	**37.5**	38	36	**37**
6	36	34	**35**	36	34	**35**	36	34	**35**
6.5	34	34	**34**	34	33	**33.5**	34	33	**33.5**
7	33	32	**32.5**	33	32	**32.5**	33	32	**32.5**
7.5	32	32	**32**	32	32	**32**	32	32	**32**
8	32	32	**32**	32	32	**32**	32	32	**32**

Table 1 shows that the burning process was completed at the average temperature of 268 °C which indicated by little amount of smoke produced. From the three experiments, it can be concluded that it took 2 hours for the candlenut shell to be burnt completely. Table 2 revealed that cooling process consuming much more time compered to burning process. All experiments confirmed that inside temperature of carbonization chamber was similar to ambient temperature, 32°C, at 7.5 hours after cooling process proceeded. Therefore, the hole carbonization process of 12 kg candlenut shell using Simple Candlenut Carbonization Tool was consumed 9.5 hours which consist of 2 hours for burning process and 7.5 hours for cooling process.

A comparison was made to see the performance of the tools, between Candlenut Carbonization Tool Using a Vertical Multi Chambers (Tool 1) and Simple Candlenut Carbonization Tool (Tool 2), by comparing the processing time and the number of repetitions of the carbonization process in one day. Table 3 displays that Tool 1 is better than Tool 2 in term of processing capacity.

Table 3. The measurement of the capacity of Simple Candlenut Shell Carbonization Tool

Description	Tool 1	Tool 2
Σ one time processed candlenut shell (kg)	24.00	24.00
Single Processsed Time (hour)	7.75	9.5
Σ Carbonization Process in one day (time)	3.10	2.52
The amount of candlenut shell that can be processed in one day (kg)	74.32	60.48

3.3 Discussion

The results confirmed that candlenut shell can be carbonized using its own energy by utilizing a simple candlenut carbonization tool. Table 3 confirmed that Tool 1 performed better than Tool 2 in term of capacity due to less processing time. In term quality of the charcoal that has been produced still remain in question. Therefore, some researches are needed to be performed to provide some more detail information regarding the quality of the charcoal that produced by those tools. The detail information will define which tool performed better in general.

4. Conclusion

A simple candlenut carbonization tool which consist of two vertical chambers can carbonize candlenut shell without any additional energy except energy that derived from candlenut shell itself. The process of carbonization consists of 2 stages, which are burning stage and cooling stage. The tool has lesser holes compared to the previous one. It is done in order to minimize the possibility of oxidation involved in cooling stage. Tool's experiments which consist of 12 kg candlenut shell in every carbonization chamber revealed the real time consumption of carbonization process. Burning stage consumed less time than cooling stage. Burning stage completed in only 2 hours and cooling stage was taking 7.5 hours to be finished. It is can be concluded that the total carbonization process time of 24 kg candlenut shell using Simple Candlenut Shell Carbonization Tool is 9.5 hours with 60.48 kg in total process capacity in one day. However, Candlenut Carbonization Tool Using a Vertical Multi Chambers can carbonize 74.32 kg in one day. Therefore, Candlenut Carbonization Tool Using a Vertical Multi Chambers is better than Simple Candlenut Shell Carbonization Tool in term of processing capacity.

References

[1] Rosman R and Djauhariya E 2005 Status dan Teknologi Budidaya Kemiri (Bogor: Balai Penelitian Obat dan aromatika)

[2] Krisnawati H Kalio M and Kanninen M 2011 Aleurites Moluccana (L) Wild (Bogor: Center for International Forestry Research)

[3] Estrada F Gusmaol R Mudjijati and Indraswati N 2007 Pengambilan minyak kemiri dengan cara pengepresan dan dilanjutkan ekstraksi cake oil *Widya Teknik* 6 2 121-130

[4] Tambunan B H Saptoadi H and Syamsiro M 2014 A Preliminary Study on Use of Candlenut Shell as a Renewable Source of Energy, Min Indonesia *Journal of Ocean, Mechanical and Aerospace -Science and Engineering* 9

[5] Romallosa ARD and Kraft E 2017 Feasibility of Biomass Briquette Production from Municipal Waste Streams by Integrating the Informal Sector in the Philippines *Resource* 6 12 1-19

[6] Demirbas A, Ahmad W, Alamoudi R and Sheikh M 2016 Sustainable charcoal production from biomass *Energy Sources, Part A: Recovery, Utilization, and Environmental Effects* 38 13 1882-1889

[7] Afriani C D, Yufita E and Nurmalita 2017 Nilai Kalor Briket Tempurung Kemiri dan Kulit Asam Jawa dengan Variasi Ukuran Partikel dan Tekanan Pengepresan *Journal of Aceh Physics Society (JAcPS)* **6** 1 6-9

[8] Lehmann J and Joseph S 2009 Bio char for Enviornmental Manganement*: Science and Technology* (UK: Earthscan)

[9] Sunarsih S Hastutiningrum S & Nisa T D 2016 Activated Carbon from Jackfruit Peel Waste as Decolouring Agent of Screen Printing Waste Water *Prosiding Seminar Nasional Teknik Kimia "Kejuangan" Pengembangan Teknologi Kimia untuk Pengolahan Sumber Daya Alam Indonesia,* Yogyakarta

[10] Habiati 2014 Kajian Potensi Produk Pirolisis Limbah Padat Kelapa Sawit *Jurnal Biologi Edukasi*

IOP Conf. Series: Materials Science and Engineering **506** (2019) 012053 doi:10.1088/1757-899X/506/1/012053

Composition on Essential Oil Extraction from Lemongrass Fragrant by Microwave Air Hydro Distillation Method to Perfume Dermatitis Production

Teuku Rihayat[1*], Suryani[1], Zaimahwati[1], Salmyah[1], Sariadi[1], Fitria[2], Satriananda[1], Alfian Putra[1], Zahra Fona[1], Juanda[1], Raudah[1], Aida Safitri[1], Mawaddah[1], Nurhanifa[1], Shafira Riskina[1], Wildan Syahputra[1]

[1] Department of Chemichal Engineering, Lhokseumawe State Polytechnic, 24301, Aceh, Indonesia
[2] Department of Dermato Venereology, Medical Faculty, University Syiah Kuala, 24311, Aceh, Indonesia

*Corresponding author: teukurihayat@yahoo.com

Abstract. Essential oils from fragrant lemongrass are generally obtained by conventional extraction methods that require large energy, a significant amount of solvents, and a fairly long processing time. This study used a method developed from microwave hydro distillation, which is a microwave water- hydro distillation to optimize the extraction of essential oils. The purpose of this study was to produce citronella essential oil with the best quality and a large amount compared to the extraction method, where the obtained fragrant citronella oil can be used as raw material for making perfume. Where extraction is carried out at 600 W power; the ratio of the raw material to be extracted and the solvent is 0.5 g mL-1 and the extraction time is 150 minutes. In extraction with microwave water-hydro distillation, the air flow rates used are 1, 2, 3, 4, and 5 L/min. The results showed that the extraction of citronella oil with microwave water-hydro distillation was easier and produced higher yields compared to extraction by microwave hydro distillation method. Testing the physical properties of fragrant citronella oil shows that essential oils obtained by microwave hydro distillation and microwave water- hydro distillation have the same quality (refractive index and specific gravity). Furthermore, testing the chemical properties of fragrant citronella oil showed that essential oils obtained by hydro distillation-air microwaves had better flavor compared to oils obtained by microwave hydro distillation.

1. Introduction

Essential oils are oils from species of aromatic plants obtained from plant parts such as roots, bark, stems, fruit, seeds or flowers through the process of hydro distillation or steam distillation. Essential oils produced are needed by various industries, such as the perfume, cosmetics, pharmaceutical / medicine industry and the food and beverage industry. In trade, essential oils are viewed as having a role to produce primary and secondary products, both for domestic and export needs [1].

Every year, consumption of essential oils increases by around 8-10%. Due to the increasing need for essential oils for the perfume, cosmetics and health care industries [2]. This is also driven by consumers to switch from the use of materials containing synthetic compounds to use natural ingredients; this is what causes the demand for essential oils. In addition, essential oil products cannot be replaced by synthetic ingredients.

One of the many essential oils in Indonesia is fragrant citronella oil. where fragrant citronella oil can be used as a natural insect repellent and very attractive to the pharmaceutical industry and fragrances [3]. Fragrant citronella oil has uses as analgesic, anticonvulsant, anxiolytic, anti-fungal, anti-bacterial, anti-parasitic and

IOP Conf. Series: Materials Science and Engineering **506** (2019) 012053 doi:10.1088/1757-899X/506/1/012053

nematicidal activity [5]. In traditional practice, citronella oil has been used as antipyretic, aromatic, vermifuge, diuretic and mental illness [6].

Indonesia places the 9th supplier position in the world as of 2011, with a value of USD 0.16 billion [7]. The biggest essential oil user industry is the flavor & fragrance industry whose global trade value is estimated at USD 18 billion per annual and Indonesia's import value is estimated at USD 400-500 million p. Aa Industry other users include Spa /Aromatherapy, pharmaceuticals, insecticides and one of the industry perfume.

Perfume is a product that is in great demand by both local and foreign people, because the use of perfume itself and the aroma produced from perfume are of particular value to consumers. Perfume is obtained from the addition of essential oils, alcohol and extracts from plants that have a distinctive aroma and interest by the public. One of the most important plants in Indonesia, especially in Aceh, is the fragrant lemongrass found in the Bener Meriah area, where the area is a strategic area for cultivating fragrant lemongrass.

At present, fragrant lemongrass is exported abroad for processing into essential oil and then used as a variety of cosmetic and perfume products [8], which is then imported back into the country to increase foreign exchange. Where the needs of the public for body deodorizers and as a treatment for dermatitis also increase. Therefore, it is needed another alternative to minimize the production of fragrant lemongrass export-import by processing the fragrant lemongrass into a product that can boast Indonesia abroad, and to improve the economic needs of the local community and can boast domestic products that come from local materials but have international quality which will then be marketed on the world market.

Therefore, in this study to produce fragrant lemongrass essential oil in greater amounts than the usual extraction method using microwave assisted air hydro distillation method, where the essential oil obtained can be used as raw material for the manufacture of perfumes that can treat disease Dermatitis.

2. Methods

2.1 Materials

The materials used in this study were fragrant lemongrass is obtained from Takengon, North Aceh. All dried plants are moistened in the semi-dark room for 10-15 days and then cut into small pieces. Plant samples are then stored in a zip plastic bag at room temperature. Aquadest and anhydrous sodium sulfate. In making perfume, the ingredients used are ethanol, locking oil (sandalwood oil or other flavorful essential oils) A subsubsection. The paragraph text follows on from the subsubsection heading but should not be in italic.

2.2 Microwave Hydro Distillation

Microwave ovens that have been modified for microwave hydro distillation operations. 30 grams of citronella powder which has been crushed which is placed in a 1 liter flask containing deionized water (500 mL). The pumpkin is arranged in a microwave oven cavity and condenser at the top of the oven to collect the essential oil extracted. Using a compressor, electric motor and maximum pressure of 3 bars. Air flow is added to the distiller which contains fragrant lemongrass powder and deionized water. Microwave ovens are operated at a power level of 600 W for a period of 2 hours. Then the oil is separated using a separating funnel. To remove water, the essential oil extracted is then dried over anhydrous sodium sulfate.

2.3 Gas Chromatography-Mass Spectrometry (GC-MS)

The qualitative and quantitative analysis of the fragrant citronella oil extracted was carried out using Shimadzu GC-MS equipment with a wavelength of 60 m and an internal diameter of 0.25 mm. The initial oven temperature was 500°C to 2600°C. Injectors and transfer line temperatures are 250°C and 200°C, respectively. The holding time is 2 minutes and the solvent delay time is 8 minutes. Helium is used as a carrier gas at 1 ml min-1. 2 μl of volume of oil sample (2% solution of citronella oil in hexane) is injected. Split ratio of 50: 1 and scanning range of 50-600 Da are used. Constituent identification retention time and search is carried out based on spectral spectrum (NIST).

2.4 Fourier Transform Infrared (FTIR)

The IR spectrum of fragrant citronella oil (obtained from the leaf portion on the highest oil yield) was obtained from the "IR Affinity 1" FTIR spectrometer (Shimadzu Corporation, Japan), measured in the range between 4000- 400 cm − 1 with a resolution of 2 cm− 1 and 30 scans per spectrum. Potassium bromide (KBr) is used for reference. KBr pellets were prepared and about 20 µl of lemongrass oil was spread on the pellets and then immediately taken for analysis. Basic correction and frequency variations are applied using the Shimadzu IR solution. Software 1.5 is included with the equipment.

3. Results and Discussion

3.1 The Effect of Water Content from Fragrance Citronella Oil

Based on the dry weight of the fragrant lemongrass plant which was 2.11% at 3 hours distillation time. These results indicate that oil is very influential by the moisture content of the plant samples. During drying of plant samples, volatile compounds may seep into the surface and evaporate together with water resulting in a decrease in oil content.

In the drying process, temperature is the most important parameter for preserving whole essential oil in plant material. The drying method significantly affects the oil content and composition. From the explanation above, it is clear that the drying process affects the active ingredients of essential oils. Therefore, in this study, shady plant materials are dried in a semi-dark room. In addition, given the traditional practice of refining lemongrass plants and the unavailability of fresh ingredients regularly, parts of dried fragrant lemongrass plants are used in optimization studies by hydro-distillation processes.

The following is research data from the results of fragrant citronella oil using the water-hydro distillation method, can be seen in table 1 below.

Table 1. Yield of Citronella Oil obtained by Microwave Air-Hydro Distillation.

Extraction Method	Air Flow Rate (L/min)	Citronella Oil Yield (%)
Microwave Air-Hydrodistilation	1	1.2074± 0.1123
	2	1.2387 ± 0.1057
	3	1.3145 ± 0.0968
	4	1.3891 ± 0.0973
	5	1.3878 ± 0.0985

Experiments were carried out to evaluate the effect of different parameters such as the fragrant citronella plant, solvent ratio and distillation time on the oil yield. The oil yield increased with an increase in extraction time, and almost all oil was extracted within 2 hours of the extraction period. Increasing extraction time up to 3 hours increases oil yield by only 1-5% of the total oil extracted and outside of 3 hours deteriorates oil quality (Ahmed, 2005).

3.2 GC-MS Analysis

The extracted oil from various parts (leaves and stems) obtained at the highest yield was analyzed using GC / MS. Percentages of areas higher than the peak indicate a higher percentage of compounds that are suitable in oil. The chromatogram plot obtained from the leaf section revealed high concentrations of three major commercially important compounds, namely citronellal (55.23%), geraniol (26.29%) and citronellol (13.41%) accounting for 94.94% of total constituents. Meanwhile, all aerial parts have 89.15% of these three compounds and the stem part only covers 69.84%.

The following is research data from citronella oil fragrance using GC / MS analysis can be seen in Figure 1 below.

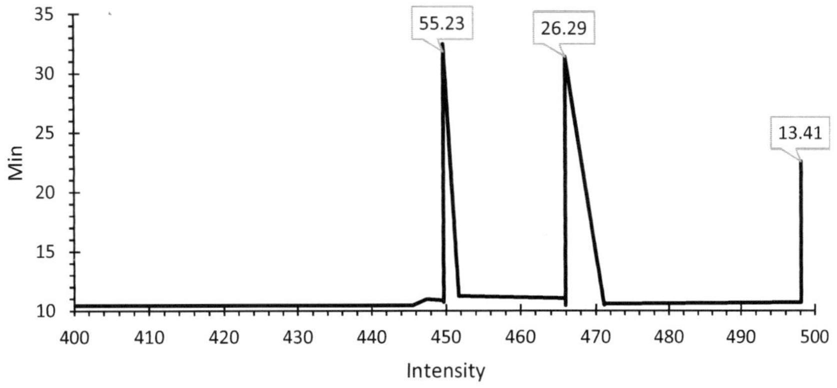

Figure 1. GC/MS Chromatogram of Air-Hydro Distillation Citronella Oil.

3.3. FTIR Analysis

Qualitative analysis of different organic compounds can be ascertained from the characteristics of the infrared spectrum at certain frequencies that are influenced by certain functional groups. The percentage of transmittance corresponding to the wave number concluded in the total infrared spectrum is weakened as shown in figure 2.

There is an intense broad peak in the range 3600–3200 cm-1 specifically at 3365.78 cm-1 according to the hydroxyl (OH) group. Another intense and branching peak in the range 2935-2915 cm-1 is the C-H methyl group and stretches of methylene, mostly aliphatic alkyl groups. The medium peak at 2719.63 cm-1 is the C-H aldehyde group of carbonyl compounds. Cluster C = O ketones at wavelength 1750-1705 cm-1. The alkene group is detected at a wavelength of 852.53 cm-1. Whereas in small vibrations on the wave of 750-660 cm-1 there is an aromatic group namely vinyl C-H.

The following is research data from citronella oil fragrance using GC / MS analysis can be seen in figure 2 below.

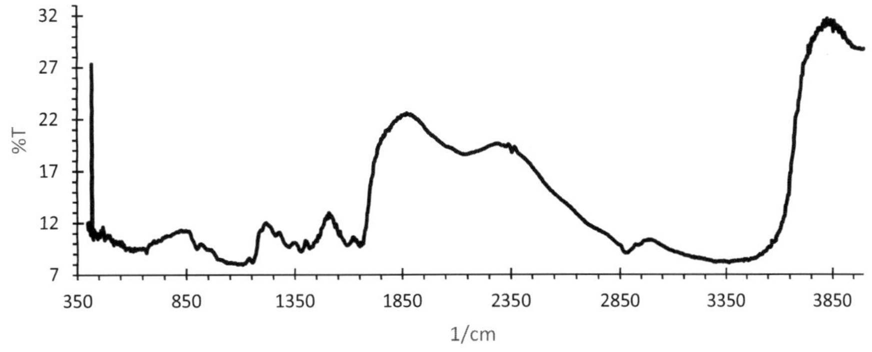

Figure 2. FTIR Spectra of Air-Hydro Distillation Citronella Oil.

4. Conclusions

Scented Lemongrass Oil is one of the many essential oils in Indonesia that can be used as natural insect repellents and are very attractive to the pharmaceutical and fragrance industries. Therefore, this study successfully to produce fragrant citronella essential oil by extraction method but is assisted by solar energy for heat needs, where the essential oil obtained can be used as raw material for making perfumes that can treat dermatitis. This is a preliminary study and is expected to be continued for further research.

References

[1] A Wany, A., Jha, S., Nigam, V.K., Pandey, D.M., 2013. Chemical Analysis and Therapeutic uses of Citronella Oil from Cymbopogon Winterianus: A Short review. Int. J. Adv. Res. 1, 504-552
[2] Wany, A., Kumar, A., Nallapeta, S., Jha, S., Nigam, V.K., Pandey, D.M., 2014. Extraction and Characterization Of Essential Oil Components Based On Geraniol and Citronellol From Java Citronella (Cymbopogon Winterianus Jowitt). Plant Growth Regul, Vol. 73, 133–145.
[3] Almeida, R.N., Navarro, D.S., Barbosa-Filho, J.M., 2001. Plants with central analgesic activity. Phytomedicine Vol. 8, 310–322.
[4] Almeida, R.N., Motta, S.C., Faturi, C.B., Catallani, B., Leite, J.R., 2004. Anxiolytic like effect of rose oil inhalation on the levated plus maze test in rats. Pharmacol. Biochem. Behav, Vol. 77, 361–364.
[5] Robinson, T., Chitta, R, B., Sukumar, P., Vaibhav, V, G., 2016. Composition and Analysis of Antibacterial Activity of Citronella Oil Obtained by Hydrodistilation: Study of Process Optimized. Industrial Crops and Products, Vol. 94,178-188.
[6] Kusuma, H.S., Mahfud, M., 2017. Kinetic Studies on Extraction of Essential Oils from Sandalwood (Santalum Album) by Microwave Air Hydrodistilation Method. Alexandria Engineering Journal, 57, 1163-1172.
[7] Market Brief, 2015. Essential Oil. Indonesian Trade Promotion Center Chennai
[8] Alberto,C., Marina, L. N., Amparo, S., 2018. Perfumes: Encyclopedia of Analytical Science 3rd Edition, https://doi.org/10.1016/B978-0-12-409547-2.14037-5.
[9] Megil, J, M., Roy, B.R.P., Lois, R., Oliver, D., Sheena, F., Nicola, L., Rupika, D., 2018. Chemical Composition and Biological Activities of The Essential Oil from Cleome Rutidosperma DC: Fitoterapia, Vol. 129, 191-197.
[10] Rihayat, Suryani, Agusnar, H., Wirjosentono, B., Nurhanifa., *et.al.* 2018. Thermal Degradation of Aceh's Bentonite Reinforced Poly Lactic Acid (PLA) Based on Renewable Resources for Packaging Application: AIP Conference Proceedings, 2049, 020040
[11] Ridwan., Wirjosentono, B., Tamrin., Rihayat, T., Nurhanifa., *et.al.* 2018. Modification of PLA/PCL/Aceh's bentonite nanocomposites as biomedical materials. AIP Conference Proceedings, 2049, 020008
[12] Zulkifli, Z., Rihayat, T., Suryani., Zaimahwati, Z., Rosalina, R., *et.al.* 2018. Purification process of jelantah oil using active chorcoal kepok's banana. AIP Conference Proceedings, 2049, 020022
[13]Rihayat,T., Suryani., Satriananda., Fitriah., Helmi, *et.al.* 2018. Poly lactic acid (PLA)/chitosan/bentonite nanocomposites based on cassava starch for materials in biomedical applications. AIP Conference Proceedings, 2049, 020021
[14] Rihayat, T., Suryani., Satriananda., Riskina, S., Khan, N.S.P., Saifuddin, *et.al.* 2018. Influence of coating polyurethane with mixture of bentonite and chitosan nanocomposites. AIP Conference Proceedings, 2049, 020020
[15] Rihayat, T., Suryani., Fauzi, T., Nurhanifa., Alam, P, N., Sami, M., *et.al.* 2018. Synthesis and innovation of PLA/clay nanocomposite characterization againts to mechanical and thermal properties. IOP Conf. Series: Materials Science and Engineering 334, 012047

Synthesis and Characterization of North Aceh CEC Bentonite Determination with Methylene Blue Method and Increased D-Spacing after Addition of Surfactants CTAB-SDS

Teuku Rihayat[1*], Suryani[1], Januar Parlaungan Siregar[3], Satriananda[1], M Yunus[1], Sariadi[1], Fitria[2], Shafira Riskina[1], Zuhra Amalia[1], Wildan Syahputra[1], Nurhanifa[1], Safari[1], Nurul fitriah[1], Mardani[1], Jamiluddin Jaafar[3]

[1]Department of Chemichal Engineering, Lhokseumawe State Polytechnic, 24301, Aceh, Indonesia
[2]Department of Dermato Venereology, Medical Faculty, University Syiah Kuala, 24311, Aceh, Indonesia
[3] Structural Materials and Degradation Focus Group, Faculty of Mechanical Engineering, University Malaysia Pahang, 26600 Pekan, Pahang, Malaysia

[*]Corresponding author: teukurihayat@yahoo.com

Abstract. Bentonite is not susceptible to polymers because it is organophilic and has low basal spacing. The main objective of this study is to increase bason spacing of bentonite and change the properties of bentonite to hydrophobic by modifying bentonite with cationic and anionic surfactants using Cetyltrimethylammonium bromide (CTAB) and sodium dodecyl sulfate (SDS). Surfactant with different CEC concentration, namely CTAB 1 CEC, 2 CEC and 3 CEC while SDS 0.5 CEC, 1 CEC and 1.5 CEC where the CEC is generated by halometric methods. From the results of the analysis with CEC bentonite value through the halometric method obtained 95.2 meq/ 100 grams clay, while the basal spacing of interlayer bentonite with modification of cationic and anionic surfactants using X-Ray Diffraction (XRD) obtained an increase in basal spacing from 14. 18393 Å to 25.08971 Å on CTAB 3 CEC and SDS 1.5 CEC. This research can produce bentonite from northern Aceh with the largest basal spacing so that it can be intercalated by other additives.

1. Introduction

Bentonite is a rock term and is part of a group of smectite minerals containing montmorillonite. The main content of bentonite is mineral montmorillonite. Bentonite application has been widely used for fillers, pellets, drilling mud, catalysts and as adsorbents. Currently, it has also been made into several high value products such as nano materials, drugs and chemicals. Bentonite is widely used for the treatment of esophagitis, gastritis, and colitis and so on. Bentonite is also used as a raw material for cosmetics due to decontamination, detoxification, as a moisturizer. In addition, with chemical adsorption and stability, bentonite has been widely used in food processing industries which are closely related to human health and life [1].

Bentonite usually contains a negative charge which allows a cation exchange reaction. This charge comes from one or more of several different reactions [2]. The main contribution of the negative charge is the isomorphic substitute and the dissociation of the open hydroxyl group. The ions that can be exchanged are different ions around the alumina silica clay minerals.

Ion exchange reactions are stoichiometric and different from absorption or sorption and desorption. Ion exchange is a process whereby the cations usually found in the crystal layer are replaced by cations from the solution. In water, the cations on the surface of the layer become more easily replaced by other cations contained in the solution, known as exchangeable cations [3].

The absorption ability is stated in milliequivalent per 100 gram dry clays called Cation Exchange Capacity (CEC) .The price of clay mineral CEC varies according to the type and amount of colloids in the clay [4]. Breaking the bond around the angle of the silica-alumina unit in montmorillonite will cause an imbalance of surface charge. Al3 + substitution replaces Si4 + in tetrahedral sheets and substitution of more refined valence ions, especially Mg2 + replacing Al3 + in octahedral sheets produces an unbalanced charge in the montmorillonite structure unit.

The use of bentonite as an adsorbent for inorganic compounds such as heavy metal ions because bentonite has cation exchange capacity and has hydrophilic properties on the surface. The chemical properties of the bentonite pore structure generally determine its adsorptive capacity [5]. Because of the ineffective hydrophilic nature of bentonite to absorb organic compounds, to increase its capacity. One of the deficiencies of clay is its hydrophilic nature which can cause agglomeration of clay minerals in a hydrophobic polymer matrix. This deficiency can be overcome by intercalating organic cations such as amino acids or alkyl ammonium to form organoclays that are hydrophobic.

Increasing the basal spacing after the intercalation process can also increase the diffusion capability of the polymer or polymer coating into clay interlayer. Intercalation is based on the exchange of cations found in the layers of clay, such as Na +, K + and Ca2 +. Intercalation into the clay structure resulted in an increase in surface area, basal spacing (spacing of montmorillonite silicate layer) and surface acidity which affected the adsorption power [6].

This intercalation process can cause clay pores to become larger and homogeneous and the layers become more stable than before being intercalated [7]. The purpose of intercalation is to: expand the interlayer distance, reduce solid-solid interactions between clays and increase the interaction between clay and matrix.

Organic compounds, bentonite is intercalated with surfactants that can interact with the negative charge present on the surface of the interlayer. Bentonite has a complex chemical surface and arises from its ability to form thixotropic water gels, high water absorption, high surface area, layered structure and high cation exchange capability, and contains crystals of quartz, cristobalite, feldspar, or other compounds [8]. Mineralogical composition will affect the CEC values obtained. Analytical problems arise from the method of determining CEC resulting from specific interactions between bentonite components and cation exchange modes which can cause an increase or decrease in the concentration of the cation index [9].

The nature of smectite is to absorb the cationic type of solution with the exchange between the cation equilibrium that is charged with a negative layer and the cation in solution is called the cation capacity exchange (CEC) and the CEC value can be expressed in centimole positive charge. per kilogram of mineral clay to dry is equal to milliequivalent per 100 g of clay (meq / 100 g) [10].

Determination of CEC value depends on the method used. However, determining the value of CEC is sometimes difficult and time consuming in the process of achieving perfect cation exchange. The researcher determined the value of Bentonite CEC in north Aceh and Kabupaten Bener Meriah with one method namely the methylene blue method.

The purpose of this study was to increase the basal spacing of north Aceh bentonite which was then characterized using X-ray Diffraction (XRD) to determine the element or compound content (qualitative analysis) and composition determination (quantitative analysis), phase change as a function of temperature and time in bentonite using Fourier Transform Infrared (FTIR) and Scanning Electron Analysis Microscope (SEM).

2. Methods

2.1 Materials

The materials used in this study were bentonite from Nisam, North Aceh, Methylene Blue (Merck, Germany), aquadest, aquabidest, Sodium Deodecyl Sulfate (SDS), Cetyltrimethylammonium Bromide (CTAB), $AgNO_3$ solution and sodium hexametaphosphate $(NaPO_3)_6$.

2.2 Determination of Cation Exchange Capacity (CEC)
Determination of CEC values by determining the specific surface of the bentonite analyzed. Bentonite is crushed first, then mashed and sieved with a size of 100 meshes. Weigh 1.25 grams of methylene blue and dissolve it with an aquadest of 125 ml of aquadest. Weigh 2.00 grams of bentonite and dissolve with distilled water 100 ml of water and stir the two solutions in a glass beaker with a stirring speed of 700 rpm for 5 minutes. Take 1 ml of the methylene solution and put it into a bentonite solution and stir it with a stirring speed of 400 rpm for 2 minutes. Note the methylene blue normality is 0.028 N. Then, drop the solution on the filter paper and then observe, if the droplets form is a blue hello, then that is the end of the titration and it hasn't formed like hello above. Then add 1 ml of methylene blue into the solution bentonite. Every change when titration is followed by dripping on filter paper, the aim is to see the blue halos around dark blue and this substitution between bentonite and methylene blue exists and continues to absorb. This is called the principle of the methylene blue method. The volume of methylene blue is added to the value of the bentonite CEC obtained where the CEC unit is milliequivalent per 100 grams of clay bentonite. This is a halometric method.

Measurement of CEC values using the following equation (1)

$$\text{CEC}\,\frac{meq}{100\;gram} = \frac{100}{F} \times V \times NMB \qquad (1)$$

Where, F is the dry sample weight, the volume of methylene is blue during the titration process with 0.028 N methylene blue concentrations.

2.3 Purification of Bentonite with Cationic and Anionic Surfactant
Bentonite with a size of 100 meshes as much as 2 grams mixed with 98 ml of aquadest with 0.5 hours at 80°C. CTAB surfactants are added with various CEC concentrations, namely 0.7 CEC, 1.0 CEC and 1.3 CEC stirred for 2 hours at 80°C. After cationic surfactant is homogeneous with bentonite, then placed the anionic surfactant into bentonite solution which has been homogeneous with cationic surfactant. Next, stir until homogeneous for 1 hour at 80°C with varying CEC concentrations 0.1 CEC, 0.2 CEC and 0.3 CEC. Furthermore, the results of the resulting samples were analyzed using XRD to determine the basal spacing of bentonite after the addition of cationic and anionic surfactants.

2.4 X-Ray Diffraction (XRD)
XRD analysis was used to identify the magnification of bentonite basal spacing after the addition of cationic surfactants and anionic surfactants. Analysis results in the form of basal spacing data with Å units.

3. Result and Discussion

3.1 Effect of Cation Exchange Capacity (CEC) Surfactant on the Bentonite Interlayer
CEC or called cation exchange capacity (CEC) is a soil component to absorb and exchange or rerelease soil solutions. One of the CEC tests is the methylene blue method. Methylene blue powder when mixed with water and identified with the chemical formula: $C_{16}H_{18}N_3SCl$. When mixed with clay (bentonite), the chloride ion in a solution of methylene blue is replaced with cation in clay. The amount of adsorbed methylene solution varies according to the number of minerals and types of clay [11].

Liao's research in 2016 stated that bentonite with the addition of cationic surfactants can improve the natural properties of bentonite itself (high hydrophobic) and with the addition of anionic surfactants can the cation driving force in the interlayer of bentonite. Where the linking of these two surfactants can increase the basal spacing of the interlayer bentonite, so that the cation can enter the interlayer of bentonite.

The mixture is made by adding 5 ml of methylene blue solution to the stirred clay solution to 200 rpm. After 1 minute, a number of mixtures is taken with a dropper pipette and dripped onto filter paper. Generally, on filter paper there is a circle but there is no blue shadow around it this means the test is negative. To get a positive test, repeat the previous step until you see the last drop has a bright blue circle around dark blue meaning the test is positive or this can be called as halometry.

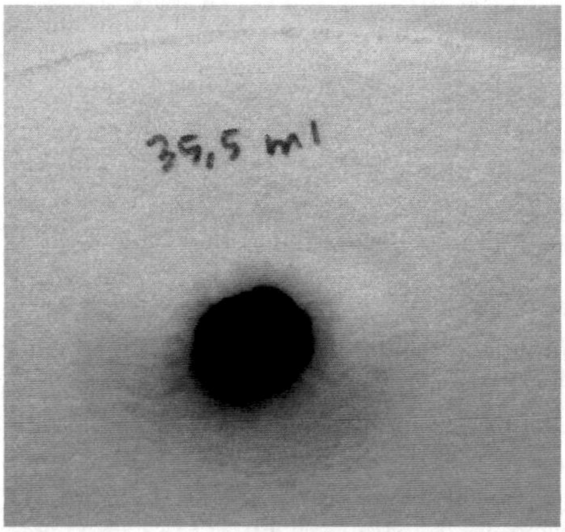

Figure 1. The acquisition of North Aceh bentonite CEC is marked by Halo metric formation on filter paper.

3.2 Bentonite Preparation with Addition of Cationic-Anionic Surfactants
Bentonite clay is widely used in paint. White Bentonite is the material of choice that is widely available around. In water-based paint, sodium or montmorillonite suspend and thicken substances. This montmorillonite is also used as an emulsifier in water and oil paint formulations. Organoclays can be made specifically with organic compounds to meet different vehicle requirements including varnishes, epoxy resins, and vinyl resins which are used in paint formulations. These organoclays improve pigment suspension, viscosity, and techotropic control and are very good in non-drip emulsion paints [12].

Bentonite used is from Nisam, north Aceh. This bentonite was previously ground and then sieved to a size of 100 meshes. Before being mixed with surfactant, raw bentonite was first analyzed using XRD to find out the spacing of the interlayer bentonite.

3.2.1 X-Ray Diffraction (XRD) Analysis. X-Ray Diffraction is one of the characterizations that use x-rays; x-ray scattering is produced if an electrous metal is fired by electrons at high speed in a vacuum tube. A crystal can be used to diffract x-ray beam due to the order of x-ray wavelengths almost the same or smaller with the order of distance between atoms in a crystal.

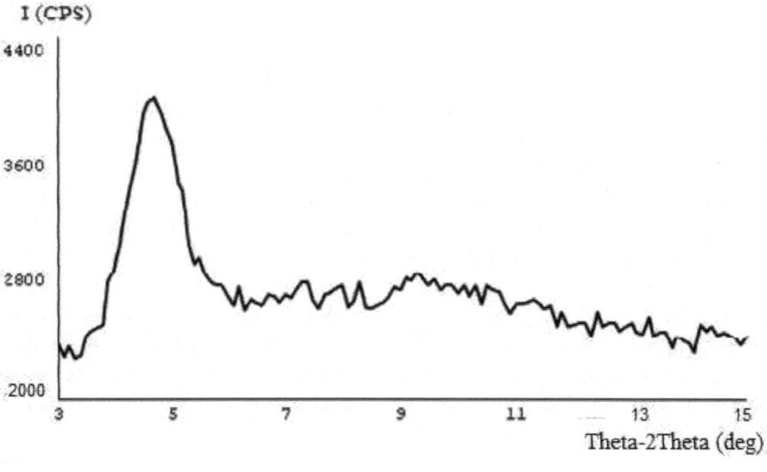

Figure 2. Graph XRD of bentonite sample without modification using surfactants

The basal spacing of raw bentonite is 14.18393 Å from figure 2. Then the raw bentonite was mixed with Cetyltrimethylammonium Bromide (CTAB) cationic surfactant stirred for 2 hours and then the mixture of Anionic Sodium Deodecyl Sulfate (SDS) surfactant was added for 1 hour. The results of these two types of surfactant were then washed to remove surfactants contained in the interlayer of bentonite using aquadest. To test whether the surfactant has been lost from the bentonite solution, $AgNO_3$ solution was used which was dripped into surfactant water. If the water turned gray then still contained surfactant, therefore it is necessary to re-wash with aquadest.

Then solid free from surfactants are filtered with a vacuum pump to separate solids with the remaining liquid. The solids are obtained then in the oven at 150°C until dry. Then it is crushed again until smooth and then re-analyzed with XRD to see the increase in d-spacing and SEM to observe the morphological structure of the bentonite obtained.

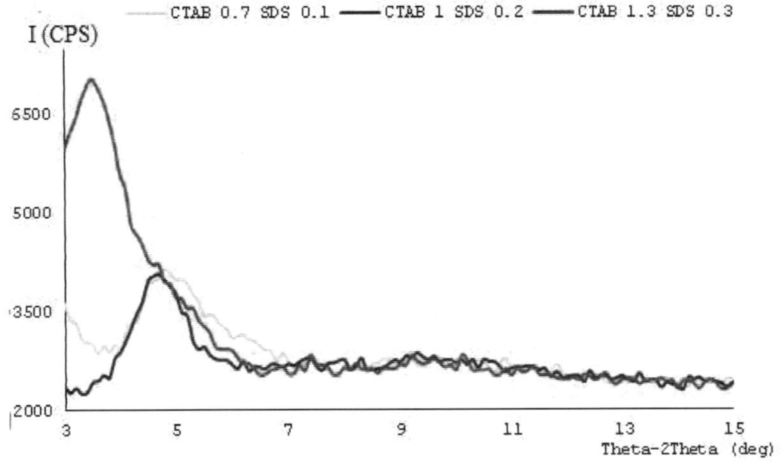

Figure 3. Graph XRD of bentonite samples after intercalation

The bentonite basal spacing has been modified using cationic and anionic surfactants on CTAB 0.7 CEC and SDS 0.1. CEC obtained spacing namely 15.11134 A on CTAB 1 CEC and SDS 0.2 CEC obtained spacing for example 19.11531 A on CTAB 1.3 CEC and SDS 0.3 CEC it is found that spacing is 23.00023A from figure 3. Thus, the highest d-spacing is the addition of CTAB 1.3 CEC and SDS 0.3. CEC surfactant which is 23.00023A. The intercalation mechanism between CTAB is positively charged and interlayer of negatively charged bentonite through cation exchange. When SDS is added, interlayer bentonite increases [6].

4. Conclusion

Bentonite is part of a group of smectite minerals containing montmorillonite. The application of bentonite has been widely used for fillers, pellets, drilling mud, catalysts and as adsorbents. This study has succeeded in determining the value of CEC using several methods. However, determining the value of CEC is sometimes difficult and time consuming in the process of achieving perfect cation each.

IOP Conf. Series: Materials Science and Engineering **506** (2019) 012054 doi:10.1088/1757-899X/506/1/012054

References

[1] Alves, Jefferson L., Paulo R., Ana R.M. 2017. *Evaluation of Organic Modification of Montmorillonit With Ionic and Nonionic Surfactants*. Applied Clay Science **150** 23-33

[2] Chatel G., Liudmila N., Sabine P. 2016. *How Efficiently Combine Sonochemistry and Clay Science*. Applied Clay Science **119** 193-201

[3] Chen D., Jian C., Xinlong L., Haipeng J., Zhiguo X. 2011. *Characterization of Anion- Cationic Surfactants Modified Montmorillonite and its Application for the Removal of Methyl Orange*. Chemical Engineering Journal **171**. 1150-1158

[4] Dewi N P W T., Nengah S., Wayan S., 2017. *Modifikasi Lempung Bentonite Teraktivasi Asam Dengan Benzalkonium Klorida Sebagai Adsorben Zat Warna Rhodamine B*. Universitas Udayana ,Bukit Jimbaran **11**. 75-81

[5] Flach C F., Chandan P., Carl J S., Erik K., Marcus O., Johan B. P, Mats T, Joakim L 2017. *Does Antifouling Paint Select for Antibiotic Resistance. Science of the Total Environment*

[6] Liu G., Guomin W., Shuping H., Can J., Zhenwu K. 2017. *Synthesis and Properties of Non-Isoyanate Polyurethane Coatings derived from Cyclic Carbonate-Functionalized Polysiloxanes. Progress in Organic Coatings* **112**. 169-175

[7] Pilch P B., Lukasz B., Bogdan M., 2017. *Synthesis and Surface Modification of Polyurethaneswith Chitosan for Antibacterial Properties. Carbohydrate Polymers* **112**. 39- 47

[8] Rahman A., Yateman A., Sutarno., Ariadne L. J., Seto R. 2015. *Sintesis dan Karakterisasi Organolempung Dari Bentonit Indonesia*. Jurnal Fisika dan Aplikasinya **16**. 1

[9] Clouttier M. D., Mantovani and Rosei F. 2015. *Antibacterial Coatings: Challenges, Perspectives, and Opportunities.Trends in Biotechnology* **33**. 11.

[10] Salwiczek M., Yue Q., James G., Richard S, Trevor L, Keith M M, Helmut T 2014. *Emerging Rules for Effective Antimicrobial Coatings*. Trends in Biotechnology **32**.2

[11] Turoz M., Hasan T. 2011. *The Use of Methylene Blue Test for Predicting Swell Parameters of Natural Clay Soils.* Scientific Research and Essay 6 **(8).** 1780-1792

[12] Martinez.J.M., Volzone C., Garrido. L.B. 2017: *Evaluation of Polymeric Al-Modified Bentonite for its Potential Application as Ceramic Coating*. Applied Clay Science **140**. 20-27

Effect of Determination Temperature on Nutrition and Organoleptic Tuna Fish Floss

Teuku Rihayat[1*], Suryani[1], Zaimahwati[1], Salmyah[1], Sariadi[1], Fitria[2], Satriananda[1], Alfian Putra[1], Zahra Fona[1], Juanda[1], Raudah[1], Mawaddah[1], Nurhanifa[1], Shafira Riskina[1], Wildan Syahputra[1], Safari[1]
[1] Department of Chemichal Engineering, Lhokseumawe State Polytechnic, 24301, Aceh, Indonesia
[2] Department of Dermato Venereology, Medical Faculty, University Syiah Kuala, 24311, Aceh, Indonesia

*Corresponding author: teukurihayat@yahoo.com

Abstract. Tuna fish is one type of marine fish that is known to have benefits that are rich in protein and vitamin B6. One effort to utilize tuna is to process it into shredded fish. This study aims to determine the effect of steaming temperature on the nutritional and organoleptic content of tuna shredded fish and to obtain the optimum temperature to produce shredded with the best nutritional and organoleptic content. The treatment in this study was the variation of steaming temperature, then an analysis was carried out on the resulting Fish Floss to the protein content, fat content, moisture content, ash content and organoleptic test (Flavor, taste, color, texture). The results of the study were processed using simple randomized design with 3 replications. The results showed that different steaming temperatures had a significant effect on the nutritional and organoleptic content of shredded fish. The best treatment was obtained at a steaming temperature of 55° C (A) with an average protein content value of 8.9812%; fat content of 1.3871%; moisture content of 4.9876%; ash content 3.6543%; organoleptic value of Flavor 9,0098; taste of organoleptic value 8,9654; color of organoleptic 8,8432 and organoleptic value of texture 8,876.

Keywords: Nutrition content, organoleptic content, tuna fish, steaming temperature variation

1. Introduction

The fisheries production in the city of Lhoseumawe in recent years has experienced fluctuations. Marine fisheries produce about 7405 tons of fish in 2014, 10066 tons in 2015 and 11526 tons in 2016 [1].

Fisheries sector are an important resource for people's life and have the potential to be the main contributors to the national economy. This is based on the fact that Indonesia owns large fisheries resources in term of quantity of diversity. In addition, fisheries sector also linked with other sectors. The national resource-based fishing industry is also known as national resources based and Indonesia also has a high comparative advantage in the fisheries sector as reflected in the potential of existing resources [2].

Processed fishery products is very important in fish commodity sector. Fish is a commodity that rapidly decays (perishable food). Decay can be caused by enzymes, whether from the fish itself and microbes, and rancidity [3]. High numbers of fresh fish water accelerate the proliferation process of decomposition by the microorganisms contained in it. The low durability of fresh fish is an obstacle in the effort to expand the market of fishery products. In fact, it is often causing large losses when there are abundant of fish available. Therefore, for a long time the community have tried to do various kinds of post-harvest fish processing to minimize these obstacles. One of strategies was making shredded tuna products [4].

Tuna is one of the highly potential fish species in Aceh, especially Lhokseumawe. Tuna is a saltwater fish that has thick flesh with a good flavour and contain 28 times more omega-3 than freshwater fish. Consumption of tuna for about 30 g a day can reduce the risk of heart disease by 50% [5].Tuna is rich in protein, including many niacin, selenium, and tryptophan. Tuna is also rich in phosphorus, potassium, omega-3 fatty acids, magnesium, thiamine, and vitamin B6 which are well suited to be made into floss products because of their high nutritional content [6]. For this reason, this study aims to determine the effect of steaming temperature on the nutrient and organoleptic content of tuna fish floss, and to obtain the optimum temperature to produce tuna shredded fish with the best nutrient and organoleptic content.

2. Methodology

2.1. Materials

The tuna fish with weigh ± 1 kg each and length of ±30 cm are obtained from the Lhokseumawe Inpres Market, together with the additional ingredients which include shallots, garlic, coriander, coconut milk, beans soil, femented bilimbi, sugar and salt. Other materials that are use for the analysis included aquades, label paper, Kjeldahl tablets, H2SO4, NaOH-thiosulfates, indicators of red metal, NaOH, n-hexane, and Na-K tartate.

2.2 Processing of Tuna Fish Floss

In general, the processing of tuna into floss includes several stages, which are the collection and selection of raw materials, separation of fish meat, making spices, boiling, cooking and packaging the processed floss. Initially tuna that were caught in the sea around Ujong Blang, Lhokseumawe were collected and sorted based on fish quality criteria, including freshness and size of fish. Freshness of the fish is determined based on the fresh colour of the fishes' eyes instead of red eyes, the appearance of fish colour that is not pale, and the aroma of fresh fish. The raw tuna's flesh is slice out from its stomach and head. Furthermore, the separation of flesh from the skin and bone is done by using a flesh separator. The flesh separator will separate the flesh from the skin and bones at low temperatures.

Afterwards, the tuna flesh is washed and cut, then the fish is boiled for about 20 minutes. The boiled tuna flesh is then drained from excess water. This is done in order to remove the excess moisture from the floss products as it will affect the durability of floss products. Fish flesh is ensured to have enough contact area of the meat with herbs so that the spices can penetrate into the flesh perfectly.

Meanwhile, shredded spices are prepared. This seasoning is the distinctive flavour of the Aceh region that will distinguish this product from other shredded tuna products. This spice consists of: onion, garlic, red chili, cayenne pepper, and fermented bilimbi. The use of this flavours makes the tuna floss product has a distinctive taste of the Aceh region. The people of Aceh always used various types of spices, which can also offer durable properties in food products. The spices are milled using a blender until smooth and evenly mixed. The spices are then cooked (sautéed) for about 15 minutes using oil until the spices are cooked and fragrant. Furthermore, tuna meat is mixed into the spices and cooked for 2 hours until the seasoning is absorbed. Floss is ready to be packaged. The tuna shreds produced were analysed chemically and undergo organoleptic tests.

2.3 Research Treatment

The treatment of this study using variations of steaming temperature which are as follows:

i. Steaming temperature A: 55.0 °C
ii. Steaming temperature B: 57.5 °C
iii. Steaming temperature C: 60.0 °C
iv. Steaming temperature D: 62.5 °C

v. Steaming temperature E: 65.0 °C

The output parameters observed in this study were protein content, fat content, moisture content of ash content and organoleptic test (aroma, taste, colour, texture).

2.4 Data Analysis
Data processing in this study uses a simple Randomized Design (CRD) model with 3 replications. The data obtained were then analyzed using Analysis of Variance (ANOVA) and continued with the LSD test.

3. Result and Discussion
In this main study the treatment used was to use a different steaming temperature for making tuna shredded fish. The steaming temperature used is 55; 57.5; 60; 52.5 and 65°C. The maximum temperature in the main study is 65°C. this is because at a temperature of 70°C it has been damaged because the use of steaming temperature is too high which is then fried. The results of the study of the effect of steaming temperature on the nutrient and organoleptic content of tuna shredded consisted of chemical parameters (protein content, fat content, ash content, moisture content) and organoleptic parameters (aroma, color, texture, taste). The average value of the main research results with successive chemical and organoleptic parameters can be seen in table 1 and table 2.

Table 1. Main Research Results of Tuna Floss on Chemical Parameters

Temperature (°C)	Parameters (%)			
	Protein Content	Fat Content	Moisture Content	Ash Content
55.0	8.9812	1.3871	4.9876	3.6543
57.5	7.1393	2.1543	4.9768	4.7839
60.0	6.8007	2.4763	5.6147	5.3195
62.5	6.9987	2.7629	5.8453	5.8253
65.0	5.9567	1.1819	4.2987	4.5524

Table 2. The Main Results of Tuna Floss on Organoleptic Parameters

Temperature (°C)	Organoleptic Parameters			
	Aroma	Taste	Color	Texture
55.0	9.0098	8.9654	8.8432	8.8765
57.5	8.6745	8.6355	8.7969	8.5181
60.0	8.5675	8.2908	8.6762	8.4557
62.5	8.6748	8.6252	8.6315	8.8471
65.0	8.6084	8.6752	8.9853	8.6879

3.1. Chemical Parameters
3.1.1 Protein Content. Proteins are macro molecules that have a molecular weight of between 5000 and several million. Proteins consist of long chains of amino acids, which are bound to one another in peptide bonds. Nitrogen is the main element of protein, because it is present in all proteins, which has a proportion of 16% of total protein [5].

The purpose of analyzing protein in food is to reduce the amount of protein in food ingredients; determine the level of protein quality in terms of nutrition; and examine protein as a chemical [7], the calculated protein content is the level of crude protein (crude protein). This is because nitrogen contained in food is not only derived from protein amino acids, but also from other nitrogen compounds that can / cannot be used as a source of nitrogen in the body. In fish, in one part of nitrogen there are free amino acids and peptides namely volatile nitrogen bases and amino-metal compounds.

The results of the protein content test on tuna floss ranged from 5.9567% to 8.9812%. The results of analysis of variance (ANOVA) showed that steaming treatment with different temperatures had a very significant effect on protein content parameters. This can be seen from the F count > F table 5%, then to find out the differences from each treatment followed by the LSD test. The average protein content in tuna fihs floss can be seen in table 3.

Table 3. Average Protein Content in Tuna Fish Floss

No.	Temperature (°C)	Protein Content (%)	
		Average ± St. Dev	Notation
1.	55.0	8.9812 ± 0.8892	d
2.	57.5	7.1393 ± 0.6831	c
3.	60.0	6.8007 ± 0.7526	b
4.	62.5	6.9987 ± 0.6869	a
5.	65.0	5.9567 ± 0.6163	a

Based on the data in table 3 above, it can be seen that the highest protein content is found in treatment A, which is steaming at 55°C, which is 8.9812%, while the lowest protein content is in treatment E, which is steaming at 65°C which is 5.9567%. The higher the steaming temperature used, the lower the tuna protein content in tuna fish. In addition to the higher steaming temperature, a decrease in the amount of protein is also caused by the temperature of the frying pan. This is presumed, the protein content in the material is denatured due to repeated cooking, namely steaming and frying. processing protein foods that are not well controlled can cause a decrease in nutritional value. The most widely used processing is processing using heating such as sterilization, cooking and drying. Conversely, the lower the steaming temperature used, the higher the protein content in tuna fish floss.

The requirements for floss quality standards in general the value of protein content is at least 15%, and the highest protein content in tuna tuna is 8.9812% so that the protein content in tuna tuna has not met the requirements of floss quality standards. This is due to the repeated processing of heat, which is the steaming and frying process which causes damage to the protein in tuna meat.

3.1.2 Fat Content. Fat is a food substance that is important for the health of the human body. In addition fat is also found in almost all food ingredients with different contents [8]. According to [9], fat consists of mixed triglycerides, which are esters of glycerol and long chain fatty acids. If the fat is hydrolyzed, it will produce 3 long chain fatty acid molecules and 1 glycerol molecule.

Determination of fat content of an ingredient can be done using the soxhlet apparatus. This method can be used for oil extraction from oil-containing materials [10], fat extraction from dry matter can be carried out intermittently or continuously.

The results of the fat content test on tuna floss ranged from 1.1819% to 2.1543%. The results of analysis of variance (ANOVA) showed that steaming treatment with different temperatures had a significant effect on fat content parameters. This can be seen from the value of F count > F table 5%, then to find out the differences of each treatment carried out LSD test. The average fat content in tuna fish shredded can be seen in table 4.

Table 4. Average Fat Content in Tuna Fish Floss

No.	Temperature (°C)	Protein Content (%)	
		Average ± St. Dev	Notation
1.	55.0	1.3871 ± 0,1392	b
2.	57.5	2.1543 ± 0,3543	c
3.	60.0	2.4763 ± 0,4928	c
4.	62.5	1.7629 ± 0,2839	a
5.	65.0	1.1819 ± 0.1579	a

Based on the data in Table 4 above, it can be seen that the highest fat content is B treatment with a steaming temperature of 60°C and an average fat content of 2.4763% while the lowest fat content is in treatment E with a steaming temperature of 65°C, the average value average fat content of 1.1819%. This shows a decrease in the value of the average fat content of the material. The decrease was caused by an increase in steaming temperature which caused fat to be damaged and the amount decreased. The level of fat damage varies depending on the temperature used and processing time [11]. The higher the temperature used, the fat damage will increase. Essential fatty acids are isomerized when heated in alkaline solutions and sensitive to light, temperature and oxygen. The process of fat oxidation can cause inactivation of its biological functions and can even be toxic. In addition to fat damaged by oxidation, fat can also be damaged by hydrolysis.

The requirements for floss quality standards in general are maximal fat content values of 30%, and the highest fat content values in tuna shredded fish are 2.4763% so that the fat content in tuna floss meets the requirements of floss quality standards.

3.1.3 Moisture Content. The principle of determining the water content by the Thermogravimetric method is to evaporate the water in the food by heating then weighing the material to a constant weight which means that all the water has been evaporated.

The water content test results on tuna fish floss ranged from 4.2987% to 5.8453%. While the results of analysis of variance (ANOVA) showed that steaming treatment with different temperatures did not have a significant effect on fat content parameters. This can be seen from the value of F count <F table 5%. The average water content in tuna fish floss can be seen in table 5.

Table 5. Average Moisture Content in Tuna Fish Floss

No.	Temperature (°C)	Protein Content (%)	
		Average ± St. Dev	Notation
1.	55.0	4.9876 ± 3.5879	a
2.	57.5	4.9768 ± 0.5601	a
3.	60.0	5.6147 ± 1.1752	a
4.	62.5	5.8453 ± 3.0837	a
5.	65.0	4.2987 ± 2.9593	a

Based on the data in Table 6 above, it can be seen that at a temperature of 60°C the highest average water content is 5.8453%, and at a temperature of 65°C the lowest average water content is 4.2987%. The highest water content is found in treatment D with an average water content value of 5.8453%. This is due to the steaming process with higher temperatures causing bonds between broken food components such as carbohydrates, fats and proteins, so that the water will bind to the material and cause the water content to increase. While the lowest water content is at steaming temperature of 65°C, this is due to the higher steaming temperature used, the fish meat will be more mature and the shredding process will be easier and more perfect so that the resulting texture can be softer so that the frying process ingredients can evaporate perfectly.

The requirements for floss quality standards in general have a maximum water content value of 7%, and the highest water content value in tuna floss is 5.8453% so that the moisture content in cork floss meets the requirements of floss quality standards. The absorption of isothermic water content by materials is strongly influenced by the temperature when drying and the higher the drying temperature, the lower the ability to absorb water. Furthermore, it was found that heating at temperatures of 65, 80 and 90°C showed that the higher the heating temperature, the ability of sorbsi decreased. In addition, an increase in processing temperature will reduce the ability to hold water which may be caused by changes due to heating that cannot return.

3.1.4 Ash Content. Ash is an inorganic substance left over from the combustion of an organic material. The ash content and composition depend on the type of material and the method of ignition.

Ash content has to do with minerals in a material. The purpose of determining total ash is to determine whether or not a processing process; to find out the type of material used and the determination of total ash useful as a parameter of the nutritional value of food ingredients.

The ash content test on tuna shredded fish ranged from 3.6290% to 5.5789%. While the results of analysis of variance (ANOVA) showed that the steaming treatment with different temperatures gave a very significant effect on the ash content parameters. This can be seen from the F count> F table 5%, then to find out the differences in each treatment carried out LSD test. The average ash content in floss from residual meat of cork fish albumin extraction can be seen in table 6.

Table 6. Swelling Properties of Wound Dressing

No.	Temperature (°C)	Protein Content (%) Average ± St. Dev	No. Notation
1.	55.0	3.6543 ± 0.3055	a
2.	57.5	4.7839 ± 0.2460	d
3.	60.0	5.3195 ± 0.5054	c
4.	62.5	5.8253 ± 0.0947	a
5.	65.0	4.5524 ± 0.5086	b

Based on the data in table 6 above, it can be seen that at a temperature of 62.5°C the highest average ash content was 5.8253%, and at 55°C the lowest average value of ash was 3.6543%. The highest ash content is in treatment D with steaming temperature 62.5°C with an average value of ash content of 5.8253%, this is thought to be a long floss frying process so that the water content in the shredded is low and leaves minerals in high floss so the ash content increases. While the lowest ash content is in treatment A with a temperature treatment of 55°C with an average value of ash content of 3.6543%. This is presumably due to the high moisture content in the ingredients and the floss texture that has not been smooth so that the ash content is low. The effect of processing on materials can affect the availability of minerals for the body. The use of water in the process of washing, soaking and boiling can reduce the availability of minerals because minerals will dissolve by the water used.

The requirements for floss quality standards in general the value of ash content is a maximum of 7%, and the highest ash content in tuna tuna is 5.8253% so that the ash content of tuna shredded meets floss quality standards.

3.2. Organoleptic Parameters

3.2.1 Aroma. The aroma of food that is in the mouth is captured by the sense of smell through a channel that connects the mouth and nose. The number of volatile components released by a product is influenced by the temperature and its natural components. Food that is brought to the mouth is felt by the senses of taste and smell which are then continued to be accepted and interpreted by the brain [10]. The results of the organoleptic scent on tuna fish floss ranged from 8,5675 to 9,0098. The average organoletic aroma results in tuna fish shredded can be seen in table 7.

Table 7. Average Aroma Organoleptic Test on Tuna Fish Floss

No.	Temperature (°C)	Average ± St. Dev
1.	55.0	9.0098 ± 0.1453
2.	57.5	8.6745 ± 0.0694
3.	60.0	8.5675 ± 0.0962
4.	62.5	8.6748 ± 0.1347
5.	65.0	8.6084 ± 0.0385

Based on the data in table 7 above, it can be seen that at a temperature of 55°C the highest average organoleptic aroma value is 9,0098, and at a temperature of 60°C the lowest organoleptic aroma value is 8,5675. The highest organoleptic value of aroma is in treatment A, which is treatment with a steaming temperature of 55°C which is equal to 9,0098, this is presumably because it is influenced by the maturity level of steamed meat. So that the distinctive aroma of fish with spices is still felt. While the lowest aroma organoleptic value is in treatment C, which is treatment with steaming temperature of 60°C. This is suspected to be fish floss, no longer typical of floss odor.

Based on the calculation of consumer acceptance of the organoleptic aroma of tuna shredded showed that the best P value is in treatment A with a value of 9,0000 then rounded to 9. Descriptively, at this value it can be concluded that the aroma of cork fish on treatment A favored panelists.

3.2.2 *Taste.* Taste is something that is received by the tongue. In sensing the gecko is divided into four main gases, namely sweet, bitter, sour and salty and there is an additional response if done modification [11]. Taste is influenced by several components, namely chemical compounds, temperature, concentration and interaction with other taste components. The increase in temperature will increase the stimulation of sweet taste but will reduce the stimulation of salty and bitter taste [12]. The organoleptic test results of taste on tuna fish floss ranged from 8,2908 to 8,9654. The average yield of organoletic taste in tuna fish floss can be seen in table 8.

Table 8. Average Taste Organoleptic Test on Tuna Fish Floss

No.	Temperature (°C)	Average ± St. Dev
1.	55.0	8.9654 ± 0.1347
2.	57.5	8.6355 ± 0.1678
3.	60.0	8.2908 ± 0.2457
4.	62.5	8.6252 ± 0.0509
5.	65.0	8.6752 ± 0.1018

Based on the data in table 8 above it can be seen that at a temperature of 55°C the highest organoleptic taste value was 8,9654, and at a temperature of 60°C it had the lowest organoleptic taste value of 8,2908. The organoleptic value of taste in tuna floss with different steaming does not give a different value. The resulting shred has almost the same value. This is because the formulation of making floss is used constantly, so that the taste produced is almost the same.

Based on the calculation of consumer acceptance of organoleptic taste shows that the best P value is in treatment A and E with a value of 8,9654 then rounded to 9. Descriptively, at this value it can be concluded that the tuna fish floss A and E are preferred by panelists.

3.2.3 *Color.* Color is one parameter besides taste, texture and nutritional value that determines consumer perceptions of a food ingredient. Consumer preferences are often determined based on the appearance of a food product. Bright food colors provide more appeal to consumers. Color in food products has several functions, among others, as an indicator of maturity, especially for fresh food products such as fruits, as an indicator of freshness for example in vegetable and meat products and as an indicator of perfection of food processing processes such as frying, brown used as the final indicator of food product maturity [12].

The color organoleptic test results on tuna fish floss ranged from 8,6315 to 8,9853. The average yield of color organoletics in tuna fish floss can be seen in table 9.

Table 9. Average Color Organoleptic Test on Tuna Fish Floss

No.	Temperature (°C)	Average ± St. Dev
1.	55.0	8.8432 ± 0.3533
2.	57.5	8.7949 ± 0.0192
3.	60.0	8.6762 ± 0.3977
4.	62.5	8.6315 ± 0.2589
5.	65.0	8.9853 ± 0.0667

Based on the data in table 9 above, it can be seen that at 65°C the highest color organoleptic value was 8,9853, and at 62.5°C the lowest color organoleptic value was 8,6315. The color organoleptic values of the tuna shredded with different broccoli do not give different values. The resulting shred has almost the same value.

Based on the calculation of consumer acceptance of the organoleptic color shows that the best P value is in treatment E and A with a value of 9.1000 then rounded to 9. Descriptively, at this value it can be concluded that the color of tuna fish floss E and A is preferred by panelists.

3.2.4 Texture. Observation of the texture of fish floss is very important. This is because texture is one of the things that distinguishes fish floss from other fishery products, namely soft fibers. The texture of meat is very influential on the final product produced and determines the level of consumer preference for the product [12].

The organoleptic test results of the texture on tuna fish floss ranged from 8,4557 to 8,8765. The average yield of organoletic textures on tuna fish floss can be seen in table 10.

Table 10. Average Organoleptic Texture Test on Tuna Fish Floss

No.	Temperature (°C)	Average ± St. Dev
1.	55.0	8.8765 ± 0.1202
2.	57.5	8.5181 ± 0.1732
3.	60.0	8.4557 ± 0.2143
4.	62.5	8.8471 ± 0.0509
5.	65.0	8.6879 ± 0.0770

Based on the data in Table 10 above, it can be seen that at a temperature of 55°C the highest average organoleptic texture value was 8,8765, and at a temperature of 60°C it had the lowest color organoleptic value of 8,4557.

Based on the calculation of consumer acceptance of organoleptic colors shows that the best P value is in treatment A with a value of 8.9000 then rounded to 9. Descriptively, at this value it can be concluded that the color of tuna fish floss A is favored by panellists.

4. Conclusion

From this study, we can conclude that different steaming temperatures had a significant effect on the nutritional and organoleptic content of shredded fish. The best treatment was obtained at a steaming temperature of 55° C (A) with an average protein content value of 8.9812%; fat content of 1.3871%; moisture content of 4.9876%; ash content 3.6543%; organoleptic value of aroma 9,0098; taste of organoleptic value 8,9654; color of organoleptic 8,8432 and organoleptic value of texture 8,8765.

5. References

[1] Almatsier S 2009 Prinsip Dasar Ilmu Gizi *PT Gramedia Pustaka Utama Jakarta* pp 337
[2] Andarwulan N F Kusnandar D Herawati 2011 Analisis Pangan *Dian Rakyat* (Jakarta). 328 hlm
[3] Chayati I and Andian A A 2008 Diktat Kimia Pangan *Universitas Negeri Yogyakarta* (Yogyakarta) pp 62
[4] De Garmo E P W G Sullivan J R Canada 1984 Engineering Economy *Mac Millan Publishing Company* (New York)
[5] De Man J M 1997 Kimia Makanan *Alih Bahasa: Kosasih P. Institut Teknologi Bandung* (Bandung) pp 550
[6] Kusnandar F 2010 Kimia Pangan Komponen Makro *Dian Rakyat* (Jakarta) pp 264
[7] Muchtadi D 2010 Teknik Evaluasi Nilai Gizi Protein *Penerbit Alfabeta* (Bandung) pp 190
[8] Nazir M 2005 Metode Penelitian. Ghalia Indonesia (Bogor) pp 58-59
[9] Purnomo H 1995 Aktivitas Air dan Peranannya dalam Pengawetan Pangan *UI Press* (Jakarta)
[10] SNI 01-3707-1995. Floss. http://sisni.bsn.go.id/index.php?/sni_main/sni/detail_sni /4128.
[11] Sudarmadji S B Haryono Suhardi 2007 Analisa Bahan Makanan dan Pertanian *Penerbit Liberty* (Yogyakarta)
[12] Sulistiyati T D 2011 Pengaruh Suhu dan Lama Pemanasan dengan Menggunakan Ekstraktor Vakum terhadap Crude Albumin Ikan Gabus (Ophiocephalus striatus) (Malang, Jawa Timur)
[13] Suprayitno E A Chamidah and Carvallo 1998 Studi Profil Asam Amino Albumin dan Seng pada Ikan Gabus (Ophiocephalus striatus) *Fakultas Perikanan dan Ilmu Kelautan* (Universitas Brawijaya. Malang)
[14] Ulandari A D Kurniawan and A S Putri 2011 Potensi Protein Ikan Gabus dalam Mencegah Kwashiorkor pada Balita di Provinsi Jambi *Universitas Jambi* (Jambi) pp 6
[15] Winarno F G 2002 Kimia Pangan dan Gizi *PT Gramedia* (Jakarta)
[16] Zura C F 2006 Cita Rasa (Flavor) *Departemen Kimia FMIPA* (Universitas Sumatera Utara, Medan)
[17] Rihayat Suryani Agusnar H Wirjosentono B and Nurhanifa 2018 Thermal Degradation of Aceh's Bentonite Reinforced Poly Lactic Acid (PLA) Based on Renewable Resources for Packaging Application *AIP Conference Proceedings* vol 2049
[18] Ridwan Wirjosentono B Tamrin Rihayat T Nurhanifa 2018 Modification of PLA/PCL/Aceh's Bentonite Nanocomposites as Biomedical Materials *AIP Conference Proceedings* vol 2049
[19] Zulkifli Z Rihayat T Suryani Zaimahwati Z and Rosalina R 2018 Purification process of Jelantah Oil using Active Chorcoal Kepok's Banana *AIP Conference Proceedings* vol 2049
[20] Rihayat T Suryani Satriananda Fitriah and Helmi 2018 Poly Lactic Acid (PLA)/Chitosan/Bentonite Nanocomposites Based on Cassava Starch for Materials in Biomedical Applications *AIP Conference Proceedings* vol 2049
[21] Rihayat T Suryani Satriananda Riskina S Khan N S P and Saifuddin 2018 Influence of Coating Polyurethane with Mixture of Bentonite and Chitosan Nanocomposites *AIP Conference Proceedings* vol 2049
[22] Rihayat T Suryani Fauzi T Nurhanifa Alam P N and Sami M 2018 Synthesis and Innovation of PLA/clay Nanocomposite Characterization Againts to Mechanical and Thermal Properties *IOP Conf. Series: Materials Science and Engineering* vol 334

Wound Dressing Based on Banana Peels Waste and Chitosan by Strengthening Lignin as Wound Healing Medicine

Teuku Rihayat[1], Suryani[1], Januar Parlaungan Siregar[3], Zaimahwati[1], Salmyah[1], Helmi[1], Sariadi[1], Fitria[2], Satriananda[1], Alfian Putra[1], Zahra Fona[1], Juanda[1], Raudah[1], Mawaddah[1], Nurhanifa[1], Shafira Riskina[1], Wildan Syahputra[1], Jamiluddin Jaafar[3]

[1] Department of Chemical Engineering, Lhokseumawe State Polytechnic, 24301, Aceh, Indonesia
[2] Department of Dermato Venereology, Medical Faculty, University Syiah Kuala, 24311, Aceh, Indonesia
[3] Structural Materials and Degradation Focus Group, Faculty of Mechanical Engineering, Universiti Malasyia Pahang, 26600 Pekan, Pahang, Malaysia

Corresponding author: teukurihayat@yahoo.com

Abstract. Recent studies in new applications of wound dressing offer and promote the process of wound healing. The purpose of this study was to develop lignin-based wound dressing on banana peel and then put it into chitosan film for wound dressing applications. The use of banana peels helps the formation of skin compounds so that they can be used to treat better wounds. Banana peel added to chitosan as matrix filler with concentration (0, 1, 3, 5, 7, 10% wt), then the dressing membrane swelling in 48 hours is at 0% wt 120.5931%, at 1% wt 99. 9981%, 3% wt 79.2916%, 5% wt is 68.1819%, 7% wt is 61.9173% and 10% wt is 45.3981%. Fourier Transform Infrared (FTIR) test, Scanning Electron Microscopy (SEM), swelling properties were carried out to characterize the prepared film. Antimicrobial tests were carried out by disc diffusion method and film growth inhibitory effects including different amounts studied in Escherichia coli and Staphylococcus aureus. FTIR at a concentration of 7% wt showed that there was an interaction between banana peel and chitosan on the absorption band 3185,728 cm^{-1}. In SEM the best sample morphology structure at 5% wt shows a good interface interface. Addition of banana peel as lignin decreases the level of swelling of water in wound dressing. In addition, Staphylococcus aureus is the most sensitive strain recorded in wound dressing. Banana skin which is included in chitosan film seems to be a potential and new biomaterial for wound healing applications.

Keyword : *chitosan, banana peels powder, fourier transform infrared scanning electron microscope*

1. Introduction

Wound infections are serious infections that have occurred in all parts of the world [1]. Many solutions have been studied and found to be able to cure wound infections. Wound healing requires a complex method and certainly does not restore the skin as usual. Severe wounds can even die from dark scars due to damage to skin tissue [2,3].

At present there are many multifunctional biomaterials for cell scaffolding, one of them is chitosan which is highly recommended in the engineering of skin tissue because it has structural characteristics similar to glucosamine from the natural extracellular matrix [4]. Chitosan has good biocompatibility, low toxicity, anti-infection activity and biodegradability [5]. Judging from the physical-chemical properties of chitosan, it can be used for various cell forms such as wound dressing, membranes, coatings, fibers, and sponges. Several studies have shown that higher molecular weight chitosan has better film-forming properties in skin tissue engineering [6]. This is the reason for the use of chitosan as a cell scaffold, specifically in making wound dressing.

Up until now, there have been many studies and development of wound dressing as an approach to the field of tissue engineering to deal with the effects of injuries in a faster period of time [7].

Several studies have shown that chitosan has a weakness if it is used as a wound dressing, this is due to the low mechanical strength of chitosan. In [8], trying to add lignin to chitosan, the results show that the tensile strength, storage modulus, thermal degradation temperature and chitosan can be increased by adding lignin.

Lignin is considered a raw material with high recovery potential, widely available in nature, cheap and environmentally friendly [9]. Lignin is almost present in all types of wood fiber plants [10], such as bananas, sugar cane, sengon and so on.

Especially bananas, containing about 10% lignin [11]. For making wound dressing, the banana part used is the skin, this is because besides lignin, the banana skin also helps the formation of skin compounds so that it can be used to treat bruises, burns, and other wound infections [12]. In addition, the use of banana peel as a wound dressing is supported by the availability of banana skin as waste [13]. The reuse of banana peel waste into a product can increase its economic value.

The development of chitosan-based wound dressing and banana peel waste in Indonesia tends to be lacking, due to limited information and research conducted. This is a good opportunity, especially in the medical field in Indonesia, which generally relies on healing methods using chemicals. The incorporation of natural polymers namely chitosan with banana peel is expected to be able to create wound dressing products with characteristics that are appropriate and provide benefits, both in terms of science, health, and in terms of economics. The purpose of this study was to make chitosan-based wound dressing and banana peel waste which were characterized using Fourier Transform Infrared (FTIR), Scanning Electron Microscopy (SEM), Swelling Properties and antibacterial susceptibility to Escherichia coli and Staphylococcus aureus.

2. Materials and Methods

2.1 Materials

Chitosan banana peel waste, glycerol, acetic acid, Escherichia coli bacteria and Staphylococcus aureus.

2.2 Processing of Banana Peels Powder

Banana Peels that have been collected, washed with water to remove dirt, and cut into small pieces and then dried in an oven at 120°C for 24 hours. After drying, the banana peel is ground with micro milling. Then the skin is crushed and sifted so that the particle size remains stable and stored at room temperature in a plastic container until it is used. Furthermore, banana peel powder was carried out by morphological testing using Scanning Electron Microscope (SEM) and functional group testing using the Fourier Transform Infrared (FTIR).

2.3 Synthesis of Membrane Wound Dressing Chitosan and Banana Peels Powder

Prepared as much as 1 gram of chitosan powder, dissolved in 2% acetic acid solution and stirred for 4 hours at room temperature. Prepared 1 ml of glycerol and added as a crosslinker and plasticizer. Chitosan solution was put into Teflon mold and mixed with banana peel powder with different concentrations (0, 1, 3, 5, 7 and 10% wt), dried at 40°C for 24 hours to evaporate the solvent and form a membrane. The prepared membrane is gently peeled, and further dried by keeping the oven 40°C for 4 hours. To give a better sight, the membrane wound dressing sample were depicted in Figure 1.

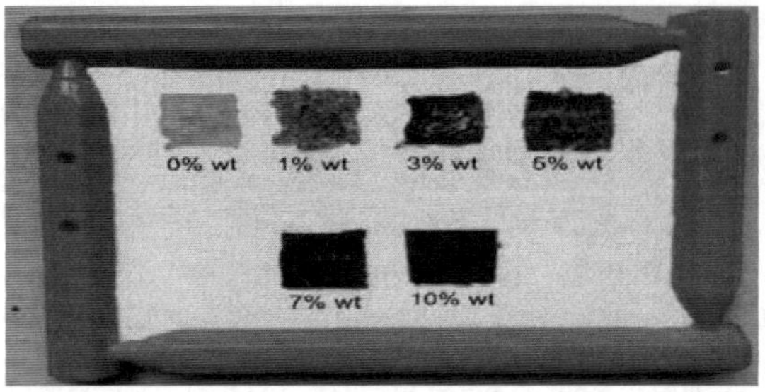

Figure 1. Membrane Wound dressing sample

2.4 Measuring Techniques

2.4.1 Fourier Transform Infrared (FTIR)
The testing of synthesized chitosan functional groups was characterized using Fourier Transform Infrared (FTIR) on wave absorption of 4000-500 cm^{-1}. For the wound dressing sample a thin and clear sized layer is formed. After that the sample is inserted into the FTIR device tube to get the functional group contained in the sample. The spectrum will be seen in the range 4000-500 cm^{-1}.

2.4.2 Scanning Electron Microscopy (SEM)
The homogeneity of the morphological structure of banana skin powder and wound dressing membrane as distribution of chitosan matrix fillers was seen by Scanning Electron Microscope (SEM).

2.4.3 Swelling Properties
Weight specimens were immersed in distilled water at 37°C. After 24 hours of immersion, the specimen is removed from the water, dried with a paper filter and weighed. The sample was soaked again for another 24 hours and then removed from the water, dried with filter paper and weighed.

$$\% \text{ Swelling} = \frac{(Mw - Md)}{Md \times 100} \quad (1)$$

where, *Mw* is the wet weight and *Md* is the dry weight of the sample.

2.4.4 Antimicrobial Susceptibilty
The antibacterial activity of membrane chitosan-banana powder wound dressing was investigated using the disc diffusion method on a petri dish [14]. Bacterial cultures that grow in the mid-logarithmic phase are placed in agar media. Escherichia coli and Staphylococcus aureus are injected into agar media. After compaction to coat the agar, the chitosan-banana powder (15 mm diameter) membrane with different concentrations (0, 1, 3, 5, 7 and 10% wt) was placed on the agar surface. The layers were incubated at 37°C for 24 hours and at 28°C for the next 72 hours.

3. Result and Discussion
3.1 Fourier Transform Infrared (FTIR)
The spectra of FTIR membrane chitosan wound dressing-waste banana peel with different concentrations (0, 1, 3, 5, 7 and 10% wt) has been recorded in wave spectrum of 4000-500 cm^{-1}. The chitosan spectrum and banana peel powder are shown in Figures 2 and 3, respectively.

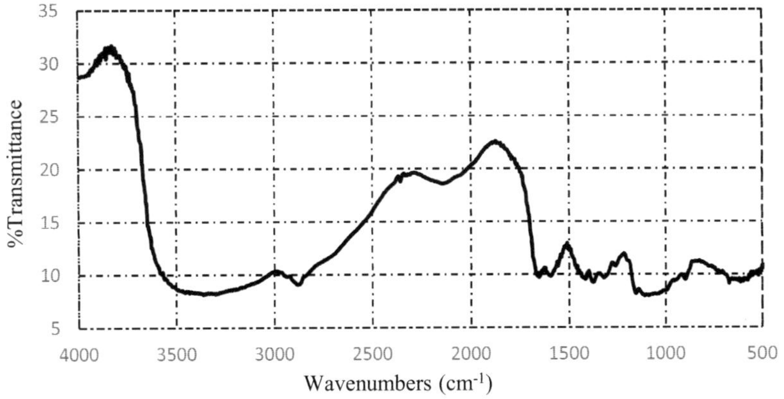

Figure 2. FTIR Spectrum of Chitosan

From Figure 2 it is shown that chitosan contains many functional groups (hydroxyl, carbonyl, carboxyl, amine, and amide). Please note that adding glycerol to chitosan does not provide a new band in the chitosan spectrum. Previous studies on banana peel powder revealed that banana peel powder contains polymers (pectin, hemicellulose and lignin) and they are rich in organic functional groups such as -COOH, $-NH_2$ and O-H [15]. Characteristics of absorption of chitosan and banana peel bands are similar to those previously studied [16]. FT-IR on banana skin chitosan (10% by weight), as shown in the picture there is a clear composite difference between the ingredients compared to the pure chitosan spectrum. Its peak at 3520 cm^{-1} corresponds to the expanding O-H stretch wave and shifts to 3628 cm^{-1}. Increasing the intensity of the band 2898 cm^{-1} which corresponds to the C-H wave. The new band appears at 1760 cm^{-1} which corresponds to the carboxyl group.

FTIR results also show the role of acid in chitosan. With FTIR profiles that show absorbance similar to each other, this shows the role of acids in the reaction of chitosan nanoparticles preparation. Acids are not intra- and intermolecular-bound with di- and tri- carboxylic acids [17]. The FTIR spectrum also does not show the formation of new polymers. This shows that chitosan does not polymerize with acid. It can be concluded that acids only act as proton donors which can dissolve chitosan [18].

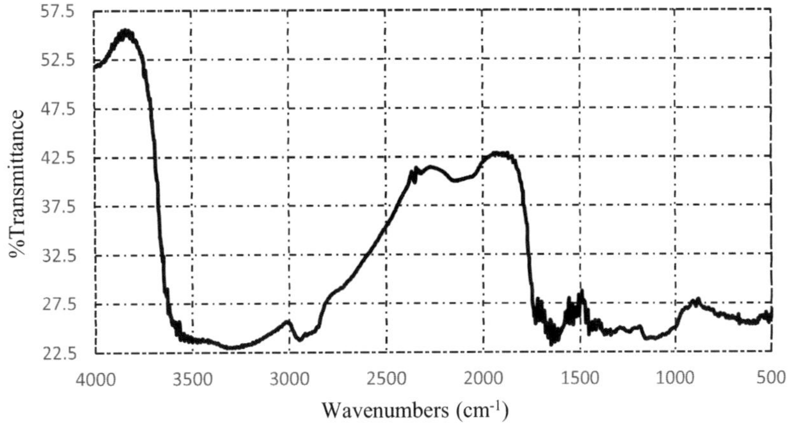

Figure 3. FTIR Spectrum of Banana Peels Powder

Figure 3 shows the membrane membrane of wound dressing containing banana peel powder with different concentrations (0, 1, 3, 5, 7 and 10% wt). From the picture it was noted that there was an increase in amide intensity which reached its peak at 1638 cm^{-1} with the addition of fillers. Bands on the absorption band of 2890 cm^{-1} and 2,900 cm^{-1} have increased as indicated by the characteristics for C-H due to the addition of fillers. This band appears on the absorption band 1760 cm^{-1} which corresponds to the carboxyl group in addition to increasing the intensity of the absorption band 1440 cm^{-1} which occurs an interaction between the positive charge of the amine chitosan group and the negatively charged carboxyl group on banana skin which makes good interactions filler and matrix.

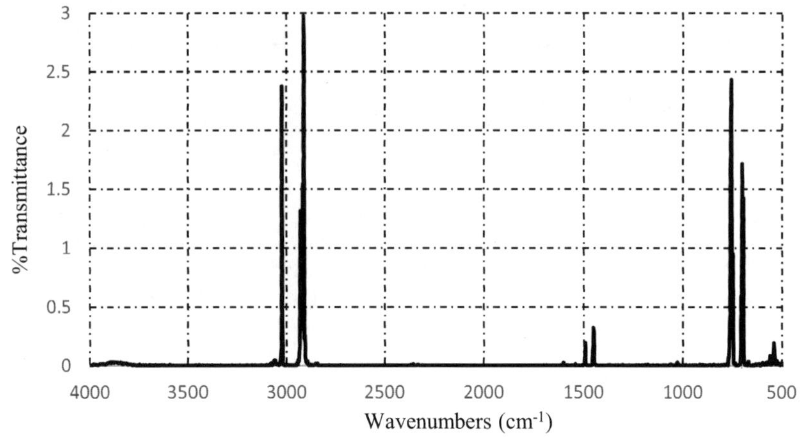

Figure 4. The FTIR of the membrane wound dressing of 7% wt

3.2 Scanning Electron Microscope (SEM)

In Figure 5 shows the SEM morphological structure of banana skin powder particles. The particle size is in the range of 20-80 μm in length and 20-30 μm in width.

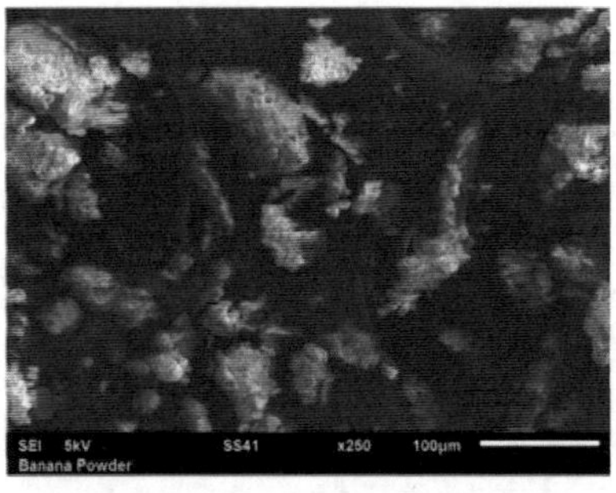

Figure 5. Morphological structure of banana peels powder

The morphological structure of SEM with concentrations (0, 1, 3, 5, 7 and 10% wt) of chitosan and banana peel powder is shown in Figure 5. Morphological analysis of SEM showed that uniform wound dressing membrane fillers were distributed in a 5% wt concentration chitosan matrix, with the increase in filler concentration in irregular distribution with several filler aggregates.

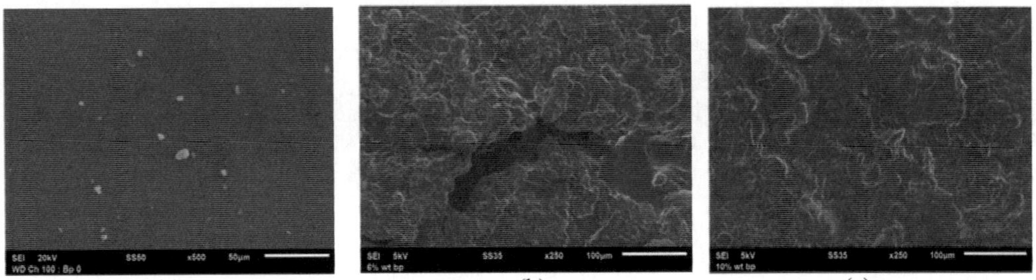

(a) (b) (c)

Figure 6. Membrane morphological structure of wound dressing at concentrations (a) 0% wt; (b) 5% wt; and (c) 10% wt

3.3 Swelling Properties

It is important to determine the swelling properties of water for biomaterials if they are used as wound cover material. This is a measure of the film's capacity to absorb wound exudates. The ideal environment for good wound healing is to keep the wound moist [19].

Table 1. Swelling Properties of Wound Dressing

Sample	*Swelling* (%)		
	0 h	24 h	48 h
0% wt	57.94	93.14	120.59
1% wt	33.06	60.17	100
3% wt	19.22	50.94	79.29
5% wt	10.18	40.94	68.18
7% wt	59.72	33.73	61.92
10% wt	29.92	24.18	45.4

Water absorption from chitosan is related to the hydrophilic group (hydroxyl and amino groups) of the polysaccharide. Membrane swelling of chitosan-banana powder wound dressing after 24 hours and 48 hours of immersion in distilled water were shown in Figure 7.

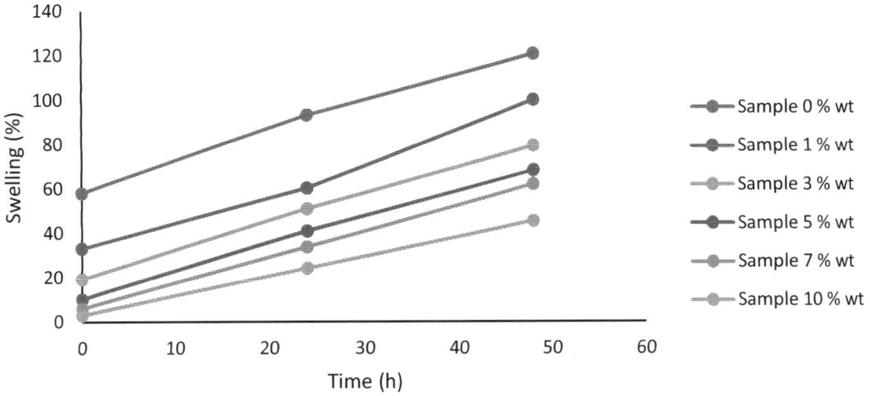

Figure 7. Swelling properties on membranes wound dressing chitosan-banana peels powder

It is clear from the figure that the 0% wt sample shows a higher swelling rate compared to the added filler, adding banana skin powder to chitosan results in a decrease in swelling and hence increases water resistance This behavior is caused by a kind of interaction between negatively charged compounds in banana peel powder such as carboxylic acids and positively charged groups on the chitosan backbone chain. This interaction limits the mobility of the chitosan chain which reduces water uptake. To better understand the effect of banana peel powder on chitosan swelling behavior, the plot of water uptake versus time is illustrated in Figure 7.

3.4 Antimicrobial Activity

The point of this study shows that the combination of chitosan with banana peel fillers has a synergistic interaction with a broad antimicrobial spectrum against gram negative, gram positive bacteria and even against strains of yeast culture that show the ability of biofilm formation. The results in Figure 8 show that the chitosan-sanitary membrane as a banana peel has a synergistic action with the highest activity of 10% wt. In addition, Staphylococcus aureus is the most sensitive strain recorded for this membrane.

Chitosan-banana peel powder dressing membrane also showed higher activity against Escherichia coli strains (gram negative bacteria) than Staphylococcus aureus strains (gram positive bacteria). On that basis, several studies note that chitosan affects gram negative and positive but rather like argumentative effects, some findings indicate more effectiveness against gram negative compared to gram positive [20].

However, this finding regarding the surface polarity of bacteria, the outer membrane of gram-negative bacteria has a lipopolysaccharide, very negatively charged, which allows attachment to polycational chitosan relative to gram-positive, which consists of peptidoglycan associated with polysaccharides and teicoic acid [21]. which in turn supports glucose [22]. Effects similar to fungal and yeast cells that can interfere with growth [23].

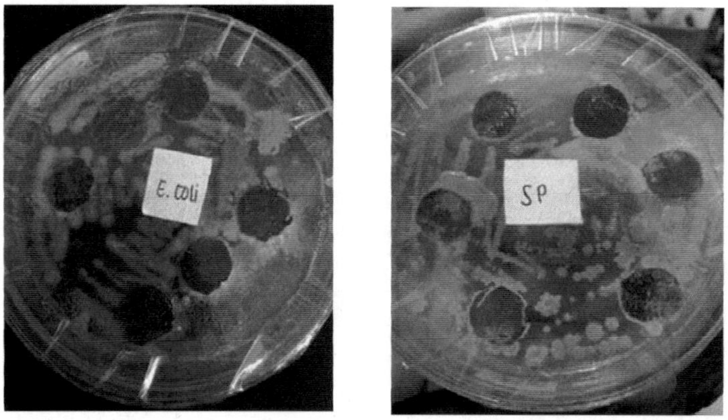

(a) (b)

Figure 8. Effect of antibacterial activity to membrane wound dressing chitosan and banana peel powder on (a) Escherichia coli and (b) Staphylococcus aureus

Scientists support another mechanism, the antimicrobial activity of chitosan which is associated with metal chelation, where chitosan has excellent metal binding ability because the amine group takes cationic metals with chelation [24]. Here, we hypothesize that such a relationship is one of the new approaches to improve the properties of these composites [25] such as increasing biodegradability and antimicrobial activity, moreover banana peels contain polymers such as lignin, hemicellulose and pectin [26] which can help improve composite properties newly formed. Therefore, we can use it for medical purposes as wound dressing. In the food industry, banana flour/chitosan is applied to preserve new vegetables, showing antimicrobial properties against bacteria when used [27].

Although in this study banana peels were used as they are, the results of antibacterial studies were comparable to those studied by P.B. Franco and his co-authors studied the antibacterial activity of chitosan membranes associated with active compounds extracted from banana peels [28].

4. Conclusion

The use of banana skin helps the formation of skin compounds so that it can be used to treat better wounds. Banana skin is added to chitosan as a matrix filler with concentrations (0, 1, 3, 5, 7, 10% wt), then the swelling of the swelling membrane in 48 hours is at 0% wt 120.5931%, at 1% wt 99. 9981 %, 3% with 79.2916%, 5% with 68.1819%, 7% with 61.9173% and 10% with 45.3981%. FTIR at a concentration of 7% wt showed that there was an interaction between banana peel and chitosan on the absorption band 3185,728 cm-1. In SEM the best morphological sample structure at 5% wt shows a good interface. Addition of banana skin because lignin reduces the level of swelling of water in wound dressing. In addition, Staphylococcus aureus is the most sensitive type recorded in wound dressing. Banana peels included in chitosan films seem to be potential and new biomaterials for wound healing applications.

References

[1] Kamel N A Salwa L A E Neveen M.S. 2017. Chitosan/banana peel powder nanocomposites for wound dressing application: Preparation and characterization *Materials Science and Engineering* vol 72 pp 543-550

[2] Guarino V C Altobelli T R Ambrosio L 2015 Degradation Properties and Metabolic Activity of Alginate and Chitosan Polyelectrolytes for Drug Delivery and Tissue Engineering Applications *AIMS Materials Science* vol 2 (4) pp 497-502.

[3] Güneş S Funda Tıhmınlıoğlu 2017 Hypericum perforatum Incorporated Chitosan Films as Potential Bioactive Wound Dressing Material *International Journal of Biological Macromolecules*

[4] Kumar M.N.V Ravi 2000 A review of chitin and chitosan applications. Reactive & Functional Polymers vol 46 pp 1-27.

[5] Ma Y Lian X H T Ming F Jianliang L Yang J Zhonghua L. Yong C Xiaohong H 2017. Chitosan membrane dressings toughened by glycerol to load antibacterial drugs for wound healing. *Materials Science & Engineering* vol 81 pp 522-531.

[6] Tan H Constance R C Karin A P Kacey G M 2009 Injectable in Situ Forming Biodegradable Chitosan-Hyaluronic Acid Based Hydrogels for Cartilage Tissue *Engineering. Biomaterials* vol 30 pp 2499-2506.

[7] Ishihara M Kuniaki N Katsuaki O Masato S Makoto K Yoshio S Hirofumi Y Takemi M Hidemi H Maki U Akira K 2002 Photocrosslinkable chitosan as a dressing for wound occlusion and accelerator in healing process. *Biomaterials* vol 23 pp 833-840.

[8] Chen L Chang Y T Nan Y N Chao Y W Qiang F Qin Z 2009 Preparation and Properties of Chitosan/Lignin Composite Films *Chinese Journal of Polymer Science* vol 27 (5) pp 739-746.

[9] Popa V I Adina M C Silvia G Teodor M 2011 Nanoparticles Based on Modified Lignins With Biocide Properties *Cellulose Chemistry and Technology* vol 45 (3-4) pp 221-226.

[10] Bahri S 2015 Pembuatan Pulp dari Batang Pisang. Jurnal Teknologi Kimia Unimal vol 4 (2) pp 36-50.

[11] Wina E 2001 Tanaman Pisang Sebagai Pakan Ternak Ruminansia Balai Penelitian Ternak. Wartazoa vol 11 1.

[12] Rangan A Manjula V.M Rajendran M.T Satyanarayana G.K Reghu M 2017 Novel method for the preparation of lignin-rich nanoparticles from lignocellulosic fibers. Industrial Crops and Products vol 103 pp 152-160.

[13] Vu H T Scarlett C & Vuong Q V 2018 Phenolic Compounds Within Banana Peel and Their Potential Uses: A Review. *Journal of Functional Foods* vol 40 pp 238-248.

[14] C Deng L He M Zhao D Yang Y Liu 2007 Biological properties of the chitosan-gelatin sponge wound dressing. Carbohydr. Polym vol 69 pp 83-589.

[15] H Zheng L Wang 2013 Banana peel carbon that containing functional groups applied to the selective adsorption of Au (III) from waste printed circuit boards Soft Nanosci Lett vol 3 (2) pp 29-36.

[16] Castro R S D Cetano L Ferreira G Padilha P M Saeki M J Zara L F Martines M A U and Castro

G R 2011. Banana peel applied to the solid phase extraction of copper and lead from river water: preconcentration of metal ions with a fruit waste *Ind. Eng. Chem. Res.* Vol 50 (6) pp 3446-3451.

[17] Bodnar M Hartmann J F Borbely J 2005 Preparation and Characterization of Chitosan-Based Nanoparticles *Biomacromolecules* vol 6 pp 2521-2527.

[18] Moura M R & Aouada F A 2008 Preparation of Chitosan Nanoparticles Using Methacrylic Acid *Journal of Colloid and Interface Science* vol 321(2) pp 477-483.

[19] W H Eaglstein et al Optimal use of an occlusive dressing to enhance healing: effect of delayed application and early removal on wound healing Arch Dermatol vol 124 (3) pp 392–395.

[20] Papineau A M Hoover D G Knorr D & Farkas D F 1991 Antimicrobial Effect of Water-Soluble Chitosans With High Hydrostatic Pressure *Food Biotechnol*. Vol 5 (1) pp 45-57.

[21] Huang J C 2002 Carbon Black Filled Conducting Polymers and Polymer Blends. Adv. Polym. Technol vol 21 (4) pp 299-313.

[22] T H Emaga Robert C Ronkart S N Wathelet B Paquot M 2008 Dietary fibre components and pectin chemical features of peels during ripening in banana and plantain varieties *Bioresour. Technol*. Vol 99 (10) pp 4346–4354.

[23] N. Pitak S.K. Rakshit 2011 Physical and antimicrobial properties of banana flour/chitosan biodegradable and self-sealing films used for preserving Fresh-cut vegetables *LWT-Food Sci. Technol*. Vol 44 (10) pp 2310-2315.

[24] P Battaglini Franco L A de Almeida R F C Marques G Brucha M G N Campos 2016 Evaluation of antibacterial activity of chitosan membranes associated to unripe banana peel *Mater. Sci. Forum* vol 869 pp 859-863.

[25] Zaimahwati Harry A Rihayat, T Deni R Saharman G 2015. The Manufacture of Palm Oil-Based Polyurethane Nanocomposite with Organic Montmorillonite Nanoparticle as Paint Coatings *International Journal of Chem Tech Research* vol 7 pp 2537-2544.

[26] Rihayat, T Suryani Fauzi T Agusnar H Syafruddin Helmi Zulkifli Alam P N Sami M 2017 Mechanical Properties Evaluation of Single and Hybrid Composite Polyester Reinforced Bamboo *PALF and Coir Fiber* vol 334 pp 1-8.

[27] Rihayat, T Suryani Arlina A Fonna Z Jalal R Alam P N Zaimahwati Sami M Syarif J Juhan N 2017 Determination of CEC Value *(Cation Exchange Capacity)* of Bentonite from North Aceh and Bener Meriah, Aceh Province, Indonesia Using Three Methods vol 334 pp 1-7

[28] Suryani Agusnar H Wirjosentono B Rihayat T Nurhanifa 2017 Improving the Quality of Biopolymer (Poly Lactid Acid) With the Addition of Bentonite as Filler vol 222 pp 1-7

[29] Rihayat T Suryani Zimahwati 2014 Preparation and Properties and Application of Renewable Source (Palm Oil Polyol) Based Polyurethanes Coatings and its Thermal Stability Improvement by Clay Nanocomposite *Advanced Materials Research* vol 887-888 pp 566-569.

[30] Rihayat T Suryani Zaimahwati 2014 Effect of Heat Treatment on The Properties of Polyurethane/Clay Nanocomposite Paint *Advanced Materials Research* vol 525 pp 97-100.

[31] Rihayat T Saari M Hilmi M M Wan Yunus W M Z Suraya A R Dahlan K Z H M Sapuan S M 2007 Mechanical Characterisation of Polyurethane/Clay Nanocomposite *Polymer and Polymer Composites* vol 15 pp 647-652

[32] Rihayat T Suryani 2010 Synthesis and Properties of Biobased Polyurethane/Montmorillonite Nanocomposites *World Academy of Science Engineering and Technology* vol 4 pp 714-718

IOP Conf. Series: Materials Science and Engineering **506** (2019) 012057 doi:10.1088/1757-899X/506/1/012057

The Benefit of Infrastructure Development: An Analysis

Khairuman Khairuman[1], E S Barus[2,3], Niskarto Zendrato[3], Diana Alemin Barus[4], Jijon Rapitha Sagala[5], Andi Elhanafi[6], Jenni Veronika Ginting[2]

[1] Department of Teknik Komputer, Politeknik Aceh Selatan, Indonesia
[2]Department of Sistem Informasi, STMIK Kristen Neumann Indonesia, Jalan Jamin Ginting Km 10.5, Medan, Indonesia
[3] Faculty of Science Computer and Information Technology, Universitas Sumatera Utara, Medan, Indonesia
[4] Faculty of MIPA Universitas Sumatera Utara, Medan, Indonesia
[5] Department of Teknik Informatika, STMIK Pelita Nusantara, Medan, Indonesia
[6] Department of Teknik Informatika, Universitas Harapan Medan, Medan, Indonesia

*Corresponding author: khairuman.poltas@gmail.com

Abstract. In analyzing the benefits of infrastructure development, several rules of economics and feasibility studies for infrastructure development are used, namely aspects of benefits, effectiveness and efficiency. These rules are applied to the results of the benefits data when infrastructure development is carried out in the first year and the results of the benefits data are processed using mamdani fuzzy logic reasoning which consists of 2 inference processes. In processing fuzzy input data produces output from the inference process which is then classified in 5 feasibility conditions, namely, low, normal, high, very high and not feasible which is used as a support facility in making infrastructure development decisions.

1. Introduction

The increasing rapidly development of the era triggered the government's performance in terms of equitable development in each region in Indonesia. Infrastructure development in each region is one of the government's efforts to improve the welfare of the Indonesian people [1]. So that to see an infrastructure development in an area that is really beneficial to improve the welfare of the people around, it is necessary to design a computer application that can analyze the benefits of infrastructure development that has been implemented in area that is in accordance with government objectives and shows a percentage value that represents the level of prosperity of the local people after infrastructure development was carried out.

This system is an analysis of the benefits of infrastructure development based on fuzzy logic that shows the level of feasibility of infrastructure development in an area, so that it is expected to facilitate officers working to evaluate infrastructure development in an area in determining which areas can be used as development priorities and which areas need to be evaluated re-building the project. Thus, the initial goal of the government to improve people's welfare and equitable development through infrastructure development in the regions can be achieved.

The purpose of this research is to design an analysis of the benefits of infrastructure development based on fuzzy logic in an area that can analyze the increase in the level of people's welfare in an area after infrastructure development is carried out.

To avoid extensive discussion, the author will limit the discussion of this Final Project with riset Analysis of people's prosperity in terms of economics and infrastructure development feasibility studies. Rules are determined based on the experience of the PNPM PISEW economic expert team in analyzing the benefits of infrastructure development. Analyzing the prosperity of the people only in Sitinjo sub-district, Dairi Regency, where infrastructure development has been carried out by PNPM PISEW. Analysis of the system model mamdani method fuzzy inference system

2. Methodology

2.1 System Application Thinking Framework

The first step is to develop the benefits of infrastructure development variables according to figure 1, where the benefit variable consists of 3 aspects, namely aspects of efficiency, aspects of effectiveness and aspects of benefits. These three aspects are measured from the following parameters, namely: the benefits of rolling out funds, saving benefits, benefits of increasing production, BC Ratio, increasing access to economic business productivity, facilitating social relations of citizens, increasing accessibility of the poor and opening isolation among citizens. The following is a system application framework. The benefits of developing fuzzy logic base on infrastructure development.

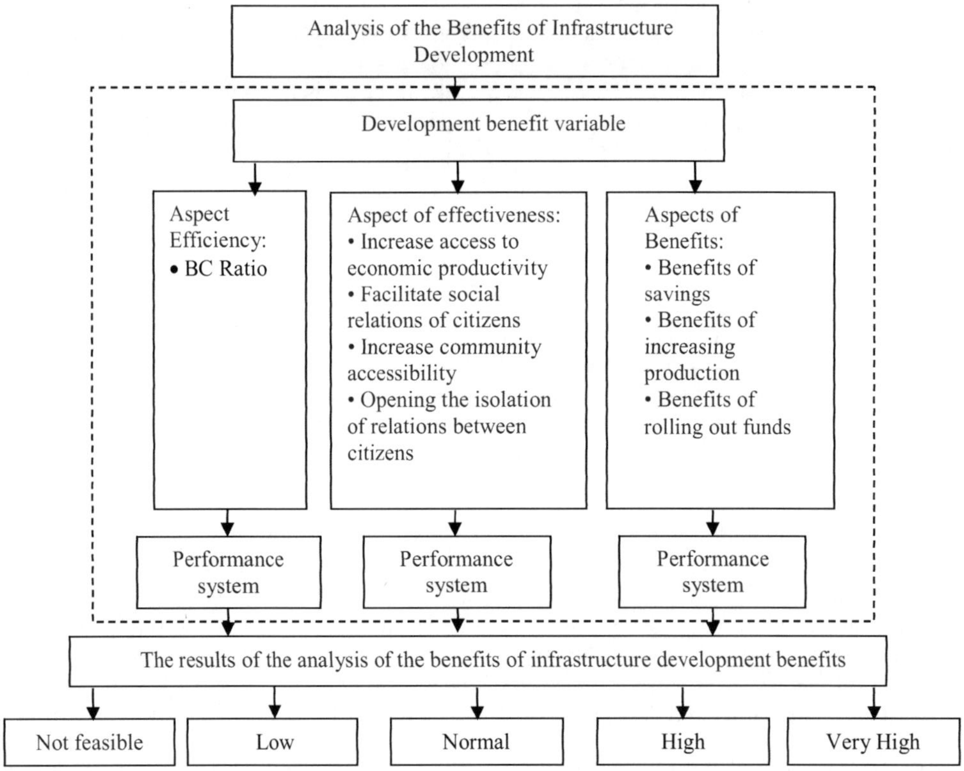

Figure 1. Framework system thinking.

For sensitivity analysis calculations to see how many percent increase and decrease the factors that cause changes in the benefits of infrastructure development in each aspect, namely from proper, normal or improper so that it needs to be re-evaluated against the process of infrastructure development. Figure 1 is the framework of the application system that will be built.

2.2 Flowchart Research

Flowchart is a scheme that describes the sequence of activities from beginning to end. Flowcharts are used to describe a program algorithm in an easier and simpler way. The process that occurs in this system can be described in the flowchart as shown in figure 2 below:

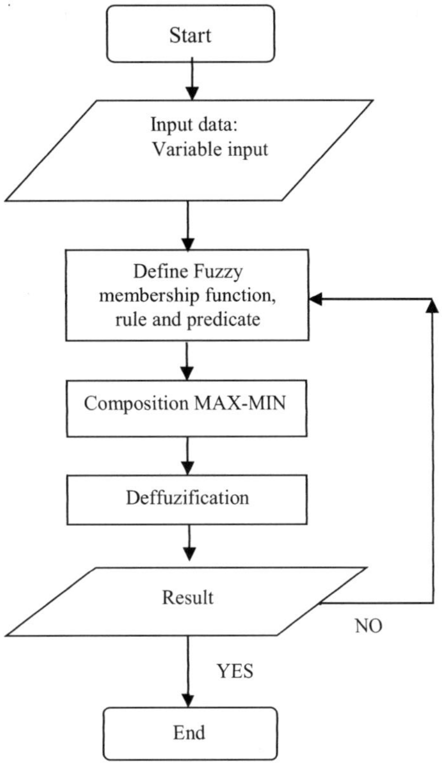

Figure 2. Flowchart of the FIS problem solving process using Mamdani Method.

Based on the figure 2 can be explained that the data training is the first data must be there and stored in the computer. Then the training data must first be normalized using the formula contained in equation (1). The normalized data will then be processed using backpropagation algorithm parameter using the bipolar sigmoid activation function. The next stage of the network will train the data training based on the parameters that have been determined. After all the steps are done, it will get the best testing results that will be used to predict.

2.3 Determine fuzzy sets and inputs

Based on figure 1 can be determined there are 10 fuzzy variables that can be model are namely:

i. BC Ratio (BcR) consists of 4 fuzzy sets, namely: low, normal, high and very high.
ii. Increasing access to economic productivity (Pr) consists of 6 fuzzy sets, namely: very low, low, normal, very normal, high and very high.
iii. Facilitating citizens' social relations (SR) consists of 3 fuzzy sets, namely: Low, Medium and High.
iv. Increasing community accessibility (Acc) consists of 6 fuzzy sets, namely: very low, low, normal, very normal, high and very high.
v. Opening isolation between citizens (Is); consists of 3 fuzzy sets, namely: low, medium and high.

vi. Effectiveness aspects (Ef) consist of 4 fuzzy sets, namely: very low, low, normal and high
vii. The saving benefit (Sa) consists of 4 fuzzy sets, namely: low, normal, high and very high.
viii. The benefits of increasing income (Inc) consist of 4 fuzzy sets, namely: low, normal, high and very high.
ix. The benefits of rolling out funds (Fu) consist of 4 fuzzy sets, namely: low, normal, high and very high.
x. Benefits (B) aspects consist of 4 fuzzy sets, namely: low, normal, high and very high.

2.4 Membership function

Facilitating citizens' social relations (Sr) is shown in figure 3 below.

Membership function:

$$\mu HuS\ Low\ [h] = \begin{cases} 1; & h \leq 30 \\ (50-h)/20; & 30 \leq h \leq 50 \\ 0; & h \geq 50 \end{cases} \quad (1)$$

$$\mu HuS\ Normal\ [h] = \begin{cases} 0; & h \leq 40 \text{ or } h \geq 60 \\ (h-40)/10; & 40 \leq h \leq 50 \\ (60-h)/10; & 50 \leq h \leq 60 \end{cases} \quad (2)$$

$$\mu HuS\ High\ [h] = \begin{cases} 0; & h \leq 50 \\ (h-50)/30; & 50 \leq h \leq 80 \\ 1; & h \geq 80 \end{cases} \quad (3)$$

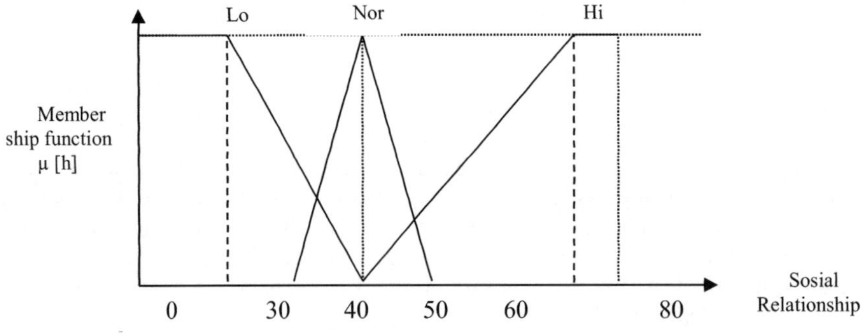

Figure 3. Fuzzy set of levels of citizens Social Relations (Sr).

3. Results And Discussion

3.1 Display of effectiveness aspect input

The following is a membership function of the productivity aspect of the productivity variable input which consists of 5 conditions, namely low, medium, very medium, ordinary, very ordinary and high according to formula (1), (2) dan (3).

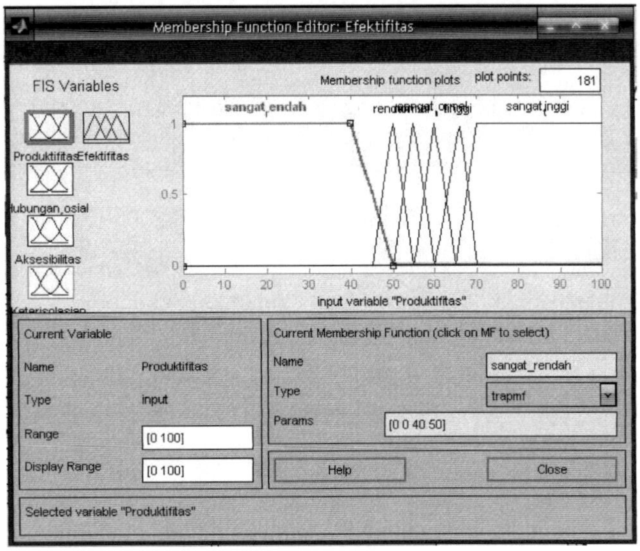

Figure 4. Display productivity membership functions.

Next is the display of the aspects of effectiveness of the input variables of social relations where this input has the appearance of a membership function consisting of 3 conditions namely low, medium and high. Figure 4 shows a display of the effectiveness aspects of the input accessibility variable of the community where this input has the appearance of a membership function consisting of 6 conditions namely low, medium, very medium, ordinary, very ordinary and high. The following is a display of the accessibility input membership function.

Figure 5 shows is a display of the effectiveness aspects of the community isolation input variable where this input has the appearance of a membership function consisting of 3 conditions, namely low, medium and high following is a display of the community isolation input membership function. It can be seen in the image display that the low and high conditions use the tramp curve type while the conditions using the trim curve type (triangle), the parameter values. These can be seen in Figure.

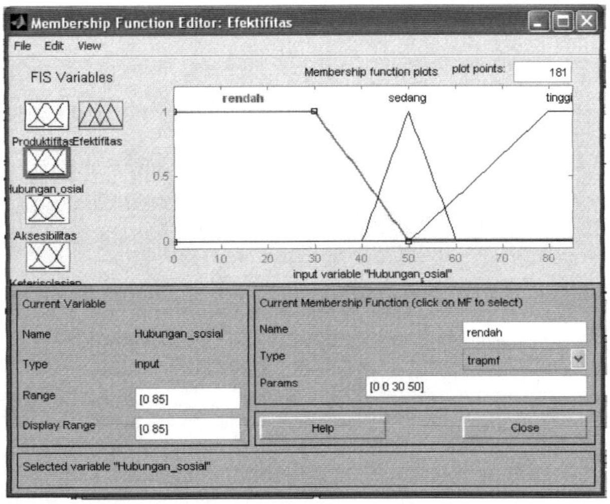

Figure 5. Display of isolation membership functions.

Figure 6 is a display of the effectiveness aspects of the community isolation input variable where this input has the appearance of a membership function consisting of 3 conditions, namely low, medium and high according to equations 3.20, 3.21 and 3.22. The following is a display of the community isolation input membership function. It can be seen in the image display that the low and high conditions use the tramp curve type while the conditions using the trim curve type (triangle), the parameter values shown are adjusted to equations 3.20, 3.21 and 3.22. These can be seen in Figure 7.

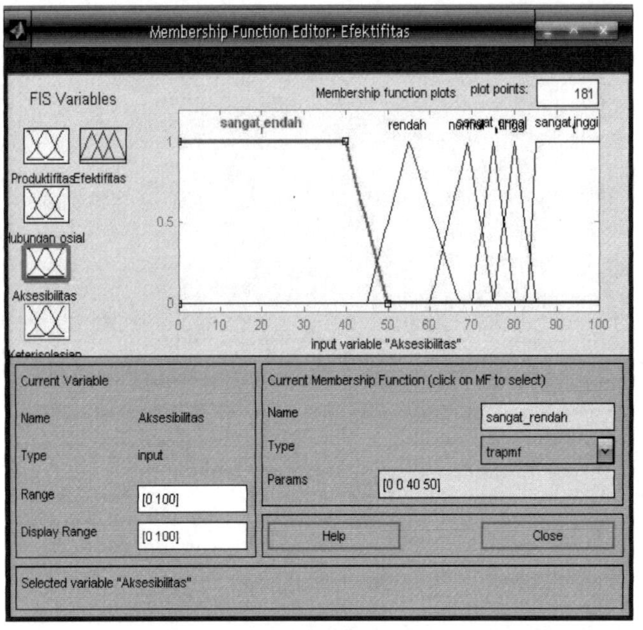

Figure 6 Display accessibility membership functions.

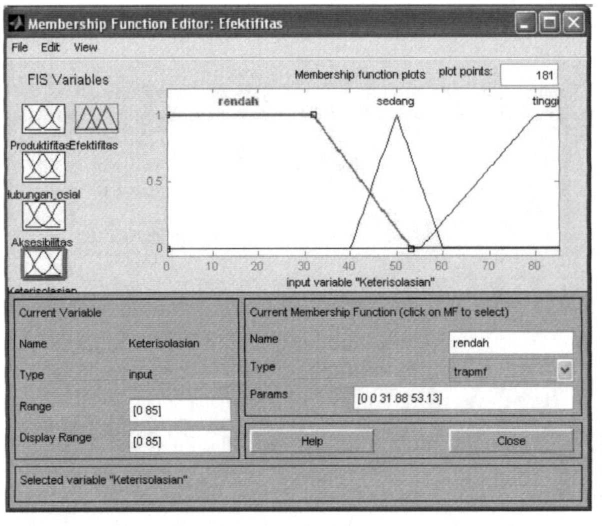

Figure 7. Display of isolation membership functions.

3.2. Display of Benefit Inference Results

After all membership functions of each input are filled. Then the rule button contains a rule that is used for the benefit aspects in accordance with the rule in figure 8.

Then the view rules button will display the results of the centroid benefit analysis as seen in figure 7. Similarly, the results of the analysis of effectiveness in the display of results can also be changed in the input values so as to produce a variable output value. To change the input value can be done by changing the value in the text input in the lower left panel or it can also be done by sliding the red line on each input curve to the left or right. The shape of the blue curve is the output curve and the output value on this aspect of benefit, which is what appears in the description of benefits, while the yellow curve is the input curves whose values can be changed so as to produce varying output values.

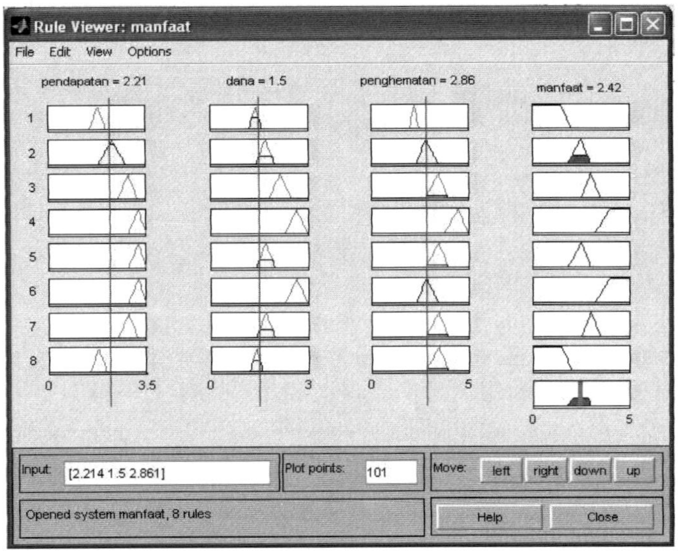

Figure 8. Display of functionality function.

In Figure 8, the input value can be changed in the input text on the lower left, by changing the input values, we can obtain a varied output value. Furthermore, the changes made in the input column can be seen the results in the testing table 1 Every changing made to the input value will result in a significant change in the output, then the output is then classified into 5 conditions of the feasibility analysis system of benefits.

3.3 Testing

From the tests performed, where the user enters fuzzy input as follows:

- BC Ratio = 2.8
- Productivity access = 53
- Social Citizens = 49
- Community Accessibility = 61.4
- Inter-citizenship isolation = 41.2
- Savings = 2.47
- Income Increase = 1.9
- Revolving of Funds = 3.22

Fuzzy input will go through the inference process 1, then the inference 2 process in the form of classification of feasibility in this case consists of 5, namely feasible, very feasible, normal, very normal and not feasible. Following are the details of the process. For the calculation of inference 1 aspect of effectiveness begins with the calculation of the implications of the implications.

3.4 Application function implications
Based on the rules in figure 8, the rules of the predicate α-pred can be determined as following:

Inference 1 effectiveness
Pers 3.5; Pers 3.10; Pers 3.10; Pers 3.18;
R1 = α-pred = min (µAkP Normal, HuS Low1, AkM Low2, IsW normal3)
= min (0,0,0,0)
= 0
Pers 3.6; Pers 3.11; Pers 3.14; Pers 3.19;
R2 = α-pred = min (µAkP very normal, HuS normal1, AkM very normal2, IsW normal3)
= min (0,0,0,0)
= 0
Pers 3.7; Pers 3.12; Pers 3.15; Pers 3.20;
R3 = α-pred = min (µAkP normal, HuS high1, AkM normal2, IsW high3)
= min (0,0.67,0.8,0.33)
= 0.33
Pers 3.8; Pers 3.12; Pers 3.16; Pers 3.20;
R4 = α-pred = min (µAkP very normal, HuS normal1, AkM very normal2, IsW normal3)
= min (0.8,0.67,0,0.33)
= 0.33
Pers 3.9; Pers 3.10; Pers 3.17; Pers 3.18;
R5 = α-pred = min (µAkP low, HuS low1, AkM low2, IsW normal3)
= min (0,0,0,0)

3.5 Rule composition
Inference 1 effectiveness from the rules of the existing predicate, the fuzzy boundary region is generated as follows:

$$(e1-65)/20 = 0.33$$
$$e1 = 71.67$$
$$(85-e)/20 = 0.33$$
$$e2 = 78.33$$

Inference 1 effectiveness from the rules of the existing predicate, the fuzzy boundary region is generated as follows:

$$\mu[x] = (x-65)/20;\ x \leq 71.67$$
$$0.33;\ 71.67 \leq x \leq 78.33$$
$$(85-x)/20;\ x \geq 78.33$$

3.6 Defuzzy
Inference 1 effectiveness can be calculated using the centroid method, the calculation of the moment is as follows:

$$M1 = \int_{65}^{71.67} (0.05x^2 - 3.25x)\, dx$$
$$= 0.06167\, x^3 - 1.625x^2$$
$$= -2199.0303 + 2279.3875$$
$$= 80.3572$$

$$M2 = \int_{71.67}^{73.33} (0.33x)\, dx$$

$= 0.165x^2$
$= 887.25267-847.53717$
$= 39.7255$

$$M23= \int_{73.33}^{85} (4.25x- 0.05x^2)\, dx$$

$= 2.125x^2 – 0.0167x^3$
$= 5097.237-4841.6518$
$= 255.5852$

Then calculate the area of each region:
A1 = ((71.67-65) *0.33)/2= 1.10055
A2 = (73.33-71.67) *0.33= 0.5478
A3 = (85-73.33) *0.33= 3.8511

Result

$$= \frac{M1 + M2 + M3}{L1 + L2 + L3}$$

$$= \frac{80.3572 + 39.7255 + 255.5852}{L1 + L2 + L1.10055 + 0.5478 + 3.85113}$$

$= 67.85$

The value of 67.85 is the value of the aspect of effectiveness obtained from the input of productivity access inputs, social relations, community accessibility and the isolation of citizens. Value 67, 85 then classified where the value is included in the category of infrastructure development in the Normal category. Thus, the calculation of the effectiveness inference process manually, as well as the calculation of the manual inference benefit calculation process. After calculation, the results will be classified into 5 conditions. The input values entered in the program, the output of the analysis results is obtained as 1 shows 10 conditions for changes in input values that are different so that it can be seen the form of output variations generated from the simulation program that is adjusted to the input.

4. Conclusions

Analysis of the benefits of infrastructure development based on fuzzy logic can be used as one of the references in decision-making on infrastructure development in an area, supported by fuzzy logic reasoning is expected to produce accurate data. The more rules used in the inference process will produce better output. The fuzzy inference process in this application is used to determine the value of the benefit aspects and aspects of effectiveness then classify them into the boundary values of the 5 standard feasibility conditions, namely low, normal, high, very high and not feasible.

References

[1] Sri Hartati, Imas S Sitanggang, A Fuzzy Based Decision Support System for Evaluating Land Suitability and Selecting Crops *Journal of computer science* 6(4):417-424 2010
[2] Panduan Pelaksana PNPM PISEW TAHUN 2010
[3] Kecerdasan Buatan http://idhaclassroom.com/2007/09/15/./ kecerdasan buatan.html.
[4] Suparman 2007 Komputer Masa Depan Pengenalan Artificial Intelligence
[5] Kusumadewi S 2002 Analisis Desain Sistem Fuzzy Menggunakan Tool Box Matlab, Penerbit Graha Ilmu, Yogyakarta
[6] Kusumadewi S, dan Purnomo H 2004 Aplikasi Logika Fuzzy untuk Pendukung Keputusan, Penerbit Graha Ilmu, Yogyakarta
[7] IGunaidi Abdia Away 2010 MATLAB Programming, Informatika, Bandung
[8] Kusumadewi S 2002 Analisis Desain Sistem Fuzzy Menggunakan Tool Box Matlab, Penerbit

Graha Ilmu, Yogyakarta
[9] Efraim Turban, Jay E Aronson, Ting Peng Liang 2005 Dicision Support System and Intelligent Systems (Sistem Pendukung Keputusan dan Sistem Cerdas) ANDI, Yogyakarta
[10] Agus Naba 2009 Belajar Cepat Fuzzy Logic menggunakan MATLAB, ANDI, Yogyakarta

The Shelf-life Prediction of Sweet Orange Based on Its Total Soluble Solid by Using Arrhenius and Q 10 Approach

Rita Khathir[1*], Ria Yuliana[1], Raida Agustina[1] and Bambang Sukarno Putra[1]

[1] Department of Agricultural Engineering, Faculty of Agriculture, University of Syiah Kuala

*Corresponding author: rkhathir@unsyiah.ac.id

Abstract. The information about shelf-life of sweet orange is necessary for its post-harvest handling to reduce the loss facing by farmers and sellers. The study aimed to observe if the Arrhenius and Q 10 models can be used to predict the shelf-life of sweet orange based on its total soluble solid (TSS). The fresh sweet oranges obtained from the market were stored at three extreme temperatures i.e. 55, 65, and 70°C. During the storage, the TSS was analyzed every hour, until the sweet oranges quality had decreased. The TSS was observed by using a hand digital refractometer. The TSS change in linear model was chosen to perform the Arrhenius model since its R-square value was higher than that of exponential model. The study had proved that there was a possibility of using the Arrhenius and Q 10 models to predict the shelf-life of sweet oranges based on its TSS contents. The Arrhenius model of TSS change was and the Q 10 value was 1.92. The real shelf-life of sweet orange at storage temperature 10°C was 18 days. The shelf-life of sweet orange at any storage temperatures can be predicted by, and as the results of drawing this prediction, the shelf-life of sweet orange can be simply calculated by, for more comfortable and easy purposes. However, the shelf-life prediction model of sweet orange should be improved by further study since it contains a relative error above 2%. For further study it is suggested to assure the TSS change at temperature 55°C as well as the ascorbic acid change in order to improve the shelf-life prediction of sweet oranges.

Keywords: Sweet orange, Shelf-life, Total soluble solid, Arrhenius, Q10

1. Introduction

The sweet orange (Citrus sinensis) is one of potential fruits produced in Aceh Province, Indonesia. The central production of sweet orange is Bireuen Distric with share about 77% of whole production in Aceh Province (Zikria, 2015). Therefore, it is necessary to predict the shelf-life of this product with the objective of to reduce the loss facing by farmers and sellers. The information of shelf-life of sweet orange can be used to organize its post-harvest handling method, and maintain its quality. The farmers can use this information to maintain the production while the sellers can use this information for controlling their stock in the market. method (simulation). One of the simulation methods that can be used to predict the shelf-life of the product is the Arrhenius Model [3-6]. The Arrhenius Model is based on some assumptions i.e. the quality change is caused by a specific reaction during constant temperature without the influence of previous process [7]. The further study of Arrhenius Model is the Q_{10} model provided the more convenient results of the prediction [8]. Therefore, the combination use of Arrhenius and Q_{10} Models is applied to predict the shelf-life of sweet orange.

Since the change of total soluble solid (TSS) values were indicated to the decreased quality of sweet orange [9-11], the study aimed to predict the shelf-life of sweet orange by using Arrhenius and Q_{10} models based on its TSS contents.

2. Methodology

The study was conducted at Laboratory of Postharvest Technology, Department of Agricultural Engineering, Faculty of Agriculture, University of Syiah Kuala, Banda Aceh, Indonesia from Mei to October, 2018. The instruments used were refractometer DBR 85 and oven. About 10 kg of sweet oranges were observed.

The fruits were selected based on its size and colour to have the same maturity index. The selected sweet oranges were stored at three extreme temperatures i.e. 55, 65, and 70 °C. During the storage, the total soluble solid (TSS) was observed hourly until the quality of the fruit had decreased indicated by the colour and texture changes. The TSS was measured by using a hand digital refractometer. After calibrated to zero by using distilled water at 20°C, the juice of sweet oranges was placed a few drops on the prism window. Then, by a single press on "meas" button, the instrument starts to test in a few second.

The TSS changes were analysed by using linear model in equation 1 and exponential model using equation 2. The model with the higher R-square was chosen for the next calculation. The rate constant of TSS changes was determined as the slope of the model (k). Furthermore, the slope was calculated as the natural logarithm (ln k) and the temperature in degree Centigrade was transformed to Kelvin as 1/T. The trend line of this curve was used to determine the Arrhenius model. The Arrhenius model is written as equation 3 and the Q_{10} is written as equation 4. Finally, the shelf-life of sweet orange was predicted by using equation 5. Previously, another separated experiment by storing the fruits at temperatures 10°C was applied to find out the real shelf-life of this fruits as ts_{T2}.

$$A = A_0 - kt \tag{1}$$
$$A = A_0 \times e^{-kt} \tag{2}$$
$$k = k_0 \times e^{-E/RT} \tag{3}$$
$$Q_{10}^{\delta T/10} = \frac{k_{T+10}}{k} = \frac{ts_T}{ts_{T+10}} \tag{4}$$
$$ts_{T1} = ts_{T2} \times Q_{10}^{\delta T/10} \tag{5}$$

Where; A is the TSS content of sweet oranges (% brix), A_0 is the initial TSS content of sweet oranges (% brix), t is Storage time (hour), k is rate constant of quality change, k_o is rate constant without temperature influence, E is the activation energy (cal/mol), T is temperature (K), R is Gas constant (1.986 kal/mol K), ts_{T1} is the shelf-life of sweet oranges at estimated temperature (day), ts_{T2} is the shelf-life of sweet oranges at temperature basis (day) and T+10 is the storage temperature at 10°C higher.

3. Results and Discussion

3.1. The TSS change of sweet oranges at extreme temperature

Results of TSS changes of sweet oranges during storage at three extreme temperatures can be seen in figure 1-3. The initial TSS content was about 9 to 10 %brix. During the storage, the TSS content had fluctuated irregulary. The final TSS content was between 3 to 5%brix at temperature 65 and 70°C, which cut off more than 50% of its initial values. However, the final TSS content at 55°C was approximately 7% brix. The linear and exponential models of TSS changes are shown in table 1.

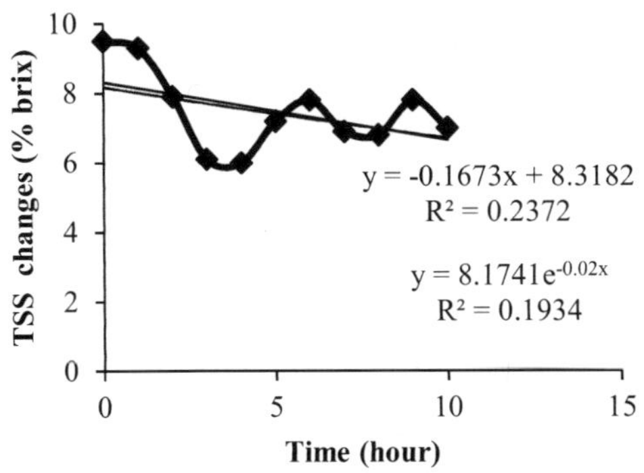

Figure 1. The TSS change of sweet oranges at temperature 55 °C.

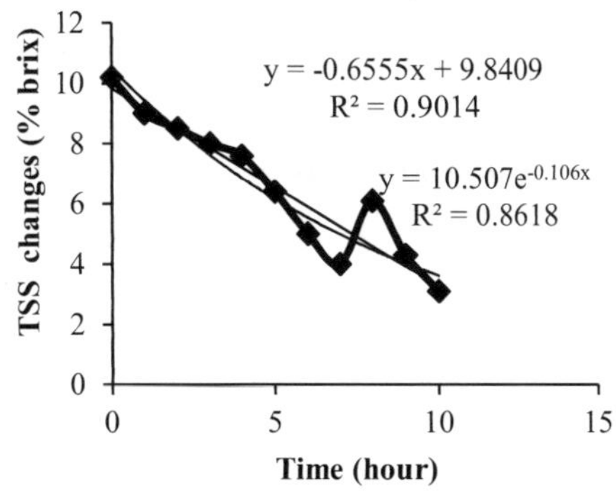

Figure 2. The TSS change of sweet oranges at temperature 65 °C.

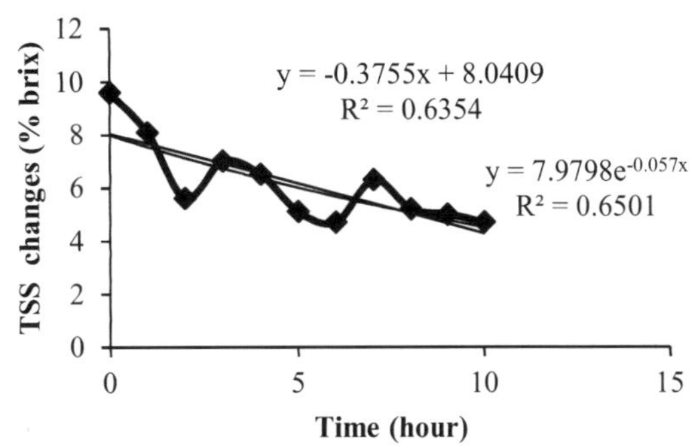

Figure 3. The TSS change of sweet oranges at temperature 70 °C.

Generally, the figures 1-3 explain that during storage, there was a decrease trend of TSS values. The highest rate constant of this changes was observed at temperature 65°C, while the lowest rate constant was found at temperature 55°C. According to the R-square values, the models of TSS changes in sweet oranges at temperature 65 and 70 °C are better than the model of TSS change in sweet oranges at temperature 55 °C. On the other word, it is important to repeat the experiment at temperature 55°C to improve the model. In fact, the linear model of TSS changes of Sweet oranges is better than that of the exponential model recognized by the higher R-square. Therefore, the linear model of TSS changes was used to determine the Arrhenius model.

Table 1. The linear and exponential model of TSS changes of sweet oranges.

No	Storage Temperature (°C)	Linear Model	R-square	Exponential Model	R-square
1.	55	y=8.32-0.167x	0.237	$y=8.174e^{-0.02x}$	0.19
2.	65	y=9.84-0.655x	0.901	$y=10.5e^{-0.1x}$	0.86
3.	70	y=8.04-0.375x	0.635	$y=7.979e^{-0.05x}$	0.65
	Average		0.591		0.57

However, the results were in contrast as reported by Adnan. He had found that the slight increase of TSS content of sweet oranges during storage had contributed to their broken quality [10]. Similarly,

Aborisade and Sakhale had also confirmed about a slight increase of the TSS content at ambient temperature [9,12]. This difference probably was caused by the extreme temperature applied in this study leaded to the higher TSS changes.

3.2. The Arrhenius Model Based on the TSS changes

The rate constant (k-value) of TSS changes of sweet oranges was recognized from the slope values of those models (figure 1-3). The lists of rate constants can be seen in table 2. The rate constant of TSS change at temperature 55°C was 0.167. As the temperature increased to 65°C, the rate constant also increased to 0.665, corresponded to almost 4 times higher. However, at temperature 70°C, the rate constant of TSS change was lower than that of at temperature 65°C, but it was almost double if it was compared to the rate constant of TSS change at temperature 55°C.

Table 2. The rate constant value and its natural logarithm of TSS changes of sweet oranges.

No	Storage Temperature (°C)	Storage Temperature (K)	1/T (K^{-1})	k	ln k
1.	55	328	0.003049	0.167	-1.78976
2.	65	338	0.002959	0.655	-0.42312
3.	70	343	0.002915	0.375	-0.98083

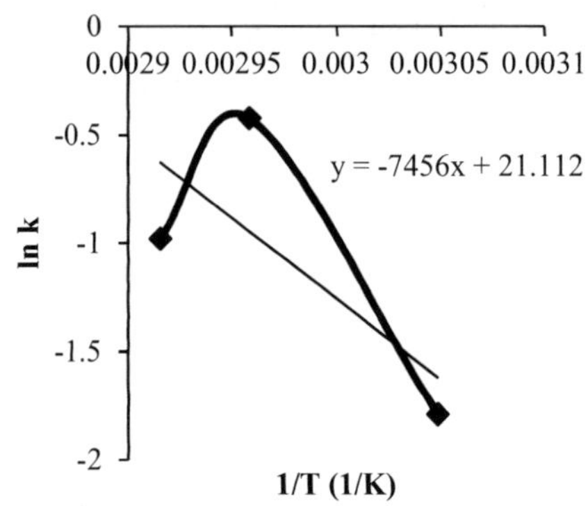

Figure 4. The plot of ln k and 1/T from the linear TSS model of sweet oranges.

$$y = 21{,}11 - 7456x$$

$$\ln k = \ln k_0 - \frac{E}{R}\left(\frac{1}{T}\right)$$

$$k_0 = \exp(21{,}11) = 1{,}4 \times 10^9$$

$$k = 1.47 \times 10^9 \times e^{-7{,}456/T} \quad (6)$$

By plotting the natural logarithm of rate constant value (ln k) and 1/T in degrees of K on a graph (figure 4), the Arrhenius model is determined as equation 6. Later, the Arrhenius model can be used to determine the rate constant of TSS changes at any storage temperatures. In addition, the model also explains about the activation energy needed to this reaction. The activation energy of the TSS changes was 14,807 cal/mol. The activation energy of TSS changes in soft drink was also confirmed about 13,030 cal/mol [12].

The comparison of rate constant values between the experiment and the prediction results are shown in table 3. It can be seen that the predicted rate constant follows the theory that the higher the storage temperatures the higher the rate constant values. Now, it can be inferred that if it is compared to the rate constant at temperature 55°C, the rate constant at temperature 65°C is two-fold, whereas at temperature 70°C it is almost three-fold.

Table 3. The comparison of rate constant value between the experiment and the prediction results

No	k experiment	k^ prediction
1.	0.167	0.197
2.	0.655	0.386
3.	0.375	0.533

3.3. The Q_{10} approach and shelf-life prediction
By using equation 4, it was found that the Q_{10} value was 1.92. The real shelf-life of sweet oranges at storage temperature 10°C was found about 18 days (according to the real observation through direct method). Recently, the equation 5 can be written as equation 7. The results of shelf-life prediction of sweet orange can be seen in table 4 estimated by using equation 7. Sakhale and Kapse had determined that the shelf-life of sweet oranges at temperature 30°C was about 5-7 days, whereas the shelf-life prediction of sweet oranges at temperature 30°C was 4.9 days [11]. It means that there is a relative error between 2 and 30 %.

$$ts_{T1} = 18 \times 1.92^{(10-T_1)/10} \quad (7)$$

Table 4. The shelf-life prediction (days) of sweet oranges based on its TSS contents.

No	Storage Temperature (°C)	ts (day)
1	35	3.5
2	30	4.9
3	25	6.8
4	20	9.4
5	15	13.0
6	10	18.0

The prediction of sweet oranges shelf-life by using this approach is very important in corresponding to the increase average daily temperature due to the climate change since the prediction can be done at any storage temperature. The relationship of shelf-life and temperature as the results of prediction by using equation 7 is shown in figure 5. It explains that the shelf-life of sweet orange follows the logarithmic model and now it can be written as equation 8. It is suggested that the equation 8 can be used to simply predict the shelf-life of sweet orange rather than equation 7, for more comfortable and easy use purposes.

There is a possibility to extend the shelf-life of sweet orange by the application of technology such as the use of 100 ppm of giberelic acid, 500 ppm of bavistin and wrapped in LDPE bags of 20% vents [11], curing in hot air at 37°C [9] and dipping in hot water at 45°C for 15 minutes [13]. The prediction for shelf-life model in this study only can be used for the untreated fruits. In the other words, there is a chance to develop the prediction model for the manipulated fruits following the same approach.

$$ts = 34{,}56 \times e^{-0.06T} \quad (8)$$

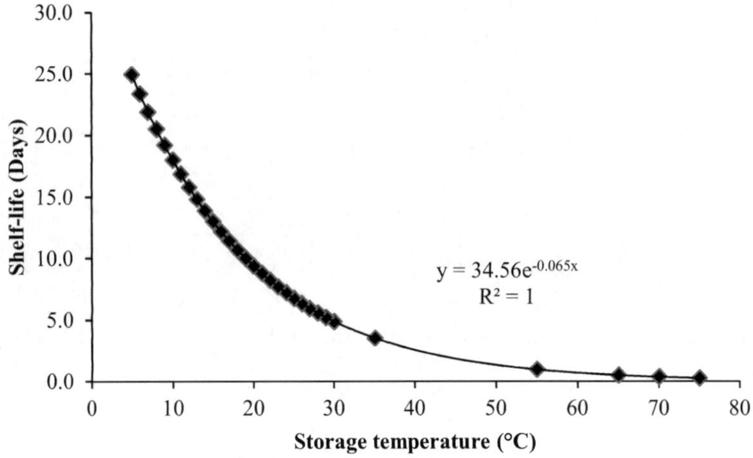

Figure 5. The predicted shelf-life of sweet orange in corresponding the temperature change.

4. Conclusion

The study had proved that there was a possibility of using the Arrhenius and Q_{10} models to predict the shelf-life of sweet oranges based on its TSS contents. The Arrhenius model was performed by using the linear model of TSS change because it was better than that of the exponential model recognized from its high R-square value. The Arrhenius model was $k = 1.4 \times 10^9 \times e^{-7{,}456/T}$ and the Q_{10} value was 1.92. The real shelf-life of sweet oranges at storage temperature 10°C was 18 days. The activation energy of TSS change was about 14,807 cal/mol. The shelf-life of sweet oranges can be predicted at any storage temperatures by equation 7, $ts_{T1} = 18 \times 1.92^{(10-T_1)/10}$, and it is suggested to use the equation 8, ts = 34,56 × $e^{-0.06T}$, for more comfortable and easy purposes. However, the equation 7 should be improved by further study since it contains a relative error above 2%. For further study, it is suggested to assure the TSS change at temperature 55°C as well as the ascorbic acid change in order to improve the shelf-life prediction of sweet oranges.

Acknowledgments

We highly acknowledge the University of Syiah Kuala for financing the study through Hibah Laboratory Programme.

References

[1] Zikria R 2015 Outlook jeruk (Jakarta : Pusat Data dan Sistem Informasi Kementerian Pertanian) p 88
[2] Wilbey R A 1997 *International Journal of Dairy Technology* 50(2) pp 64-67
[3] Zanoni B, Pagliarini E, Galli A, Laureati M 2005 *Journal of Food Engineering* 70 pp 512-517
[4] Hong W, Jianfeng S, Qian L, Yahui P, Ziyang J, Jie W, Shujun L 2015 *Journal of Chemical and Pharmaceutical Research* 7(4) pp 484-489

[5] Kurniadi M, Salam N, Kusumaningrum A, Nursiwi A, Angwar M, Susanto A, Nurhikmat A, Triwiyono, Ferdiansyah A 2016 *AIP Conf. Proc. 1788*
[6] Pulungan MH, Sukmana AD, Dewi IA 2018 *IOP Conf. Series: Earth and Environmental Sicience* 131
[7] Syarief R, Halid H 1993 Teknologi Penyimpanan Pangan (Jakarta: Arcan) pp 185-208
[8] Khathir R, Ratna, Puri M A 2015 *J. Agritech* 35(2) pp 200-204
[9] Aborisade AT and Ajibade AA 2010 *Rev. Bras. Frutic. Jaboticabal – SP* 32(3) pp 910-915
[10] Adnan 2011 *Proc. Seminar National on Post Harvest Agriculture Innovation Technology III* (Jakarta: Research and Development Division, Ministry of Agrculture) pp 55-61
[11] Sakhale B K, Kapse B M 2012 *J International Food Research* 19(2) pp 779-781
[12] Swadana AW and Yuwono SS 2014 *Jurnal Pangan dan Agroindustri* 2(3) pp 203-213
[13] Ullah F, Sajid M, Gul SL, Zainub B, Khan M 2018 *Sarhad Journal of Agriculture* 34(1) pp 220-224

IOP Conf. Series: Materials Science and Engineering **506** (2019) 012059 doi:10.1088/1757-899X/506/1/012059

Assessing local vulnerability to climate change by using Livelihood Vulnerability Index: Case study in Pahang region, Malaysia

M I Nor Diana[1*], S Chamburi[1], T Mohd. Raihan[1] and A Nurul Ashikin[1]

[1]Institute for Environment and Development (LESTARI), Universiti Kebangsaan Malaysia, 43600 Bangi, Selangor, Malaysia.

[*]Corresponding author: nordiana@ukm.edu.my

Abstract. The aim of this research is to test the Livelihood Vulnerability Index (LVI) in order to estimate climate change vulnerability to communities in Pahang. It will focus in particularly on two areas that are severely impacted by flooding namely Pekan and Temerloh. The data sample is taken from a survey of 402 households and it is based on their socio-demographics, livelihoods, social network, knowledge and skill, health, food, water security, housing, land, finances and income. This approach estimates vulnerabilities in term of exposure, sensitivity and adaptive capacity. While Pekan is more vulnerable in terms of social network, knowledge and skill but less adaptive in capacity to cope with climate and in contrast, Temerloh is the most vulnerable to natural disaster and climate variability, and sensitivity in term of health, food, finance and income and livelihood strategy. The research finding will allow a better planning of resources in terms of assistance in areas with limited support, enhanced evaluation for potential programs or current policy effectiveness, and provide access to reliable data by incorporating scenarios into the LVI model as a baseline comparison for future design management.

1. Introduction

Malaysia has had an annual increase in temperature per decade by 0.15-0.25°C which is lower than the predicated global average of 6°C by 2100 [1], but the rates of warming for the last 40 years were as high as 4°C per decade for several locations in Malaysia [2]. In addition, floods are associated with climate variability and will become more serious in the future, increasing in duration, number and frequency. A million people live in river basins world-wide and rivers flooding present a serious menace, often with the destruction of assets and the significant loss of human life due to the impact of climate change [3]. Increasing temperatures due to high rates of melting ice, glacial retreat, drought and floods, have all had a big impact on the livelihood of locals. If we look back at the historical records of Malaysia's climate change which is related to disasters and shocks it's even more prominent (Kelantan, Terengganu and Pahang) and it is often still unreported. It is now essential to identify socio-economic, health and well-being impacts on vulnerable places, communities and the magnitude and aspects of livelihood vulnerability in Malaysia. In order to do that, the research needs to encompass and study the various adaptations and vulnerability reducing projects in Malaysia. Flooding is a recurring seasonal event in Malaysia. About 29,800 KM2 or 9% of the land mass is located in the flood-plain and is affecting almost 4.82 million people which is circa 22% of the total population [4]. The most immediate and serious consequence of heavy rain is the flooding of river basins through both inundation and erosion. This not only displaces those locals affected, but it has an impact on their health and well-being as well as damaging the eco-system. Climate change not only effects a gradual change to the normal conditions it also has an effect on the frequency and intensity of extreme events such as heavy rainfall or drought or periods of extreme cold or heat [5].

The Pahang river Basin experiences flooding almost every year, but the 2014 flood was one of the worst flood disasters ever and caused the most devastation to Malaysia by a natural disaster. Flooding also exposes those affected by the event to stress and health related illness such as physical and mental, health problems and its related to negative impact. Very few qualitative vulnerability assessments or integrated vulnerability assessments have been done in Malaysia and is very much a new approach. This study aims to estimate the effect on people's livelihoods and their vulnerability in the flood-plains of Pekan and Temerloh districts using the Livelihood Vulnerability Index.

Vulnerability is also a function of the character, magnitude, and rate of climate change and the variation to which a system is exposed its sensitivity to climate change and its adaptive capacity. The IPCC definition views vulnerability as a function of exposure, sensitivity and adaptive capacity so the second category is largely associated with the Third and Fourth Assessment Reports (TAR and AR4, respectively) of the Intergovernmental Panel on Climate Change [5-6]. The approach differs from the natural hazards approach as it views vulnerability as a function of both 'internal' factors (sensitivity and adaptive capacity) and "external" factors (exposure to shocks and stresses). The latter are the various climate hazards associated with climate change and variability to which a system or population is exposed. The IPCC defines exposure as the nature and degree to which a system is exposed to significant climate variation [6], and sensitivity as the degree to which a system is affected either adversely or beneficially by climate related stimuli [6]. Adaptive capacity is defined as the ability of a system to adjust to climate change including climate variability and extremes to moderate potential damages to take advantage of opportunities or to cope with the consequences [6]. Vulnerability has traditionally had been determined through indices. For example, climate change vulnerability indices focused on agriculture [7-8], tourism [9-10] and health [11]. Most of the indices develop by socio-economic and biophysical indicator and transform to components of exposure, sensitivity and adaptive capacity. Recent studies point to the importance of vulnerability assessment when determining adaptation strategies for climate change adaptation strategies for climate change. Finally, the concept of vulnerability may include local geographical and environmental factors that mediate risks and outcomes. It is strongly rooted in social and political processes and tends to take an actor-oriented approach [12-14]. Consequently, the interpretation of an external hazard might be considered an undesirable outcome such as complex disaster) but is does however result from the interaction of, hazard and vulnerability. The people risk of flooding will also depend on its intrinsic ability to tolerate changing climate and its sensitivity.

This analysis was used to identify sources and forms of vulnerability that are specific to this context in order to design context-specific resilience measures which may help the governments to better understand communities' vulnerabilities in order to establish their resilience. This will provide development organizations and local policy makers with practical tools to understand demographics, social and other related factors contributing to the flood.

2. Material and method

2.1 Study area and household survey

The reference period for most of the survey is between 2007 and 2014 and the primary source of the data came from people residing in the Pekan and Temerloh districts. A questionnaire covering 42 key variables used in calculating the LVI was designed, tested and administrated at the household level. Pekan and Temerloh were selected for the study (locations in figure 1), as being the most flood affected area in Pahang. The Temerloh district is located in middle of Pahang and up stream of the Pahang river basin and is normally flooded from October. Whereas the Pekan district is located down-stream of the Pahang river basin and close to the South China Sea. Normally both areas face long term flooding because of the slow water to flow out from the river. A total of 402 of households were interviewed in June 2015 with approximately 200 residents in Pekan and 202 within Temerloh, respectively. The reference period for the climate events data was 2007 to 2014. A further criterion of the research was that the residents were living in the area most likely to flood and had a different social group.

2.2 Data analysis

To guide the assessment of livelihood vulnerability to floods, the study used the sustainable Livelihood framework. The impact of flood and climate variability is considered as a vulnerability, which is a major indicator of sustainability of livelihoods and assets which influence the livelihood strategies, the institutional process, and the livelihoods outcome of a community [15].This can be show in equation (1) below:

$$Index_{S_d} = \frac{S_d - S_{min}}{S_{max} - S_{min}} \tag{1}$$

Where S_d is the dimension value of each indicator, and S_{min} and S_{max} are the minimum and maximum sub-dimension values determined dimension respectively. After standardization, the values of each dimension are averaged using equations (2):

$$M_d = \frac{\sum_{i=1}^{n} index_{sdi}}{n} \tag{2}$$

Where M_d one of the ten major is dimensions for region d, $Index_{sdi}$ represents the sub-dimension value of each indicator i of major dimension M_d and n is the number of sub-dimensions in major dimension of M_d. The 11 major components are Socio-demographic profiles (SDP), Social networks (SN), Livelihood strategies (LS), Knowledge and Skill (KS), Health (H), Food (F), Water (W), Land (L), Housing (Hou), Finance and Income (FI) and Natural disaster and climate variability (NDCV). Once values for each of the 11 major components for a region are calculated , they are directly used in equation (3) to obtain the district level LVI:

$$LVI_d = \frac{\sum_{i=1}^{11} W_{Mi} M_{di}}{\sum_{i=1}^{11} W_{Mi}} \tag{3}$$

This can also be shown as follows:

$$LVI_d = \frac{w_{SDP}SDP_d + w_{SN}SN_d + w_{LS}LS_d + w_{KS}KS_d + w_H H_d + w_F F_d + w_W W_d + w_L L_d + w_{Hou}Hou_d + w_{FI}FI_d + w_{NDCV}NDCV_d}{w_{SDP} + w_{SN} + w_{LS} + w_{KS} + w_H + w_F + w_W + w_L + w_{Hou} + w_{FI} + w_{NDCV}} \tag{4}$$

Where LVI$_d$ is the livelihood vulnerability index for the district *d*, and the weightings of the 11 major components, W$_{MI,}$ are determinants by number of subcomponents that each of the components, contribute equally to the overall LVI. In this study, the LVI is scaled from 0 (least vulnerable) to 1 (most vulnerable). Following the equations (1), (4), and [16] also calculated the new equation of vulnerability LVI-IPCC. This takes consideration of the IPCC vulnerability definition as a function of system expose, sensitivity and adaptive capacity. These approaches utilise household level primary data to quantify the subcomponents. Using the same data, the LVI is based on the IPCC vulnerability definition by grouping the eleven major components into each of these three categories. The three IPCC factors are calculated based on the equation (5):

$$CF_d = \frac{\sum_{i=1}^{n} W_{Mi} M_{di}}{\sum_{i=1}^{n} W_{Mi}} \tag{5}$$

Where district d CF_d is an IPCC-defined contribution factor (exposure, sensitivity or adaptive capacity) for the district d, M_{di} is the major components for the district *d*, indexed by i, W_{Mi} is the weightage of each major components, and n is the number of major components in each contribution factor. Once exposure, sensitivity and adaptive capacity were calculated, the three contributing factors were combined using equation (6):

$$LVI - IPCC_d = (e_d - a_d) * s_d \tag{5}$$

Where $LVI - IPCC_d$ is the LVI for the district d expressed using the IPCC vulnerability framework, e_d is the calculated exposure score for the district d (equivalent to the natural disaster and climate variability major component) a_d is the calculated adaptation capacity score for district d (weighted average of the Socio-Demographic, Social Networks, Livelihood Strategies and Knowledge and Skill major components) and s is the calculated sensitivity score for district d (weighted average of Health, Food, Water, Land, Housing and Finance and Income major components). The LVI-IPCC is scaled for -1 (least vulnerable) to +1 (most vulnerable) and is the best understood as an estimate of the relative vulnerability of compared populations. A total of 37 component values were computed from scored responses of individuals in each community was treated scores taken from a set of scales ranging from minimum to maximum. All individual score was used for computing the variable components values for the community. The foregoing to this, scores were checked with their respective values and were all found significant at $p < 0.05$ using the three-standard deviation rule and a standard error of 0.1.

3. Result and Discussions

The initial readings of the finding suggest little difference between the two districts. Further detailed reading does however reveal several important differences between Pekan and Temerloh. These nuances are now discussed in greater detail by using the vulnerabilities indices of the major components which range from 0.10 to 0.85 as shown in table 1. The value index for the socio-demographic profile component of the LVI shows Temerloh to be the most vulnerable (0.257) and the Pekan district to be the least vulnerable (0.178). Furthermore, the Temerloh districts recorded the highest percentage of female headed households, lowest household income which is below the Poverty Line Income (PLI) of Malaysia in fact; they are also reported taking care of at least one orphan. The starts show that this area and its inhabitants are the most vulnerable to flooding. The indices are relative values and are compared across two districts in Pahang within the study sample only. The major component of Social Networks is made up of two sub-components which are borrowing money and do not involved in any organization. In term of social networks, Pekan is the most vulnerable (0.462) whereas Temerloh is (0.348). Households in Pekan reported receiving more help than giving help to others compared to households in the other districts. Help is a classed as small funding programs for small scale agricultural or other small business projects. This activity is recorded between themselves and their social network and this also includes assistance in kind. Households that borrow more money than they lend are more vulnerable. If the subcomponents were added together, the most reliant on their social network was the Pekan districts. These results indicated a need for strengthening community networks and local organizations such as Farmer Associations, Business Associations, and Youth Associations at village level to reduce social vulnerability. Most of the households are comfortable with borrowing from relatives and close friends rather than from community. Good social networking seems to lessen the impact of climatic stresses on individual households [17]. The third major component is livelihood strategies which are made up of three sub-components. Due to individual experiences, knowledge and exposure to natural disasters [18], the livelihood strategies of households were diverse. These strategies include no members migrating to others areas, depending solely on agriculture income and the diversification of agriculture. When the results of all the sub-components where aggregated, it was found that the Temerloh area was the most vulnerable (0.412) and Pekan slightly less (0.356). The results also revealed that Temerloh retained their members within the community and was less likely to diversify away from agriculture. The majority of them remained in catfish farming and thus they had the highest indices of vulnerability. In contrast, although Pekan households reported relying solely on agriculture income too, they were more likely to have diversification strategies such as growing crops, animal husbandry and farming natural resources which shows their livelihood diversification. When the three sub-components were averaged, the overall Livelihood Strategies vulnerability score was higher for Temerloh than for the Pekan districts. The finding also shows that the inhabitants of the Temerloh district where more likely to suffer health issues than the Pekan district. This is mainly due to issues arising from flooding such as cough and colds, flu, fever, sore throats and headaches and consequently it was taking an average of two weeks to return to work or school. Another impact was the time to reach a health facility due to flood related illness which was 13 minutes. Inadequate access to the health services tends to decrease the health status of households, thereby increasing their vulnerability to extreme climatic stresses. Sickness increases the vulnerability of households to other external stresses it was also revealed that 5.5% of households in the study area have post-traumatic stress disorder (PTSD).

The Pekan districts have a higher knowledge and skills than Temerloh. This shows that areas with lower levels of education and especially flood training are more vulnerable to disaster. Lower adaptation strategy without any training is less likely to cope with stress after a flood for the Pekan households. A Temerloh household also without enough skill to cope with shock and just recently left primary school. The food major component is made up by three sub components show that Temerloh is the highest again in food storage and grow the own crops and diversify rather than Pekan The overall food vulnerability score for Temerloh (0.745) was higher than Pekan (0.648). This is important as food security enhances a households' resilience to external stress and shock due to extreme climatic events.

Temerloh also had a lower vulnerability score (0.235) for the water component than Pekan (0.243). It was seen that they were much better prepared storing fresh water, and the Temerloh households reported less conflicts of water supply and accessibility of water resources compared to Pekan. The average number of liters water storage by Pekan residents during the flood is the lowest, coupled with the utilization of natural water resources which is more likely to expose them to the symptom of water borne disease. Land is the most important assets of inhabitants in the area and is a measurement of wealth. This major segment is made up of two sub components; one being landless and the other small farms linked or owned by the households. Nonetheless, they are considered more susceptible to flood damage and climate change. The land at Pekan registers as slightly more vulnerable (0.650) and more than fifty percent of households are reported are landlessness 57% and the percentage of small land packages is 73%. This was constructed by raising the vulnerability value for Pekan rather than Temerloh (0.625). Most of the residents rent or live in their ancestral home. The research also discovered that non-landowners have limited to access the formal safety nets and other entitlements which are seriously vulnerable.

In terms of natural resources, Temerloh has a higher index than Pekan (0.100) mainly because a larger percent of household does not receive a flood warning and a larger percent of injury or death due to flood. Additionally, the last seven years have seen severe flooding every year which also contributes in a small way to the higher index. The survey results also show the average water level of seven feet with sub component indices of about 0.347 and households affected by flood, partially to totally submerge was 68.0% and shows a higher percentage of households with non-solid houses in Temerloh district 18.0% and that the water levels that engulfed the houses is average. The combinations of these three sub-components provide a higher housing vulnerability index for Temerloh districts (0.403) than Pekan (0.378). In the Temerloh districts, households were affected by income loss during the flood and it was these households that needed to borrow money during the crisis. Furthermore, these residents are the ones on the lowest incomes RM 520 and are in the lowest band of 2%. When the results of all three sub-components are aggregated, the Temerloh district is found to be the most vulnerable (0.158) in term of finance and income. Overall, the aggregated score in table 1 shows that Temerloh was more vulnerable to climate change than Pekan, LVIs of 0.308 compared to 0.278. The results also suggest that Temerloh was more vulnerable in term of SDP, Health, NDCV, Housing and F&I. In contrast, Pekan was more vulnerable in SN (0.462), Knowledge and Skill (0.207), Water (0.243) and Land (0.650). Therefore, a further model 2, LVI-IPCC where the findings show that all of communities in all of these districts share a very similar degree of vulnerability -0.053 and -0.052, respectively and could be described as "mid-range" on the scale -1 to +1. The results do suggest that Temerloh is more sensitive (0.357) and was most affected by food, health, land and housing and that Pekan (0.326) which is less exposed than Temerloh 0.100 vs. 0.141, respectively. The adaptive capacity of the communities was most affected by their livelihood strategies which was 0.288 in Temerloh and 0.263 in Pekan were also the elements of social networks (0.462) were most affected and the indices is the lowest at 0.263.

Table 1 LVI-IPCC contributing factors calculation for Pekan and Temerloh districts, Pahang, Malaysia (IPCC, 2001)

Contributing factors	Major components	Major component values	Number of subcomponent per major component	Contributing factor values	LVI- IPCC value
Adaptive capacity	Socio-demographic profile	0.178 (0.257)	6	0.236 (0.288)	
	Livelihood strategies	0.356 (0.412)	3		
	Social Network	0.462 (0.348)	2		
	Knowledge and skill	0.207 (0.187)	3		
Sensitivity	Health	0.107 (0.170)	4	0.326 (0.357)	-0.053(-0.052)
	Food	0.648 (0.745)	3		
	Water	0.243 (0.235)	4		
	Housing	0.378 (0.402)	3		
	Land	0.650 (0.625)	2		
	Finance and income	0.140 (0.158)	3		
Exposure	Natural Disaster and Climate variability	0.100 (0.141)	4	0.100 (0.141)	

Noted: Values in the parenthesis are for Temerloh district.MIndex value should be interpreted as relative values to be compared within the study sample only. The LVI-IPCC is on a scale from -1(least vulnerable) to 1 (most vulnerable) [16].

Next, figure 1 below show the triangle diagram of the contributing factors at related places.

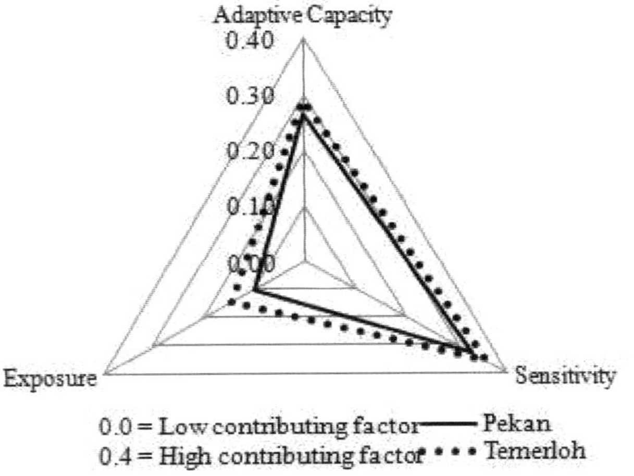

Figure 1: Vulnerability triangle diagram of the contributing factors of the Intergovernmental Panel on Climate Change livelihood vulnerability index (LVI-IPCC) for districts

The LVI-IPCC was calculated by grouping the 11 major components into three categories, namely, adaptive capacity, sensitivity, and exposure, which all differ across districts. The calculations show the high levels of exposure relative to adaptive capacity yield positive vulnerability scores while low values of exposure relative to adaptive capacity, yield negative vulnerability scores. Sensitivity roles as a multiplier such as that high sensitivity in a district for which exposure exceeds adaptive capacity will result in a large positive for example high vulnerability LVI-IPCC scores and vulnerability to other areas such as its socio-demographic profile, health , natural disaster and climate control variability components are impending. There was also found to be a disproportionate number of households with orphans, female breadwinners and a sole dependence on agriculture with little diversification with little awareness towards climate variability. It is also this socio-group which is impacted more with health problems consequently missed school and work due to flooding.

The majority of these groups received no warning about flooding and where the most severely affected by the loss of property including equipment. Over the last seven years, they were also the most exposed to injury and death, natural hazards and climate variability. It was revealing that those are living in natural disaster areas such as a river basin affected by flooding are exposed to the hazard [19]. Furthermore, flooding is one of the environmental shocks that are associated with poverty drivers [20] Thus; the government had to rescue the community in this district to reduce the risk of climate change. Medical assistance and visits from various health teams after the flooding should be enhanced so those affected can get back to normality and health as quickly as possible after the disaster. Additionally, Pekan was found to be most vulnerable districts in term of its adaptation capacity and this could be a major factor due to its enhanced social network. Furthermore, the results highlighted a definite need to strengthen community networks and associations such as farming, business and youth groups in an effort to reduce social vulnerability. This will also reduce the impact of climatic stress [17] and could also facilitate economic well-being and security for coping with climate variability and hazard [21]. The LVI is designed to provide development organization, policy-makers and practitioners with a practical tool to understand various livelihood assets contributing towards climate vulnerability. The index-scale can help development planners refine and focus their analysis of geography units.

4. Conclusion

The research performed some pragmatic analysis to assess and understand the local vulnerability to climate change and variability in two specific areas by using the livelihood and vulnerability index (LVI) as developed [16]. In addition, it also investigated the significant contributions of the index to the development and understanding of vulnerability indicators and the assessment tool for policy analysis. Additionally, it is a systematic means for comparing livelihood vulnerability in different socio-ecological circumstance. The presenting of LVI-IPCC as alternative methods for assessing relative vulnerability of community to climate variability impacts shows that Pahang river basin is most vulnerable to climate change due to the high exposure and relatively low level of adaptive capacity than Temerloh. Flood preparedness is a good strategy in order to limit the impact of extreme flooding in the future. In addition, the upgrading of communication system and strengthening of social institution networks are highly recommended to a community to deal with climate change and climate variability and can help identify and determine actions that can ameliorate adverse impacts of disaster, such as flooding. If the community improves their coping capacity, it could help to reduce the damage caused by extreme events or environmental degradation, especially in the Temerloh districts. In contrast, the community in the Pekan district was the most vulnerable and had the lowest financial support. It is recommended that they are the most likely to need assistance for small and medium enterprises, smallholders and producers to start-up production or enterprises to improve their living conditions.

Acknowledgments

The authors acknowledge the support from the Ministry of Education Malaysia for funding this study on Application of Geographical Information System (GIS) to mapping the Livelihood Vulnerability Index (LVI) in Pahang River Basin" under the DPP-2018-008 project.

References

[1]. Tangang F.T., Juneng L., Salimun E., Sei K.M., Le L.J. and Muhamad H. 2012. *Climate change and variability over Malaysia*: Gaps in Science and Research Information. Sains Malaysiana. **41 11** 1355-1366

[2]. Tangang F.T., Xia C., Qiao F., Juneng L. and Shan F. 2011. *Seasonal circulations in the Malay Peninsula Eastern continental shelf from a wave-tide-circulation coupled model.* Ocean Dynamic. **61** 1317

[3]. Chan N.W. 2012. *Impacts of Disaster & Disaster Risk Management in Malaysia: The case of floods.* pp.503-551

[4]. Sani G., Gasim M.B., Toriman M.E., Abdullahi M.G. 2012. *Floods in Malaysia: Historical reviews, causes, effects and mitigation approach. International Journal of Interdisciplinary Research and Innovations.* **2 4** 59-65

[5]. IPCC 2007. Climate change 2007: *The physical science basis. Contribution of working group II to the Forth Assessment Report of the Intergovernmental Panel on Climate Change (Ch.9).* Cambridge University Press, Cambridge, UK

[6]. IPCC,2001. Climate change 2001: *Impacts, adaptation and vulnerability. Contribution of working group II to the Third Assessment Report.* Cambridge University Press, Cambridge, UK.

[7]. Tao S., Xu Y., Liu K, Pan J. & Gou S. 2011. *Research progress in agricultural vulnerability to climate change. Advances in Climate Change Research.* **2** 4, 203-210

[8]. Millari A.E. 2016. *Climate change vulnerability assessment in the Agriculture sector: Typhoon Santi Experience. Procedia-Social and Behavioral Sciences.* **216** 440-451

[9]. Sabine L.P.-N. 2010. The *vulnerability of beach tourism to climate change- an index approach. Climatic Change.* **100** 3-4, 579-606

[10]. Raquel S.L., Clave S. A., Saladie O. 2017. *The vulnerability of coastal tourism destination to climate change: The usefulness of policy analysis. Sustainability.* **9** 11 2062

[11]. Malik S.M., Awan H., Khan N. 2012. *Mapping vulnerability to climate change and its repercussions on human health in Pakistan. Globalization and Health.* **8** 31

[12]. Cannon T., Muller-Mahn D. 2010. *Vulnerability, resilience and development discourses in context of climate change. Nat Hazards.* **55** 3 621-635

[13]. Miller F. H., Osbahr E., Boyd F., Thomalla S., Bharwani G., Ziervogel B., Walker J., Birkmann.S., Van der Leeuw J., Rockström J., Hinkel T, Downing C., Folke and D. Nelson. 2010. *Resilience and vulnerability: complementary or conflicting concepts?. Ecologyand Society.* **15** 3, 11

[14]. Wisner B., Blaikie P., Cannon T., Davis I. 2004. *At Risk: Natural Hazards, people's vulnerability and disasters.* London: Routledge

[15]. Chamber R., Conway G.R. 1992. *Sustainable rural livelihood: Practical concepts for the 21st century* discussion paper 296 Institute of Development Studies, Brighton

[16]. Hanh M.B., Riederer A.M., Foster S.O. 2009. *The livelihood vulnerability index: A pragmatic approach to assessing risks from climate variability and change-A case study in Mozambique. Global Environmental Change.* **19** 1, 17-88

[17]. Madhuri Tewar H.R., Bhowmick P.K. 2014. *Livelihood vulnerability index analysis: An approach to study vulnerability in the context of Bihar. Jamba: Journal of Disaster Risk Studies.* **6** 1, 1-13

[18]. Etwire P.M., Al-Hassan R.M., Kuwornu J.K.M., Owusu O.Y. 2010. *Application of livelihoodvulnerability index in assessing vulnerability to climate change and variability in Northern Ghana. Journal of Environment and Earth Science.* **3** 2 157-170

[19]. Fussel H.M. 2007. *Vulnerability: A generally applicable conceptual framework for climate change research. Global Environment Change.* **17** 2 155-167

[20]. Leichenko R., Julie A. S. 2014. *Climate change and poverty: vulnerability, impacts and alleviation strategies. WIREs Climate Change.* **5** 4 539-556

[21]. Adger W.N., Huq S., Brown K., Conway D., Hulme M. 2003. *Adaptation to climate change in the developing world. Progress in Development Studies.* **3** 3 179-195

IOP Conf. Series: Materials Science and Engineering **506** (2019) 012060 doi:10.1088/1757-899X/506/1/012060

Sound Intensity Mapping of Two Stroke Engine by Using Hemispherical Surface Coordinate Arrangement

M S M Sani[1,2*], M A F A Rahimi[1] and D Mohamed[2]
[1]*Advanced Structural Integrity and Vibration Research (ASIVR), Faculty of Mechanical Engineering, Universiti Malaysia Pahang, 26600 Pekan, Pahang, Malaysia*
[2]*Automotive Engineering Centre (AEC), Universiti Malaysia Pahang, 26600 Pekan, Pahang, Malaysia*

[*]Corresponding author: mshahrir@ump.edu.my

Abstract. This paper presents to investigate the noise source identification of two stroke engine for different engine speed and different general case in two stroke engine without rear cover, with rear cover, with rear cover (composite fibre), and with rear cover (acoustic foam). The experiment carried out utilizing by sound intensity in hemispherical surface arrangement and sound intensity mapping method for engine speed at range 800 to 4800 revolution per minute (rpm). A single module of microphone is used to identify sound power level in hemispherical surface coordinate arrangement while a pair microphone in sound intensity mapping. This method can be applied on-site rather than in an anechoic chamber as long as the background noises are stationary. The sound power level is within 73.7 dB to 81.3 dB in the range of engine speed. The noise level is increasing with engine speed. Rear cover should be applied in order to reduce the sound power level. The project delivers the reliable input for the engineering practice to reduce engine noise level.

1. Introduction

Sound is such a common part of everyday life that are rarely appreciate all of its functions. It is also defined as the pressure variation in air, water and other medium that are detected by human ear. When sound produced, energy have transferred from sound source to the surrounding air molecules takes place. The energy transfer is called Sound Power that have unit of Sound Power is W (Watt). The audible range of sound power is from 10-9 W to 1000 x 10-9 W is the lowest level which can heard by people that are closed to the source and 1000 W can immediately create hearing damage. Lower level of sound power also can create hearing damage but in long period of time. Sound have affected human in positive and negative way. In positive way, sound can be defined as pleasant and relaxing. In other way, the unpleasant and unwanted sound is called noise which are occurred in high decibel (dB) frequency. Noise affected people in everyday life in different way depending on the type of noise, noise level, frequency characteristics, time of day and variance over time [1-2].

When a source produces sound power (P) it have created a certain Sound intensity (I) at distance away from the source. The intensity is a measure for the amount of power through a certain area of the distance. Sound intensity is the rate of acoustic energy flow per unit area. It is also a current of acoustic energy through medium that have magnitude and direction. Sound Intensity Mapping is by integrated over enclosing surface of an object and its sound power. Sound intensity measurement is able to determine the sound directional and localise noise hot spot. Mapping software have provided a contour map of noise level that are provided intuitive documentation of noise problems.

Noise measuring devices are typically used a sensor to receive noise signals from source. However, sensor not only detect noise from the source but it is also detected any ambient background noise. Sound measurement for machine is running, it is important to make the background noise in lower level to ensure any influence on the result. Therefore, the anechoic chamber is needed to measure the source of noise. Noise source identification (NSI) is a technique to study noise control program. Noise source have two categories which are vibration-induced noise and flow-induced noise. First category has included noise from rotating machinery, noise due to structural resonance, impact noise, etc., then second category included fan noise, pump noise, jet noise, etc. Other than that, NSI is sound field visualization techniques and find applications in non-destructive evaluation, underwater imaging and machine diagnosis [3-4]. NSI techniques is useful in estimating the sources position and sound strength. NSI is also to identify the major sources of noise radiation. The best reduction in noise can be achieved by reducing the noise that are radiated from the noisiest sources. The reduction noise is required not only the treatment of many different sources but also called for different approaches in each particular case depending upon the degree of the contribution of the various individual sources to the total noise. Noise of vehicles and machines is mainly defined by noise generated by the power part which are an engine. The main part that are highly produced engine noise is the noise that are emitted by engine surfaces [5-8].

Overall, this paper aims to investigate noise source from two stroke engine by using sound intensity mapping. A single module of microphone has been applied to identify sound power level in hemispherical surface coordinate arrangement. This study also suggests to design rear cover in order to reduce sound power level.

2. Experimental set up

In conducting this experiment, it has studied in four general cases which are the condition of two stroke engine without rear cover, with rear cover, cover by composite fibre and acoustic foam. Acoustic foam and composite fibre have used in this study as noise reduction for two stroke engine.

The two stroke engine needs a proper setup before the experiment performed. Figure 1 shows the apparatus setup used in experiment which are Prepolarized Free – Field ½ Microphone type 4189, Output Generator Module 50kHz, Hand – held analyzer, and Sound Intensity Microphone Pair. The coordinates in hemispherical surface arrangement is according to ISO 3745 [9], Sound power levels (sound pressure – free field) as shown in Figure 2. Furthermore, Figure 3 shows the microphone positions on equal areas on the surface of a hemisphere. Sound power of the source is then calculated from sound pressure measurements and knowledge of acoustic environment.

Figure 1. Apparatus of Sound Intensity

Figure 2. Hemispherical arrangement

IOP Conf. Series: Materials Science and Engineering **506** (2019) 012060 doi:10.1088/1757-899X/506/1/012060

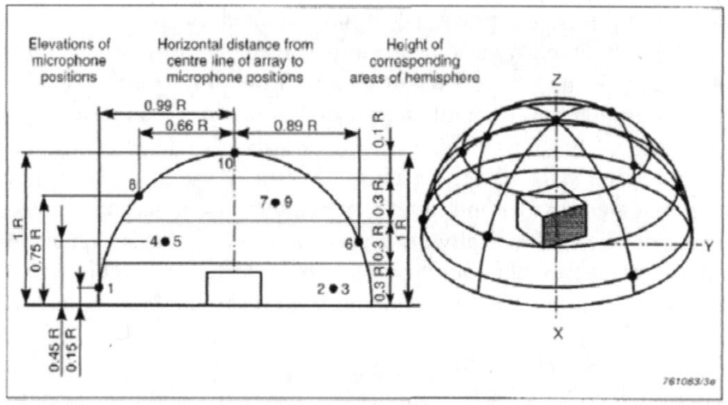

Figure 3. Microphone positions on equal areas on the surface of a hemisphere [6].

The noise on two stroke engine needs proper set up in the semi anechoic chamber before the experiment can be performed. In investigate the noise source identification at free field, four parameters, (800,1800,3600 and 4800 rpm) are setup in this experiment. The calibration for microphone is followed calibration chart from Bruel & Kjaer type 4297. The microphone has calibrated until it reaches sound level is around 94.0 dB. In Noise Source Identification experiment, Prepolarized Free-field ½ Microphone type 4189 is used. It has connected by wire to the Bruel & Kjaer Type 3160 A 4/2 Output Generator Module and the data will be transfer directly to the laptop by cable wire. The single microphone is setup by following the coordinates point on hemispherical arragement. When the test was running, data is generated and analyse by Bruel & Kjaer – PULSE LabShop Version 16.1.0 in the laptop to detect the higher noise level in the hemispherical arrangement. Graph of decible (dB) versus frequency are used to find which point of location that detected high level of decibel reading.

After the experiment of noise source identification by single microphone conducted, it is followed by using the Bruel & Kjaer Hand-held analyser Type 2270 and Sound Intensity microphone pair type 4197 to get the sound intensity mapping. Figure 4 shows the investigations that were carried out using Sound Intensity technique for each side or view of two stroke engine. The noise analysis can be interpreted by looking at the colour of the contour graph mapping to determine which part that produces major source of noise by using post processing software, PULSE LabShop Version 16.1.0. The noise measurement was carried out in Noise Source identification at 40 point of coordinates in hemispherical arrangement for single microphone than Sound Intensity Mapping were carried out at the front, rear, and at both side of the engine according to the SAE recommendation for microphone positions. The noise measurement was repeated in five level of parameter in rpm.

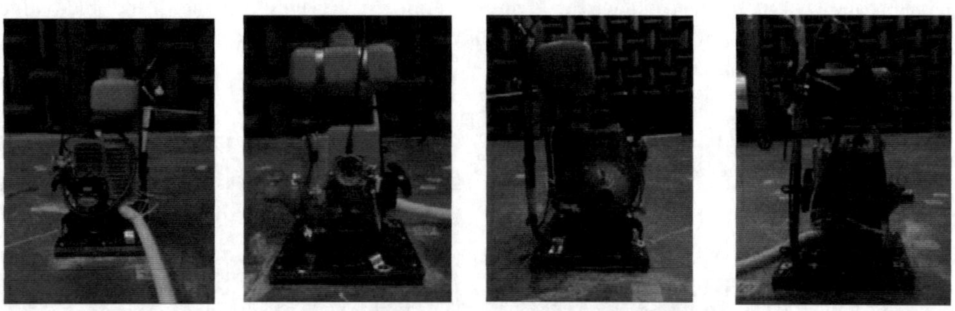

(a) (b) (c) (d)

Figure 4. Two stroke engine a) Right view, b) Front view, c) Left view, d) Rear view

3. Mathematical Modelling

Sound intensity at any point, in a sound field in an interest direction x can be written as

$$SI_x = \frac{1}{T}\int_0^T h(t)\, w_x(t) dt \tag{1}$$

where T is a sufficiently long averaging time, $h(t)$ and $w_x(t)$ are instantaneous sound pressures and particle velocity in the direction of x, respectively. The sound pressure at the midpoints of two microphones can be determined by taking the average of two sound pressure signals as Eq. (2):

$$h(t) = \frac{h_1(t) + h_2(t)}{2} \tag{2}$$

The particles velocity at midpoint can be obtained by Euler's equation for a zero mean velocity medium as Eq. (3):

$$w_x(t) = -\int \nabla h(t) dt \approx -\frac{h_2(t) - h_1(t)}{j\omega\rho\Delta x} \tag{3}$$

By using Fourier transforms, sound intensity can be summarized as Eq. (4):

$$SI_x(\omega) = \frac{\mathrm{Im}\,[K_{12}(\omega)]}{\omega\rho\Delta x} \tag{4}$$

where $K_{12}(\omega)$ is the one sided cross power spectrum density function between microphone channels 1 and 2, ω is the angular frequency, ρ is the air density and Δx is the separation distance between the two microphones.

4. Result and Discussion

4.1 Engine Speed 800 rpm

Investigation NSI for two stroke engine were carried out using sound intensity technique at 40 points of coordinates in hemispherical arrangement and sound intensity mapping technique for each side of two stroke engine. The maximum noise level for two stroke engine without cover at speed 800 rpm is 80.3 dB at point 5 in hemispherical arrangement. The other sound power level for engine speed 800 rpm is at Table 1. Result shown the higher noise level is occurring at the rear part of two stroke engine. Then, noise reduction is suggested at rear part of the engine.

Table 1. Maximum Noise Level at engine speed 800 rpm

Point	Maximum Noise Level (dB)	Point	Maximum Noise Level (dB)	Point	Maximum Noise Level (dB)	Point	Maximum Noise Level (dB)
1	77.0	11	79.9	21	79.1	31	78.2
2	77.3	12	78.6	22	78.7	32	76.7
3	76.8	13	78.4	23	78.9	33	77.3
4	78.9	14	79.1	24	78.6	34	78.4
5	80.3	15	78.4	25	77.8	35	77.0
6	77.5	16	79.3	26	79.9	36	75.1
7	78.9	17	77.7	27	77.8	37	78.3
8	79.0	18	78.8	28	77.4	38	75.9
9	77.8	19	78.3	29	77.5	39	77.3
10	77.5	20	77.0	30	80.0	40	75.3

Sound Power level two stroke engine with rear cover, cover by composite fibre and acoustic foam is 79.9 dB at point 5, 81.3 dB and 81.5 at point 7. Point 5 is located at the rear part while point 7 is located at front part of two stroke engine. Based on trend data that are plotted in graph in Figure 5, several point in hemispherical arrangement shows the decreases of sound power level. Point 5 shows the decreases from 80.3 dB for two stroke engine without cover to 79.9 dB, 77.6 dB and 73.9 dB for two stroke engine with rear cover, cover by composite fibre and acoustic foam. Other point that are shows the decreases of sound power level is point 1, 2, 8, 9, 12, 15, 16, 18, 21, 23, 24, 26, 27, 30, 31, 34, 37, and 40. All presented spectra are averages of four measurements of each cooling fan speed variations.

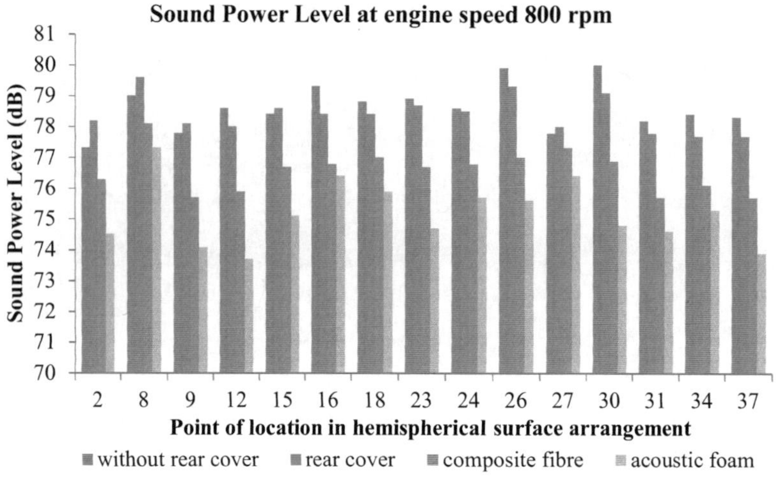

Figure 5. Graph of comparison Sound power level at engine speed 800 rpm

Then, investigations were carried out using sound intensity mapping technique for each side of two stroke engine had been made. Thus, in this section the results of the experiment are showed. In order to analyzed and interpreted the results produced by CPB analyzer that presented on the screen of a Hand-Held analyzer, sound power level also can be determined by looking at the colour of the contour graph to know the part that produces major source of noise. As shown by the colour indicator at the right side of the picture, sound power level that produced ranging from 75 dB to 116 dB. Instead of positive values, negative values also take as account in these discussion where the negatives intensity showed that the sound coming from the back of the probe. The highest of sound power level produce during the test have been conducted is 93.89 dB at rear view, 99.54 dB at right view, 95.15 dB at front view and 95.75 dB at left view. Other sound power level ay engine speed 800 rpm is shown in Table 2. The yellow shape indicates the engine part that are produced the highest noise level while the green shape indicates the lower noise level of two stroke engine.

4.2 Engine Speed 1800 rpm

Sound power level (dB) at engine speed 1800 rpm was recorded is 83.3 dB as the maximum noise level at frequency 800 Hz and located at point 25 in hemispherical arrangement. Others for maximum noise level (dB) at engine speed 1800 rpm is shows in Table 2.

IOP Conf. Series: Materials Science and Engineering **506** (2019) 012060 doi:10.1088/1757-899X/506/1/012060

Table 2. Maximum Noise Level at engine speed 1800 rpm

The maximum noise level (dB) for two stroke engine without rear cover is 85.1 dB at point 1 in

Point	Maximum Noise Level (dB)	Point	Maximum Noise Level (dB)	point	Maximum Noise Level (dB)	point	Maximum Noise Level (dB)
1	77.3	11	79.0	21	77.6	31	77.8
2	78.2	12	78.0	22	79.1	32	77.2
3	77.5	13	78.4	23	78.7	33	78.4
4	79.2	14	79.3	24	78.5	34	77.7
5	79.9	15	78.6	25	78.1	35	77.6
6	78.6	16	78.4	26	79.3	36	78.1
7	79.5	17	78.0	27	78.0	37	77.7
8	79.6	18	78.4	28	77.6	38	76.8
9	78.1	19	78.9	29	78.1	39	77.4
10	78.1	20	77.9	30	79.1	40	76.7

hemispherical surface while for two stroke engine with rear cover, cover by composite fibre and acoustic foam is 83.3 dB, 84.1 dB and 84.9 dB at point 25. Point 1 is located at the right part while point 25 is located at front part of two stroke engine. Based on trend data that are plotted in graph in Figure 6, several point in hemispherical arrangement shows the decreases of sound power level. Point 37 shows the decreases from 81.6 dB for two stroke engine without cover to 81.2 dB, 80.7 dB and 79.4 dB for two stroke engine with rear cover, cover by composite fibre and acoustic foam. Other points that are shows the decreases of sound power level is point 1, 2, 8, 9, 12, 15, 16, 18, 21, 23, 24, 26, 27, 30, 31, 34, 37, and 40.

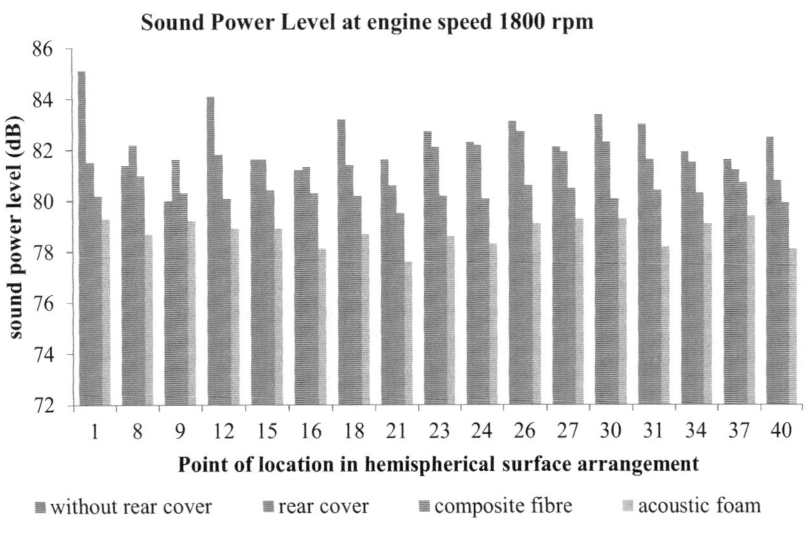

Figure 6. Graph of comparison Sound power level at engine speed 1800 rpm

4.3 Engine Speed 3600 rpm

The maximum noise level at engine speed 3600 rpm is 87.2 dB at frequency 100 Hz and located at point 4 and 7. Table 3 shows the other maximum noise level at engine speed 3600 rpm.

Table 3. Maximum Noise Level at engine speed 3600 rpm

Point	Maximum Noise Level (dB)	Point	Maximum Noise Level (dB)	point	Maximum Noise Level (dB)	point	Maximum Noise Level (dB)
1	78.1	11	77.1	21	76.1	31	75.7
2	76.3	12	75.9	22	79.9	32	77.9
3	79.5	13	78.8	23	76.7	33	80.1
4	80.3	14	80.4	24	76.8	34	76.1
5	77.6	15	76.7	25	79.3	35	78.9
6	79.9	16	76.8	26	77.0	36	79.2
7	81.3	17	79.7	27	77.3	37	75.7
8	78.1	18	77.0	28	77.9	38	78.8
9	75.7	19	81.2	29	79.2	39	78.7
10	79.3	20	79.3	30	76.9	40	76.9

The maximum noise level (dB) for two stroke engine without rear cover is 86.9 dB at point 11 in hemispherical surface while for two stroke engine with rear cover, cover by composite fibre and acoustic foam is 86.4 dB at point 11, 87.2 dB at point 4 and 7 and 87.5 dB at point 7. Point 4, 7, and 11 is located at the right and front part of two stroke engine. Based on trend data that are plotted in graph in Figure 7, several point in hemispherical arrangement shows the decreases of sound power level. Point 12 shows the decreases from 86.9 dB for two stroke engine without cover to 84.4 dB, 82.5 dB and 81.7 dB for two stroke engine with rear cover, cover by composite fibre and acoustic foam. Other points that are show the decreases of sound power level is point 1, 2, 8, 9, 12, 15, 16, 18, 21, 23, 24, 26, 27, 30, 31, 34, 37, and 40.

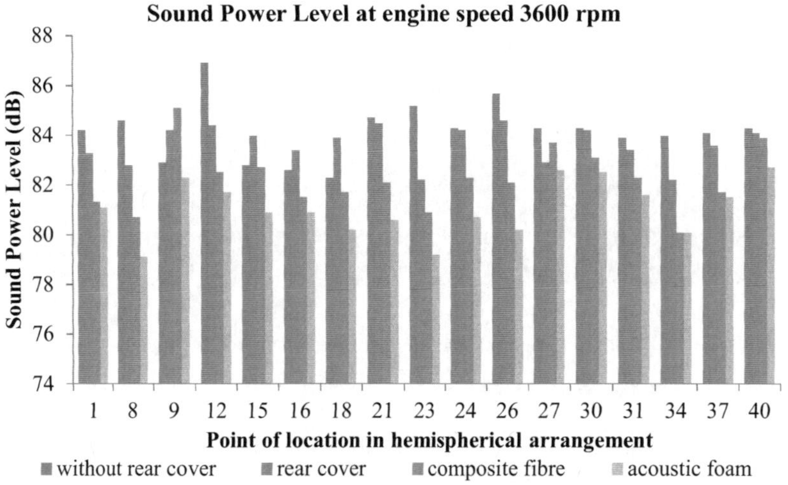

Figure 7. Graph of comparison Sound power level at engine speed 3600 rpm

IOP Conf. Series: Materials Science and Engineering **506** (2019) 012060 doi:10.1088/1757-899X/506/1/012060

4.4 Engine Speed 4800 rpm

The maximum noise level at engine speed 4800 rpm is 93.3 dB at frequency 125 Hz and located at point 14. The other maximum noise level and graph of sound power level at engine speed 4800 rpm as shown in Table 4.

Table 4. Maximum Noise Level at engine speed 4800 rpm

Point	Maximum Noise Level (dB)	Point	Maximum Noise Level (dB)	Point	Maximum Noise Level (dB)	Point	Maximum Noise Level (dB)
1	78.0	11	78.1	21	76.3	31	74.6
2	74.5	12	73.7	22	79.3	32	78.2
3	79.9	13	79.2	23	74.7	33	81.0
4	80.5	14	80.3	24	75.7	34	75.3
5	73.9	15	75.1	25	79.8	35	79.6
6	81.4	16	76.4	26	75.6	36	78.7
7	81.5	17	79.5	27	76.4	37	73.9
8	77.3	18	75.9	28	77.1	38	79.7
9	74.1	19	80.7	29	79.4	39	79.8
10	80.1	20	79.7	30	74.8	40	76.6

The maximum noise level (dB) for two stroke engine without rear cover is 92.9 dB at point 5 and 26 in hemispherical surface while for two stroke engine with rear cover, cover by composite fibre and acoustic foam is 92.5 dB at point 1, 92.7 dB at point 22, 32, and 38 and 93.3 dB at point 14. Point 5 is located at the rear part while point 1, 22, 32, and 38 is located at right and front part of two stroke engine. Based on trend data that are plotted in graph in Figure 8, several point in hemispherical arrangement shows the decreases of sound power level. Point 21 shows the decreases from 90.9 dB for two stroke engine without cover to 89.6 dB, 88.1 dB and 87.7 dB for two stroke engine with rear cover, cover by composite fibre and acoustic foam. Other points that are show the decreases of sound power level is at point 1, 2, 8, 9, 12, 15, 16, 18, 21, 23, 24, 26, 27, 30, 31, 34, 37, and 40.

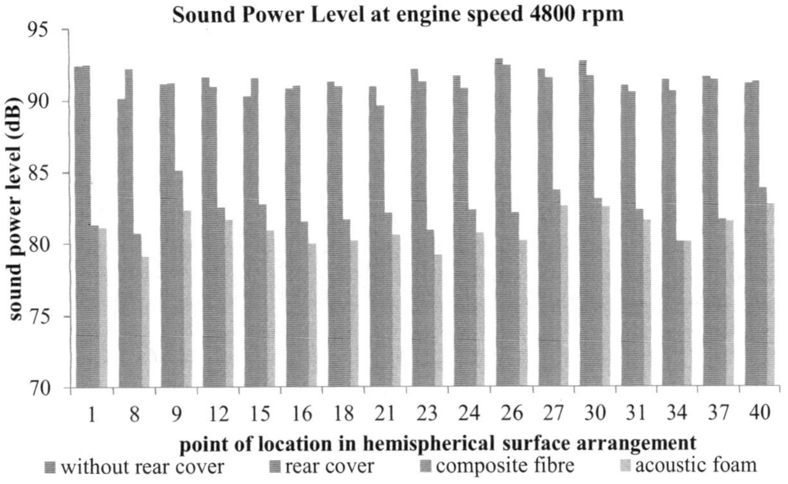

Figure 8. Graph of comparison Sound power level at engine speed 4800 rpm

5. Conclusion

Sound intensity mapping is an effective technique to investigate the maximum noise level that are produced by the two stroke engine. Data have been analysed by using hemispherical surface arrangement that are plotted 40 points in this area and sound intensity mapping that are detected noise at rear, front, right and left view of two stroke engine. Higher engine speeds contribute more noise level. Maximum sound level has increases 6 dB in every increases of engine speed. The results of different side at different engines speed present the different contribution to the engine noise. Thus, results in the sound intensity mapping shows the decreases from two stroke engine without cover 4 dB, 6dB and 7 dB at two stroke engine with rear cover, rear cover (composite fibre) and rear cover (acoustic foam).

Acknowledgments

The author would like to greatly acknowledge the support by focus group of Advanced Structural Integrity of Vibration Research (ASIVR) and Universiti Malaysia Pahang for providing all the equipment used for this project under UMP research grant scheme – RDU 130392.

References

[1] S Gustafson, R MacCreecry, B Hoover, J G Kopun and P Stelmachowicz 2014 Listening effort and perceived clarity for normal-hearing children with the use of digital noise reduction *Ear and hearing* vol 35(2) pp 1-13

[2] R B Mingsian and L Jia-Hong 2007 Source identification system based on the time-domain nearfield equivalence source imaging: Fundamental theory and implementation *Jounal Sound and Vibration* Vol 307(1-2) pp 202-225

[3] N Razak, Sani M S M, W Azmi and B Zhang 2017 Noise and vibration analysis for automotive radiator cooling fan in *IOP Conference Series: Materials Science and Engineering* p 012083

[4] Sani M S M, Zaman I, Rahman M 2015 *Journal of Advanced Research in Fluid Mechanics* vol 9 pp 28-33

[5] Sani M S M, Rahman M M, Baharom M Z and Zaman I 2015 Sound intensity mapping of an engine dynamometer *International Journal of Automotive and Mechanical Engineering* Vol 12 pp 2820-2828

[6] W J P Casas, E P Cordeiro, T C Mello and P H T Zannin 2014 Noise mapping as a tool for controlling industrial noise pollution *Journal of Scientific & Industrial Research* vol. 73 pp. 262-266

[7] J H Zhang and B Han 2005 Analysis of engine front noise using sound intensity techniques *Mechanical Systems and Signal Processing* vol.19 pp 213-221

[8] X Jian, F Zengming, W Guoqiang and W Fei 2015 Vibration and noise source identification methods for a diesel engine *Journal of Mechanical Science and Technology* vol. 29 pp 181-189

[9] ISO 3745:2003(E) Acoustics: Determination of sound power levels of noise sources using sound pressure: Precision methods for anechoic and hemi-anechoic rooms

The Utilization of Agricultural Waste Biochar and Straw Compost Fertilizer on Paddy Plant Growth

K Nisa[1], L Siringo-ringo[1], Zaitun [2] and Muyassir [2]

[1] Department of Agriculture Agribusiness, Faculty of Agriculture, Teuku Umar University Meulaboh, 23681, West Aceh, Indonesia
[2]Department of Agro technology, Faculty of Agriculture, Syiah Kuala University Banda Aceh ,Indonesia.

*Corresponding author: khr.nisa@gmail.com

Abstract. The agricultural environment continues to be used for inappropriate technology, reduced agricultural land, insufficient inputs (chemical fertilizers and inorganic pesticides), and air. Rice is the most important food crop in Indonesia because almost all residents use rice as a staple food. Rice straw is a source of organic material that is available after harvesting rice with a large enough amounts but the use of rice straw has only been used in paddy fields. This research is a field experiment followed by laboratory research with the following stages: (a) analysis of soil samples before the research was conducted at the Soil Chemistry Laboratory, Faculty of Agriculture, Syiah Kuala University, (b) field experiments by planting Inpari 30 variety rice, biochar treatment and straw compost treated according to the combination of each plot treatment, and soil sample analysis at the end of the study to re-test the soil chemical properties after conducting research with biochar and straw compost. This research method uses factorial randomized block design (RBD) consisting of two factors, namely: Biochar and Straw Compost. The results of this study indicate that biochar space has an influence on plant growth, namely plant height and number of paddy tillers. Observations on rice growth were 28 day after planting (DAP), 35(DAP) and (DAP) to find out the effects of giving biochar and straw compost, it is necessary to conduct further research on the next planting season so that it can be recognized and applied to save the use of chemical fertilizers. The administration of biochar and straw compost affects plant growth, namely plant height and the number of tillers in each rice with higher yields. It would be even better by giving biochar and straw compost together with higher yields. Thus, it is hoped that further research will be carried out in the next rice planting season to see how much residue is giving biochar and straw compost to improve rice yields. The results of the variance analysis showed that plant height 28 HST was significantly affected by biochar treatment with a significance value of 0.033. The results of the variance analysis showed that plant height 35 HST was significantly affected by the treatment of straw compost with a significance value of 0.018. The results of the variance analysis showed that plant height of 45 DAP was significantly affected by the treatment of biochar with a significance value of 0.019 while the treatment of straw compost had a very significant effect on plant height 45 DAP with a significance value of 0.001. The results of the variance analysis showed that the numbers of tillers 28 DAP were significantly affected by the treatment of biochar with a significance value of 0.013. The results of the variance analysis showed that the number of tillers 35 DAP and 45 DAP all had no significant effect by the treatment of biochar, straw compost, and interaction of biochar and straw compost because at plant age 35 DAP and 45 DAP the significance values were above 0.05.

IOP Conf. Series: Materials Science and Engineering **506** (2019) 012061 doi:10.1088/1757-899X/506/1/012061

1. Introduction

Technology, reduced agricultural land, insufficient inputs (inorganic chemical fertilizers and pesticides), and water. Concerns about environmental pollution and degradation, the impact of the global economy, and food needs continue to increase, resulting in the agricultural environment continuing to experience changes in the future.

According to Fitri [1] each country has a goal to improve the development of the economy including Indonesia. Indonesia is an agricultural country with a large portion of its population living from the agricultural sector. Agriculture is a very important sector in Indonesia's economic growth where agriculture has a contribution to both the economy and the fulfillment of basic needs of the community. Agricultural development aims to provide food supply. One way to achieve this goal is by increasing production where increasing production is a prerequisite for meeting people's food needs especially rice. Rice is the most important food crop in Indonesia because almost all residents use rice as a staple food. Rice is also a strategic food commodity that has a considerable influence on economic stability especially the level of inflation, social stability and political stability.

Biochar contains high carbon (C) which is more than 30%. Biochar does not experience further weathering so that when applied in the soil it can last for a long time to millions of years [2]. Research using biochar with various age levels shows that biochar has greater adsorption properties over cations through surface oxidation than through adsorption by ordinary organic matter. Although the new biochar has a low adsorption capacity, the old ones show very high CEC [3]. Asai [4] in the 2007, rainy season has also tested the effect of biochar administration on soil properties and grain yield of upland rice in northern Laos. Biochar administration of 16 that -1 increases the hydraulic conductivity of the top soil. In soils with low P availability biochar can increase grain yield. Besides, the response to *N* fertilization increases with the addition of biochar. Rice straw is a source of organic material that is available after harvesting rice with a large enough amounts, but the use of rice straw has only been used in paddy fields only [5-9]. The provision of rice straw compost can increase C-organic and P-available Ultisol soil, plant height, plant dry weight, N uptake and P uptake on maize plants. The administration of rice husk ash can increase C-organic Ultisol soil and N uptake of corn crops. The interaction of compost and rice husk ash had a significant effect on all variables of growth and production of sweet potato which included variable stem length, number of branches, number of tuber leaves / plants, tuber weight or tuber and tuber starch content [10-13].

2. Research Methodology

This research is a field experiment followed by laboratory research with the following stages:

i. Analysis of soil samples before the research was conducted at the Soil Chemistry Laboratory, Faculty of Agriculture, Syiah Kuala University
ii. Field experiments by planting Inpari 30 variety rice, biochar treatment and straw compost treated according to the combination of each plot treatment, and soil sample analysis at the end of the study to re-test the soil chemical properties after conducting research with biochar and straw compost.

This research method uses factorial randomized block design (RBD) consisting of two factors, namely: Biochar and Straw Compost.

i. Biochar Factors: The biochar factor consists of 4 different doses: treatment without biochar, biochar 5 tons / ha, biochar 10 tons / ha, and biochar 15 tons / ha.
ii. Straw Compost Factors: These consists of 3 different measures, namely treatment without biochar, compost 10 tons / ha and compost 20 tons / ha. From the table, 12 combinations of treatments were repeated (3) times, becoming 36 unit test units.

2.1 Data analysis. This can be analyzing by using the statistical models for factorial randomized block design (RBD) are as equation (1) follows:

$$Y_{ijk} = \mu + K_k + F_i + B_j + (FB)_{ij} + \varepsilon_{ijk} \quad (1)$$

Where Y_{ijk} is the observation value in the experimental group to those who received a combination of ij treatment (level i of biochar administration factor and j-level of straw compost), μ the average value (mean) of hope. Next, K_k is the value of observing the influence of the group. FP_i is the value of observing the effect of straw composting on the level of I whereas F_j is a value of observing the effect of biochar administration on the j-level. $(FB)_{ij}$ is the value of interaction observation of giving straw compost at level i and giving biochar at level j and ε_{ijk} is a randomized effect of experiment on the experimental group that received a combination of biochar treatment at the level and the provision of straw compost at the level of.

2.2 Data Collection Methods and Procedures

2.2.1 Preliminary Analysis of Soil Examples. Soil collection was carried out in a composite manner, each composite consisting of 5 sample points taken diagonally at a depth of 0-20 cm (using a drill). Analysis of initial soil samples prior to the study, was carried out on soil chemical properties.

2.2.2 Soil Processing. The first treatment carried out three weeks before planting using a rotary with a depth of 20-25 cm. The second treatment is done one week before planting with rotary.

2.2.3. Planting Preparation. Before planting rice seeds, the experimental plot with a size of 5 m x 5 m was made beforehand, this plot was made before NPK fertilizer was applied (15:15:15). To prevent fertilizer seepage between treatment plots, each treatment plot is separated from one treatment to another by using a dike. The water channel is made on the edge of the plot in such a way that the entrance water gate is separated by the exit gate.

2.2.4 Rice Nursery. Seed treatment is carried out by means of seeds soaked in salt water (5 liters) of 3% salt for 30 minutes; the seeds that float are discarded and then soaked with fresh water for 12 hours, then pressed 2 x 24 hours. Nursery area of 5.0% of planted area.

2.2.5Planting. After the soil is processed to a suitable condition for rice cultivation, the Legowo 2: 1 system is planted with a distance of 20 cm x 40 cm x 20 cm, which is the distance according to the path 20.0 cm and distance by row 20.0 cm. Each hole is planted with 2 new rice stems removed from the nursery. The age of rice planting is carried out 15 days after seedling (DAS). Then, plant simultaneously in one day from all experimental plots.

2.2.6 Provision of Biochar, Compost Straw and Fertilization. The biochar treatment is given at the time of final soil treatment, dried , sprinkled and immediately immersed in the soil evenly on the 20 cm processing layer. Biochar and straw compost are given at the earliest before planting. NPK Basic Fertilization is given according to the results of balanced fertilizer recommendations using PUTS carried out with regard to fertilizer time and dosage, a week after planting. Then given additional Urea fertilizer at 30 days after planting (DAP) and 45 DAP.

2.3 Maintenance
Maintenance of rice plants includes; control of plant pest organisms (OPT), weed control and water regulation. To prevent pest attacks pesticides are used. Weed control is done manually by removing weeds by pulling them by hand Weed weeding is done when monitoring weed growth, usually when the rice is 25 DAP.

3. Result and Discussion

3.1 Products from Research
Before weaving rice, the product made in this study is first made. The activities that have been carried out in this study are as follows:

3.1.1 Making Biochar
The making of biochar is done using a tool in the form of a modified drum for the combustion using a little oxygen or the pyrolysis process and is assisted by using fan angina to speed up the process of cooking into biochar husk charcoal. So, this can be show in figure 1 below:

Figure 1. Making of biochar from rice husk

3.1.2 Making Prebiotic Decomposer
The making of the prebiotic decomposer is done by using waste around the research location with the aim of being easily available and easy to use by farmers at relatively low prices compared to having to buy on the market. The prebiotic decomposer making materials use coconut water, fish head or fish waste remnants decayed straw, decomposed sawdust, palm sugar residual production and palm oil bunch waste. The production of prebiotic has been done 100% to speed up the process of making straw compost and this composer can be visualized in figure 2.

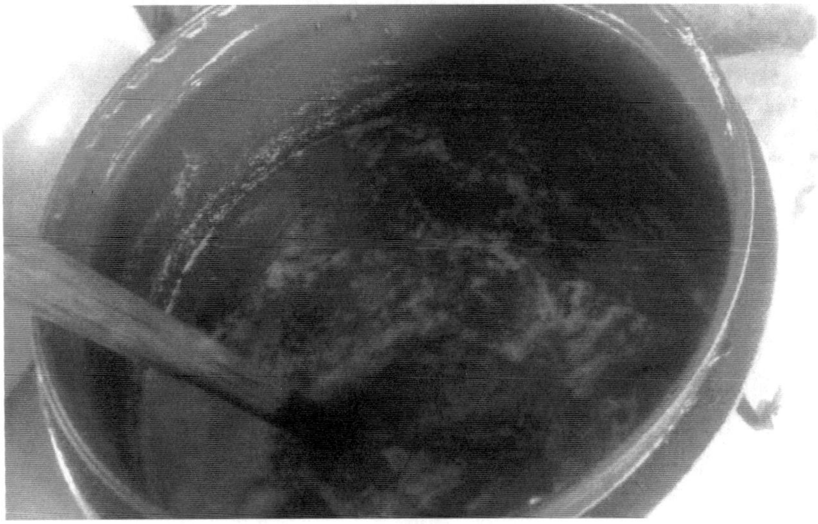

Figure 2. Making of prebiotic of agricultural waste

3.1.3 Making Straw Compost

Making straw compost is done by mixing 50% of the basic ingredients of straw from the total straw compost material and this can show in figure 3 until figure 7. The other supporting materials are rice husk, husk ash, water hyacinth, sawdust, prebiotic decomposer and water. The manufacturing process is carried out with each layer of material and given every layer of water and prebiotic decomposer. At this time the manufacturing process is ongoing 50% waiting for the reversal process and perfect decomposition process.

Figure 3. Making of straw compost from agricultural waste

Figure 4. Place for paddy nursery in the fields

Figure 5. Planting paddy after giving biochar and straw compost

Figure 6. Observation of paddy plant growth

Figure 7. Plant paddy 28 days after planting

After that, this plant result can be tabulated as in table 1.

Table 1: Average Plant Height from 12 treatment combinations

No	treatment combinations	Replication I	Replication II	Replication III	Average (cm)
1	B_0 K_0	54.10	54.57	57.41	55.36
2	B_0 K_1	62.77	55.55	56.10	58.14
3	B_0 K_2	58.58	56.62	63.20	59.47
4	B_1 K_0	59.79	55.03	60.34	58.39
5	B_1 K_1	62.37	54.58	58.00	58.32
6	B_1 K^2	59.68	58.57	62.44	60.23
7	B_2 K_0	55.42	52.88	59.66	55.99
8	B_2 K_1	56.78	57.23	60.48	58.16
9	B_2 K_2	56.63	57.54	58.24	57.47
10	B_3 K_0	60.18	54.40	59.16	57.91
11	B_3 K_1	59.07	59.74	61.39	60.07
12	B_3 K_2	63.05	59.35	64.43	62.28

Table 1 shows that the highest value of plant height in the treatment of biochar as much as 15 tons ha^{-1} and the provision of straw compost as much as 20 tons ha^{-1} were given together with an average plant height of 62.28 cm. After that the high plant height was followed by the treatment of 5 tons ha^{-1} of biochar and 20 tons ha^{-1} of straw compost together with an average plant height of 60.23 cm. The average value of plant height without the administration of 55.36 cm of biochar and straw compost was the control that had not been carried out at the research site. This can show in figure 8 and the results tabulated in table 2.

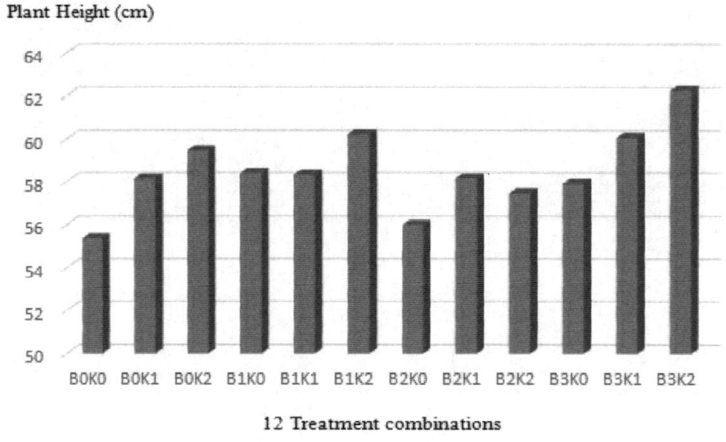

Figure 8. Plant paddy 28 days after planting

Table 2: Average number of tillers from 12 treatment combinations.

No	treatment combinations	Replication I	Replication II	Replication III	Average (tillers)
1	B_0 K_0	15.30	15.40	15.73	15.48
2	B_0 K_1	16.07	15.70	16.13	15.97
3	B_0 K_2	17.33	15.10	17.40	16.61
4	B_1 K_0	16.57	16.93	17.20	16.90
5	B_1 K_1	16.93	17.43	16.83	17.06
6	B_1 K^2	15.20	15.33	19.43	16.65
7	B_2 K_0	17.00	16.50	18.97	17.49
8	B_2 K_1	17.33	16.70	19.30	17.78
9	B_2 K_2	17.30	16.40	19.90	17.87
10	B_3 K_0	16.93	14.73	19.83	17.16
11	B_3 K_1	18.03	20.00	17.50	18.51
12	B_3 K_2	17.83	19.13	18.70	18.55

Table 2 shows that the highest value of number of tillers in the treatment of biochar as much as 15 tons ha^{-1} and the provision of straw compost as much as 20 tons ha^{-1} was given together with an average of number of tillers of 18.55 tillers. After that the high of number of tillers was followed by the treatment of 15 tons ha^{-1} of biochar and 10 tons ha^{-1} of straw compost together with an average plant height of 18.55 cm. The average value of number of tillers without the administration of 15.48 tillers of biochar and straw compost was the control that had not been carried out at the research site. So, it can picturized as figure 9 while table 3 until table 9 results for tests of between-subjects effects according to the tillers and so on.

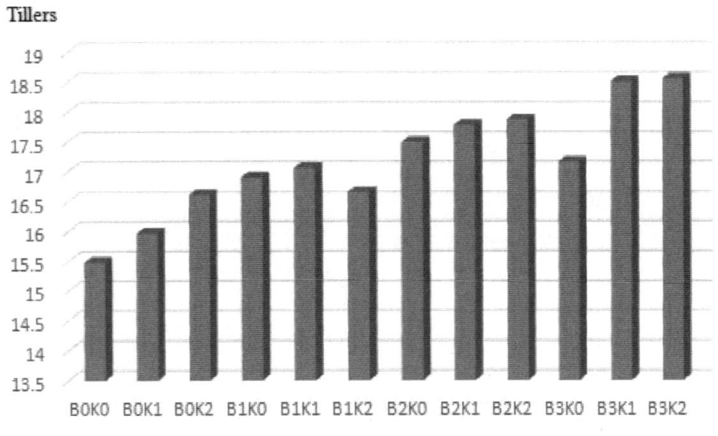

Figure 9. Plant growth with paddy number of tillers from 12 treatment combinations

Table 3: Dependent Variable Plant Height 28 DAP

Source	Type III Sum of Squares	df	Mean Square	F	Sig.
Corrected Model	184,874[a]	13	14,221	1,534	,182
Intercept	83508,477	1	83508,477	9009,846	,000
Replication	36,386	2	18,193	1,963	,164
Biochar	96,785	3	32,262	3,481	,033
Straw Compost	32,105	2	16,053	1,732	,200
Biochar * Straw Compost	19,598	6	3,266	,352	,901
Error	203,909	22	9,269		
Total	83897,260	36			
Corrected Total	388,783	35			

a. R Squared = ,476 (Adjusted R Squared = ,166)

Table 4: Dependent Variable: Plant Height 35 DAP

Source	Type III Sum of Squares	df	Mean Square	F	Sig.
Corrected Model	171,115[a]	13	13,163	2,141	056
Intercept	119965,250	1	119965,250	19517,462	000
Replication	25,802	2	12,901	2,099	146
Biochar	39,152	3	13,051	2,123	126
Straw Compost	59,760	2	29,880	4,861	018
Biochar * Straw Compost	46,402	6	7,734	1,258	316
Error	135,224	22	6,147		
Total	120271,589	36			
Corrected Total	306,340	35			

a. R Squared = ,559 (Adjusted R Squared = ,298)

Table 5. Dependent Variable: Plant Height 45 DAP

Source	Type III Sum of Squares	df	Mean Square	F	Sig.
Corrected Model	577,702[a]	13	44,439	12,917	000
Intercept	174160,156	1	174160,156	50624,886	000
Replication	415,890	2	207,945	60,446	000
Biochar	41,922	3	13,974	4,062	019
Straw Compost	75,305	2	37,653	10,945	001
Biochar * Straw Compost	44,584	6	7,431	2,160	087
Error	75,685	22	3,440		
Total	174813,543	36			
Corrected Total	653,387	35			

a. R Squared = ,884 (Adjusted R Squared = ,816)

Table 6. Dependent Variable: Number of tillers 28 DAP

Source	Type III Sum of Squares	df	Mean Square	F	Sig.
Corrected Model	38,058[a]	13	2,928	2,109	059
Intercept	5083,690	1	5083,690	3663,121	000
Replication	10,715	2	5,357	3,860	037
Biochar	18,677	3	6,226	4,486	013
Straw Compost	1,312	2	,656	473	630
Biochar * Straw Compost	7,355	6	1,226	883	523
Error	30,532	22	1,388		
Total	5152,280	36			
Corrected Total	68,590	35			

a. R Squared = ,555 (Adjusted R Squared = ,292)

IOP Conf. Series: Materials Science and Engineering **506** (2019) 012061 doi:10.1088/1757-899X/506/1/012061

Table 7: Dependent Variable:Number of tillers 35 DAP

Source	Type III Sum of Squares	df	Mean Square	F	Sig.
Corrected Model	30,939[a]	13	2,380	812	644
Intercept	10096,900	1	10096,900	3443,311	000
Replication	20,169	2	10,084	3,439	050
Biochar	7,536	3	2,512	857	478
Straw Compost	751	2	375	128	881
Biochar * Straw Compost	2,483	6	414	141	989
Error	64,511	22	2,932		
Total	10192,350	36			
Corrected Total	95,450	35			

a. R Squared = ,324 (Adjusted R Squared = -,075)

Table 8. Dependent Variable: Number of tillers 45 HST

Source	Type III Sum of Squares	df	Mean Square	F	Sig.
Corrected Model	198,817[a]	13	15,294	2,040	068
Intercept	17897,980	1	17897,980	2387,949	000
Replication	143,674	2	71,837	9,584	001
Biochar	26,381	3	8,794	1,173	343
Straw Compost	10,704	2	5,352	714	501
Biochar * Straw Compost	18,058	6	3,010	402	870
Error	164,893	22	7,495		
Total	18261,690	36			
Corrected Total	363,710	35			

a. R Squared = ,547 (Adjusted R Squared = ,279)

4. Conclusion

This study can be concluded as follows;

i. Preparation of prebiotic decomposers in the form of liquid requires ingredients containing nitrogen, phosphorus, potassium and other elements by using bacterial decomposers to accelerate the decomposition process.
ii. The composition of straw compost requires other ingredients besides straw, namely husk ash, sawdust and prebiotic decomposers, water hyacinth. So that the elements needed by plants can be fulfilled.
iii. The administration of biochar and straw compost affects plant growth, namely plant height and the number of tillers in each rice with higher yields. It would be even better by giving biochar

and straw compost together with higher yields. Thus, it is hoped that further research will be carried out in the next rice planting season to see how much residue is giving biochar and straw compost to improve rice yields.

iv. The results of the variance analysis showed that plant height of 45 HST was significantly affected by the treatment of biochar with a significance value of 0.019, while the treatment of straw compost had a very significant effect on plant height 45 HST with a significance value of 0.001. The results of the variance analysis showed that the number of tillers 28 HST was significantly affected by the treatment of biochar with a significance value of 0.013. The results of the variance analysis showed that the number of tillers 35 HST and 45 HST all had no significant effect by the treatment of biochar, straw compost, and interaction of biochar and straw compost because at plant age 35 HST and 45 HST the significance values were above 0.05.

References

[1] Fitri Nurul.2015. *Analisis faktor-faktor yang Mempengaruhi Produksi Padi di Propinsi*
[2] *Aceh.*Jurnal Ilmu Ekonomi ISSN 2302-0172 **Vol.3** pp. 81- 95.
[3] Gani, A. 2009. *Iptek Tanaman Pangan* (ISSN 1907-4263) **Vol. 4 No. 1**, Juli 2009. p:33-48.
[4] Cheng, C.H., Lehmann J. and Engelhard M. 2008. *Natural oxidation of black carbon in soils: changes in molecular form and surface charge along a climosequence.* Geochimica et Cosmochimica Acta **72**, 1598-1610.
[5] Asai H B.K., Samson H. M. Stephan K. Songyikhangsuthor K., Homma Y., Kiyono Y. Inoue T. Shiraiwa and T. Horie. 2009. *Biochar amendment techniques for upland rice production in Northern Laos: 1. Soil physical properties leaf SPAD and grain yield.* Field Crops Research **Volume 111 (1-2):** 81-84.
[6] Anas I., O.P. Rupela T.M.Thiyagarajana N. Uphoff. 2011. *A review of studies on SRI effects on beneficial organism*. Paddy Water Environ. **9**:53-64.
[7] Chan K.Y., Van Z.L. Meszaros I, Downie A., and Joseph S. 2007. *Agronomic values of greenwaste biochar as a soil amendment.* Australian Journal of Soil Research **45 (8):** 629-634.
[8] Chan K.Y., Van Z. L., Meszaros I., Downie A., and Joseph S. 2008. *Using poultry litter biochars as soil amendments. Australian Journal of Soil Research* **46 (5):** 437-444.
[9] FAO. 2009. *Climate change mitigation and adaptation in agriculture, forestry and fisheries. Office of the Assistant Director-General Natural Resources Management and Environment Department Food and Agriculture Organization of the United Nations* Viale delle Terme di Caracalla 00153 Rome, Italy.
[10] Gadde,B.,C.Menke, Werner Siemers, and Suneerat Pipatmanomai. 2007. *Technologies for energy use of rice straw:* International Rice Research Notes **32(2):** 5-14
[11] Gunther, F. 2007. *Nutrient adsorption by fresh charcoal.* Submitted by Folkeg on Sat.
[12] Kimetu, J., H.J. Lehmann, S. Ngoze, D. Mugendi, J. Kinyangi, S. Riha, L. Vercot, J. Rencha, and A. Pell. 2008. *Reversibility of soil productivity decline with organic matter of differing quality along a degradation gradient. Ecosystems in press.*
[13] Sintia Megi. 2011. Jurnal Tanaman Pangan. *Pengaruh Beberapa Dosis Kompos Jerami Padi dan Pupuk Nitrogen terhadap Pertumbuhan dan Hasil Jagung Manis (Zea mays saccharata Sturt.).*

Heat Transfer Enhancement with Considering Pressure Loss Penalty of Airflow through Heated Plate Mounted by Perforated Concave Rectangular Winglet Vortex Generators

Syaiful[1*], Ganesha Rachmandala[1], Bambang Yunianto[1], MSK Tony SU[1] and M. Ilham Maulana[2]

[1]Mechanical Engineering Department of Diponegoro University, Semarang, Indonesia
[2]Mechanical Engineering Department Syiah Kuala University, Banda Aceh, Indonesia

*Corresponding author: syaiful.undip2011@gmail.com

Abstract. The use of longitudinal vortex generator results in decreased hydrodynamic flow performance by increasing pressure drop values. The purpose of this research is to obtain the improvement of heat transfer with pressure drop as low as possible. In order to achieve this purpose, experiment was performed on perforated concave rectangular winglet pair vortex generators (CRWP VGs). The test results show that the CRWP VGs without hole increase thermal performance by 24.02% up to 129.75% against baseline (without using any vortex generators) and by 12,21% up to 19,41% against rectangular winglet pair vortex generator (RWP VG). The pressure drop value for the CRWP VGs case increased by more than 10 times against the baseline case and increased by almost 100% against RWP VGs case. By using the perforated CRWP VGs, pressure drop value can be reduced by 15.07% to 72.73% followed by a slight decrease in thermal performance of 5.50% to 5.73% against the CRWP VGs without hole. Based on the evaluation using London goodness factor criteria, perforated CRWP VGs is better than the CRWP VGs without hole.

Keyword : *pressure drop, concave rectangular winglet pair vortex generator, rectangular winglet pair vortex generator*

1. Introduction

Compact heat exchangers have been widely applied in areas such as radiators in the automotive industry, condensation and evaporation in refrigeration, power plant cooling systems, solar-powered air/water heaters, exhaust gas recovery, cogeneration, steam generators and the pharmaceutical industry [1]. The most common type of heat exchanger is the fin-and-tube heat exchanger where heat from the hot fluid on the tube side is transferred to the gas using the extended surface, commonly referred to as fin. The performance of this type of heat exchanger is often limited by the thermal resistance on the gas side because the heat transfer coefficient of the gas is lower than the liquid fluid [2]. It has been known that 90% of the total thermal resistance of the fin-and-tube heat exchanger found on the gas side [3]. The development of fin with high effectiveness needs to be done to improve convection heat transfer on the gas side.

Convective heat transfer enhancement can be done by active and passive methods [4-5]. The active method is complicated because this method requires some external source input to increase the rate of heat transfer [4]. Instead, passive methods do not require the supply of external resources. The passive method uses a modified heat transfer surface and/or uses a turbulence generator element. The passive method is very popular in the industry because it has a simple design, cheaper production cost, and high reliability.

The passive method for convection heat transfer cases in the fin-and-tube heat exchanger can be done with two general concepts. The first concept is to enlarge the area of surface heat transfer such as fin. The second concept is to increase the convection heat transfer coefficient using a vortex generator (VG). VG is a protrusion in the flow wall and serves to generate the vortex in the flow [1]. Vortex can be divided into two types: longitudinal vortex (LV) and transverse vortex (TV). LV is a vortex that has a rotation axis with the main flow, while the TV is a vortex that has the direction of the rotation axis normal to the main flow direction.LV is considered to be more effective because LV is capable of covering three basic mechanisms for increasing the convection heat transfer coefficient: reducing the boundary layer thickness, swirl flow, and enhancing turbulence intensity. While the TV only covers the mechanism of increasing turbulence intensity. Based on its shape, there are four basic types of VG namely rectangular wings, delta wings, rectangular winglets, and delta winglets [6]. Winglets are better than wings because of winglets able to enhance heat transfer by generating more intensive LV with lower pressure drop [7].

The study of VG has been widely investigated because VG has an excellent heat transfer performance. L. H. Tang et al. [8] have conducted numerical studies using 2 types of VG namely rectangular winglet pairs (RWP) and delta winglet pairs (DWP) with common flow up (CFU) and common flow down (CFD) configurations. Each of VG surface is combined with an elliptical pole. The test results show that the use of DWP with the elliptical pole can improve heat transfer with high effectiveness. This is indicated by the angle between the velocity and the temperature gradient smaller than the baseline case (without VGs). Based on j/f factor, DWP-CFU with the elliptical pole is the best VG followed by DWP-CFU, RWP-CFD, and DWP-CFD.

Cheng-Hung Huang and Po-Ching Chiang [9] carried out a numerical three-dimensional study using the CFD-ACE+ commercial package and Lavenberg-Marquardt Method (LMM) methods to estimate the optimal position and shape of the delta winglet vortex generators on the pin-fin heat exchanger. The purpose of their research was to minimize the average temperature of the heat exchanger base plate. The thermal performance of the heat sinks that they observe can be increased with the help of vortex generators. The numerical test results show that the pin-fin heat sink has the best thermal performance when using the optimal vortex generator shape and position.

Anupam Sinha et al.[10] has numerically tested the fin-and-tube heat exchangers using a rectangular winglet pair (RWP) with a common-flow-up (CFU) configuration. Tests were carried out to determine the effect of heat transfer using variations in tube arrangement (inline and staggered) and variations in VG attack angle on each variation of tube arrangement. From the test results, it was found that the thermal performance of the inline tube arrangement is better than the staggered arrangement. From the results of this study observed that the performance of heat exchanger increases with increasing angle of attack. U. Kashyap et al.[11] conducted a numerical study of the impact of modified cross-section surface shape on the RWP on the heat transfer rate. From the test results, it was found that the surface modification can help in determining the position of the main vortex behind the VG which resulted in an increase in the strength of the main vortex core behind the VG. The use of curved profiles on the VG front surface is judged to be appropriate to increase the average Nusselt number.

In addition to increasing the rate of heat transfer, the use of VG results in an increase in pressure drop in the flow [4]. Flow with a large pressure drop requires a large pumping power as well. Many researchers have developed the concept of VG that can increase the rate of heat transfer with a low-pressure drop. Anupam Sinha et al. [12] has numerically tested the effect of using two rows of delta winglet pairs (DWPs) using a combination of VG orientation (CFU-CFU, CFD-CFD, and CFD-CFU) as well as VG line arrangement configurations (inline and staggered)for thermal performance and quality factor. From the test results found that DWP with CFU-CFU configuration has the best thermal performance and quality factor.

P. Saha et al. [13] performed the numerical study to determine the effects of RWP and DWP prepared with CFD and CFU configurations for heat transfer and pressure drop. From the test results, it is observed that the pressure drop in RWP is greater than DWP. Performance analysis shows that RWP provides better thermal performance than DWP.CFD configuration is better than CFU configuration.

Azita Abdollahidan M Shams [7] conducted a numerical study to determine the effect of VG (rectangular winglets, trapezoidal winglets, and delta winglets) and the angle of attack on thermal and pressure drop performance. Their study is carried out by using Pareto optimal strategy to get an increase in maximum heat transfer and a low-pressure drop. From the test results found that rectangular winglet with 45 ° attack angle has the best heat transfer performance and highest pressure drop.

Li Li et al. [14] performed numerical testing to determine the effect of RWP and DWP on thermal and pressure drop performance. Their test results show that RWP and DWP can improve thermal performance compared to baseline. Thermal performance enhancement on RWP is better than DWP.Y. He et al. [6] performed a numerical study to determine the effect of RWP on heat transfer and pressure drop. The test is performed using variations of the number of VG rows, angle of attack, Reynolds number, and configuration of the RWP arrangement. The test results show that RWP can increase the heat transfer coefficient. The heat transfer coefficient also increases as the number of RWP rows increases. Increased heat transfer coefficient in this case is accompanied by an increase in pressure drop value. From the results of their study, it was found that the pressure drop on the staggered configuration was lower than the inline configuration.

S. Skullong et al. [15] performed an experimental study to determine the effect of winglet perforated tape (WPT) and non-perforated winglet tape (WTT) on heat transfer and friction factor. The Reynolds numbers varied from 4,000 to 30,000. The test results show that WPT and WTT produce higher heat transfer performance than baseline. WTT provides a higher heat transfer rate than the WPT because the holes in the winglet cause weakening of the vortex in the flow. The weakness of vortex strength in the flow can reduce the temperature gradient. Friction factor for WPT case is lower than WTT.

Syaiful et al. [16] conducted experimental and numerical study using the concave delta winglet pair vortex generator (CDWP VG). The variations used in their study are the number of VG rows. From their study results, it was found that the thermal performance for the CDWP VG case was greater than the DWP VG due to the instability of the centrifugal force when the fluid passed through the concave surface. This causes the LV to become stronger so that mixing between hot fluid and cold fluid becomes better which ultimately results in improvement of heat transfer. The results of their analysis also showed a greater increase in pressure drop in the case of CDWP than DWP. Their test results observed that thermal and pressure drop performance increased with increasing number of VG rows. When compared with the results of experimental tests, the numerical test results have a relatively small difference of less than 6%. Hemant Naik et al. [17] performed numerical testing to determine the effect of convex and concave RWP VG on thermal and pressure drop performance. From the test results, it was found that the highest thermal performance and pressure drop were owned by concave RWP VG. S. K. Sarangi and D. P. Mishra [18] performed a numerical study to determine the effect of RWP rows, attack angles, and RWP positions on thermal and pressure drop performance in fin-and-tube heat exchangers. Their test results show that thermal and pressure drop performance increases with increasing number of RWP lines and increasing the angle of attack. Each number of RWP rows each has an optimal position on the tube.

Then, Syaiful et al. [19] performed numerical study using RWP VG and concave rectangular winglet pair vortex generator (CRWP VG).Variations used are the number of VG rows and Reynolds numbers. Their test results show that the thermal performance in the CRWP VG case is better than RWP. This is because CRWP VG produces a stronger LV due to centrifugal instability when the fluid flows through the concave wall. From the test results, it was found that thermal performance increased with increasing number of VG rows.

According to a study conducted by Syaiful et al., thermal performance in the CRWP VG case is better than RWP [19]. However, the use of CRWP VG results in a higher pressure drop than RWP VG. Therefore, the current study is focused on reducing the pressure drop resulting from the use of the CRWP VG by giving holes on VG. For each RWP VG and CRWP VG cases, the variations used are inlet air velocity, number of VG rows, and number of holes. The attack angle used in this study is set to 45 ° from the direction of the main flow.

2. Materials and Methods

2.1. Experimental apparatus

This study was carried out in a rectangular air duct made of glass with a thickness of 10 mm. This air duct has a length of 370 cm, width 8 cm, and height H = 8 cm. The air was sucked by a centrifugal blower and enters through the inlet side. Then, the air was flowed through a straightener consisting of pipes arrangement with a diameter of 5 mm and a screen so that the flow becomes uniform. The inverter is used to adjust the rotating speed of the electric motor in the blower so that the inlet air velocity can be adjusted. The inlet air velocity was varied from 0.4 m/s to 2.0 m/s with an interval of 0.2 m/s. Inlet air velocity was measured using a hot wire anemometer (Lutron type AM-4204 with accuracy of ± 0.1). To determine the effect of VG on the rate of heat transfer, air was passed through a flat plate with/without VG. This flat plate was heated at a constant rate, Q, of 35 Watt by a heater whose was regulated using a heater regulator. Wattmeter (Lutron DW-6060 with an accuracy of ± 1.0) was used to monitor the heat rate into the plate. Inlet, outlet, and plate surface temperatures were measured using K-type thermocouples. Thermocouples were connected with data acquisition (Advantech type USB-4718 which has ± 0.001 accuracy) and computer CPU. Then, in order to determine the effect of the vortex generators on the pressure drop value, the pitot tubes were installed on the inlet and outlet sides of the test section. The pitot tubes were connected with micromanometer (Fluke) type 922 which has an accuracy of ± 0.001. In order to observe the characteristics of the vortex due to the use of VGs in the flow, visualization of the flow was carried out using smoke as a medium. The smoke was produced from heated oil in the oil heater. A compressor was used to push smoke from inside the oil heater to the air duct through the capillary tube. The capillary tube was mounted on the inlet side of the test specimen. For the vortex to be seen clearly in the flow, three green lasers were used and arranged in the direction of the main flow. The laser beam is refracted so that the beam illuminates the flow cross-section. The schematic diagram of this testing apparatus can be seen in Figure 1.

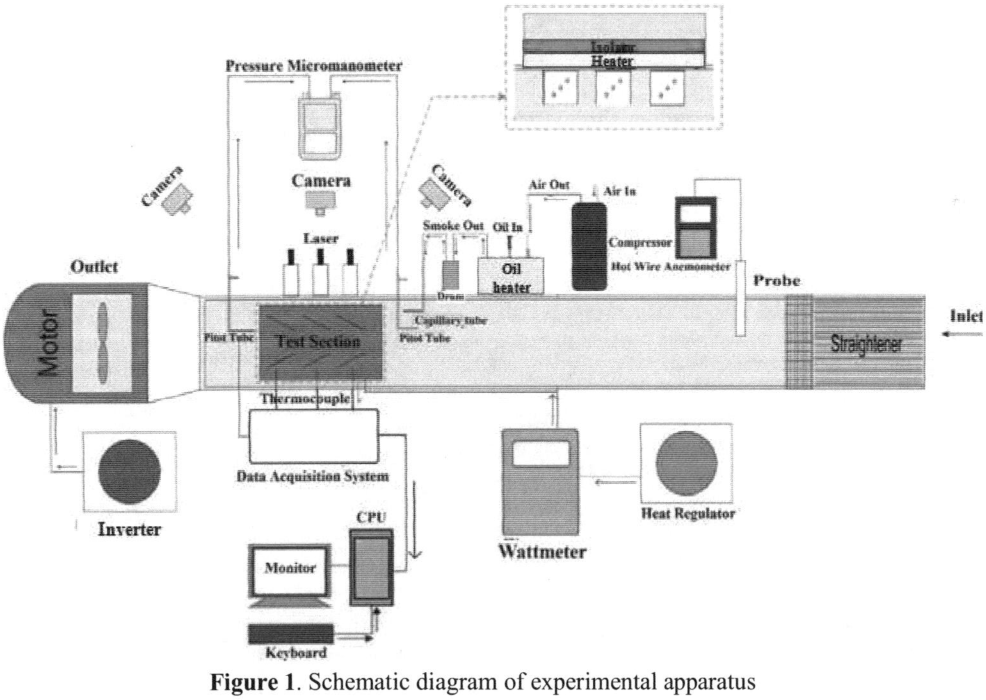

Figure 1. Schematic diagram of experimental apparatus

2.2. Test specimens

The test specimen used in this study was the winglet type. The type of winglet geometry used was rectangular winglet pair vortex generators (RWP VGs) and concave rectangular winglet pair vortex generators (CRWP VGs). Both test specimens were mounted on an aluminium plate that has a length of 500 mm, a width of 155 mm, and a thickness of 1 mm. The flow configuration used was common-flow-down (CFD) with the distance between the leading edge is 20 mm. The attack angle was set at 45° from the direction of the main flow. The specifications of the test specimen can be seen in Figure 2.

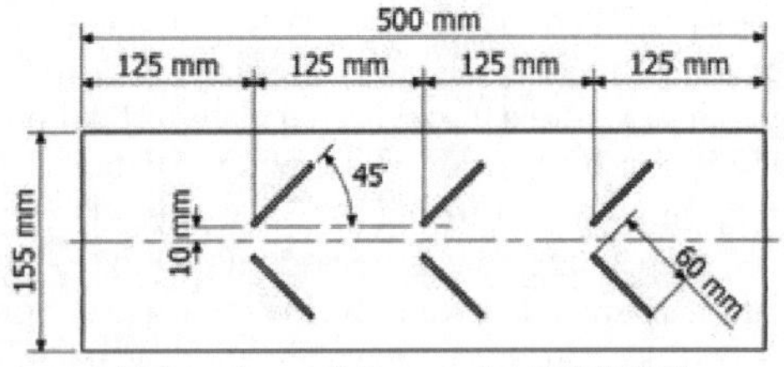

(a) Top view of three pairs DWP VGs

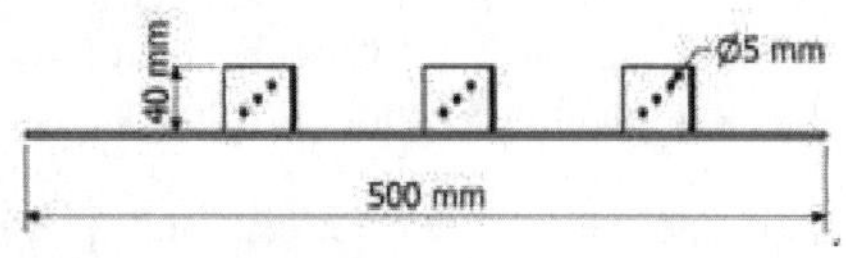

(b) Side view of three pairs DWP VGs

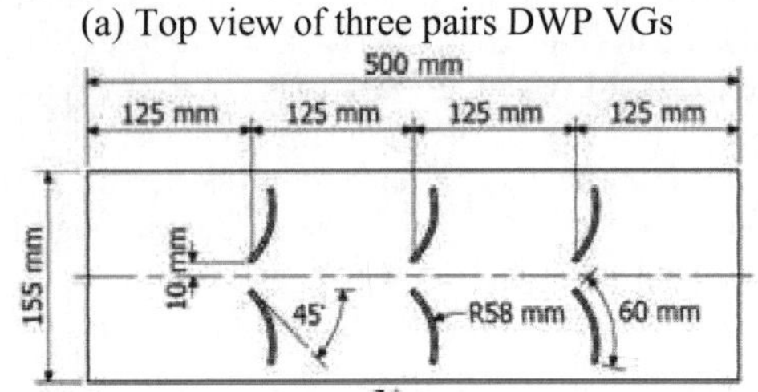

(c) Top view of three pairs CDWP VGs

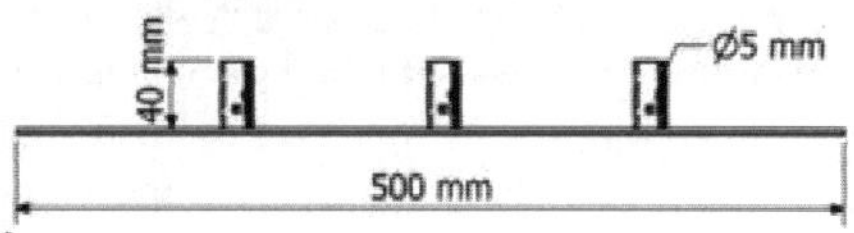

(d) Side view of three pairs CDWP VGs

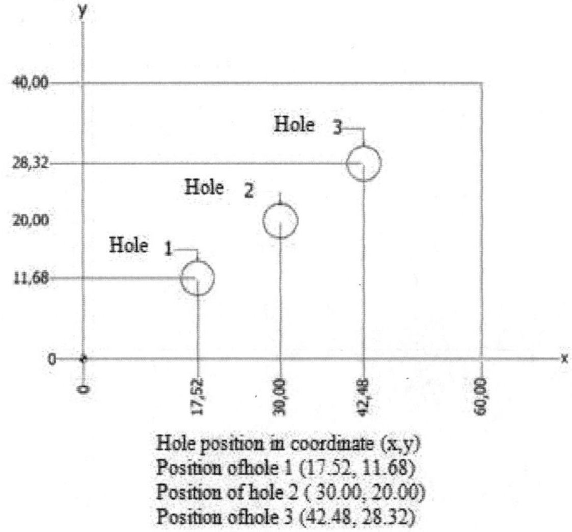

(e) Detail holes location in x-y coordinates

Figure 2. Geometry detail of test specimens

In the variation of the number of VG rows, the test was carried out using one, two, and three pairs of RWP VGs and CRWP VGs which are arranged inline. The distance of the leading edge between one pair of VGs and another VGs pair was 125 mm. The top view specifications of the test specimen can be seen in Figure 2 (a) for three pairs RWP VGs and Figure 2 (c) for the CRWP VGs. Figure 2 (b) and (d) show the side view of geometry detail of three pairs DW and CDW VGs, respectively.

In variations in the number of holes, the test was carried out using VG with one, two, three holes, and VGs without hole. Holes with a diameter of 5 mm were positioned as can be seen in Figure 2 (e). Position 2 is the hole location of the VG test with a variation of one hole. Positions 2 and 3 are the holes position of VG test with variation of two holes. Whereas, the three-hole vortex generator variation uses all hole positions.

Figure 3 shows an example of a test specimen of perforated RWP and CRWP VG that has been mounted on an aluminium plate with variations of one, two, and three rows of VGs. The test specimen of RWP VG case can be seen in Figure 3 (a). While the test specimen of CRWP VG case can be seen in Figure 3 (b).

(a)

(b)

Figure 3. Test specimens for (a) RWP VGs and (b) CRWP VGs

2.3. Experimental method

For thermal performance experiment, the parameters measured were the temperature on the inlet, outlet, and flat surfaces of the plate. The temperature was measured using K type thermocouples which were connected with data acquisition. The data acquisition was connected to a computer device so that the temperature read by the thermocouples can be monitored directly and recapitulated into the table. The test specimen was heated at a constant heat rate, Q, of 35 W using a heater placed under the plate as shown in Figure 1. A heater regulator was used to adjust the rate of heating on the heater. The test specimen was heated to a steady temperature of 54°C-55°C. After steady was reached, the test can be started by turning on the blower and adjusting the frequency of the inverter until the reading of the air flow velocity on the hot wire anemometer reaches the desired velocity. The flow velocity range in this experiment was 0.4 m/s to 2.0 m/s with an interval of 0.2 m/s.

To calculate the convection heat transfer coefficient (h), the formula used refers to the Wu & Tao study [20]. The value of h can be calculated using the following equation:

$$h = \frac{\overline{Nu}k}{2H} \quad (1)$$

where $\overline{Nu}$ is the Nusselt number, k is the thermal conductivity, and H is the height of the duct. The Nusselt number value can be calculated using the following equation:

$$\overline{Nu} = \frac{Q}{A(T_w - T_{mb})} \frac{2H}{k} \quad (2)$$

where A is the heat transfer surface area, T_w is the average temperature of the heat transfer surface area and T_{mb} is the bulk temperature of the working fluid, $T_{mb} = 1/2\,(\bar{T}_{out} + \bar{T}_{in})$.

For hydrodynamic performance experiment, the pressure drop, ΔP measurement was carried out by measuring the pressure difference on the inlet side (P_{inlet}) and outlet side (P_{outlet}) of the test specimen in the test section so that:

$$\Delta P = P_{inlet} - P_{outlet} \quad (3)$$

Two pitot tubes were placed on the inlet and outlet sides of the test section. The two pitot tubes were connected with a micromanometer to monitor the pressure difference. The pressure drop data retrieval was carried out at 5-second intervals as much as 30 times at each inlet velocity variation. Similar with thermal performance experiment, the inlet velocity was varied in the range of 0.4 m/s to 2.0 m/s with an interval of 0.2 m/s.

Analysis of thermal and hydrodynamic performances was carried out using the London goodness factor criteria (j/f). Each thermal and hydrodynamic parameter was evaluated using the Colburn factor (j) for thermal parameter and friction factor (f) for hydrodynamic parameter. Colburn factor is calculated using the equation:

$$j = \frac{\overline{Nu}}{Re.Pr^{1/2}} \tag{4}$$

where Re is the Reynolds number, Pr is the Prandtl number. While friction factor is calculated using the equation:

$$f = \frac{2H}{L}\frac{\Delta P}{\frac{1}{2}V_{in}^2} \tag{5}$$

where L is the length of the test specimen plate and ρ is the density of the working fluid.

For the flow visualization, oil was put into the oil heater and heated until it evaporates. Then the smoke was pushed out of the oil heater using a compressor. The smoke entered the rectangular channel through the capillary tube placed in the inlet of the test section. To observe the vortex structures in the flow, three lasers were arranged in the direction of the main flow. The laser beam was refracted by using a transparent cylinder so that the area of cross-section laser beam was formed to capture the LV structures. A digital camera was used to capture and record the LV structures.

The thermal and hydrodynamic performance experiments were carried out on each type of VG (RWP VGs and CRWP VGs), the number of rows (one, two, and three rows), as well as the number of holes (one, two and three holes). As a comparison, a plate without VG was also tested for its thermal and hydrodynamic performance which is then referred to as the "baseline" case. For flow visualization, the test specimen used was CRWP VG with a variation of one row and three holes.

2.4. Data validation

Data validation was done to know that the data taken in the experiment are correct. The data in the present experiment were compared with the experiment data from Wu & Tao study [21] because of the present experiment is referred to their study. The data that is compared was found from testing with an identical VG configuration. In this case, an identical configuration is a baseline case (without VG). Comparison of data from the baseline case between the present and Wu & Tao studies can be seen in Figure 4.

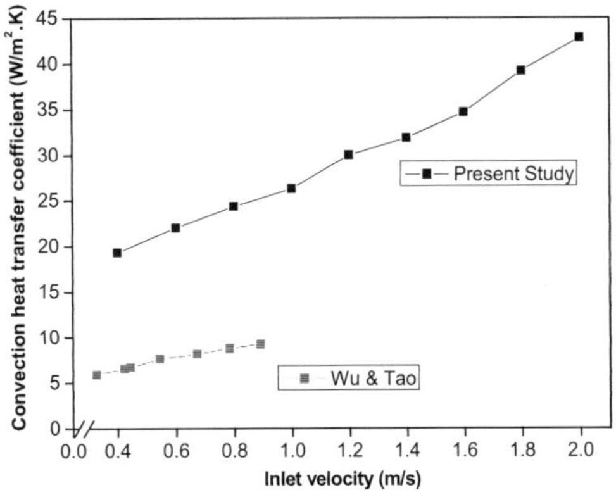

Figure 4. Comparison of convection heat transfer coefficient between present study and Wu and Tao study [20] for baseline case

From Figure 4, it is known that the two data have the same tendency, the convection heat transfer coefficient increases with increasing inlet air velocity. However, the results of the baseline data in the present study are greater than the results of the baseline on Wu & Tao's study due to smaller heat transfer to the test plate. In the Wu & Tao study, VG was mounted in the middle of the air duct. Whereas the present study, VG was placed on the channel wall. In other words, the hydraulic diameter in the present study is greater so that the convection heat transfer coefficient value becomes larger.

3. Results and Discussion

3.1. Effect of vortex generators on convection heat transfer coefficient

The effect of inlet air velocity, number of VG rows, and number of holes on convection heat transfer coefficients have been experimentally conducted. The convection heat transfer coefficient for each variation of the inlet velocity, number of VG rows, and number of holes for the RWP and CRWP VGs can be seen in Figures 5 and 6, respectively. The convection heat transfer coefficient was calculated by referring to the Wu and Tao study [20]. From Figures 5 and 6 it is observed that the use of RWP VG and CRWP VG can increase the convection heat transfer coefficient to the baseline case [19].

Generally, the convection heat transfer coefficient increases with increasing air inlet velocity because the LV produced by VG becomes stronger so that the intensity of mixing between hot fluid and cold fluid in the flow becomes better [16, 21, 22]. The convection heat transfer coefficient also increases with increasing number of VG rows [16, 18] because the resulting LV will continue to be strengthened until the last VG row. However, the convection heat transfer coefficient decreases slightly with increasing number of holes of the VG surface. Holes on the surface of the VG cause weakening of the vortex strength in the flow which results the reduction of the intensity of mixing between hot fluid and cold fluid [15]. The convection heat transfer coefficient for CRWP VGs case is greater than that of RWP VGs at the same of inlet velocity and the number of rows because the fluid experiences centrifugal force instability when the fluid passes through the concave wall producing stronger vortices [16, 19]. The LV produced by CRWP VGs becomes stronger and the radius becomes larger than RWP VGs causing better mixing between hot fluid and cold fluid.

As seen in Figure 5, convection heat transfer coefficient for RWP VGs without hole cases with variations of one, two, and three VG rows are increased by 35.73%, 65.55%, and 92.40%, respectively, against the baseline at the inlet velocity of 2.0 m/s. For perforated RWP VGs cases, the convection heat transfer coefficient decreases slightly with increasing number of holes. As shown in Figure 5 (a), the convection heat transfer coefficient for one pair RW VGs case with variation of one, two, and three holes decreases 0.58%, 3.48%, and 4.35%, respectively, against one pair RW VGs without hole. In

Figure 5 (b), a decrease in convection heat transfer coefficient on two pairs RW VGs with variation of one, two, and three holes are found by 2.95%, 4.76%, and 6.67% against two pairs RW VGs without holes case. Whereas in Figure 5 (c) it is observed that convection heat transfer coefficient for the three pairs RW VGs case with variations of one, two, and three holes decrease 0.65%, 3.33%, and 7.47%, respectively, against three pairs RW VGs without hole.

Figure 6 shows a comparison of the convection heat transfer coefficient for the CRWP VG case. For the case of CRWP VGs without hole, convection heat transfer coefficients with variations of one, two, and three VG pairs increase 54.01%, 89.53%, and 129.75%, respectively, against the baseline or 13.47%, 14, 48%, and 19.41% of RWP VGs without hole with the same number of pairs at the inlet velocity of 2 m/s. As in the case of RWP VGs, the convection heat transfer coefficient on the use of perforated CRWP VGs decreases slightly compared to those of case without holes. In the case of a single pair of CRW VGs with one, two, and three-holes as shown in Figure 6 (a), it is observed that the convection heat transfer coefficient decreased by 0.38%, 2.18%, and 5.12%, respectively, against CRWP VGs without hole. In the case of two pairs of CRW VGs as seen in Figure 6 (b), the convection heat transfer coefficients with one, two, and three holes decreases 2.53%, 3.28%, and 4.06%, respectively, against the CRWP VGs without hole. Whereas in the variation of the three-pairs of CRW VGs, the decrease in convection heat transfer coefficient is 2.15%, 3.95%, and 5.49% for the variations of one, two and three holes, respectively, against CRWP VG without hole as observed in Figure 6 (c).

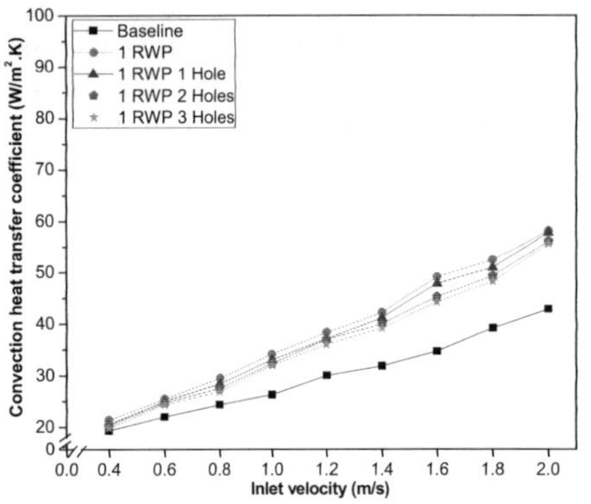

(a)

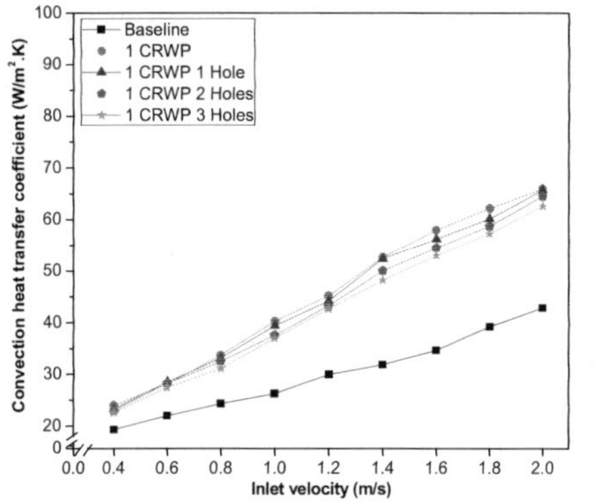

(a)

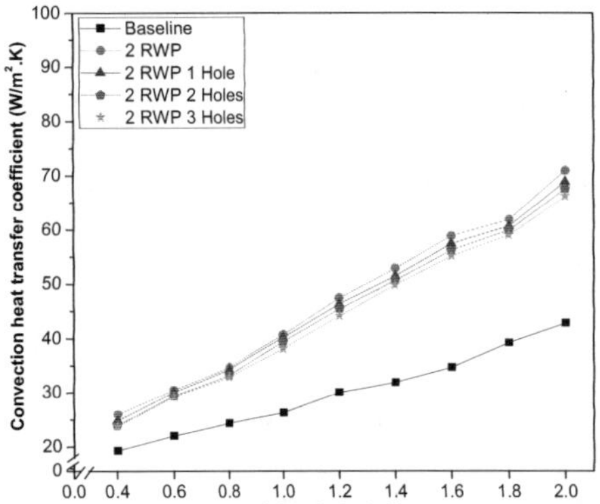

(b)

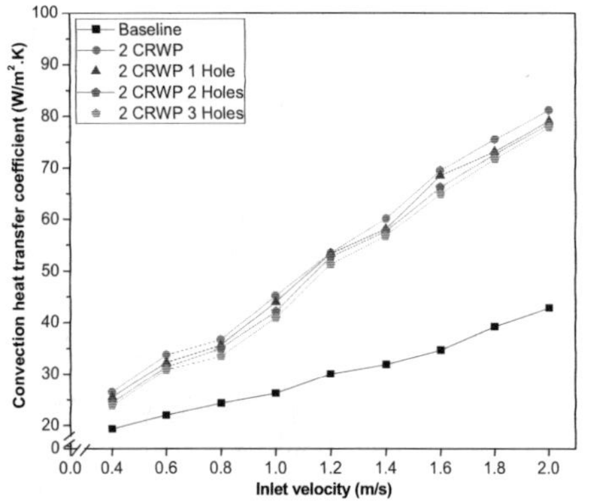

(b)

IOP Conf. Series: Materials Science and Engineering **506** (2019) 012062 doi:10.1088/1757-899X/506/1/012062

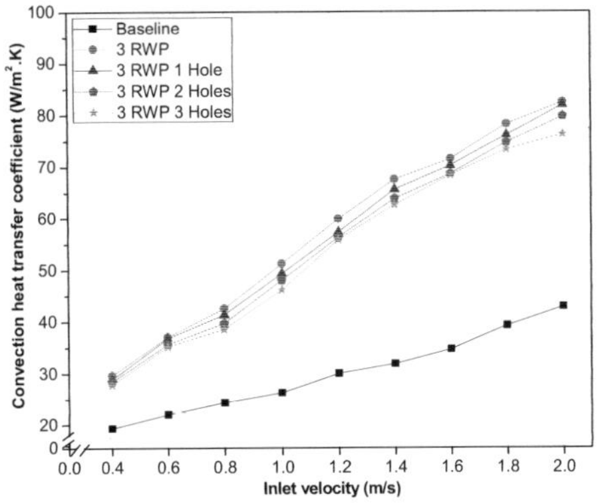

(c)

Figure 5. Comparison of convection heat transfer coefficients with variations of inlet velocity and number of holes for different number of pairs on (a) one pair, (b) two pairs, and (c) three pairs of RW VGs

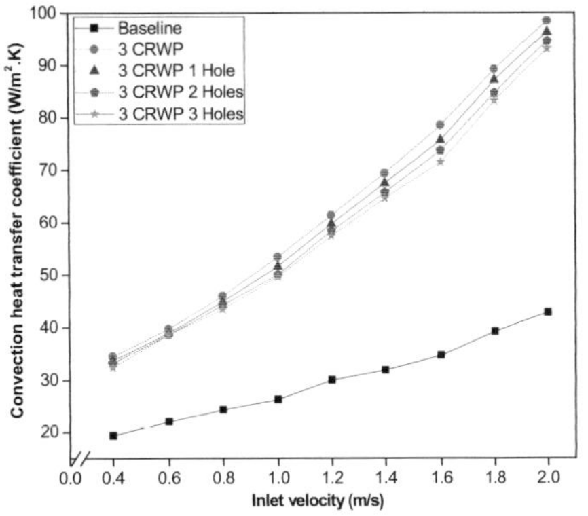

(c)

Figure 6. Comparison of convection heat transfer coefficients with variations of inlet velocity and number of holes for different number of pairs on (a) one pair, (b) two pairs, and (c) three pairs of CRW VGs

3.2. Effect of perforated vortex generator on pressure drop

The effect of VG on the pressure drop can be seen in Figure 7 for the RWP VGs case and Figure 8 for the CRWP VGs case. In general, the pressure drops increase with increasing inlet velocity because the formed drag increases with increasing turbulent intensity. As shown in Figures 7 and 8 it is observed that the baseline case has the lowest pressure drop in each variation of the inlet velocity and the variation in the number of pairs because the baseline does not have VG that inhibits the flow. With the same number of pairs and number of holes, the pressure drop for the CRWP VGs case is greater than that of RWP VGs due to larger frontal area of CRWP VGs producing higher formed drag [16, 19].

As can be expressed in Figure 7, it is observed that the pressure drops for RWP VGs without hole case at the inlet velocity of 2 m/s increases to 158.67%, 355.33%, and 460% of the baseline case for RWP VGs with one, two and three pairs, respectively. The pressure drop can be reduced by using the perforated VGs because the longitudinal vortices (LVs) that are generated are weakened due to leakage of flow by the existence of holes [15]. At one pair RW VGs as described in Figure 7 (a), pressure drop decreases 7.22%, 14.95%, and 21.13% against RWP VGs without hole for one, two and three holes, respectively. The similar tendency is observed in Figure 7 (b), the pressure drops for the case of two pairs RW VGs with one, two and three holes is decreased by 8.20%, 11.86%, and 25.04% against RWP VG without hole, respectively. Then Figure 7 (c) expresses that the pressure drops for three pairs RW VGs with one, two, and three holes decreases 6.31%, 10.36%, and 13.93% against RWP VG without hole, respectively.

For the case of CRWP VGs without hole as shown in Figure 8, pressure drop with one, two, and three pairs increases 372.67%, 660%, and 1014.67%, respectively, against the baseline or increases 82.73%, 66.91 %, and 99.05% of RWP VGs, respectively, at 2 m/s inlet velocity. As in the case of RWP VGs, the pressure drop for the perforated CRWP VG case decreases with increasing number of holes. The longitudinal vortices produced by VG are weakened with the presence of hole on VGs due to flow leakage [15]. As seen in Figure 8, the pressure drops decrease with increasing number of holes for each variation in the number of pairs. In the one pair CRW VGs as expressed in Figure 8 (a), the pressure drops for one, two, and three holes decreases 7.89%, 12.27%, and 15.09%, respectively, against the CRWP VGs without hole. For the case of one, two and three holes of two pairs CRW VGs, the pressure drop decreases 10.44%, 14.65%, and 20.96%, respectively, against CRWP VG without hole as can be seen in Figure 8 (b). Whereas for the three pairs RW VGs case as stated in Figure 8 (c), the pressure drops for variations of one, two, and three holes is decreased by 4.19%, 10.29%, and 15.07%, respectively, against CRWP VGs without hole.

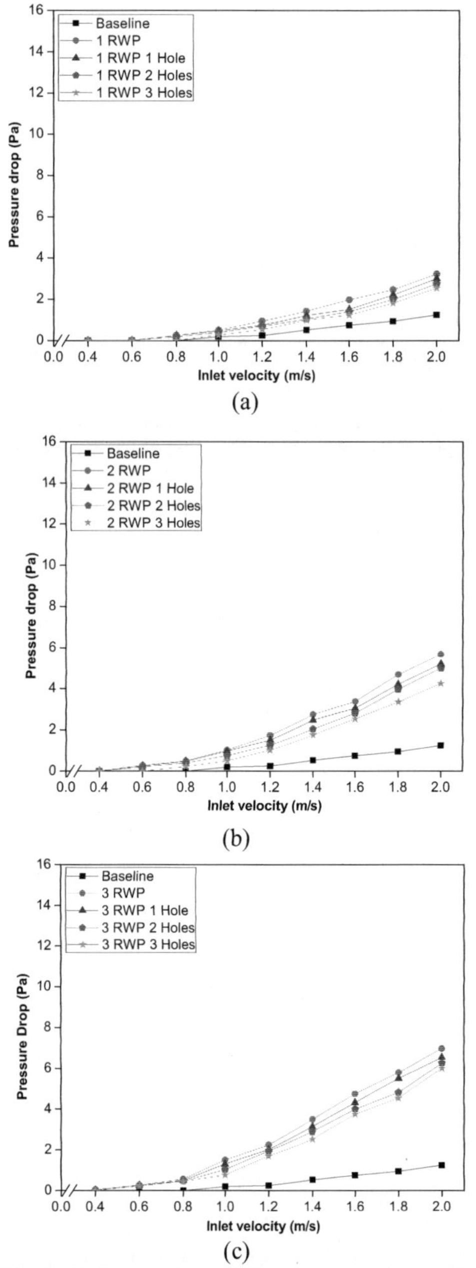

Figure 7. Comparison of flow pressure drop with variations of inlet velocity and number of holes for different number of pairs on (a) one pair, (b) two pairs, and (c) three pairs of RW VGs

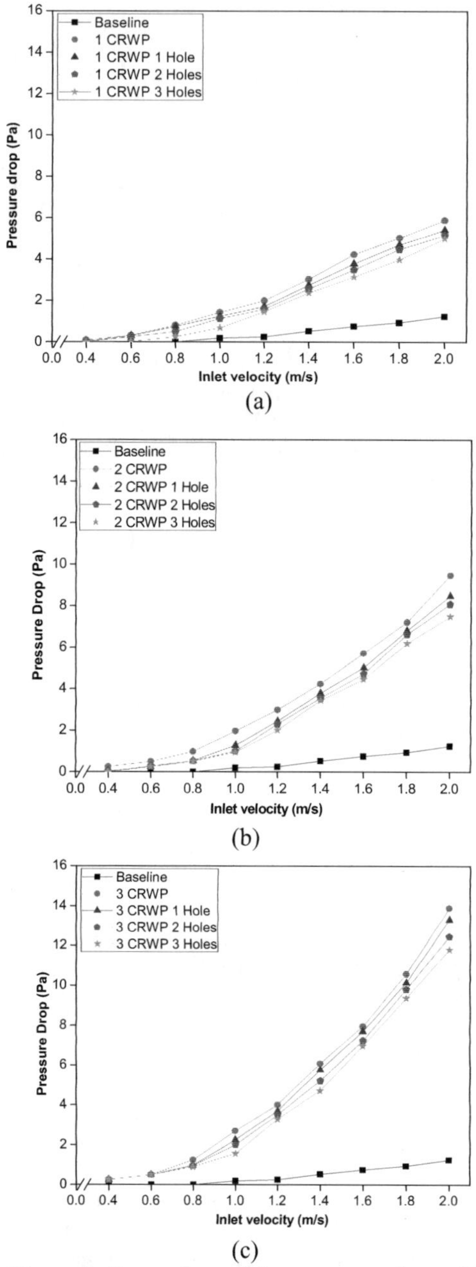

Figure 8. Comparison of flow pressure drop with variations of inlet velocity and number of holes for different number of pairs on (a) one pair, (b) two pairs, and (c) three pairs of CRW VGs

3.3. Thermal-hydrodynamic performances with presence of perforated vortex generators

Thermal and hydrodynamic performances are the main parameter components used to analyse heat exchanger performance. In order to determine the highest heat transfer and the lowest pressure drop, the criteria for London goodness factor (j/f) was used in this study. In general, the value of j/f will decrease with increasing inlet velocity [23]. At a certain inlet velocity, the difference in the value of j/f for each variation in the number of holes in the same number of VG rows tends to be insignificant [23, 24].

Figure 9 shows a comparison of the j/f criteria for the RWP VGs with and without cases with variations in the number of pairs. The variations of the j/f value for the RWP VG case become insignificant when the inlet velocity is more than 1.4 m/s. This is because the increase in pressure drop is greater than the increase in convection heat transfer performance at the highest velocity resulting low value of the j/f. At certain inlet velocity, the j/f value is unreadable in the image because the pressure drop value is outside of the micromanometer accuracy limit. Thus, the value of the criteria to be discussed in the RWP VGs case is the value found at the inlet velocity of 1.4 m/s. For RWP VG without hole cases, the j/f values for variations of one, two, and, three pairs with an inlet velocity of 1.4 m/s are 0.12, 0.08, and 0.08, respectively.

Figure 9 (a) shows a comparison of the j/f value for the RWP VGs one pair case with variations in the number of holes. By comparing London goodness between RWP VGs with and without holes, RWP VG with one, two, and three holes increases the j/f values of 14.15%, 29.46%, and 33.16%, respectively. The value of j/f in the case of inlet velocity less than 0.8 m/s is unreadable (for RWP VGs cases of one pair with one, two and RWP VGs without hole) and below 1 m/s (for RWP VGs one pair with three holes). This is caused by the pressure drop at the incoming velocity which is very low because the value is outside the micromanometer accuracy limit.

Furthermore, for two pairs RWP VGs case as shown in Figure 9 (b), an increase of 7.77%, 29.05%, and 47.31% is observed in a pair with the use of RWP VG with one, two, and three holes, respectively, against RWP VGs without hole. The value of j/f for this case is unreadable at inlet velocities less than 0.6 m/s (for two pairs RW VGs cases with one, two and without hole) and less than 0.8 m/s (for two pairs RW VGs case with three holes) due to limitation of micromanometer accuracy. Whereas for the case of three pairs RW VGs with one, two and three holes, the j/f value increases 6.34%, 8.55%, and 23.46%, respectively, against RWP VGs without hole as seen in Figure 9 (c). In this case, the j/f value is unreadable at the inlet velocity less than 0.6 m/s for all variations in the number of holes. This is because the pressure drop at velocity less than 0.6 m/s is very low.

From the evaluation results using the London goodness factor criteria in the RWP VGs case, the best configuration is found for the case of RWP VGs with a variation of three holes. This is expressed by the j/f value for the RWP VGs case with three holes in each variation in the number of VG pairs is the highest for almost all variations of the inlet velocity. The second bests configuration was found for RWP VGs with a variation of two holes. Then RWP VG with three holes and without holes is in the third and fourth position, respectively.

For the CRWP VGs case, the comparison of the j/f value between with and without holes can be seen in Figure 10. In this case, the change in the j/f value becomes insignificant when the inlet velocity is more than 1.2 m/s. This is because the increase in pressure drop is greater than the increase in convection heat transfer performance at the highest inlet velocity yielding the small j/f value. At certain inlet velocity, the j/f value is unreadable because the pressure drop value is outside of the micromanometer accuracy limit. In the case of CRWP VGs without hole at 1.0 m/s inlet velocity, the j/f values for one, two, and three pairs of VGs are 0.07, 0.06, and 0.05, respectively.

As observed in Figure 10 (a), the value of j/f for the case of one, two, and three holes of one pair CRW VGs are increased by 10.04%, 12.19%, and 17.20% against CRWP VG without hole. The value of j/f at this case is unreadable at the inlet velocity below 0.8 m/s (for one pair RW VGs case with one, two and RWP VGs without hole) and less than 1.0 m/s (for one pair RW VGs with three holes case). Then for the case of two pairs CRW VGs with one, two, and three holes, the value of j/f increases 8.23%, 14.30%, and 16.37%, respectively, against CRWP VG without hole as shown in Figure 10 (b). In the case of a two pairs CRW VGs, the j/f value is unreadable at inlet velocity less than 0.6 m/s (for two pairs CRW VGs case with one, two and CRWP VGs without hole) and less than 0.8 m/s (for the two pairs CRW VGs with three holes). Whereas for the three pairs CRW VGs case as shown in Figure 10 (c), the j/f value for one, two and three holes increases 2.75%, 10.61%, and 20.36% against CRWP VGs without hole. As it is observed in the case of CRWP VGs with one and two pairs, the j/f value is not readable at the inlet velocity less than 0.6 m/s for all variations of the number of holes.

The best thermal and hydrodynamic performances which are evaluated are found in the CRWP VGs case with three holes. This is because the value of j/f for the case of the CRWP VGs with the three holes is the highest than the other variation of holes number at the same VG pairs and the same inlet velocity. Then, the second bests configuration was found by CRWP VGs with two holes and followed by CRWP with three holes and no holes.

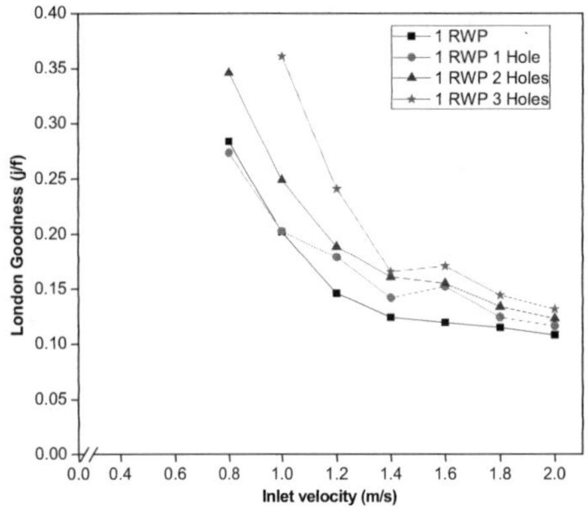

(a)

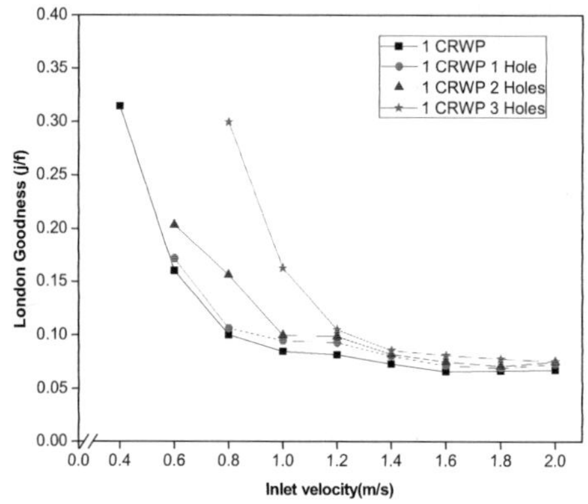

(a)

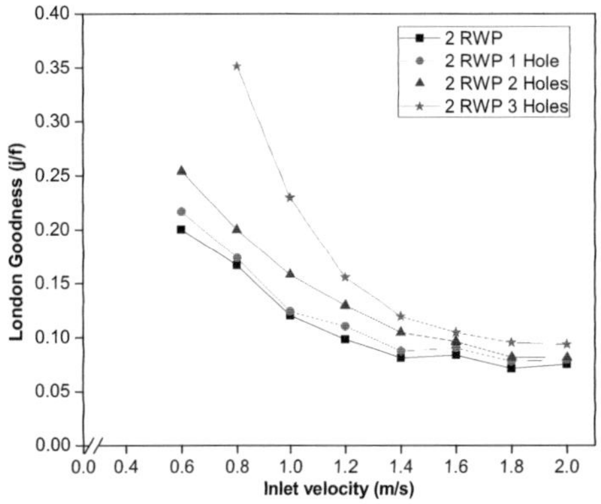

(b)

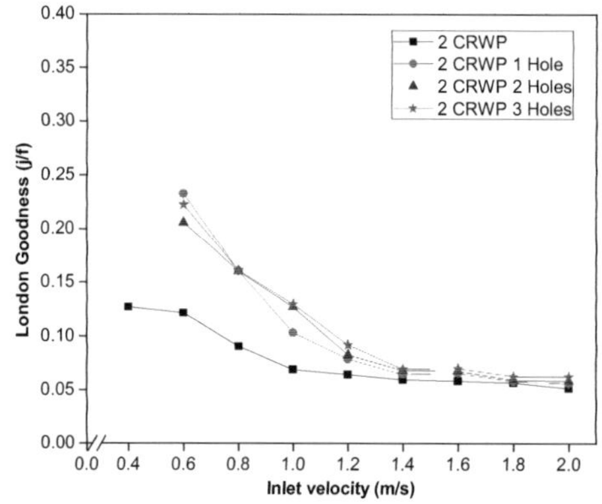

(b)

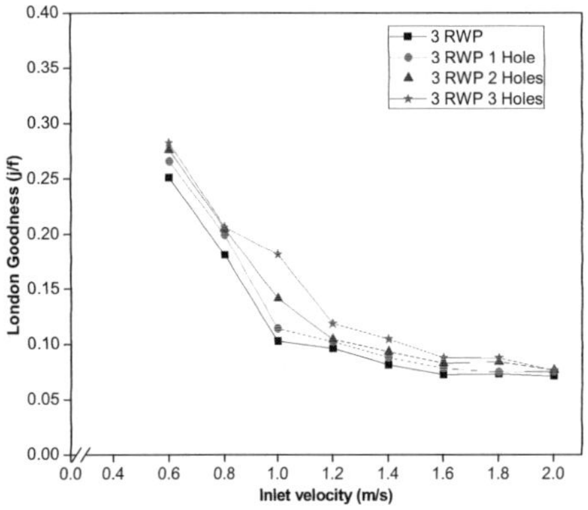

(c)

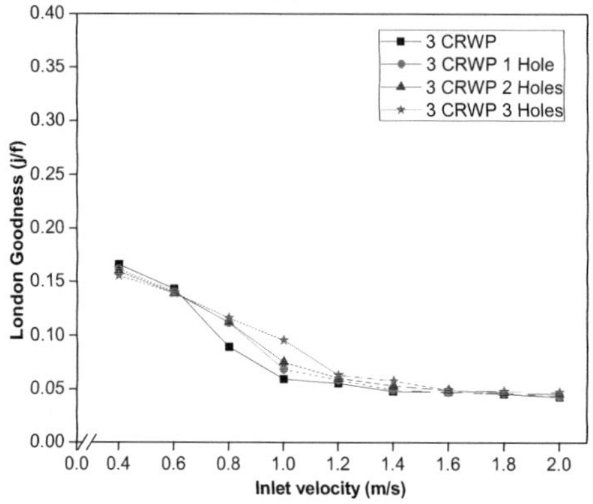

(c)

Figure 9. Comparison of flow London goodness with variations of inlet velocity and number of holes for different number of pairs on (a) one pair, (b) two pairs, and (c) three pairs of RW VGs

Figure 10. Comparison of flow London goodness with variations of inlet velocity and number of holes for different number of pairs on (a) one pair, (b) two pairs, and (c) three pairs of CRW VGs

3.4. Flow visualization

Flow visualization was performed to determine the LV structure produced by VG in the flow. The test was carried out using smoke from heated oil. The smoke was projected on cross-section planes in the flow obtained from laser beam induced to transparent cylinder. This visualization test used one pair CRW VGs with one pair and three holes at the inlet velocity of 0.6 m/s. The result of the visualization can be seen in Figure 11. From the Figure 11, it is observed that the pair winglet produces two LVs in the flow. Both LVs have opposite rotation directions. When it has been viewed from the inlet side, the LV to the left of the flow rotates clockwise while the LV to the right of the flow rotates counter-clockwise.

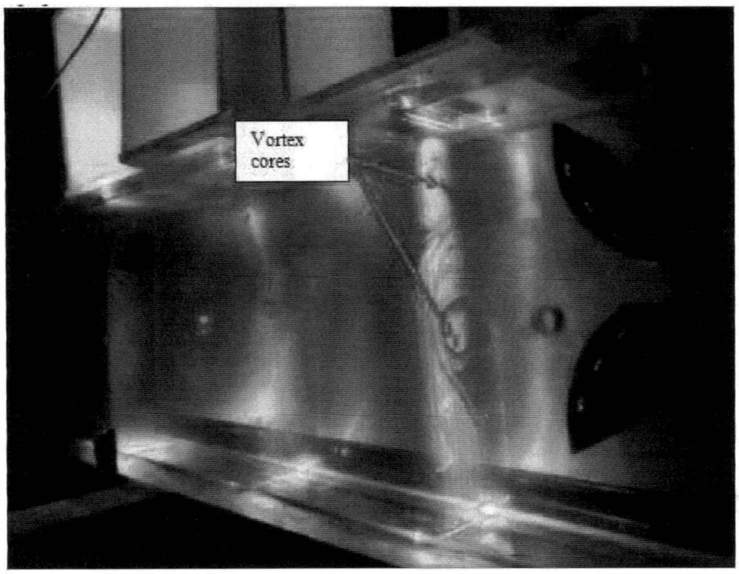

Figure 11 Visualization of longitudinal vortices generated by CRWP VG with various of one pair CRW VGs with three holes in the inlet velocity of 0.6 m/s.

3.5. Error calculation in experiment results

When the physical quantity measurement is performed using a measuring instrument and obtains numeric values, an evaluation of how close the measurement results to the true value needs to be performed. This is because it is very impossible to get physical data measurement results that are very precise in the measurement process. A good measurement process is carried out in accordance with the instructions for making measurements (measurement methods) that are truly implemented as best as possible. When this process is carried out repeatedly with the assumption of identical conditions, large amounts of measurement data has been found. Usually, the measurement data in each measurement process is not the same. By taking large amounts of physical data, therefore, requires scientific data processing to find out the errors in the data obtained from the actual value [25].

The calculation method used in processing data to determine deviations that occur is the Mean, Standard Deviation, Standard Deviation of the Mean, and Overall Error. Calculation of the error used in this test refers to Ernest O. Doebelin [25]. From the calculation, it is found that the average deviation of the calculation of convection heat transfer and pressure drop coefficients are 1.55% and 3.73%, respectively.

4. Conclusion

In the present study, the effect of using perforated RWP and CRWP VGs in thermal and hydrodynamic performances has been experimentally tested. The test results can be summarized as follows:

i. Thermal performance for the CRWP VGs case is better than RWP VGs. This was indicated by the convection heat transfer coefficients in the case of the CRWP VGs increased up to 19.41% against the RWP VGs. When using perforated VGs, thermal performance experienced a slight decrease of 7.47% against the VGs without hole.
ii. By judging from the hydrodynamic performance, the pressure drops in the case of the CRWP VGs increased to 99.05% against RWP VGs. However, the use of perforated VGs can reduce pressure drop up to 25.04% against VGs without hole.
iii. To determine the variation of VG with the best thermal and hydrodynamic performance, evaluation using the London goodness factor criteria (j/f) was required. The evaluation results showed that RWP VG and CRWP VG with 3 holes have the best thermal and hydrodynamic performances for each variation in the number of VG pair. This was indicated by the value of j/f for VG with three holes increased up to 47.31% against VG without hole.
iv. From the results of flow visualization, it was observed that the winglet pair VG produced two LVs with opposite rotation direction.

Acknowledgement

This work was supported by the Riset Publikasi Internasional (RPI Batch II) of Indonesia (Universitas Diponegoro Number:474-94/UN7.P4.3/PP/2018). The authors are grateful to all research members, especially Lab. Thermofluid of Mechanical Engineering of Diponegoro University and Mechanical Engineering Department of Syiah Kuala University, Indonesia.

References

[1] M Awais and A.A.Bhuiyan 2018 Heat transfer enhancement using different types of vortex generators (VGs): A review on experimental and numerical activities *Thermal Science and Engineering Progress* vol. 5 pp 524-545.
[2] P Wais 2016 Correlation and numerical study of heat transfer for single row cross-flow heat exchangers with different fin thickness *Procedia Engineering* vol 157 pp 177–184.
[3] M J Li H Zhang J Zhang Y T Mu E Tian D Dan X Zhang and W Tao 2018 Experimental and numerical study and comparison of performance for wavy fin and a plain fin with radiantly arranged winglets around each tube in fin-and-tube heat exchangers *Applied Thermal Engineering* vol 133 pp 298–307.
[4] T Alam and M H Kim 2018 A comprehensive review on single phase heat transfer enhancement techniques in heat exchanger applications *Renewable and Sustainable Energy Reviews* vol. 81 pp 813-839.
[5] Mostafa Mirzaei Majid Saffar-Avval 2018 Enhancement of convection heat transfer using EHD conduction method *Experimental Thermal and Fluid Science* vol 93 pp 108-118.
[6] He Y L Chu P Tao W Q Zhang Y W Xie T 2013, Analysis of heat transfer and pressure drop for fin-and-tube heat exchangers with rectangular winglet-type vortex generators *Applied Thermal Engineering* vol 61 pp 770-783.
[7] A Abdollahi and M Shams 2015 Optimization of shape and angle of attack of winglet vortex generator in a rectangular channel for heat transfer enhancement *Applied Thermal Engineering*, pp 376-387.
[8] L H Tang W Chu N Ahmed and M Zeng 2016 A new configuration of winglet longitudinal vortex generator to enhance heat transfer in a rectangular channel *Applied Thermal Engineering* vol 104 pp 74-84.

[9] C-H Huang and P C Chiang 2016 An inverse study to design the optimal shape and position for delta winglet vortex generators of pin-fin heat sinks *International Journal of Thermal Sciences* vol 109 pp 374-385.

[10] A Sinha H Chattopadhyay A K Iyengar and G Biswas 2016 Enhancement of heat transfer in a fin-tube heat exchanger using rectangular winglet type vortex generators *International Journal of Heat and Mass Transfer* vol 101 pp 667–681.

[11] U Kashyap K Das and B K Debnath 2018 Effect of surface modification of a rectangular vortex generator on heat transfer rate from a surface to fluid *International Journal of Thermal Sciences* vol 127 pp 61-78.

[12] A Sinha K A Raman H Chattopadhyay and G Biswas 2013 Effects of different orientations of winglet arrays on the performance of plate-fin heat exchangers *International Journal of Heat and Mass Transfer* vol 57 pp 202-214.

[13] P Saha G Biswas and S Sarkar 2014 Comparison of winglet-type vortex generators periodically deployed in a plate-fin heat exchanger A synergy based analysis *International Journal of Heat and Mass Transfer* vol 74 pp 292–305.

[14] L Li X Du Y Zhang L Yang and Y Yang 2015 Numerical simulation on flow and heat transfer of fin-and-tube heat exchanger with longitudinal vortex generators *International Journal of Thermal Sciences* vol 92 pp 85-96.

[15] S Skullong P Promvonge C Thianpong and M Pimsarn 2016 Heat transfer and turbulent flow friction in a round tube with staggered-winglet perforated-tapes *International Journal of Heat and Mass Transfer* vol 95 pp 230–242.

[16] Syaiful A Ayutasari M F Soetanto A I Siswantara and M-W. Bae 2017 Thermo-hydrodynamics performance analysis of fluid flow through concave delta winglet vortex generators by numerical simulation *International Journal of Technology* vol 7 pp 1276-1285.

[17] S H S T Hemant Naik 2018 Numerical investigations on heat transfer characteristics of curved rectangular winglet placed in a channel *International Journal of Thermal Sciences* vol 129 pp 489–503.

[18] S K Sarangi and D P Mishra 2017 Effect of winglet location on heat transfer of a fin-and-tube heat exchanger *Applied Thermal Engineering* vol 116 pp 528-540.

[19] Syaiful I Syarifudin M F Soetanto and M W Bae 2018 Numerical simulation of heat transfer augmentation in fin-and-tube heat exchanger with various number of rows of concave rectangular winglet vortex generator *Matec Web of Conferences* vol 159.

[20] J M Wu and W Q Tao 2008 Numerical study on laminar convection heat transfer in a rectangular channel with longitudinal vortex generator Part A: Verification of field synergy principle *International Journal of Heat and Mass Transfer* vol 51 pp 1179-1191.

[21] R K Ali H A Refaey and M R Salem 2018 Effect of package spacing on convective heat transfer from thermal sources mounted on a horizontal surface *Applied Thermal Engineering* vol 132 pp 676–685.

[22] A Esmaeilzadeh N Amanifard and H Deylami 2017 Comparison of simple and curved trapezoidal longitudinal vortex generators for optimum flow characteristics and heat transfer augmentation in a heat exchanger *Applied Thermal Engineering* vol 125 pp 1414–1425.

[23] Y G Lei Y L He L T Tian P Chu and W Q Tao 2010 Hydrodynamics and heat transfer characteristics of a novel heat exchanger with delta-winglet vortex generators *Chemical Engineering Science* vol 65 pp 1551–1562.

[24] M Zeeshan S Nath D Bhanja and A Das 2018 Numerical investigation for the optimal placements of rectangular vortex generators for improved thermal performance of fin-and-tube heat exchangers *Applied Thermal Engineering* vol 136 pp 589–601.

[25] E O Doebelin 1990 Measurement Systems: Application and Design New York: McGraw-Hill Publishing Company.

IOP Conf. Series: Materials Science and Engineering **506** (2019) 012063 doi:10.1088/1757-899X/506/1/012063

Unmanned Aerial Vehicle Applications In Agriculture

C Y N Norasma[1*], M A Fadzilah[2], N A Roslin[2], Z W N Zanariah [3], Z Tarmidi[4], F S Candra[5]

[1]Department of Agriculture Technology, Faculty of Agriculture, Universiti Putra Malaysia, Malaysia

[2]Institute of Tropical Agriculture and Food Security (ITAFoS), Universiti Putra Malaysia

[3]Department of Basic Science and Engineering, Faculty of Agriculture and Food Sciences
Universiti Putra Malaysia Bintulu Sarawak Campus, 97008 Bintulu, Sarawak, Malaysia

[4]Dep. of Geoinformation, Fac.of Geoinformation and Real Estate, Uni. Teknologi Malaysia

[5]CV. Mitra Geotama, Gg. Kalimantan G36, Sono, Sinduadi, Sleman, Yogyakarta, Indonesia

*Corresponding author: niknorasma@upm.edu.my

Abstract. Agriculture becomes extremely important as a main source of food production to feed the population in this planet. On the other hand, agriculture provides a lot of benefits to the country such as food and non-food product, transportation and balancing the environment. The demand for food security creates pressure to the decision maker to ensure our world has sufficient food for the entire world. Thus, the usage of the unmanned aerial vehicle (UAV) is an alternative to manage a farm properly to increase its yield. In order to promote the use of UAV in agriculture to support its sustainability, this paper provides an understanding towards the usage of UAV and its applications in agriculture. The objective of this paper is to review the usage of UAV in agriculture applications. Based on the literature, we found that a lot of agriculture applications can be done by using UAV. In the methodology, we used a comprehensive review from other researches in this world. As a result, from the revision, we found that different sensors give different analysis to the agriculture applications. Thus, the purpose of the project needs to be investigated before using the UAV technology for better data quality and analysis. As a conclusion, a suitable sensor and UAV need to be identified before using UAV to gather accurate data and precise analysis.

1. Introduction

Agriculture can be defined as cultivation of animals, plants, fungi and other life forms for food, fiber, biofuel, medicinal and other products used to sustain and enhance human [1]. It is the most comprehensive word used to show the many ways crop plants and domestic animals contribute in order to sustain the global human population especially in providing food and also other products [2]. Agriculture development is important especially for social and economic benefit, where agriculture increases food production, increases net income and improves family productivity, reduces income imbalance between rural agricultural employment and urban income of factory workers, and improves

IOP Conf. Series: Materials Science and Engineering **506** (2019) 012063 doi:10.1088/1757-899X/506/1/012063

farmers' living condition [3]. Malaysia has introduced Agrofood Policy 2011-2020, as a continuity from Third National Agriculture Policy (DPN3). This policy aims to outline a strategic planning to improve in the food value chain as well as increasing food production in sustainable manner [4]. From the latest knowledge and technologies in agriculture development, Department of Agriculture (DOA) will then spread the knowledge to farmers and give proper training to improve the workforce in agriculture sector [5].

Government have initiatives to improve development in agriculture by using technologies. One of the technologies is unmanned aerial vehicle (UAV) or also known as drone. It was first introduced in military for their surveillance during war. However now there are wide applications such as the forestry industry [6] rangeland and ecology research [7] as well as agriculture [8, 9, 10, 11, and 12]. Small or medium sized plantation or farm are suitable to use UAV for data collection and field monitoring effectively. UAV could be one of the solutions in getting higher resolution at a reasonable price, using high-end systems, fulfilling the requirements of the spatial, spectral, and temporal resolution needed, real-time processing capability as well as reducing the influence of environmental conditions [13, 14, 15, 16 and 6]. UAV can provide high resolution imagery in contrast with satellite imagery, which has certain date and day to obtain the image. This paper discusses the definition of UAV, Precision Agriculture, the usage of UAV in agriculture fields, and how to improve agriculture development via using UAV technology.

1.1 Unmanned Aerial Vehicle (UAV)
Unmanned Aerial Vehicle is a low-cost alternative in sensing technology and data analysis technique in the recent years. Basically, remote sensing using electromagnetic energy to determine the properties of targeted object from a distance and has the advantages of extensiveness, non-invasiveness, timeliness, and flexibility. However, remote sensing measures soil properties in the farm is far from actual data due to the complex natures of remote sensing, agricultural production and soils [17]. According to [17], some early studies using multispectral satellite images for soil surveys and soil mapping were done, however the images were generally unable to provide quantitative information concerning specific soil properties. Thus, UAV can be an option to collect accurate data in the field. It had been identified as a potential technology that can generate high spatial resolution imagery (< 1m) and at temporal frequency appropriate for timely responses in the production of actionable information about crop and field status [18]. The small UAV belongs to Low Altitude Remote Sensing (LARS), which is less as costly compared to traditionally manned aircraft is probably one of the main reason why the UAV industry had been exceeding the market demand [19].

The advantages of UAV are, it can capture the images of the farmer's crop with a variety of camera filters that can provide the farmers with multiple spectral imaging, allow the image processing and analysis which gives better information on their crop's health and at the same time identifying areas of the crops that require specific forms of attention. The small UAV can be easily flown and maintained with little training, making it a great option for farmers looking to advance their farming by merging agriculture with the technology of remote sensing [20]. UAV can be categorized in two types which is fixed wing and multirotor. It is a potential substitute for satellite-based remote sensing and it had been identified to generate high spatial resolution imagery and temporal frequency appropriate for timely responses in the production of actionable information about crops and field status [18]. Table 1 shows the comparison between multirotor and fixed wing UAV.

Table 1. Comparison between multirotor and fixed wing UAV [21]

Type	Pay-load	Flight time	Benefits	Limitation	Examples
Multirotor UAV	0.8–8.0 kg	8 – 120 min	• Applicable with waypoint navigation • Hovering capabilities • Can hold range of sensors from thermal, multispectral to hyperspectral cameras	• Payload may limit battery usage and flight time	DJI Inspire, Mikrocopter ARK OktoXL 6S12, Yamaha RMAX
Fixed wing UAV	1.0–10 lg	30 – 240 min	• Better flight time • Multiple sensors can be mounted • Limited hovering capacity	• Lower speed is required for image stitching	Landcaster Precision Hawk, senseFly eBee

1.2 Sensor and type of UAV

Good imagery depends on the type of UAV, sensor types and flight plan [22]. The type of UAV and sensors depend on the applications. Table 2 shows the different type of sensors used in agriculture. The applications of UAV will lead the type of the sensors. For example, to detect disease the suitable sensor is multispectral because it has many bands that can detect the sensitivity of the symptom. However, for agriculture mapping the RGB camera should be enough for data collection.

Table 2. Different types of sensors used in plant phenotype characterization

UAV and Sensor type	Applications	Photo	Ref.
3DR Iris+ (a) and DJI Phantom 2 (b), camera gimbal and GoPro camera	Agriculture field monitoring, autonomous navigation	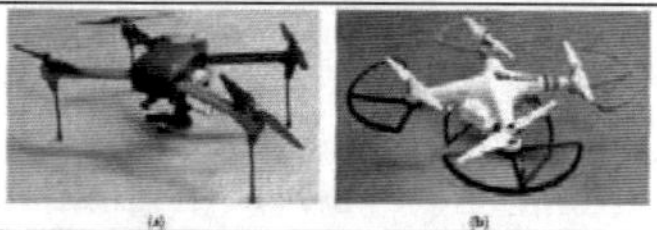	[20]
Turnigy 9XR Octocopter UAV, Digital camera (RGB)	Basal Stem Rot (BSR) disease in oil palm		[23]
Sensefly eBee UAV, 16-megapixel digital camera	Mapping changes to land cover, transmission of infectious diseases	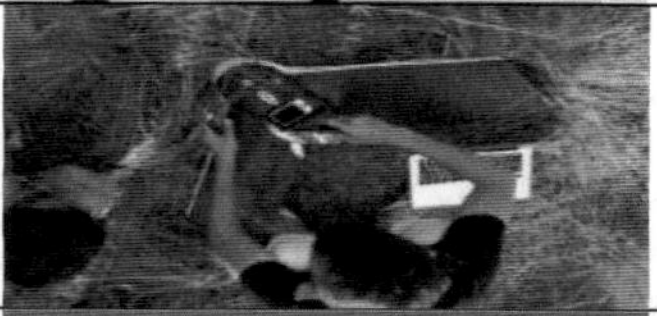	[24]
(HiSystems GmbH Mikrokopter, Germany, RGB camera	Develop a new estimation technique for disease severity		[25]
Microdrones MD4-200, A Tetracam ADC Lite digital camera	NDVI and grain yield, aerial biomass and nitrogen content	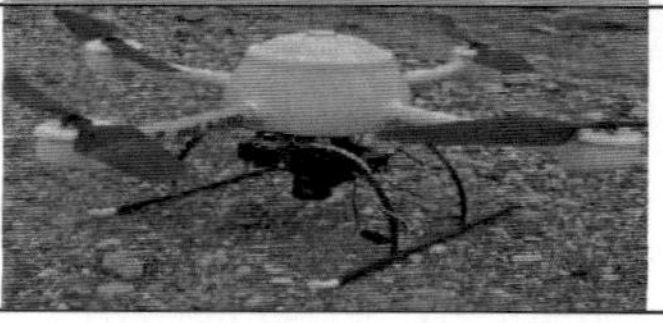	[26]

All aspects need to be considered before flying any UAV applications. In addition, geometric correction and geocoding are also important in producing an accurate weed map [27]. This is considered high accuracy for site-specific application. The accuracy of the geo-referencing of the images depends on the altitude of the UAV. The spectral and spatial resolution of the images depends on the sensors carried by the UAV. It is not necessary to use expensive sensors to map an area because it can be done using a typical digital camera such as a Canon 550D 15 Megapixel, DSLR (Digital Single-Lens Reflex), with a Canon EF-S 18-55 mm F/3.5 – 5.6 IS lens [28]. However, to detect the occurrence of specific diseases, pests and weeds, it is necessary to use a multispectral or thermal sensor camera [9]. The type of sensor should be matched to the type of information to be collected. The sensors can collect higher spatial resolution imagery than satellite sensors and at a lower cost [29].

1.3 Precision Agriculture

Precision agriculture (PA) concept was established in the late 1970s where the global positioning system (GPS) was developed and this system provides its own potential to determine accurate position (latitude, longitude and altitude) anywhere and at any time [30]. It is a holistic farm system to monitor and manage the field by using many components such as high-resolution data and high technology to maximize the output and minimize the input [20]. Furthermore, precision agriculture also referred to the application of "the right input, at the right place, at the right time with the right amount" [31] by using an integrated agricultural management system incorporating several technologies. The technological tools often include the GPS, geographical information system (GIS), yield monitor, variable rate technology (VRT), and remote sensing. Over the years, remote sensing approaches place remote sensors on towers over the crop field (thermal imagery, multi and hyper-spectral cameras, fluorometers) where the fixed positions were the main limitation for data collection [32].

These technologies can maximize the crop output of the farmers [20]. Satellite Imaging has been used before in crop monitoring, however there are several problematic issues on the satellite usage such as the prohibitive cost, the low image resolution, and its low sampling frequency [20]. Precision farming, PA, site-specific crop management (SSCM) or even site-specific farming basically use sensors that are capable of detecting field variability, such as VRT and grain yield monitor have been used in combination with high position accuracy GPS. The main concern of PA is to provide an alternative and realistic ways in order to promote a healthier environment for humans and in the meantime to reduce and optimize the usage of potentially harmful compounds [33] by using the spatial analysis like GIS tools. The flow of GIS tool purpose in agriculture is shown in figure 1.

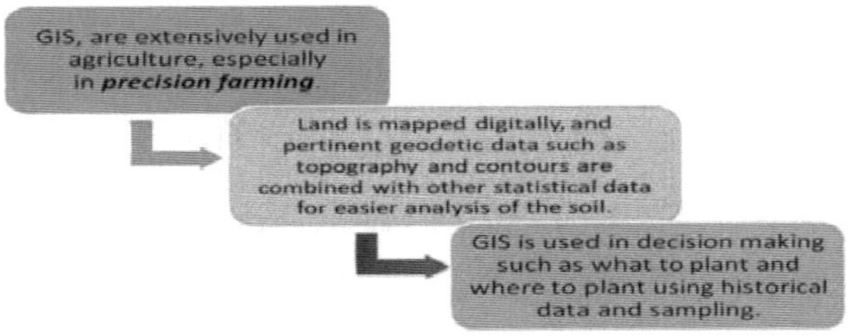

Figure 1. GIS flow for decision making in agriculture application.

The main key point of PA refers to the way farmers manage their crops to make sure there is continuous supply of water and fertilizer, and maximizing its productivity, quality and yield. In addition, it also involves minimizing pests, unwanted flooding and disease to the crops area. The invention of highly technical images had also lead in practicing agriculture by using information technologies with the approaches of looking at the whole system [34]. According to [35], most of the farmers face huge financial loss due to wrong weather prediction and poor irrigation method for crops. Some of the major problems faced by farmers was lack of knowledge on soil content, types of soil, lack of knowledge on which fertilizer is suitable for certain areas and the correct amount of irrigation according to the areas. Thus, the usage of UAV will allow farmers to constantly monitoring their crops' condition by air to find problems quickly and at the same time reduce time consumed for ground-level spot checks. For instance, a farmer might find through time-lapse UAV photography that part of their crop is low in nutrient or not well irrigated in some areas that are difficult to access quickly.

Precision agriculture can reduce the cost of input such as the fertilizer, herbicides, and pesticides and at the same time maximize the output. For a farmer, knowing the health and current state of the year's crop is essential for effective crop management and this is where the main role of PA especially the usage of UAV take place [20]. The aim of these PA is parallel with the Fourth Industrial Revolution that brings technological revolution that will fundamentally alter the way we live, work and relate to one another. The First Industrial Revolution used water and steam power to mechanize production. The Second and Third used electric power to create mass production and used electronics and information technology to automate production respectively. Now a Fourth Industrial Revolution is building on the Third, the digital revolution that has occurred since the middle of the last century. The Fourth revolution is characterized by a fusion of technologies that is blurring the lines between the physical, digital and biological spheres [21].

However, there is some policy requirements before the UAV can be departed. It is the policy of the Department of Civil Aviation, Malaysia (DCA) that UAVs operating in Malaysia must meet or exceed the safety and operational standards as those for manned aircraft. Thus, UAV operations must be as safe as manned aircraft, for example the UAV operations must not present or create hazard to person or property in the air or on the ground. UAV must not present or create a hazard to persons or property in the air or on the ground greater than that attributable to the operations of manned aircraft of equivalent class or category. Moreover, without obtaining prior relevant DCA approval, UAVs shall not be flown [36]. Standard and operating guideline of the UAV in Malaysia is not yet well established although the DCA office had release a minimum requirement for the safety such as the weight of the instrument should not be more than 2 kg of the total carried weight, unless obtained required official approval from the DCA office. The rules were more for the public safety and there is no guideline for agriculture practices.

2. Application of UAV in Agriculture

This paper reviewed the UAV applications and the sensors usage in the applications. There are many agriculture applications that use UAV technologies nowadays such as AeroVironment [37], X-Copter [38], VIPtero [39], Fieldcopter [40] and senseFly eBee. The imagery from the sensors contain different types of format that need to be processed using the correct software. Mosaic and orthorectification are a technique to combine each piece of the photos, which are captured using UAV to produce a complete set of maps. ENVI 4.5 (Exelis Visual Information Solutions Inc., Boulder, CO, USA) and ArcGIS 9.3 (ESRI Inc., Redlands, CA, USA) are the common softwares used for orthorectification and geo-referencing the image. One of the applications using UAV in Sydney for surveillance and management of aquatic weeds [37]. They used rotary UAV, which was cost-effective and very suitable for surveillance and management in aquatic weed. They collected the data and weed experts analyzed the data and produced a map. They used Support Vector Machine (SVM) in image processing analysis. It shows that SVM was successful in identifying the weeds in natural environments.

Oil palm is the golden crop in tropical region especially Malaysia. The number of oil palm trees in the area is important information to predict the yield, monitoring the oil palm growth and increase their productivity [41]. In South-East Asia especially Malaysia, Indonesia and Thailand where oil palm (*Elaeis guineensis*) are mostly planted. The oil from oil palm is the most consumed vegetable oil in the world about 35% compared to other oil crops such as soybean and sunflower. The demand from the consumer for vegetable oil had converted most area into plantation area for oil palm. The oil palm needs proper management for planting operations on a daily basis such as weeding, pruning, harvesting and manuring [42]. Unmanned aerial vehicle is a flight with no pilot on-board and controlled by a remote or autonomous self-control. The UAV is used to capture geospatial orthophotos to monitor oil palm growth from the early planting and identify which points at the plantation area is fertile to plant crop. Therefore, UAV is used to get the information quickly [43]. A system to detect palm oil nutrient stress had been developed and UAV images used to detect the oil palm rods [44, 45]. From the orthophoto, user can match with other maps that has agriculture elements. The UAV captured the images and then images were analysed to observe the growth in the oil palm plantation [43].

The UAV used in oil palm plantation to capture images that were processed with photogrammetry software. The photogrammetry software then were created orthomosaics of the interest zones and obtains an automatic census of the palm by using GIS software. The DJI Phantom 2 Vision+ can be controlled using mobile app iOS operating system. The Pix4D mapper Capture was used to mosaic the oil palm plantation to monitor the whole area in a faster way [46].

Rubber plantations (*Hevea brasiliensis*) have increased rapidly in tropical and non-tropical regions a few years ago in order to meet the demand for natural rubber [47]. According to report by Food and Agriculture Organization (FAO) in 2010 the rubber plantation has increased by 25% for past two decades [48]. The expansion of rubber industry has created problems between the production, forest management and conservation [49, 50, 51, 52, 53, and 54]. A better understanding of its management can aide a proper decision making for its management [55]. Drone used in precision agriculture has a lot of benefits to farmers which allows them to save their money in managing their crop by increasing the yield and the productivity [56]. The drone can be used to count the plants, nutrient applicator sprayer, mapping and NDVI plant health indexing [57, 58].

The application of UAV in rice is really important in classification, phenological behaviour, yield estimation and triggering various researchers in deriving a rice database at global, national and regional scales [38]. [39] developed a UAV for rice monitoring and management. It is known as X-Copter and can carry about 30 kg payload and fly for more than an hour. The interesting part of their research is the X-Copter, which not only capture the images but apply fertilizer in real time. This is a very good opportunity for farmers to apply precise input. In Italy, the first research using UAV in vigor mapping was based on Normalized Differentiated Vegetation Indices (NDVI) known as "VIPtero" [59]. It is a six-rotor aerial platform and flies autonomously. Multispectral camera and spectroradiometer been used for capturing and sampling the data respectively. "VIPtero" produced high-resolution imagery up to 5 cm per pixel. It definitely helps in vineyard management and is very suitable for precision agriculture in medium size.

Meanwhile, for large area rice mapping generally performed at low spatial resolution due to insufficient availability of high temporal data [59]. Image fusion or blending algorithms that combs high and low spatial resolution images is commonly applied in order to obtain the temporal data at high spatial resolution. Moreover, from the studies that were done by [60] one of the methods suggested is using spectral features along with phenological behaviors, this is because when using spectral features alone the grassland and other crops can be misclassified as rice due to similar spectral properties with rice. By combining the spectral features with phenological behavior, different crop type has different seasonal behavior that is captured through phenological behavior. This method is usually done by calculating, integrating, averaging, differencing or taking the maximum value of annual NDVI [61] found out that there is a high correlation between uncalibrated RGBVI (UAV-based RGB images) and calibrated NDVI images from their findings also, RGBVI could serve as a strong predictor for qualitative differences in crop growth for any captured growth stage.

AeroVironment is one of the UAV applications in agriculture using solar-powered and PathFinder Plus [37]. It is a high-altitude flying UAV with both multispectral imaging and hyperspectral imaging systems. It has been demonstrated in Kauai Coffee Plantations with a focus on surveillance of large rangelands. Another research has been done in a coffee plantation in Indonesia that used UAV with RGB digital camera [62]. The research investigated the usage of RGB camera as low-cost system for assessing Chlorophyll a and b, Carotenoid, and Nitrogen in coffee plantation with the combination ground sensor such as spectrometer (Ocean Optic USB2000, VIS–NIR series of portable spectrometer (Ocean optics, USA) and SPAD-502 Chlorophyll meter. Despite using spectrometer and SPAD-502 Chlorophyll meter in estimating Chlorophyll a and b, Carotenoids, and Nitrogen, they successfully measured the Chlorophyll a and b, Carotenoid, and Nitrogen in coffee plantation without any multispectral camera. They also developed, modified, and compared the VIs with the previous research.

3. Advantages and Limitations

UAVs are capable of capturing imagery at high risk situations. They can be flown below clouds and in light rain [9]. They are not limited by physiological conditions that would affect human pilots of light planes. Field Copter is a UAV that can carry multispectral sensors for soil and crop analysis [40]. It is capable of flying and capturing imagery in more than 70% weather conditions compared to satellite imagery. There are numerous advantages and benefits using UAV in various applications. However, it depends on the aircraft, sensor types, mission objectives and their platforms [63]. All these aspects need to be considered before implementing any UAV applications. The advantages of UAV in various aspects such as it can be used in high risk situations at low altitude, in cloudy and drizzly weather conditions, not burdened with the physiological limitations and economic expenses, real-time capability and the ability for fast data acquisition, can fly below the clouds, imagery is available 'on-demand', cost effective, images are geo-referenced, enabling direct links with GIS packages, low maintenance and safety.

The limitations of the UAV are the weight and dimension of the sensors of low-cost UAV. The small or medium amateur format chosen are normally less stable and not accurate. Low-cost UAV have limitations in reaching a certain altitude due to a less powerful engine [52]. There are some problems that need to be considered when using UAV, such as the path-planning system does not use professional pilot, the high-speed ultra-low situation, data downloading function during real-time application, the size and payload to avoid bottleneck and the software for its automatic processing [64].

4. Conclusions

The production of agriculture is very important for food security in the world. The field management can be improved by using the UAV and increase the yield of productivity in order to feed the increasing population in the world. The sensor and type of UAV depends on the objectives of the research. The usage of the UAV cannot solve all the problems in the field but it can help some specific issues in agriculture. The integrated technology will help farmers in field operations. This paper only reviewed some of the researches in agriculture that used UAV and more study need to be investigated in the future such as image processing, cost of the UAV and the accuracy of data.

Acknowledgements

Authors wishing to acknowledge the financial support from Ministry of Higher Education, Malaysia under HICoE and Transnational Research grant (Vote No: 5526500) and Universiti Putra Malaysia, Grant Putra-IPM (Vote No: 9611400).

References

[1] Singh A and Masuku M (2012). An Insight in Statistical Techniques and Design In Agricultural and Applied Research. World Journal of Agricultural Sciences, 568-584.
[2] Harris D, and Fuller D (2014). Agriculture: Definition and Overview. In Smith, & Claire, Encyclopedia of Global Archaeology. New York: Springer.
[3] Othman K (2004). Integrated Farming System and Multifunctionality of Agriculture in Malaysia. XV International Symposium on Horticultural Economics and Management (pp. 291-296). Malaysia: ISHS Acta Horticulturae 655.
[4] DOA. (2017a). Department of Agriculture Malaysia. Retrieved from Dasar Agro Makanan (2011-2010): http://pertanian.johor.gov.my/ms/mengenai-kami/dasar-agro-makanan-2011-2020
[5] DOA. (2017b). Department of Agriculture. Retrieved from Objektif dan Fungsi Jabatan Pertanian: http://pertanian.johor.gov.my/ms/mengenai-kami/objektif-dan-fungsi-jabatan-pertanian,
[6] Wallace L, Lucieer A, Watson C, and Turner D (2012). Development of a UAV-LiDAR System with Application to Forest Inventory. Remote Sensing, 1519-1543.

[7] Chang J, Clay S, Clay D, and Dalsted K (2004). Detecting weed-free and weed-infested areas of a soybean field using near-infrared spectral data. Weed Science, 642-648.
[8] Nebiker S, Annen A, Scherrer M, and Oesch D (2008). A light-weight multispectral sensor for micro UAV-opportunities for very high resolution airborne remote sensing. The International Archives of the Photogrammetry, Remote Sensing and Spatial Information Sciences, (pp. 1193-1200). Beijing.
[9] Berni J, Zarco-Tejada P, Suárez L, González-Dugo V, and Fereres E (2009). Remote Sensing of Vegetation from UAV Platforms using Lightweight Multispectral and Thermal Imaging Sensors. Int.Arch.Photogramm.Remote Sen.Spatial Inform Sci, p. 38(6).
[10] Gay A, Stewart T, Angel R, Easey M, Adrian J E, Thomas N, and Kemp A (2009). Developing Unmanned Aerial Vehicles for Local and Flexible Environmental and Agricultural Monitoring. Proceedings of RSPSoc 2009 Annual Conference, (pp. 471-476). Leicester, UK.
[11] Delegido J, Verreist J, Meza C, Rivera J, Alonso L, and Moreno J (2013). A red-edge spectral index for remote sensing estimation of green LAI over agroecosystems. European Journal of Agronomy, 42-52.
[12] Leon J, and Woodroffe C (2013). Morphological characterisation of reef types in Torres Strait and an assessment of their carbonate production. Marine Geology, 64-75.
[13] Knipling E (1970). Physical and Physiological basis for the reflectance of visible and near-infrared radiation from vegetation. Remote Sensing of Environment, 155-159.
[14] Swain K, Jayasuriya H, and Salokhe V (2007). Low-Altitude Remote Sensing with Unmanned Radio-Controlled Helicopter Platforms: A Potential Substitution to Satellite-based Systems for Precision Agriculture Adoption under Farming Conditions in Developing Countries. International Commission of Agricultural Engineering, 1-16.
[15] Berni J, Zarco-Tejada P, Suárez L, González-Dugo V, and Fereres E (2009). Remote Sensing of Vegetation from UAV Platforms using Lightweight Multispectral and Thermal Imaging Sensors. Int.Arch.Photogramm.Remote Sen.Spatial Inform Sci, p. 38(6).
[16] Elsenbeiss H, and Sauerbier M (2011). Investigation of uav systems and flight modes for photogrammetric applications. The Photogrammetric Record, 400-421.
[17] Thomasson J A, Sui R, and GE Y (2011). Remote sensing of soil properties in precision agriculture:a review. Front Earth Sci, 229-238.
[18] Elarab M, Ticlavilca A M, Torres-Rua A F, Maslova I, and McKee M (2015). Estimating chlorophyll with thermal and broadband multispectral high resolution imagery from an unmanned aerial system using relevance vector machines for precision agriculture. International Journal of Applied Earth Observation and Geoinformation, 32-42.
[19] Zhang Y, Wang L, and Duan Y (2016). Agricultural information dissemination using ICTs: A review and analysis of information dissemination models in China. Information Processing in Agriculture, 17-29.
[20] Cano E, Horton R, Liljegren C, and Bulanon D (2017). Comparison of Small Unmanned Aerial Vehicles Performance Using Image Processing. Journal of Imaging, 1-14.
[21] Schawab K (2016). The Fourth Industrial Revolution: what it means, how to respond. Retrieved from Fourth Industrial Revolution: https://www.weforum.org/agenda/2016/01/the-fourth-industrial-revolution-what-it-means-and-how-to-respond/
[22] Sankaran S, Khot L, Espinoza C, Jarolmasjed S, Sathuvalli V, Vandemark G, and Pavek M (2015). Low-altitude, high-resolution aerial imaging systems for row and field crop phenotyping: A review. European Journal of Agronomy, 112-123.
[23] Khairunniza-Bejo S, Jaleni M, Husin M, Khosrokhani M, Muharam F, Seman I, and Anuar M (2018). Basal Stem Rot (BSR) detection using iextural analysis of Unmanned Aerial Vehicle (UAV)image. eProceedings Chemistry, 40-45.
[24] Fornace K, Drakeley C, William T, Espino F, and Cox J (2014). Mapping infectious disease landscapes: unmanned aerial vehicles and epidemiology. Trends in Parasitology, 514-519.
[25] Sugiura R, Tsuda S, Tamiya S, Itoh A, Nishiwaki K, Murakami N, and Nuske S (2016). Field

phenotyping system for the assessment of potato late blight resistance using RGB imagery from an unmanned aerial vehicle. Biosystems Engineering, 1-10.

[26] Vega F A, Ramírez F C, Saiz M P, and Rosúa R O (2015). Multi-temporal imaging using an unmanned aerial vehicle for monitoring a sunflower crop. Biosystem Engineering, 19-27.

[27] Xiang H, and Tian L (2011). Method for automatic georeferencing aerial remote sensing (RS) images from an unmanned aerial vehicle (UAV) platform. Biosystems Engineering, 104-113.

[28] Turner D, Lucieer A, and Wallace L (2014). Direct Georeferencing of Ultrahigh-Resolution UAV Imagery. IEEE Transactions on Geoscience and Remote Sensing (pp. 2738-2745). IEEE.

[29] Peña J, Torres-Sánchez J, Isabel de Castro A, Kelly M, and López-Granado F (2013). Weed Mapping in Early-Season Maize Fields Using Object-Based Analysis of Unmanned Aerial Vehicle (UAV) Images. Retrieved from PLOS ONE : https://doi.org/10.1371/journal.pone.0077151

[30] Stafford J V (2000). Implementing Precision Agriculture in the 21st Century. J. agric. Engng, 267.

[31] Bongiovanni R, and Lowenberg-Deboer J (2004). Precision Agriculture and Sustainability. Precision Agriculture, 359-387.

[32] Gago J, Douthe C, Coopman R, Gallego P, Ribas-Carbo M, Flexas J, and Medrano H (2015). UASs challenge to assess water stress for sustainable agriculture. Agricultural Water Management, 9-19.

[33] Chunhua Z and John M (2012). The application of small unmanned aerial systems for precision agriculture:a review . Precision Agric, 694.

[34] Shibusawa S (2002). Precision Farming Approaches for Small Scale Farms. 22.

[35] Chakane S, Chaskar H, Patil P, Shelar P, and Godse P (2017). Automated Information System for Improved Crop Management. International Journal of Agriculture Innovations and Research, 740.

[36] AIS. (2008, 18/FEB). Unmanned Aerial Vehicle (UAV) Operations in Malaysia. Retrieved from Drone Laws in Malaysia: http://aip.dca.gov.my/aip%20pdf%20new/AIC/AIC%20200804.pdf

[37] Göktoğan A, Sukkarieh S, Bryson M, Randle J, Lupton T, and Hung C (2010). A Rotary-wing Unmanned Air Vehicle for Aquatic Weed Surveillance and Management. Journal of Intelligent and Robotic Systems, 467-484.

[38] Shim D, Han J S, and Yeo H T (2009). A Development of Unmanned Helicopters for Industrial Applications. Journal of Intelligent and Robotic Systems, 407-421.

[39] Primicerio J, Gennaro S, Fiorillo E, Genesio L, Lugato E, Matese A, and Vaccari F (2012). A flexible unmanned aerial vehicle for precision agriculture. Precision Agriculture, 517-523.

[40] Wal T, Abma B, Viguria A, Prévinaire E, Zarco-Tejada P, Serruys P, and Voet P (2013). Fieldcopter: unmanned aerial systems for crop monitoring services. Precision agriculture, 169-175.

[41] Li W, Fu H, Yu L, and Cracknell A (2017). Deep Learning Based Oil Palm Tree Detection and Counting for High-Resolution Remote Sensing Images. Remote Sensing, 1-13.

[42] Chong K, Kanniah K, Pohl C, and Tan K (2017). A review of remote sensing applications for oil palm studies. Geo-spatial Information Science, 184-200.

[43] Fahmi F, Trianda D, Andayani U, and Siregar B (2017). Image processing analysis of geospatial uav orthophotos for palm oil plantation monitoring. 2nd International Conference on Computing and Applied Informatics 2017 (pp. 1-7). IOP Conf. Series: Journal of Physics.

[44] Guldogan O, Rotola-Pukkila J, Balasundaram U, Le T H, Mannar K, Chrisna T, and Gabbouj M (2016). Automated Tree Detection and Density Calculation using Unmanned Aerial Vehicles. Visual Communications and Image Processing (VCIP). Chengdu, China: IEEE.

[45] Moranduzzo T, and Melgani, F (2014). Monitoring structural damages in big industrial plants with UAV images. Geoscience and Remote Sensing Symposium (IGARSS). Quebec City, QC, Canada: IEEE.

[46] Castillo J, Diaz R, and Guzmán C (2015). Development of an aerial counting system in oil palm plantations. IOP Conference Series: Materials Science and Engineering (pp. 1-16). IOP Publishing.

[47] Dong J, Xiao X, Chen B, Torbick N, Jin C, Zhang G, and Biradar C. (2013). Mapping deciduous rubber plantations through integration of PALSAR and multi-temporal Landsat imagery. Remote Sensing of Environment, 392-402.

[48] FAO. (2010). Global Forest Resources Assesment. Rome: Food and Agriculture Organization of

the United Nations.
[49] Guardiola-Claramonte M, Troch P, Ziegler A, Giambelluca T, Durcik M, Vogler J, and Nullet M (2010). Hydrologic effects of the expansion of rubber (Heveabrasiliensis) in a tropical catchment. Ecohydrology, 306-314.
[50] Li H, Ma Y, Aide T, and Liu W (2008). Past, present and future land-use in Xishuangbanna, China and the implications for carbon dynamics. Forest Ecology and Management, 16-24.
[51] Blécourt M, Brumme R, Xu J, Corre M, and Veldkamp E (2013). Soil Carbon Stocks Decrease Following Conversion of Secondary Forests to Rubber (Hevea brasiliensis) Plantations. Forest-to-Rubber Conversion Reduces Soil C Stocks, 1-9.
[52] Romeo J, Pajares G, Montalvo M, Guerrero J, Guijarro M, and Ribeiro A (2012). Crop Row Detection in Maize Fields Inspired on the Human Visual Perception. The Scientific World Journal, 1-10.
[53] Xu J, Grumbine R and Beckschäfer P (2014). Landscape transformation through the use of ecological and socioeconomic indicators in Xishuangbanna, Southwest China, Mekong Region. Ecological Indicators, 749-756.
[54] Ziegler l, Fox J, and Xu J (2009, May 22). The Rubber Juggernaut. Retrieved from www.sciencemag.org: http://science.sciencemag.org/content/sci/324/5930/1024.full.pd
[55] Chen B, Li X, Xiao X, Zhao B, Dong J, Kou W, and Xie G (2016). Mapping tropical forests and deciduous rubber plantations in Hainan Island, China by integrating PALSAR 25-m and multi-temporal Landsat images. International Journal of Applied Earth Observation and Geoinformation, 117-130.
[56] Makhdzir A (2017). Application of Artificial Intelligence and Drone Technology in Malaysian Plantation. International Proceedings of Irc 2017, (pp. 146-164).
[57] Khorramnia K, Shariff A, Rahim A, and Mansor S (2014). Toward malaysian sustainable agriculture in 21st century. 8th International Symposium of the Digital Earth (ISDE8) (pp. 1-5). IOP Conference Series: Earth and Environmental Science.
[58] Makhdzir A (2017). Application of Artificial Intelligence and Drone Technology in Malaysian Plantation. International Proceedings of Irc 2017, (pp. 146-164).
[59] Xiao X, Boles S, Frolking S, Li C Y, Babu J, Salas W, Moore III B (2006). Mapping Paddy Rice Agriculture in South and SouthEast Asia Using Multi-temporal MODIS Images. Remote Sensing of Environment, 95-113.
[60] Singha M, Wu B, and Zhang M (2016). An Object-Based Paddy Rice Classification Using Multi-Spectral Data and Crop Phenology in Assam,Northest India. Remote Sensing, 1-20.
[61] Bareth G, Bolten A, Gnyp M, Reusch S, and Jasper J (2016). Comparison of uncalibrated RGBVI with spectrometer-based NDVI derived from uav sensing systems on field scale. The International Archives of thr Photogrammetry,Remote Sensing and Spatial Information Sciences, 837-843.
[62] Putra, Bayu Taruna Widjaja, and Peeyush Soni. "Enhanced Broadband Greenness in Assessing Chlorophyll a and B, Carotenoid, and Nitrogen in Robusta Coffee Plantations Using a Digital Camera." Precision Agriculture 19, no. 2 (2018): 238-56.
[63] Watts A, Ambrosia V and Hinkley E (2012). Unmanned Aircraft Systems in Remote Sensing and Scientific Research: Classification and Considerations of Use. Remote Sensing, 1671-1692.
[64] Zhang C, and Kovacs J M (2012). The Application of UAV. Precision Agric, 693-712.

IOP Conf. Series: Materials Science and Engineering **506** (2019) 012064 doi:10.1088/1757-899X/506/1/012064

Optimization of Flow Shop Scheduling Problem using Tabu Search Algorithm: A Case Study

SS Dewi[1*], Andriansyah[1, 2] and Syahriza[1]

[1]Industrial Engineering Department, Faculty of Engineering, Universitas Syiah Kuala, Darussalam, Banda Aceh 23114, Indonesia
[2]Laboratory of Industrial Computation and Optimization, Universitas Syiah Kuala, Darussalam, Banda Aceh 23114, Indonesia

[*]Corresponding author: susysusiladewi@gmail.com

Abstract. This study discusses about scheduling of dining chair processing in a company. Products have been completed according to customer wishes. The problem is classified as flow shop scheduling problem. This paper considers a special case of a six machines of general flow shop scheduling problem which processing time of each job has the longest idle time and waiting time which causes makespan of dinning chair to be high. The goal of this problem is to minimize makespan of product processing. tabu search (TS) algorithm is used to solve this problem. Then performance of tabu search algorithm is compared with Campbell Dudek Smith (CDS) algorithm from literatures. Result shows that tabu search algorithm can improve the makespan in 1968 seconds. This result more effective compare to CDS algorithm that takes longer times 2064 seconds. Tabu search algorithm saving 4.6% makespan with the fastest computing time.

1. Introduction

Flowshop scheduling problem with the minimization makespan criterion has always been a problem that most attracted the attention of researchers. This is a due to the issues that concerns how to produce products with right on time so that it is able to fulfill customer satisfaction [1]. Companies that do not consider schedule production will be reduce customer trust then switch to another competitor firms. Many activities are carried out using the same machine, the completion of work that does not use the right pattern causes problems when the resources work on other tasks, the problem usually triggers a lot of raw materials waiting for the work process which results in too large makespan [2,3].

Many researchers have conducted a study of this scheduling, with the aim of managing flow time, makespan, machine waiting time using the heuristic method using the Campell Dudek Smith (CDS) method which uses the concept of trial and error, solving problems like this usually tends to be fast but the quality of the solution produced low and less optimal [4]. To overcome this limitation, many researchers eventually turned to methods such as the Metaheuristic Algorithms Tabu Search (TS), where this method was inspired by natural events. This method of use is easier because it is applied to computer programming languages so as to provide a better solution than the heuristic method [5-8].

The case in this paper discusses PT XYZ which is a manufacturing company engaged in the manufacture of furniture, the company produces various kinds of component products such as chair dining products and doing a lot of work using the same machine. production at this company has a waste of time which causes the machine schedule to be less efficient, so the total effect of the completion time

IOP Conf. Series: Materials Science and Engineering **506** (2019) 012064 doi:10.1088/1757-899X/506/1/012064

of all work is late from 1 to 5 working days[9]. This flowshop scheduling problem will be solved by searching for algorithms.

2. Research Methodology

In this research conducted, there are several steps that will be shown in figure 1 below:

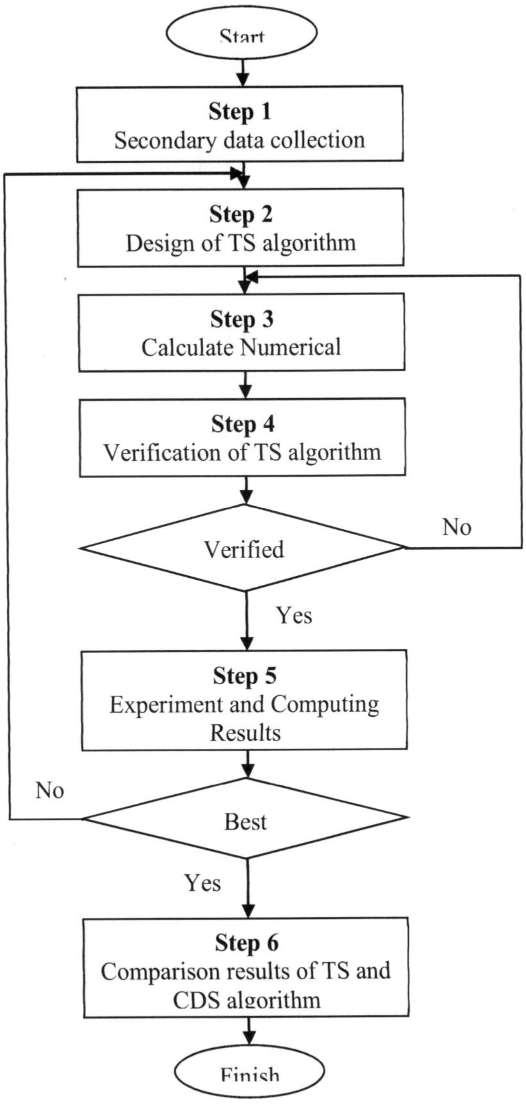

Figure 1. Flowchart on research methodology tabu search TS algorithm

IOP Conf. Series: Materials Science and Engineering **506** (2019) 012064 doi:10.1088/1757-899X/506/1/012064

Based on flowchart, this can explained clearly as details below:

i. Step 1 is the secondary data collection which obtained from previous research using CDS.
ii. Step 2 is the design of TS algorithm. TS algorithm is designed using MATLAB 2013a programming language by determination the initial solution set several parameters in solution search and continued with calculate aspiration criteria.
iii. Step 3 is the numerical example. A numerical example is done manually on the TS algorithm using initial solution from CDS.
iv. Step 4 is the verification of the TS Algorithm. TS algorithm is verified if the solution obtained does not change and the combination of parameters changes. Verification is done to check the results obtained from the TS algorithm meet the parameters and according to the manual calculation.
v. Step 5 is the experiment and computing results: The experimental results are carried out by running the programs for 20 times to get the best solution and calculating the standard deviation to see the performance of the TS algorithm and see the computation time.
vi. Step 6 is the comparison of results of TS and CDS algorithms: Results done to find out how big the levels of reliability of the method inside produce solutions.

3. Numerical Examples and Computational Analysis

3.1 Design of TS algorithm

TS algorithm is an optimization method based on local search where the solution of the problem is using the history solution which is used as a search neigborhood Search to get a solution better in the next search. Besides, TS algorithm also applies restricted area system so that the solution produced not repeated a second time and the solution produced does not produce a solution that is worse than the value for initial solution. The parameters of TS algorithm are shown in table 1.

Table 1. Parameters of TS Algorithms

Parameter	Value
Tabu List	10
Iteration	100
Makespan Early	2064 Seconds
Early job sequence solutions	g-d-c-l-i-j-a-b-k-h-e-f

3.2 Numerical Examples

Numerical examples that are performed by using data from previous studies. As for TS design steps are as follows:

i. Input cycle time data used by the job when passing through each machine.
ii. Generate the initial solution in the form of a sequence the initial job obtained uses CDS. The job order is g-d-c-l-i-j-a-b-k-h-e-f.
iii. Determine the cost of early makespan solution is 2064 seconds.
iv. Determine maximum iteration is for 100 times.
v. Tabu list is 10.

Hence, calculating aspiration criteria as shown in table 2.

Table 2. Calculation of the TS Algorithm Manual

Machine/Work Center		Job sequencing (seconds)											
		G	**d**	**c**	**L**	**i**	**j**	**A**	**b**	**k**	**h**	**e**	**f**
Cutting machine	Start	0	119	243	375	503	678	853	1120	1271	1400	1535	1662
	Finish	119	243	375	503	678	853	1120	1271	1400	1535	1662	1797
Planer machine elbow	Start	119	243	375	503	678	869	1120	1319	1456	1586	1705	1816
	Finish	227	354	491	629	869	1086	1319	1456	1586	1705	1816	1922
Planers Machines (Press)	Start	-	-	-	-	-	-	1319	1477	-	-	-	-
	Finish	-	-	-	-	-	-	1477	1614	-	-	-	-
Spindle Machine	Start	-	-	-	-	-	1086	1477	-		-	-	-
	Finish	-	-	-	-	-	1303	1632	-		-	-	-
Grinding Machine	Start	227	354	491	629	1086	1303	1632	1710	1774	1859	1932	1999
	Finish	292	421	559	715	1180	1397	1710	1774	1859	1932	1999	2064
Drilling Machine	Start	-	-	-	-	-	-	1710	1852	-	-	-	-
	Finish	-	-	-	-	-	-	1852	1978	-	-	-	-

vii. Take an alternative step, move the alternative to do by moving one or two work sequences alternately as many as the specified iteration. But what must be remembered in the TS algorithm is that there is no possibility of repeating the sequence of work that occurs.

viii. Determining the candidate list the solution used, the candidate solution is the result of an alternative step that has a sequence of jobs with the lowest makespan. How it be processed is given in table 3 below.

Table 3. List of Candidate Solutions

Iteration	Job sequencing	Makespan
1	g-d-c-l-i-j-a-b-k-h-e-f	2064
2	d-g-c-l-i-j-a-b-k-h-e-f	2064
3	c-d-g-l-i-j-a-b-k-h-e-f	2064
4	l-c-d-g-i-j-a-b-k-h-e-f	2064
5	i-l-c-d-g-j-a-b-k-h-e-f	2064
6	j-i-l-c-d-g-a-b-k-h-e-f	2064
7	a-j-i-l-c-d-g-b-k-h-e-f	1977
8	g-c-d-l-i-j-a-b-k-h-e-f	2064
9	g-l-c-d-i-j-a-b-k-h-e-f	2064
10	g-i-l-c-d-j-a-b-k-h-e-f	2064

ix. Choose the order of jobs that have the lowest makespan that is the best solution.*3.3 Verification of TS algorithm results*

The results of TS algorithm solutions obtained from MATLAB will check whether the solution does not violate. The Tabu list and the iteration determined and calculated according to the method manually. Based on table 4, the results obtained from design do not violate the boundaries specified parameters.

Table 4. Verification of TS Algorithm

Parameter	Value	Unfulfilled Constrain
Tabu list	10	X
Iteration	100	X
Early Makespan	2064 seconds	X

3.3 Results of experiment and computing

Table 5 below is the experimental results from TS stochastic and 20 attempts were made to find out level of TS algorithm data dissemination.

Table 5. Recapitulation of TS Algorithm Experiments

Experiments	Makespan (seconds)
1	1970
2	1970
3	1968
4	1968
5	1968
6	1975
7	1970
8	1968
9	1970
10	1968
11	1968
12	1968
13	1970
14	1968
15	1968
16	1968
17	1968
18	1968
19	1968
20	1968
Mean	1968.85
Standard Deviation	1.69

The standard deviated values are obtained to find out the level of data dissemination as shown in figure 2. The smaller the standard deviation obtained, the results of the experiment are the best.

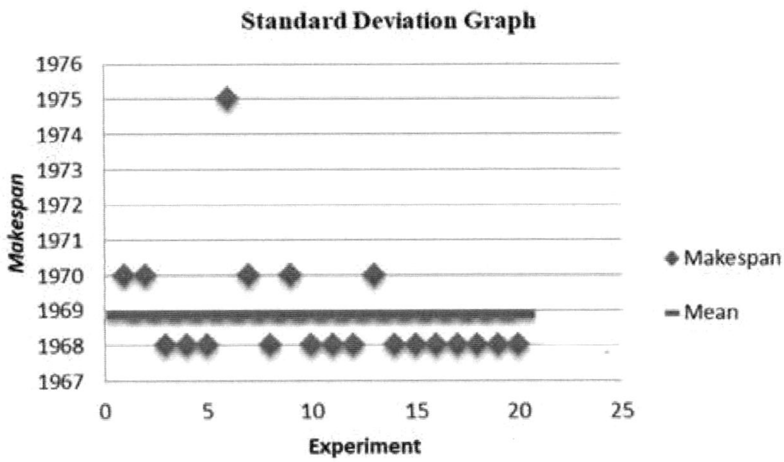

Figure 2. Standard deviation graph

The graph that shows the value of these 20 experiments spread evenly and there was no significants change from the solution what is obtained means that the algorithm has have good performance in producing solution. In each experiments, also presented the best solution chart used to find out in what iterations solutions that have the best value found based on the graph, the job sequence is selected produced to be the best solution as shown in figure 3.

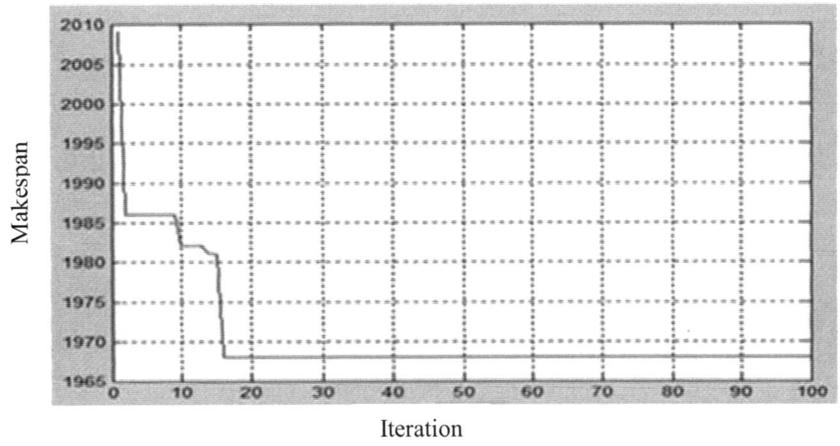

Figure 3. Best solution graph.

The graph shows that the data are in the 16 to 100 constant range and no more changes. That is, at the iteration value of the solution obtained has been convergent. The appearance of the job sequence and makespan of TS algorithm will be presented in figure 4.

The best makespan =

1968

The best sequences =

12 9 1 7 11 5 10 2 3 8 4 6

Figure 4. The best makespan and job sequences.

Next, is the scheduling repair Gantt chart using the TS algorithm So, this can be show in figure 5 as visualize.

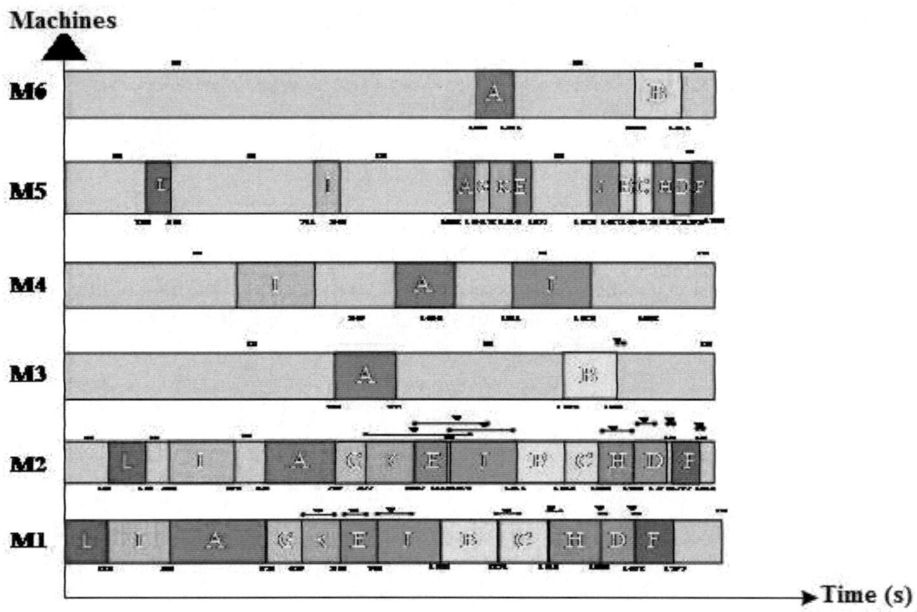

Figure 5. Gantt chart scheduling repair using TS algorithm.

Figure 5, on the improved Gantt chart done using TS looks to occur the initial reduction in makespan value 2064 seconds to 1968 seconds.

IOP Conf. Series: Materials Science and Engineering **506** (2019) 012064 doi:10.1088/1757-899X/506/1/012064

3.4 Comparison of CDS Result and TS algorithm

Table 6. Comparison of CDS Result and TS Algorithms

Method	Job Sequence	Makespan
CDS	g-d-c-l-i-j-a-b-k-h-e-f	2064 Seconds
TS	l-i-a-g-k-e-j-b-c-h-d-f	1968 Seconds

From table 6, it can be concluded that scheduling completion using TS algorithms is much better than the CDS method. TS algorithm can be shortening the completion time by 4.6%.

4. Conclusion

This paper aims to reduce the makespan of dinning chair and compare between the TS method with CDS method. The result of this paper is scheduling using the TS method produces makespan about 1968 seconds, where the makespan value is 96 seconds or 4.6% smaller than compared to results scheduling using the CDS method. Also this shows that the scheduling dining chair use tabu search (TS) algorithm is better than use CDS method.

References

[1]. Siburian R. and Ginting A. SS. 2013 *Penjadwalan produksi job shop dengan menggunakan algoritma Tabu Search pada PT . XYZ.* **8(2)** pp. 1–5

[2]. Nakandhrakumar R.S, Seralathan S, Azarudeen A and Narendran V.2014 *Optimization of Job Shop Scheduling Problem using Tabu Search Optimization Technique*. **3(3)** pp. 1241–1244

[3]. Ekşioglu, B., Ekşioglu S. D. and Jain P. 2008 *A Tabu Search algorithm for the flowshop scheduling problem with changing neighborhoods.*Computers & Industrial Engineering, pp. 1–11

[4]. Solimanpur M., Vrat P. and Shankar R.2004 *A neuro-tabu search heuristic for the flow shop scheduling problem.*Computers and Operations Research.**31(13)** pp. 2151–2164

[5]. Baker K. R. and Dan T.2009 *Principles Of Sequencing and Scheduling*. New Jersey:John Wiley & Sons Inc. Hoboken

[6]. Ben-Daya M. and Al-Fawzan M. 1998 *A tabu search approach for the flowshop scheduling problem. European Journal of Operational Research.* **109(1)** pp. 88–95

[7]. Cesaret B., Oğuz C. and Sibel Salman F.2012 *A tabu search algorithm for order acceptance and scheduling. Computers and Operations Research.* **39(6)** pp. 1197–1205

[8]. Wang C., Lig X. and Wang Q.2010 *Accelerated tabu search for no-wait flowshop scheduling problem with maximum lateness criterion. European Journal of Operational Research.* **206(1)** pp. 64–72

[9]. Hilmi T. M. J. 2017 *Identifikasi dan Minimasi Time Waste dengan Pendekatan Value Stream Mapping dan Optimasi Penjadwalan Mesin Menggunakan Metode CDS pada Proses Produksi Dining Chair*

Finite Element Modelling and Analysis for Dynamic Investigation of a Laser Spot Welded Hat-Plate Structure under Initial Stress Influence

C Peter[1], M N Abdul Rani[1*], D P Starbuck[2], M A Yunus[1], KAM Rezali [3] and M S Mohd Zin[1]

[1]Structural Dynamics Analysis & Validation (SDAV), Faculty of Mechanical Engineering, Universiti Teknologi MARA (UiTM), 40450 Shah Alam, Selangor, Malaysia
[2] Opus International, 3800 Parkside, Birmingham Business Park, Birmingham UK, B37 7YG
[3] Sound and Vibration Research Group, Department of Mechanical and Manufacturing, Faculty of Engineering, Universiti Putra Malaysia (UPM), 43400 Serdang, Selangor, Malaysia

*Corresponding author: mnarani@salam.uitm.edu.my

Abstract. This paper presents the finite element modelling and analysis of the dynamic behaviour of a laser spot welded structure under the influence of initial stress. Modal based updating with a new scheme was used to improve the finite element model of the structure. Element connectors with CWELD ELPAT format were used to represent the laser spot welded in the structure. The modal parameters of the structure were measured using impact hammer testing and roving accelerometers under free-free boundary conditions. NASTRAN SOL 200 was used in updating the finite element model in light of the experimental results. In this study, it was found that modal updating with the new scheme has successfully matched the finite element model of the welded structure with the experimental model. The achievement demonstrates the accuracy and efficiency of the proposed new scheme.

Keywords: Finite Element Analysis, Model Updating, Experimental Modal Analysis, Initial Stress

1. Introduction

Large and complex engineering structures such as a car body in white, usually consists of a large number of structural components made from thin metal sheets. They are assembled together by thousands of joints which significantly influence the dynamic behaviour of the structure [1].

Laser spot welds are one type of joints and the use of laser spot welds in the automotive and aerospace industries has lately shown a significant increase. This is because laser spot welding is a cost effective and efficient method of assembling structural components. In addition, laser spot welding offers higher process speed and excellent flexibility for automation [2]. However, laser spot welded joints are found to be complex and have many uncertainties when it comes to analytical modelling [3]. One way to represent laser spot welds in a jointed structure for dynamic analysis is to use the finite element method. For example, Nurul et. al [4], showed that CWELD element connectors were used to represent laser spot welds. In the study, modal updating had been done on the hat-plate structure but there are still noticeable errors. Recently, Syazwan et. al [5], had done a study on frequency response function (FRF) based updating for the laser spot welds on a similar structure. The method had successfully shown a strong correlation between the updated and measured frequency response function. However, although modelling and analyses of the dynamic behaviour of welded structures were demonstrated, the issues of discrepancies between the FE and EMA results have remained to be unsolved. The discovery reveals that the use of the common model properties in updating the FE models are not successful and other potential updating parameters such as initial stress should be looked into.

Initial stress can arise when components are assembled. For a structure with a large surface of low thickness with initial curvature, stiffeners can be intentionally added to remove it and they may also unintentionally remove it. Initial stress is also connected with residual stress, improved understanding of the fatigue behaviour, in connection of those two subjects is extremely important [6]. Previous work [7] has compared the fatigue behaviour between resistance spot welds and laser spot welds of dissimilar materials. Thus, the effects of initial stress on structural dynamics are still completely unknown [8].

When the initial curvature is suppressed after adding of stiffeners, initial stress arises. Initial stress can also arise as a result of fabrication and heat treatment. However, such initial stress is very problematic to estimate by theoretical analysis or to measure, unless the unstressed configuration is first measured in the latter case, which is very infrequent in reality. In addition, the influence of initial curvature and initial stress on the natural frequencies of structures was investigated and was found to be noticeable in [9-12]. Furthermore, work on the effect of initial curvature was carried out in [13]. It was also pointed out in the same paper, that finite element commercial software treats membrane and bending deformation as being independent and this approximation is only practical for structures with a small initial curvature and small deflection. However, as for a moderate initial curvature and a small deflection the interaction between membrane and bending deformations should not be ignored.

This paper presents and discusses the importance of modelling initial stress in the finite element model and the use of initial stress as an updating parameter for enhancing the performance of the finite element model of the laser spot welded hat-plate structure made of thin metal sheets.

2. Description of the Jointed Structure and Experimental Set up

In this study, the structure under investigation is a laser spot welded hat-plate made from steel sheets of a uniform thickness of 1.5mm, 110mm long and 80mm wide. The structure consists of two components welded together as shown in figure 1. Component labelled number 1 is the hat shaped plate and number 2 is the flat plate. There are in total of 20 laser spot welds used in the structure.

The test structure was set up in free-free boundary conditions by suspending the structure using strings and rubber bands as shown in figure 2. The dynamic characteristics of the test structure were obtained using impact testing with two accelerometers. An impact hammer was used to excite the structure with one accelerometer was used as a reference by fixing to the predefined excitation point. The other accelerometer was roved around to measure 60 points throughout the structure. LMS SCADAS system was used to process the input and output responses of the structure under the test.

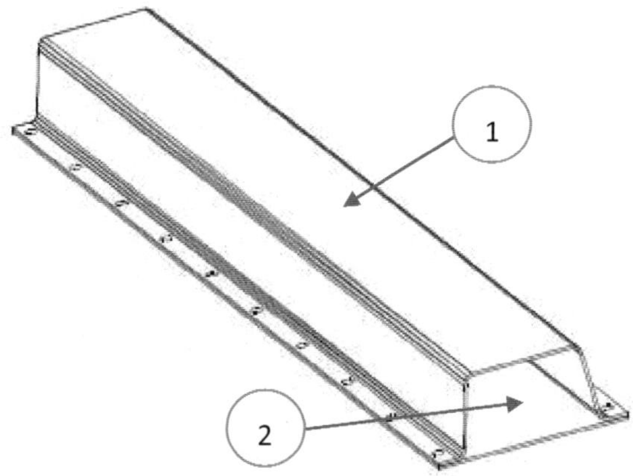

Figure 1. Hat-Plate test structure.

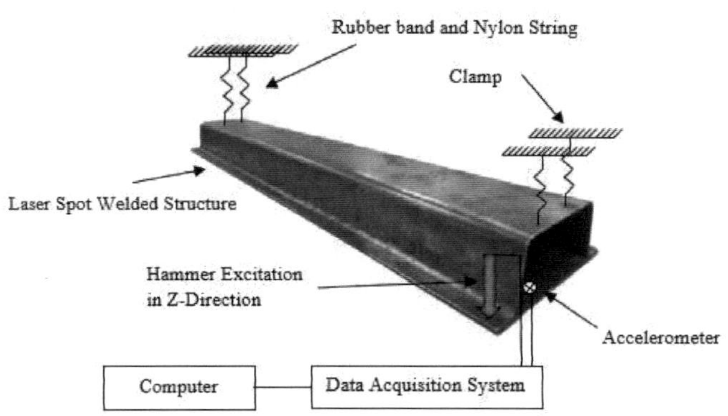

Figure 2. Experimental set-up of the test structure.

3. Finite Element Modelling and Analysis

The finite element model of the welded hat-plate structure was developed using HyperMesh software. MSC NASTRAN SOL 103 and SOL 200 were used for the prediction of the dynamic behaviour of the welded structure. The former was used for normal modes analysis and the latter for modal updating. Full details of the methods used in this study are available in [14-16].

Figure 3 shows the finite element model of the welded hat-plate. The finite element models of the hat and plate were developed based on CQUAD4 shell elements with 17673 elements and 18146 nodes. CWELD element connectors were used to represent the laser spot welds in the FE models. The actual dimensions of the test structure were measured using a coordinate measuring machine (CMM). The measurement was carried out for minimizing the uncertainties of the FE model developed. Detailed information on the model properties used in the FE model of the welded hat-plate structure is given in table 1 and table 2. In this study, initial stress exhibited in the test welded hat-plate structure was modelled in the FE model of the welded hat-plate structure. The modelling was carried out by discretizing the plate into 56 collectors mimicking the effect of the exhibited initial stress. The FE model with initial stress is shown in figure 4. By engineering judgement, the initial value of initial stress modelling was set to be 0.5 in the finite element analysis and the frequency of interest used in normal modes analysis was from 0 to

1000 Hz. Only the first six frequencies and mode shapes were investigated because there is high confidence in the modes.

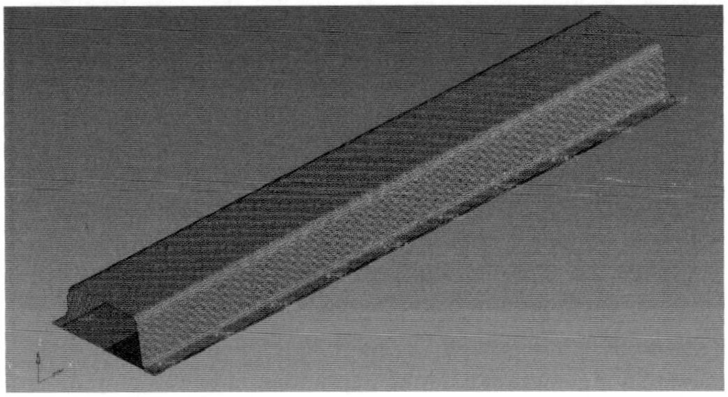

Figure 3. The finite element model of the welded hat-plate structure.

Table 1. Model properties of the FE model of the welded hat plate structure.

Property name	Nominal Value
Young's Modulus	210GPa
Poisson's Ratio	0.3
Density	7500kg/m^3

Table 2. Model properties of CWELD element connectors used in the FE model of the welded hat-plate structure.

The name of Property	Nominal Value
Young's Modulus	210GPa
Poisson's Ratio	0.3
Density	7500kg/m^3
Diameter	4.5

Figure 4. Modelling of the initial stress in the plate.

3.1 Modal Updating

Modal updating has been widely used to improve the accuracy of finite element models by correcting the invalid assumptions about the finite element models [17-19]. The improvement is carried out by systematically changing predefined potential updating parameters which have a large influence in improving the accuracy of the finite element models. The NASTRAN optimization code (SOL 200) is used to perform modal updating. Furthermore, the optimization algorithm in equation 1, which is a sensitivity based iterative-based procedure, let the objective function (J) to be minimized by correcting the eigenvalues of the initial finite element model until the objective function is converged. The objective function based on eigenvalues is defined as follows [22]:

$$J = \sum_{i=1}^{n} W_i \left(\frac{\lambda_i^{fe}}{\lambda_i^{exp}} - 1 \right)^2 \tag{1}$$

Where, λ_i^{exp} is the experimental eigenvalue and λ_i^{fe} is the predicted eigenvalue from the finite element model and n is the number of eigenvalues used in the updating procedure.

4. Results and Discussion

Previous studies [20-21] revealed that efficient and economical methods of studying the initial stress of the structure were vital to accurately determine dynamic ability which is directly related to the performance of engineering products. Model updating method has been successfully used by researchers [21-23] to accurately determine modal parameters of engineering structures in the light of experimental results which are usually obtained from experimental modal analysis.

In this study, the predicted and measured modal parameters of the test structure were obtained from the finite element method and experimental modal analysis. CQUAD4 shell elements and CWELD element connectors were used in modelling the welded hat-plate structure. The initial stress was introduced in the finite element model and modal updating scheme was then used in updating the initial finite element model of the structure to the test structure.

Table 3 tabulates the natural frequencies obtained from experimental modal analysis (column II), initial finite element modal (column III) and updated finite element model (column V). From the comparison of the natural frequencies, it was found that the errors in terms of percentage recorded in column IV are inconsistent. The largest error is shown in the 2nd mode with 2.71 percent and the smallest one is 1.02 percent from the 4th mode. The total error registered from six modes is 9.39 percent. The inconsistency in the errors suggests that the updating parameters (as stated in Tables 1 and 2) used in the updating scheme are not capable of updating the finite element model to the test structure. These findings are in fact consistent with the ones presented and discussed by [4,5,18]. Therefore, other potential updating parameters that can be identified using sensitivity analysis may be useful in updating the finite element model. Initial stress that may result from laser spot welding of the test structure was used as the potential updating parameters in the second attempt to update the finite element. In other words, initial stress with the updating parameters stated in Tables 1 and 2 were used in the attempt. Column V of table 3 presents the natural frequencies calculated from the second attempt which is the updated finite element model.

A compelling accomplishment in terms of percentage individual and total error was recorded in the updated finite element model after each of the six modes calculated from the model was compared with the experimental counterparts. An extraordinary reduction in the total error calculated from the updated model was noted in column VI with 0 percent. Another remarkable achievement recorded from the updated model is the percentage of individual error which is calculated from the direct comparison of each mode between the updated model and experimental modal analysis in which 0 percent was registered for each of the six modes. Meanwhile, figure 5 shows a graph of normalized values of the updating

IOP Conf. Series: Materials Science and Engineering **506** (2019) 012065 doi:10.1088/1757-899X/506/1/012065

parameters used in the updated finite element model of the welded hat-plate structure. Fifty-six updating parameters as shown in the figure were used in the updating. It was observed that forty-five cycles were taken to complete the updating process. In other words, the number of the cycles was required to systematically adjust the fifty-six updating parameters in the process of updating the finite element model. However, the number of the cycles could be reduced further if the 13th cycle in which all the updating parameters started converging was taken into consideration. The perturbations in the diameter and Young's modulus of the spot welds was observed in the 45th cycle. The initial diameter used has dropped from 4.5mm to 4.05mm and the initial Young's modulus defined has slid from 210GPa to 170GPa. The findings and achievement in this study have revealed that initial stress resulting from spot welding has a large influence on the accuracy of the finite element modelling and analysis of the dynamic behaviour of the laser spot welded hat-plate structure.

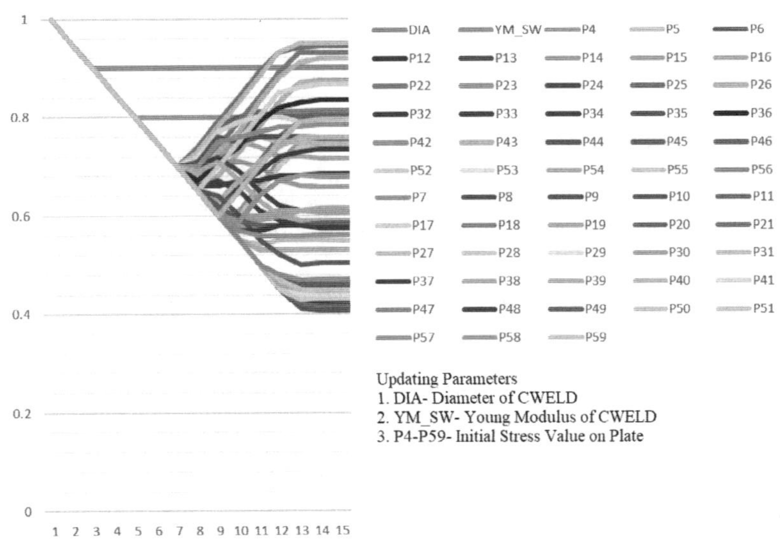

Figure 5. Graph of the normalized values of the updating parameters vs number of cycles.

Table 3. Comparison of the natural frequencies between the EMA, initial FEA and updated FE with initial stress.

Modes	EMA (Hz)	Initial FE (Hz)	Error %	Updated FEA with initial stress (Hz)	Error %
(I)	(II)	(III)	(IV)	(V)	(VI)
1	504.58	509.95	1.06	504.58	0
2	556.89	572	2.71	556.89	0
3	572.33	586.39	2.46	572.33	0
4	634.43	640.93	1.02	634.43	0
5	644.41	651.03	1.03	644.41	0
6	666.54	673.93	1.11	666.54	0
		Total error	9.39	Total error	0

5. Conclusions

The finite element modelling and analysis of the dynamic behaviour of the laser spot welded hat-plate structure under initial stress influence have been successfully investigated using experimental modal analysis, finite element method and modal based updating. From the investigation it was found that the conventional updating parameters and updating method largely used by researchers to update the finite elements models have appeared to be incapable of attaining the main objective of model updating. On the other hand, the findings from the investigation demonstrate that the proposed new updating scheme including initial stress in updating the finite element model of the structure is able to offer a superior updating capability. The proposed updating scheme has been successfully used to reduce the total error recorded in the initial finite element model of the welded structure. The reduction means that the dynamic behaviour of the initial finite element model has been perfectly matched with the test welded structure. In addition, initial stress has contributed a large of error.

References

[1] J. E. Mottershead, et al., "Experience in Mechanical Joint Model Updating," *19th International Seminar on Modal Analysis*, pp. 481-495, 1994

[2] Florea, R.S. & Hubbard, Camden & Solanki, K & Bammann, Douglas & Whittington, W.R. & Marin, Esteban. (2012). Quantifying Residual Stresses in Resistance Spot Welding of 6061-T6 Aluminum Alloy Sheets Via Neutron Diffraction Measurements. *Journal of Materials Processing Technology*. 212. 2358–2370. 10.1016/j.jmatprotec.2012.06.024.

[3] M. N. A. Rani, Stancioiu, D., Yunus, M., Ouyang, H. J., Deng, H., & James, S. (2011). Model Updating for a Welded Structure Made from Thin Steel Sheets. *Applied Mechanics and Materials*, 70, 117-122. doi:10.4028/www.scientific.net/amm.70.117

[4] Husain, N. A., Khodaparast, H. H., Snaylam, A., James, S., Dearden, G., & Ouyang, H. (2010). Finite-element modelling and updating of laser spot weld joints in a top-hat structure for dynamic analysis. *Proceedings of the Institution of Mechanical Engineers, Part C: Journal of Mechanical Engineering Science*. 224(4), 851-861. doi:10.1243/09544062jmes1787

[5] Zin, M. S., Rani, M. N., Yunus, M. A., Sani, M. S., W. I. I. Wan Iskandar Mirza, & Isa, A. A (2018). Frequency response function (FRF) based updating of a laser spot welded structure. doi:10.1063/1.5032017

[6] Martinson, P., Daneshpour, S., Koçak, M., Riekehr, S., & Staron, P. (2009). Residual stress analysis of laser spot welding of steel sheets. *Materials & Design*, 30(9), 3351-3359. doi:10.1016/j.matdes.2009.03.041

[7] Daneshpour, S., Riekehr, S., Koçak, M., & Gerritsen, C. H. (2009). Mechanical and fatigue behaviour of laser and resistance spot welds in advanced high strength steels. *Science and Technology of Welding and Joining*, 14(1), 20-25. doi:10.1179/136217108x336298

[8] A. R. de Faria and S. F. M. de Almeida (2006). The maximization of fundamental frequency of structures under arbitrary initial stress states, *International Journal for Numerical Methods in Engineering*, vol. 65, pp. 445-460.

[9] K. L. J. Fong. (2003). A Study of Curvature Effects on Guided Elastic Wave, PhD, Imperial College London, University of London.

[10] A. W. Leissa and A. S. Kadi (1971). Curvature Effects on Shallow Shell Vibrations, *Journal of Sound and Vibration*, vol. 16, pp. 173.

[11] X. H. Liu, R. Paurobally, and J. Pan (2008). The effect of prestress on vibration behavior of structures in impedance method damage detection, *Structural Integrity and Failure*, vol. 41-42, pp. 401-406.

IOP Conf. Series: Materials Science and Engineering **506** (2019) 012065 doi:10.1088/1757-899X/506/1/012065

[12] Labeas, G., Tsirkas, S., Diamantakos, J. and Kermanidis (2005). A., "Effect of Residual Stresses due to Laser Welding on the Stress Intensity Factors of Adjacent Crack," *Proc. of 11th International Conference on Fracture*.

[13] Q. L. Yu, N. Mostaghel, and K. C. Fu (1994). Effect of Initial Curvature on Natural Frequency of Thin-Plate on Hinge Supports, *Journal of Engineering Mechanics-Asce*, vol. 120, pp. 796-813.

[14] S. C. MSC: MSC. NASTRAN Version 70, Advaned Dynamic Analysis User's Guide.

[15] Abdul Rani, M. N. (2012) Structural dynamic analysis and model updating for a welded structure made from thin steel sheets. PhD Thesis, University of Liverpool.

[16] MSC.NASTRAN 2004. MSC.Nastran 2004 Rclease Guide.

[17] N.A.Z. Abdullah, M.S.M. Sani, N.A. Hussain, M.M. Rahman, and I. Zaman, "Dynamics properties of a Go-kart chassis structure and its prediction improvement using model updating approach", *International Journal of Automotive and Mechanical Engineering (IJAME),* Vol. 14, pg 3887-3897, 2017, doi: https://doi.org/10.15282/ijame.14.1.2017.6.0316

[18] M. N. Abdul Rani, S. Kasolang, and M. H. Othman, (2016). "Finite Element Modelling and Modal Based Up- Dating for the Dynamic Behaviour of a Laser Spot Welded Structure,"ICSV 2016 - *23rd International Congress on Sound and Vibration: From Ancient to Modern Acoustics. International Acoustics and Vibration*.

[19] J. E. Mottershead and M. I. Friswell, (1993). "Model Updating In Structural Dynamics: A Survey," *Journal of Sound and Vibration*, vol. 167, no. 2. pp. 347–375.

[20] Friswell, M. I. And Mottershead, J. E. (1995) Finite element model updating in structural dynamics, Kluwer Academic, Dordrecht.

[21] Anastassiou, M., Babbit, M., & Lebrun, J. (1990). Residual stresses and microstructure distribution in spot-welded steel sheets: Relation with fatigue behaviour. *Materials Science and Engineering:* A.125 (2), 141-156. doi:10.1016/0921-5093(90)90166-z

[22] M. N. A. Rani, H. Ouyang, M. A. Yunus, and B. A. Aminudin, "Model Updating For a Thin Steel Sheet Welded Structure," *20th International Congress on Sound and Vibration*, 2013.

[23] Ewins, D. J. (1990). Modal Testing Techniques. Vibration and Wear in High Speed Rotating Machinery, 299-309. doi:10.1007/978-94-009-1914-3_18

Patent Technological Field and Financial Performance of Malaysian Firms

F Ghapar
School of Business & Accountancy, Kolej Universiti Poly-Tech MARA Kuala Lumpur

Corresponding author: farha@kuptm.edu.my

Abstract. The patent specification within the patent system consists of valuable information regarding the characteristics of an invention. This includes the patent's technological fields and citations. Statistics show that Malaysia's manufacturing industries constitute its largest exporting industries compared to other industries with the U.S. being Malaysia's largest trading partner since the 1990s. The objective of this study is to examine how far does the patent technology field given impact on the financial performance at the Malaysian firm level. By segregating the patent technology field into two categories, 1) Human Necessities and Performing Operations, and 2) Mechanicals and Electronics, this study measures the financial performance based on the patent application and renewal stocks from within the patent system. This study applied a panel dataset from 1994 to 2008 and the model is estimated using panel least square, fixed effects model, and random effects model. It is found that the mechanicals and electronics technological field demonstrate more statistically significant variables compared to the human necessities and performing operations field. This findings are not surprising as Malaysia's largest export in the manufacturing industries is the mechanicals/ electronics technology field.

Keywords: Patent technological field, patent system and firm financial performance

1. Introduction

There is an ongoing debate over whether the patenting system benefits only advanced economies when compared to emerging economies and, further, whether the patent system benefits only the rich [1], [2]. However, even though Malaysia is a developing country, there is reason to be optimistic that patenting activity benefits this country as one of the emerging economies.

A patent involves many fields of technology. Some technology fields represent fast moving technology, while some do not, with the former having a higher obsolescence rate compared to the latter [3, 4]. In spite of this situation, there are firms that undertake patenting activity in both fast and slow moving technology fields.

Some firms patent products and processes either to be sold in the market or to be used in their own firms. The former may directly maximize the firm's sales revenue in order to maximize profits, while the latter is used to reduce the business costs in order to maximize profits [5, 6]. Indirectly, the latter may also maximize the firm's sales revenue once the firm's business costs have been reduced. Hence, patenting an invention represents a profit-maximizing strategy both in the short and long run.

IOP Conf. Series: Materials Science and Engineering **506** (2019) 012066 doi:10.1088/1757-899X/506/1/012066

The objective of this study is to examine how far does the patent technology field given impact on the financial performance at the Malaysian firm level. By segregating the patent technology field into two categories, 1) Human Necessities and Performing Operations, and 2) Mechanicals and Electronics, this study measures the financial performance based on the patent application and renewal stocks from within the patent system. This study applied a panel dataset from 1994 to 2008 and the model is estimated using panel least square, fixed effects model, and random effects model. The orientation of this paper introduces the institutional background, followed by literature review, hypothesis development and research method. It ends with findings and conclusion of the study.

2. Institutional background

Malaysia's economic history shows had a large quantity of manufactured products are assembled in that country from foreign components parts, and that these products are then exported to other parts of the world [7]. Statistics show that Malaysia's manufacturing industries constitute its largest exporting industries compared to other industries, with the U.S. being Malaysia's largest trading partner since the 1990s [8–10]. Furthermore, more research has been carried out into the patenting activities of manufacturing firms within the advanced economies [11–13] than into other types of industries. In fact, 73% of Malaysian firms in our sample data that have been granted patents in Malaysia and the U.S. are manufacturing firms; thus, this study would expect that Malaysian firms that have manufactured new products or processes and exported to the U.S. would also want to secure patent protection in the U.S. territory.

The Ninth Malaysia Plan reported a higher average annual growth rate of 12% of patents applied for to domestic residents compared to only 5% to non-residents in terms of the science and technology (S&T) indicator. However, as shown in Table 2.3, the number of patents filed by Malaysian residents is still much lower than for non-residents [10]. Malaysia's cumulative patent application and registration is dominated by foreigners, as demonstrated by the very large gap between residents and non-residents. In another statistic produced by MyIPO [14], of the total number of patent applications in Malaysia, only approximately 7% are applied for by Malaysian residents. On the other hand, out of the total number of patents granted in Malaysia, only 3% have been granted to Malaysian residents. Furthermore, the royalties paid to acquire foreign technological capability had an average annual growth rate of 29% compared to only 6% paid for local technology acquisition, as indicated in table 1. With patenting activity and royalty earnings dominated by foreigners, it is apparent that Malaysia is still dependent on foreign technological capability [15]

Table 1. Science and Technology (S&T) Indicators in Malaysia.

Indicators	2002	2005	Average Annual Growth Rate
Total Patents Filed (Residents)	322	522	12
Total Patents Filed (Non-Residents)	4615	5764	5
Royalties (Receipts – RM million)	74	98	6
Royalties (Payments – RM million)	2399	5851	29

*Source: Ninth Malaysia Plan [10]

In terms of technological field, Section C (chemistry; metallurgy) was the field in which the highest number of patents were granted followed by Section H (electricity), and Section B (performing operations; transporting). As discussed previously, foreigners have dominated Malaysian patent applications; hence, the technological fields were also influenced by the foreigners' technological fields. According to the Seventh and Eighth Malaysia Plan [8, 9] which reported Malaysia's progress between

1996 and 2005, technology inflows from foreigners came from electrical and electronics products; followed by chemical and chemicals products and transport equipment, 10 years in a row. The report matched with MyIPO data on patents granted based on technological field in table 2. There is no doubt that Malaysia is still dependent on foreign technological capability. This is evidenced through the technology inflows from foreigners; royalties paid to foreigners to acquire their technology; and the patenting activity undertaken by foreigners rather than Malaysian residents

Table 2. Patents Granted Based on Field of Technology in MyIPO.

Year	Section								Total
	A	B	C	D	E	F	G	H	
1993	215	169	503	15	37	52	155	138	1,284
1994	260	267	505	12	71	79	192	243	1,629
1995	336	268	542	27	48	61	194	277	1,753
1996	285	323	483	31	76	103	178	322	1,801
1997	151	138	196	13	32	45	82	132	789
1998	104	98	141	4	18	31	64	106	566
1999	132	112	191	9	21	49	68	139	721
2000	61	59	110	8	19	42	36	70	405
2001	155	233	288	18	44	102	231	399	1,470
2002	206	236	334	19	42	104	228	323	1,492
2003	224	242	396	28	38	119	190	341	1,578
2004	325	377	625	25	50	132	321	492	2,347
2005	333	452	600	30	82	164	316	531	2,508
2006	948	1,155	1,275	101	197	448	1,042	1,583	6,749
2007	1,179	1,213	1,748	109	221	407	883	1,223	6,983
2008	423	421	451	33	98	159	293	364	2,242
Total	5337	5763	8388	482	1094	2097	4473	6683	34317

*Source: MyIPO [14]

3. Literature review and hypothesis development

The patent system requires that the patent's technological fields and citations are included in the patent specification. This involves classifying the patent's technological field and citing other patents upon which the patent is built. The value of patent varies across technological fields [3, 16-17].

Griliches [18] argued that caution must be taken when studying the classification system in a patent. A patent's technology field is not the same as the industrial classification mainly categorized by economists. Thus, a researcher must have a clear understanding of the differences between these two in an attempt to answer any research questions. This is due to the fact that one technology field can be patented by many industries, and vice versa. To complicate matters, a patent can exhibit a combination of several technology fields. Some researchers have measured the technology field based on the patent's international patent classification (IPC) [3, 19] others have successfully created their own technology field [20, 21]; while others have based their classification on the industrial sector normally classified by economists [22, 23]. The most popular measure is to segregate the patents based on the technology field [3, 16]. However, other researchers have combined the technology field in an attempt to examine the dispersion of technology from its own traditional setting [19], which may create a superior technology.

Conceptually, some technology fields have a shorter technological life than others. Normally, biotechnology and pharmaceuticals have a longer technological life, compared to technical products that deal with basic human necessities which may have a shorter technological life. With regard to the former, generally at the early stage of their invention these firms would apply for a patent for defensive purposes [24, 25]. Thus, there is no commercialization of patents taking place at the early stage of the invention. Nevertheless, once the invention is ready to go into the market, it may last for a longer time period and boost the firms' sales revenues and profits. There are studies that show that pharmaceuticals, biotechnology, chemicals and other inventions classified as science-based are more valuable compared to other technological fields [16, 26]. However, the value of technological fields may also depend on the patent territory [27]. Based on Pavitt's [28], concept of the technological area which is grounded in differences in the process of innovation rather than product-based industrial classification. Greenhalgh and Rogers [21] found that the science-based technology field has the strongest effect on market value for UK patents. Schankerman [3] realized a similar result to that of Greenhalgh and Rogers [21] in his work on U.K. patents. However, he was more specific as he used the renewal behavior model [29] across the technology field, and found that pharmaceuticals and chemicals have a slow decay in private returns compared to mechanical and electronic patents which have much faster obsolescence.

In Malaysia, the pharmaceuticals and biotechnology fields are relatively new and targeted as a new source of growth [10]. The National Biotechnology Policy was launched in 2005 and divided into three main phases which end in 2020 [30]. Thus, this study expects that it is highly unlikely that Malaysian firms would have many patents in this area. Our sample contains more technical and mechanical technological area patents than pharmaceutical, biotechnology or chemical field patents. For example, even though in Table 2 it is evident that Section C (chemistry and metallurgy) has the highest granted patents compared to other technology fields, the results in show that in our sample Section C has the second lowest patents granted to Malaysian firms. This is unsurprising, as Section C in Table 2 involves all granted patents in Malaysia. Comparing with our sample, this situation obviously shows that foreign firms are dominating the Section C technology field in Malaysia. It is evident that our sample mostly comes from the human necessities, performing operations/ transporting, mechanical engineering, and electronics technology fields. This is to be expected as Malaysia has moved from post-colonial status to primary product exporter to being an industrially oriented economy [7, 31] in which these technology fields commonly rest in the aforementioned sectors.

Maintaining the unit of analysis with Malaysian firms which have been granted patent in two territories (Malaysia and the U.S.) in two technological fields (a) human necessities and performing operations and (b) mechanical and electronic, this study developed the following hypotheses.

i. There is a positive significant relationship between patents applied for and granted to Malaysian firms in human necessities and performing operations technological field and their financial performance
ii. There is a positive significant relationship between patents applied for and granted to Malaysian firms in mechanical and electronic technological field and their financial performance

4. Research method

Based on the IPC and composition of technology fields of patents granted to Malaysian firms in the two territories of Malaysia and the U.S., this study segregate the technology fields into two broad areas: 1) human necessities and performing operations, and 2) mechanicals and electronics. This study believe there is no significant difference between these two fields in terms of the impact of their patenting activity on the firms' financial performance relating to the sign and significance. This is because this study expects these two fields, both with fast moving technology, to have a short technological life. Nevertheless, the reason this study segregated these two fields is because the magnitude of their impacts may differ significantly.

In addition to segregating all the Malaysian firms which have been granted patents in the Malaysian and U.S. territories into the manufacturing industrial sector, this study were also interested in examining the impact of the technology field on the firms' financial performance. Thus, with the data in hand, this study rearranged all the firms based on the international patent classification (IPC). When segregating the firm into its technology field, this study based this on the first single alphabetical classification of its patent. For example, H01S 3/083 is classified in Section H, that is, electricity; another example C10J 3/26 is classified in Section C, that is, chemistry/ metallurgy. Nevertheless, as mentioned before, a patent can have multiple classifications within the technology field [18]. This study chose the majority number of classes in the patent specification to determine the classification of the technology field. This means that if there was a combination of two classes with only two fields, the first field classified in the patent specification was chosen. This study found the following composition of technology classes based on the broad classification in the IPC, as shown in figure 1

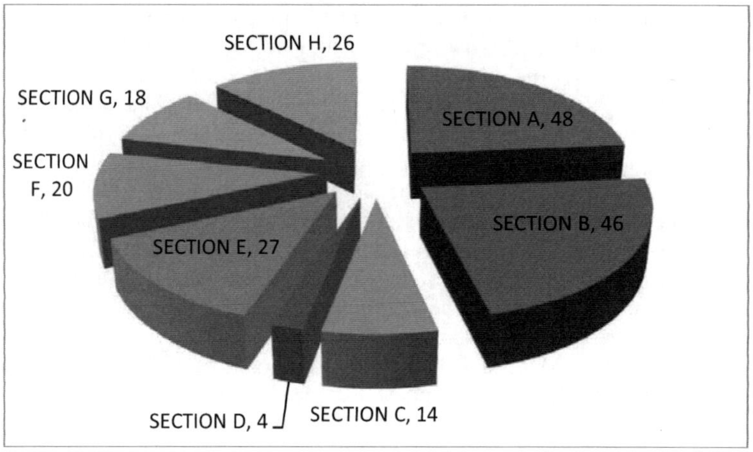

Figure 1. Patent Technology Field Based on International Patent Classification (IPC).

*Guide: Section A – Human Necessities; Section B – Performing Operations; Transporting; Section C – Chemistry; Metallurgy; Section D – Textiles; Paper; Section E – Fixed Constructions; Section F – Mechanical Engineering; Lighting; Heating; Weapons; Blasting; Section G – Physics; Section H – Electricity

Thus, to run the analysis with the patent renewal/ application, this study divided the technology field into two broad categories, that is, Section A to Section D and Section E to Section H. This study then reclassified them as Human Necessities and Performing Operations, and Mechanical and Electronics, respectively. The former comprised 112 firms with 1178 observations, while the latter was made up of 91 firms with 1047 observations for the analysis. This study applied a panel dataset from 1994 to 2008 and the model is estimated using panel least square, fixed effects model, and random effects model. This study also segregated model specification into three that Model 1 using sales, Model 2 using profits and Model 3 using profit margin as the dependent variable. Malaysian backward citations (MBWC), US backward citations (SBWC), Malaysian claims (MCL), US claims (SCL) and patent family (FAM) are the explanatory variables; tangible assets (LNTGA) and firms' age (NAGE) are the control variables

5. Findings

This study answers the research question as to the extent differences in the patent technology field impact on the financial performance of the firm at the Malaysian firm level. By segregating the patent technology field into two categories, 1) Human Necessities and Performing Operations, and 2) Mechanicals and Electronics, the measurement is based on the patent application and renewal stocks in Ghapar, Brooks and Smyth [32]. The panel model is estimated with panel least squares (PLS), fixed effects model (FEM) and random effects model (REM)

5.1. Diagnistic Test

This study tested for multicollinearity, autocorrelation, heteroskedasticity and misspecification to make sure that our results are robust, unbiased and consistent. As expected, this study found no multicollinearity problem with the patent renewal/ application measures in either of our samples as shown in table 3 and 4

Relating to autocorrelation, this study found serial correlation is present, and this study therefore corrected the first-order serial correlation with autoregressive errors of order 1 or an AR(1) model. This study also took measures to correct for the heteroskedasticity problem. This study chose the White Cross-Section coefficient covariance method to correct for heteroskedasticity in our panel model. This is due to the fact that White's coefficient covariance method gave us the majority of statistically significant explanatory variables compared to seemingly unrelated regression (SUR) and panel corrected standard error (PCSE). Finally, this study also tested for misspecification using the Ramsey [33]. Regressions Specification and Error Test (RESET) and normality of the residuals test of Jarque-Bera [34].

Interestingly, even though the number of observations in the human necessities/ performing operations sample is slightly higher than the mechanicals, none of the model specifications are found not to be misspecified. In the mechanicals/ electronics sample, all three models (models 1, 2 and 3) in this study had at least one effects specification which is not misspecified. All three models based on the patent renewal/ application measures had the REM estimations with a cross-section random effects specification, not misspecified. This situation shows that the internal factors (μ_i unobservable individual effect) in the mechanicals and electronics sample are not correlated with the explanatory variables. On the other hand, this study found the Jarque-Bera [34] normality residual test is not met in all of our specifications. Nevertheless, since our observations are large, this study followed the central limit theorem (CLT) assumptions

5.2. Panel Model Results

Our panel model is estimated using panel least squares (PLS), fixed effects model (FEM) and random effects model (REM). The model specifications were run with two separate samples which consisted of the human necessities/ performing operations technology field sample and the mechanicals/ electronics technology field sample. Both samples had similar R-squared in all model specifications to the samples discussed in the sub-topic of Research Method, even though their number of observations is much lower by nearly 50% - for example 2225 for the all firms sample and 1178 for the human necessities/ performing operations sample. On the other hand, when comparing the manufacturing firms sample with the mechanicals/ electronics technology field sample, the number of observations is nearly 40% lower - 1694 for manufacturing firms sample and 1047 for the mechanicals and electronics sample. The control variables of LNTGA and NAGE are found to have a positively statistically significant impact on the dependent variable in almost all three model specifications (see tables 5 to 10). However, this study found a near singular matrix problem in the mechanicals/ electronics sample with the FEM estimation. This study therefore omitted the NAGE variable from the FEM model in the mechanical/ electronics sample to obtain the parameter estimates.

IOP Conf. Series: Materials Science and Engineering **506** (2019) 012066 doi:10.1088/1757-899X/506/1/012066

Table 3. Correlation Matrix Based on Patent Renewal/ Application Measures – Human Necessities and Performing Operations.

	LNSALES	LNPROFITS	LNPM	APPM	GRANM	APPS	GRANS	LNTGA	NAGE
LNSALES	1								
LNPROFITS	0.881314	1							
LNPM	-0.00688	0.466454	1						
APPM	0.115663	0.110534	0.0174	1					
GRANM	0.094177	0.101364	0.038215	0.075	1				
APPS	0.067081	0.07986	0.04343	-0.00041	-0.0695	1			
GRANS	0.014987	0.020216	0.014726	-0.00897	-0.02395	0.236078	1		
LNTGA	0.898781	0.867824	0.154045	0.147512	0.145262	0.052749	-0.00701	1	
NAGE	0.263487	0.264847	0.067243	-0.01147	0.231225	-0.03744	-0.02828	0.341686	1

*Note: LNSALES, LNPROFITS and LNPM are the dependent variable of three different model specifications; APPM, GRANM, APPS and GRANS are the explanatory variables; LNTGA and NAGE are the control variables. LNSALES, LNPROFITS and LNPM are the sales, profits and profit margin. APPM and APPS are the patents applied for in Malaysia and the U.S. respectively; GRANM and GRANS are the patents granted to in Malaysia and the U.S. respectively; LNTGA is the tangible assets; NAGE is the age.

Table 4. Correlation Matrix Based on Patent Renewal/ Application Measures – Human Necessities and Performing Operations.

	LNSALES	LNPROFITS	LNPM	APPM	GRANM	APPS	GRANS	LNTGA	NAGE
LNSALES	1								
LNPROFITS	0.913438	1							
LNPM	0.010685	0.416716	1						
APPM	0.310318	0.26238	-0.04847	1					
GRANM	0.102939	0.059695	-0.08326	0.473127	1				
APPS	0.267541	0.233795	-0.02315	0.181548	0.169877	1			
GRANS	0.311796	0.235522	-0.11776	0.101896	0.098631	0.319338	1		
LNTGA	0.957354	0.931984	0.151508	0.295359	0.093642	0.251546	0.310047	1	
NAGE	0.271809	0.257042	0.024431	0.006015	0.04672	0.133979	0.084867	0.230394	1

*Note: LNSALES, LNPROFITS and LNPM are the dependent variable of three different model specifications; APPM, GRANM, APPS and GRANS are the explanatory variables; LNTGA and NAGE are the control variables. LNSALES, LNPROFITS and LNPM are the sales, profits and profit margin. APPM and APPS are the patents applied for in Malaysia and the U.S. respectively; GRANM and GRANS are the patents granted to in Malaysia and the U.S. respectively; LNTGA is the tangible assets; NAGE is the age.

IOP Conf. Series: Materials Science and Engineering **506** (2019) 012066 doi:10.1088/1757-899X/506/1/012066

Table 5. Panel Model Result for Model 1 with PLS, FEM & REM Estimations (Based on Patent Renewal/ Application Measures) – Human Necessities & Performing Operations.

Column	(1)	(2)	(3)	(4)	(5)	(6)	(7)
Estimation	PLS	PLS[(a)]	FEM	FEM[(a)]	FEM	REM	REM
Effects Specification	-	-	cross-section fixed	cross-section fixed	cross-section & period fixed	cross-section random	period random
Explanatory Variables							
C	0.320761*	5.380935***	-2.144443	-4.355525	-3.69591	0.320761**	0.32039
	(0.190116)	(1.308738)	(3.867302)	(4.086912)	(4.550238)	(0.148784)	(0.254738)
APPM	-0.000823	0.027015*	0.042356***	0.038476**	0.037124***	-0.000823	0.002084
	(0.014381)	(0.01575)	(0.009574)	(0.01596)	(0.008396)	(0.011128)	(0.019074)
GRANM	-0.118015***	0.033208	-0.03262	-0.00562	-0.13198***	-0.118015***	-0.11825**
	(0.03645)	(0.037025)	(0.039998)	(0.047017)	(0.048588)	(0.030714)	(0.05317)
APPS	0.10147***	0.07481	-0.016752	0.014308	-0.016816	0.10147***	0.108929
	(0.03053)	(0.057233)	(0.023884)	(0.029119)	(0.020388)	(0.039167)	(0.066845)
GRANS	0.058548	-0.016563	-0.027393	-0.031547	-0.063468	0.058548	0.059862
	(0.051178)	(0.097251)	(0.059673)	(0.079228)	(0.057887)	(0.037008)	(0.063346)
LNTGA	0.968115***	0.639172***	0.747681***	0.634226***	0.733988***	0.968115***	0.969077***
	(0.012863)	(0.086584)	(0.051769)	(0.078951)	(0.059125)	(0.010006)	(0.017063)
NAGE	0.006619	0.027193	0.298601	0.495044**	0.387236*	0.006619**	0.006109
	(0.00436)	(0.012897)	(0.197065)	(0.206992)	(0.223093)	(0.002603)	(0.004442)
R-Squared	0.774437	0.915117	0.929983	0.940893	0.93188	0.774437	0.775277
Number of Observations	1178	1066	1178	1066	1178	1178	1178

*Note: The dependent variable is sales. C is the constant; APPM and APPS are the patents applied for in Malaysia and the U.S. respectively; GRANM and GRANS are the patents granted to in Malaysia and the U.S. respectively; LNTGA is the tangible assets; NAGE is the age. Values shown are the coefficient estimates. Values below the coefficient estimates in the parentheses are the robust standard error. ***, **, and * indicate the significance level at 1%, 5%, and 10% respectively. [(a)] The specification is run with AR(1) correction.

Table 6. Panel Model Result for Model 2 with PLS, FEM & REM Estimations (Based on Patent Renewal/ Application Measures) – Human Necessities & Performing Operations

Column	(1)	(2)	(3)	(4)	(5)	(6)	(7)
Estimation	PLS	PLS[(a)]	FEM	FEM[(a)]	FEM	REM	REM
Effects Specification	-	-	cross-section fixed	cross-section fixed	cross-section & period fixed	cross-section random	period random
Explanatory Variables							
C	-3.650614***	-3.108045***	-4.968208	-4.624205	-6.71898*	-3.650614***	-3.650614***
	(0.345586)	(0.89281)	(4.156469)	(4.943603)	(3.579345)	(0.259522)	(0.338929)
APPM	-0.023826	0.009206	0.040296**	0.040801	0.040784*	-0.023826	-0.023826
	(0.018504)	(0.026926)	(0.019747)	(0.031309)	(0.022227)	(0.016511)	(0.021563)
GRANM	-0.058705	-0.033069	0.07936	0.082286	-0.02229	-0.058705	-0.058705
	(0.056977)	(0.095442)	(0.072229)	(0.07635)	(0.07531)	(0.048922)	(0.063891)
APPS	0.118053	0.140019*	0.063203	0.08945	0.060465	0.118053**	0.118053**
	(0.056104)	(0.082786)	(0.051815)	(0.068914)	(0.057635)	(0.058341)	(0.076192)
GRANS	0.095949	0.168685**	0.10862	0.186192	0.09724	0.095949	0.095949
	(0.083215)	(0.07773)	(0.104843)	(0.183602)	(0.101912)	(0.067229)	(0.087799)
LNTGA	1.068035***	1.007785***	0.673611***	0.56421***	0.649133***	1.068035***	1.068035***
	(0.021903)	(0.055244)	(0.072097)	(0.095901)	(0.083663)	(0.016668)	(0.021768)
NAGE	-0.009354**	0.008368	0.348112*	0.412045**	0.449205**	-0.009354**	-0.009354**
	(0.004238)	(0.010104)	(0.194367)	(0.197725)	(0.181656)	(0.004087)	(0.005337)
R-Squared	0.756237	0.858793	0.874101	0.901642	0.877828	0.756237	0.756237
Number of Observations	893	721	,893	721	893	893	893

*Note: The dependent variable is profits. C is the constant; APPM and APPS are the patents applied for in Malaysia and the U.S. respectively; GRANM and GRANS are the patents granted to in Malaysia and the U.S. respectively; LNTGA is the tangible assets; NAGE is the age. Values shown are the coefficient estimates. Values below the coefficient estimates in the parentheses are the robust standard error. ***, **, and * indicate the significance level at 1%, 5%, and 10% respectively. [(a)] The specification is run with AR(1) correction.

IOP Conf. Series: Materials Science and Engineering **506** (2019) 012066 doi:10.1088/1757-899X/506/1/012066

Table 7. Panel Model Result for Model 3 with PLS, FEM & REM Estimations (Based on Patent Renewal/ Application Measures) – Human Necessities & Performing Operations

Column	(1)	(2)	(3)	(4)	(5)	(6)	(7)
Estimation	PLS	PLS[a]	FEM	FEM[a]	FEM	REM	REM
Effects Specification	-	-	cross-section fixed	cross-section fixed	cross-section & period fixed	cross-section random	period random
Explanatory Variables							
C	-4.268543***	-5.301705***	-2.371781	-2.582953	-4.420208*	-4.268543***	-4.268543***
	(0.342554)	(0.867938)	(2.273648)	(3.181364)	(2.459335)	(0.245086)	(0.320774)
APPM	-0.003035	-0.00131	0.005989	0.002479	0.012418	-0.003035	-0.003035
	(0.012807)	(0.019541)	(0.016233)	(0.020048)	(0.018976)	(0.015592)	(0.020408)
GRANM	0.029192	0.027014	0.123288***	0.04863	0.116725*	0.029192	0.029192
	(0.042626)	(0.067346)	(0.0471)	(0.058263)	(0.062722)	(0.046201)	(0.060469)
APPS	0.074561	0.10671	0.086471	0.095933	0.09194	0.074561	0.074561
	(0.051071)	(0.07944)	(0.065132)	(0.076597)	(0.0685)	(0.055096)	(0.07211)
GRANS	0.01974	0.054234	0.050489	0.033578	0.072619	0.01974	0.01974
	(0.077904)	(0.094949)	(0.107391)	(0.199106)	(0.104242)	(0.063489)	(0.083096)
LNTGA	0.083128***	0.124954**	-0.121562***	0.048829	-0.080797*	0.083128***	0.083128***
	(0.022111)	(0.051612)	(0.037526)	(0.06663)	(0.04674)	(0.01574)	(0.020602)
NAGE	0.002125	0.01344	0.069134	-0.048805	0.132062	0.002125	0.002125
	(0.005343)	(0.010878)	(0.101837)	(0.108953)	(0.102327)	(0.003859)	(0.005051)
R-Squared	0.025627	0.426405	0.497163	0.592546	0.506541	0.025627	0.025627
Number of Observations	893	721	893	721	721	721	721

*Note: The dependent variable is profit margin. C is the constant; APPM and APPS are the patents applied for in Malaysia and the U.S. respectively; GRANM and GRANS are the patents granted to in Malaysia and the U.S. respectively; LNTGA is the tangible assets; NAGE is the age. Values shown are the coefficient estimates. Values below the coefficient estimates in the parentheses are the robust standard error. ***, **, and * indicate the significance level at 1%, 5%, and 10% respectively. [a] The specification is run with AR(1) correction.

IOP Conf. Series: Materials Science and Engineering **506** (2019) 012066 doi:10.1088/1757-899X/506/1/012066

Table 8. Panel Model Result for Model 1 with PLS, FEM & REM Estimations (Based on Patent Renewal/ Application Measures) – Mechanicals & Electronics

Column	(1)	(2)	(3)	(4)	(5)	(6)	(7)
Estimation	PLS	PLS[a]	FEM	FEM[a]	FEM	REM	REM
Effects Specification	-	-	cross-section fixed	cross-section fixed	cross-section & period fixed	cross-section random	period random
Explanatory Variables							
C	0.162634	1.95716***	3.844391***	5.427892***	4.325021***	1.779494***	0.162634
	(0.14044)	(0.619938)	(0.867848)	(1.621691)	(1.070468)	(0.430577)	(0.192207)
APPM	0.023367***	0.0205***	0.007788	0.00627	0.008811**	0.007764	0.023367**
	(0.005104)	(0.007886)	(0.004796)	(0.006333)	(0.003845)	(0.007378)	(0.009251)
GRANM	0.02458	0.025599	0.049446***	0.04126**	0.031811	0.038888	0.02458
	(0.020906)	(0.017473)	(0.01525)	(0.018772)	(0.020609)	(0.025815)	(0.037334)
APPS	-0.033032***	0.020059	0.005334	-0.007503	0.002903	0.00377	-0.033032
	(0.00971)	(0.020915)	(0.022443)	(0.025538)	(0.021373)	(0.02355)	(0.027343)
GRANS	0.035759**	0.086447***	0.125329***	0.099514***	0.115099***	0.122153***	0.035759
	(0.01766)	(0.025109)	(0.021119)	(0.029352)	(0.023516)	(0.027623)	(0.039013)
LNTGA	0.942718***	0.824395***	0.757674***	0.66741***	0.729623***	0.821316***	0.942718***
	(0.010259)	(0.04414)	(0.052169)	(0.096257)	(0.063988)	(0.024385)	(0.012102)
NAGE	0.028129***	0.037292***				0.045648***	0.028129***
	(0.003064)	(0.012823)				(0.012592)	(0.003856)
R-Squared	0.892186	0.961329	0.96432	0.973019	0.964678	0.59085	0.892186
Number of Observations	1047	956	1047	956	1047	1047	1047

*Note: The dependent variable is sales. C is the constant; APPM and APPS are the patents applied for in Malaysia and the U.S. respectively; GRANM and GRANS are the patents granted to in Malaysia and the U.S. respectively; LNTGA is the tangible assets; NAGE is the age. Values shown are the coefficient estimates. Values below the coefficient estimates in the parentheses are the robust standard error. ***, **, and * indicate the significance level at 1%, 5%, and 10% respectively. [a] The specification is run with AR(1) correction.

Table 9. Panel Model Result for Model 2 with PLS, FEM & REM Estimations (Based on Patent Renewal/ Application Measures) – Mechanicals & Electronics

Column	(1)	(2)	(3)	(4)	(5)	(6)	(7)
Estimation	PLS	PLS[(a)]	FEM	FEM[(a)]	FEM	REM	REM
Effects Specification	-	-	cross-section fixed	cross-section fixed	cross-section & period fixed	cross-section random	period random
Explanatory Variables							
C	-3.896801***	-4.424141***	0.997305	1.884711	0.212789	-2.786286***	-3.885831***
	(0.176507)	(0.710276)	(1.374049)	(2.060409)	(2.236957)	(0.620448)	(0.270922)
APPM	0.000907	-0.011711	-0.002369	-0.01192*	0.003955	-0.003404	0.002571
	(0.005644)	(0.008212)	(0.005312)	(0.007183)	(0.006715)	(0.01234)	(0.011289)
GRANM	-0.088994	0.046969	0.029238	0.077321**	0.002782	-0.00218	-0.096288**
	(0.060322)	(0.057492)	(0.03633)	(0.036676)	(0.028941)	(0.04456)	(0.047746)
APPS	0.066228*	-0.061194	-0.026928	-0.087842	-0.022026	-0.024318	0.069595
	(0.037159)	(0.073311)	(0.043364)	(0.062955)	(0.044175)	(0.056847)	(0.057494)
GRANS	-0.254154***	-0.164104	-0.208049**	-0.240723*	-0.209232**	-0.23845***	-0.253379***
	(0.046996)	(0.120565)	(0.081376)	(0.13002)	(0.085727)	(0.06061)	(0.056888)
LNTGA	1.050285***	1.068679***	0.786027***	0.740954***	0.831766***	0.960258***	1.050304***
	(0.011507)	(0.033316)	(0.079254)	(0.116184)	(0.130203)	(0.037925)	(0.015994)
NAGE	0.017801***	0.023289***				0.034232**	0.017494***
	(0.003447)	(0.00664)				(0.014296)	(0.005146)
R-Squared	0.874543	0.921384	0.92431	0.942207	0.926208	0.50365	0.875163
Number of Observations	796	645	796	645	796	796	796

*Note: The dependent variable is profits. C is the constant; APPM and APPS are the patents applied for in Malaysia and the U.S. respectively; GRANM and GRANS are the patents granted to in Malaysia and the U.S. respectively; LNTGA is the tangible assets; NAGE is the age. Values shown are the coefficient estimates. Values below the coefficient estimates in the parentheses are the robust standard error. ***, **, and * indicate the significance level at 1%, 5%, and 10% respectively. [(a)] The specification is run with AR(1) correction.

Table 10. Panel Model Result for Model 3 with PLS, FEM & REM Estimations (Based on Patent Renewal/ Application Measures) – Mechanicals & Electronics

Column	(1)	(2)	(3)	(4)	(5)	(6)	(7)
Estimation	PLS	PLS[(a)]	FEM	FEM[(a)]	FEM	REM	REM
Effects Specification	-	-	cross-section fixed	cross-section fixed	cross-section & period fixed	cross-section random	period random
Explanatory Variables							
C	-4.367013***	-4.960039***	-2.207597*	-2.659236	-3.391773*	-3.692485***	-4.367013***
	(0.156928)	(0.984108)	(1.142117)	(1.841746)	(1.871437)	(0.573087)	(0.301286)
APPM	-0.023094**	-0.02939***	-0.009435	-0.018155**	-0.003091	-0.01119	-0.023094*
	(0.009555)	(0.010612)	(0.007074)	(0.007392)	(0.007565)	(0.012359)	(0.012578)
GRANM	-0.067548	0.057477	-0.01302	0.052911	-0.033464	-0.025494	-0.067548
	(0.052598)	(0.060301)	(0.04574)	(0.037779)	(0.041301)	(0.044756)	(0.052226)
APPS	-0.003836	-0.091129	-0.007814	-0.073945	-0.001245	-0.014898	-0.003836
	(0.038693)	(0.064034)	(0.036495)	(0.049968)	(0.03904)	(0.057046)	(0.064239)
GRANS	-0.295252***	-0.197426	-0.281012***	-0.28227**	-0.277328***	-0.301479***	-0.295252***
	(0.050559)	(0.129068)	(0.081192)	(0.126812)	(0.085707)	(0.060541)	(0.063042)
LNTGA	0.107774***	0.128895***	-0.023271	0.00297	0.045518	0.058609*	0.107774***
	(0.011123)	(0.047063)	(0.066438)	(0.104606)	(0.109388)	(0.035248)	(0.017875)
NAGE	-0.001922	0.001847				0.004946	-0.001922
	(0.004372)	(0.010491)				(0.012792)	(0.005748)
R-Squared	0.064369	0.459446	0.534289	0.627632	0.541946	0.036084	0.064369
Number of Observations	796	645	796	645	796	796	796

*Note: The dependent variable is profit margin. C is the constant; APPM and APPS are the patents applied for in Malaysia and the U.S. respectively; GRANM and GRANS are the patents granted to in Malaysia and the U.S. respectively; LNTGA is the tangible assets; NAGE is the age. Values shown are the coefficient estimates. Values below the coefficient estimates in the parentheses are the robust standard error. ***, **, and * indicate the significance level at 1%, 5%, and 10% respectively.
[(a)] The specification is run with AR(1) correction.

5.2.1. Human necessities and performing operations technology fields. In the human necessities/ performing operations technology field, both APPM and APPS have a positive sign to the impact on the firm's sales, but only APPM is statistically significant. On the other hand, GRANM and GRANS both have a negatively statistically significant impact on the firm's sales. The positive impact of patents applied for may signal that the market demand and supply of the patented technology does last from when the patent is applied for until the time the patent is granted. However, the measurement of the granted patents that is based on the renewal behavior may reveal that the granted patents in the territories of both Malaysia and the U.S. are not well being demanded and supplied in the long run, which in turn leads to negative sales.

Surprisingly, however, the negative impact does not continue when this study move into model 2. All of our explanatory variables have a positive significant impact on the firm's profits. This may be due to the fact that the firms that patent in the human necessities/ performing operations technology fields patented the products or processes to be used in the firm to reduce the business costs. This contrasts with the argument in the preceding paragraph, as the patented product or process in this technology field may not be meant for sale directly to the consumers, but to be applied to their own firms. Moving into model 3 with profit margin as the dependent variable, only two variables are significantly reported with APPM having a negative impact and GRANM having a positive impact. The negative impact of APPM may be interesting, as earlier APPM gave a positive impact on both the firm's sales and profits. This may due to the lagged effects which result in a negative impact of APPM on the ratio of the firm's profits to its sales (profit margin).

5.2.2. Mechanicals and electronics technology fields. When comparing the result of the hypotheses development for mechanical/ electronics technology field sample, with the manufacturing firms sample, they are quite similar. The only difference involves one variable, that is, GRANM on the firm's sales with a positive impact in the mechanical/ electronics technology field sample, and a negative impact in the manufacturing firms sample. The significance of all variables in the two samples is also nearly the same. This is unsurprising, as pointed out in earlier that Malaysia has moved into an industrialized nation, with manufacturing industries and mechanicals/ electronics sectors providing the largest portion of manufactured products, as well as those that are exported to other parts of the world. The positive impact of GRANM on the firm's sales also shows that the firms with the patented invention in the mechanicals/ electronics technology field have a place in Malaysia until the patent is granted, and continues giving positive impact as the measurement involves the patent renewal behavior.

6. Conclusion

When comparing between the technology fields, firms that have patents in the mechanical/ electronics technology field have more statistically significant variables in model 2 and 3, compared to the human necessities/ performing operations technology field. If this study refer to the RESET test, the mechanicals/ electronics technology field has model 2 and 3 not to be misspecified compared to all models in the human necessities/ performing operations technology field that are found to be misspecified. Even though the RESET test does not tell us how to correct the misspecification, the results of this study have favor the mechanicals/ electronics technology field. That the result demonstrate more statistically significant variables in the mechanicals/ electronics technology field is unsurprising because Malaysia's largest export in the manufacturing industries is the mechanicals/ electronics technology field. Furthermore, this study also found that manufacturing firms have more statistically significant variables.

References

[1] S Yelderman 2016 Do Patent Challenges Increase Competition *University Chicago Law Reviews* vol 83 no 4 pp 1943–2026

[2] E Kranakis 2017 Patents and Power: European Patent-System Integration in the Context of Globalization *Technology and Culture* vol 48 pp 689–728

[3] M Schankerman 1998 How Valuable is Patent Protection? Estimates by Technology Field *The RAND Journal of Economics* vol 29(1) pp 77

[4] Y T Li M H Huang and D Z Chen 2011 Semiconductor industry value chain: Characters' technology evolution *Industrial Management & Data Systems* vol 111 no 3 pp 370–390

[5] S van Triest and W Vis 2007 Valuing patents on cost-reducing technology: A case study *International Journal Production Economics* vol 105 no 1 pp 282–292

[6] S S Tsang F C Chang and W C Wang 2015 A Survival Analysis on Fuel Cell Technology Patent Maintenance and Values Exploration between 1976 and 2001 *Advances in Materials Science and Engineering* vol 2015 pp 1–9

[7] V G R Chandran Govindaraju and C Y Wong 2011 Patenting activities by developing countries: The case of Malaysia *World Patent Information* vol 33 no 1 pp 51–57

[8] JPM 1995 Seventh Malaysia Plan 1996-2000 (Kuala Lumpur)

[9] Economic Planning Unit 2001 Eighth Malaysia plan (2001–2005) (Kuala Lumpur)

[10] Economic Planning Unit 2005 Ninth Malaysia Plan (2011-2015) (Kuala Lumpur)

[11] B H Hall G Thoma and S Torrisi 2007 The Market Value of Patents and R&D: Evidence From European Firms *Academy Management Proceeings* vol 2007 no 1 pp 1–6

[12] F G P Moreira A L V Torkomian and T J C C Soares 2016 Exploration and firms' innovative performance – How does this relationship work? *Reviews of Business Management* vol 18 no 61 pp 392–415

[13] A Brem P A Nylund and E L Hitchen 2017 Open innovation and intellectual property rights: How do SMEs benefit from patents, industrial designs, trademarks and copyrights? *Management Decision* vol 55 no 6 pp 1285–1306

[14] MyIPO Patent Statistics 2010 Retrieve from *http://www.myipo.gov.my/en/ip-statistics/160.html*

[15] V G R C Govindaraju F A Ghapar and V Pandiyan 2009 The role of collaboration, market and intellectual property rights awareness in university technology commercialization *International Journal of Innovation and Technology Management* vol 6 no 4

[16] J O Lanjouw and M. Schankerman 2015 Patent Quality and Research Productivity : Measuring Innovation With Multiple Indicators *The Economic Journal* vol 114 no 495 pp 441–465

[17] B Cepeda Zetter C González Brambila and M. Á. Pérez Angón 2017 Gender Desegregated Analysis of Mexican Inventors in Patent Applications Under the Patent Cooperation Treaty (Pct) *Interciencia* vol 42 no April pp 1–14

[18] Z Griliches 1990 Patent Statistics as Economic Indicators: A Survey *Journal of Economy Literature*, vol 28 no 4 pp 1661–1707

[19] J Lerner 2012 The importance of patent scope : an empirical analysis vol 25 no 2 pp 319–333

[20] M Meyer T S Pereira O Persson and O Granstrand 2004 The scientometric world of Keith Pavitt: A tribute to his contributions to research policy and patent analysis *Research Policy* vol 33 no 9 pp 1405–1417

[21] C Greenhalgh and M. Rogers 2006 The value of innovation: The interaction of competition, R&D and IP *Researvh Policy* vol 35 no 4 pp 562–580

[22] B H Hall A Jaffe and M. Trajtenberg 2005 Market Value and Patent Citations The *RAND Journal Economics* pp 16–38

[23] B H Hall 2005 Measuring the Returns to R & D : The Depreciation Problem *Annales d'Economie et de Statistique* vol July/December no June 2006 pp 341–381

[24] B Soranzo A Nosella and R Filippini 2017 Redesigning patent management process: an Action Research study *Management Decision* vol 55 no 6 pp 1100–1121

[25] L Agostini A Nosella V Lazzarotti R Manzini and L Pellegrini 2017 Introduction to the Special Issue on Intellectual Property Management: an internal and external perspective *Management Decision* vol 55 no 6 pp 1082–1086

[26] F Fai and N Von Tunzelmann 2001 Industry-specific competencies and converging technological systems: Evidence from patents *Structural Change and Economic Dynamics* vol 12 no 2 pp 141–170

[27] E Bacchiocchi and F Montobbio 2009 Knowledge diffusion from university and public research A comparison between US, Japan and Europe using patent citations *The Journal of Technology Transfer* vol 34 no 2 pp 169–181

[28] K Pavitt 1984 Sectoral patterns of technical change: Towards a taxonomy and a theory *Research Policy* vol 13 no 6 pp 343–373

[29] M Schankerman and A Pakes 1986 Estimates of the Value of Patent Rights in European Countries During the Post-1950 Period *Economic Journal* vol 96 no December pp 1052–1076

[30] Malaysia National Biotechnology Policy 2011 Retrieved from: *http://bic.org.my/?action=localscenario&do=policy*

[31] V G R C Govindaraju F A Ghapar and V Pandiyan 2009 The Role of Collaboration, Market and Intellectual Property Rights Awareness in University Technology Commercialization *International Journal of Innovation Technology Management* vol 06 no 04 pp 363–378

[32] F Ghapar R Brooks and R Smyth 2014 The impact of patenting activity on the financial performance of Malaysian firms *Journal of the Asia Pacific Economy* vol 19 no 3 pp 445–463

[33] J B Ramsey Tests for Specification Errors in Classical Linear Least-Squares Regression Analysis *Journal of the Royal Statistical Society: Series B (Methodological)* vol 31 no 2 pp 350–371

[34] C M Jarque and A K Bera 1980 Efficient tests for normality, homoscedasticity and serial independence of regression residuals *Economics Letters*